W0080841

Transcriptional Regulation by Neuronal Activity

Serena M. Dudek

Editor

Transcriptional Regulation by Neuronal Activity

To the Nucleus and Back

 Springer

Serena M. Dudek
Laboratory of Neurobiology
Synaptic & Developmental Plasticity Group
NIEHS/NIH
Research Triangle Park, NC 27709
USA
dudek@niehs.nih.gov

ISBN: 978-0-387-73608-2   e-ISBN: 978-0-387-73609-9

Library of Congress Control Number: 2007937950

© 2008 Springer Science+Business Media, LLC
All rights reserved. This work may not be translated or copied in whole or in part without the written
permission of the publisher (Springer Science+Business Media, LLC, 233 Spring Street, New York,
NY 10013, USA), except for brief excerpts in connection with reviews or scholarly analysis. Use in
connection with any form of information storage and retrieval, electronic adaptation, computer software,
or by similar or dissimilar methodology now known or hereafter developed is forbidden.
The use in this publication of trade names, trademarks, service marks and similar terms, even if they are
not identified as such, is not to be taken as an expression of opinion as to whether or not they are subject
to proprietary rights.

Cover illustration: High resolution confocal projection image from region CA3 of rat hippocampus.
This rat was exposed briefly and sequentially to "environment A" twice, once 30 min and the other
immediately before sacrifice. The red color shows the distribution of RNA to the immediate-early gene
Arc. The blue color is the nuclear counterstain DAPI. Arc mRNA can be seen in cytoplasmic and
dendritic regions, and provides a marker of neuronal activation 30 min before sacrifice. In addition, two
strong Arc transcription foci can be seen in the nucleus of this same neuron. The Arc transcription foci
provide a marker of neuronal activation from within 2 to approximately 10 min before sacrifice. Image
by John Guzowski, used with permission.

Printed on acid-free paper.

9 8 7 6 5 4 3 2 1

springer.com

Preface

As you read the chapters of this book, neurons in your central nervous system are undergoing significant changes at the nuclear level. Thanks to these changes, you may see, understand, and retain (at least in part) what you are reading. One of these nuclear events, transcriptional regulation, is important not only for restricting genes to expression in tissue- and cell-specific manners, but also for controlling genes in response to extracellular and/or environmental stimuli. In the unique case of excitable cells that include neurons, genes can be and are regulated by neuronal activity.

Activity-mediated gene regulation may occur as a result of neurotransmitter release and subsequent post- or pre-synaptic signaling, or alternatively, it may come in the form of postsynaptic action potentials and the resulting rise in intracellular calcium or signaling at the membrane due to ion channel activation. What genes are expressed in response to neuronal activation? Screens have revealed that most of the genes can be described as falling into a few major categories: 1) those that function to generally normalize the neuronal activity or otherwise facilitate homeostasis, 2) those recruited for replacement of proteins used up or secreted, or more rarely, 3) genes related to a particular function, such as for the maintenance of synaptic plasticity. Activity-regulated genes also play important roles in neuronal development that may include synapse maturation or neuronal phenotype determination.

The idea for the book came out of discussions I had with Alison Barth while we were planning a mini-symposium on the topic for the 2005 Society for Neuroscience meeting. Although most of our original speakers at the symposium were unavailable to contribute to the book, I was surprised by just how many authors did agree to participate. In hindsight, I don't know why I was so surprised. The topic is one that has blossomed into a thriving field, and I found that there were many enthusiastic would-be authors wanting to tell their stories. In this book, I have attempted to bring together chapters that cover three main topics: signal initiation and transduction to the nucleus, consequent gene regulation within the nucleus, and possible functions of the activity-regulated genes. Please bear with the unavoidable overlap in the general topics and omissions of specific ones; as the field is rapidly expanding, it is becoming increasingly difficult to cover the field in its entirety. Nevertheless, this book should provide both experts in the field and students of biology with an update on where the last 20 years of study has taken us in the nascent field of transcriptional regulation by neuronal activity.

Serena M. Dudek

Contents

List of Contributors

J. Paige Adams
Laboratory of Neurobiology
National Institute
    of Environmental Health Sciences
National Institutes of Health
111 T.W. Alexander Drive
Research Triangle Park,
NC 27709
USA

Angel Barco
Instituto de Neurociencias de Alicante
UMH-CSIC
San Juan de Alicante
Alicante, 03550
Spain

Cynthia Barber
Department of Biology
Massachusetts Institute of Technology
43 Vassar Street
Cambridge, MA 02139
USA

Alison L. Barth
Department of Biological Sciences
Carnegie Mellon University
4400 Fifth Avenue
Pittsburgh, PA 15213
USA

Azad Bonni
Department of Pathology
Harvard Medical School
77 Avenue Louis Pasteur
Boston, MA 02115
USA

Sangeeta Chawla
Department of Pharmacology
University of Cambridge
Tennis Court Road
Cambridge, CB2 1PD
UK

Hai-Ying Mary Cheng
Department of Neuroscience
The Ohio State University
333 West 10th Avenue
Columbus, OH 43210
USA

Ricardo Dolmetsch
Department of Neurobiology
Stanford University
299 Campus Drive D 229
Stanford, CA 94305
USA

Serena M. Dudek
Laboratory of Neurobiology
National Institute of Environmental
    Health Sciences
National Institutes of Health
111 T.W. Alexander Drive
Research Triangle Park,
NC 27709
USA

John S. Fitzpatrick
Department of Neurobiology
Yale University School of Medicine
P.O. Box 208001
New Haven, CT 06520
USA

Robin T. Garner
Department of Neurobiology
Yale University School of Medicine
P.O. Box 208001
New Haven, CT 06520
USA

Natalia Gomez-Ospina
Department of Molecular
     Pharmacology
Stanford University
299 Campus Drive
Stanford, CA 94305
USA

Fernando Gomez-Pinilla
Department of Physiological Sciences
University of California
621 Charles E. Young Drive
Los Angeles, CA 90095
USA

Rachel D. Groth
Department of Neuroscience,
University of Minnesota
321 Church Street, SE
Minneapolis, MN 55455
USA

John F. Guzowski
Department of Neurobiology
     and Behavior
University of California
108 Bonney Research Labs
Irvine, CA 92697
USA

Anna M. Hagenston
Department of Neurobiology
Yale University School of Medicine
P.O. Box 208001
New Haven, CT 06520
USA

Daniel N. Hertle
Department of Neurobiology
Yale University School of Medicine
P.O. Box 208001
New Haven, CT 06520
USA

Yi-Ping Hsueh
Institute of Molecular Biology
Academia Sinica
Taipei 115
Taiwan

Dragana Jancic
Instituto de Neurociencias de Alicante
UMH-CSIC San Juan de Alicante
Alicante, 03550
Spain

Bryen A. Jordan
Department of Biochemistry
School of Medicine
New York University
550 First Ave
New York, NY 10016
USA

Leszek Kaczmarek
Department of Molecular and
     Cellular Neurobiology
Nencki Institute
Pasteura 3
PL-02-093 Warsaw
Poland

Eric R. Kandel
Center for Neurobiology and Behavior
College of Physicians and Surgeons
Howard Hughes Medical Institute
    and the Kavli Institute
1051 Riverside Drive
New York, NY 10032
USA

Brian Yee Hong Lan
Department of Pharmacology
University of Cambridge
Tennis Court Road
Cambridge CB2 1PD
UK

Huan Ling Liang
Department of Cell Biology,
    Neurobiology, and Anatomy
Medical College of Wisconsin
8701 Watertown Plank Road
Milwaukee, WI 53226
USA

J. Troy Littleton
Department of Biology
    and Department of Brain
    and Cognitive Sciences
Massachusetts Institute of Technology
43 Vassar Street
Cambridge, MA 02139
USA

Claudio V. Mello
Neurological Sciences Institute
Oregon Health Sciences University
505 NW 185th Ave
Beaverton, OR 97006
USA

Paul G. Mermelstein
Department of Neuroscience
University of Minnesota
321 Church Street SE
Minneapolis, MN 55455
USA

Teiko Miyashita
Department of Neurobiology
    and Behavior
University of California
108 Bonney Research Labs
Irvine, CA 92697
USA

Ting Nie
Department of Neurobiology
    and Behavior
University of California
108 Bonney Research Labs
Irvine, CA 92697
USA

Karl Obrietan
Department of Neuroscience
The Ohio State University
333 West 10th Avenue
Columbus, OH 43210
USA

Sakkapol Ongwijitwat
Department of Cell Biology,
    Neurobiology, and Anatomy
Medical College of Wisconsin
8701 Watertown Plank Road
Milwaukee, WI 53226
USA

Rachel A. Robinson
Laboratory of Neurobiology
National Institute of Environmental
    Health Sciences
National Institutes of Health
111 T.W. Alexander Drive
Research Triangle Park, NC 27709
USA

Aryaman Shalizi
Department of Pathology
Harvard Medical School
77 Avenue Louis Pasteur
Boston, MA 02115
USA

Amanda A. Sleeper
Department of Neurobiology
Yale University School of Medicine
P.O. Box 208001
New Haven, CT 06520
USA

Thomas C. Tubon, Jr.
Department of Genetics
University of Wisconsin
425 Henry Mall
Madison, WI 53706
USA

Shoshanna Vaynman
Department of Physiological Sciences
University of California
621 Charles E. Young Drive
Los Angeles, CA 90095
USA

Tarciso A. F. Velho
Neurological Sciences Institute
Oregon Health Sciences University
505 NW 185th Ave
Beaverton, OR 97006
USA

Grzegorz M. Wilczynski
Department of Molecular and Cellular
  Neurobiology
Nencki Institute
Pasteura 3, PL-02-093
Warsaw
Poland

Anne E. West
Department of Neurobiology
Duke University Medical Center
Reseach Drive, Box 3209
Durham, NC 27710
USA

Margaret T.T. Wong-Riley
Department of Cell Biology,
  Neurobiology, and Anatomy
Medical College of Wisconsin
8701 Watertown Plank Road
Milwaukee, WI 53226
USA

Lina Yassin
Department of Biological Science
Carnegie Mellon University
4400 Fifth Avenue
Pittsburgh, PA 15213
USA

Mark F. Yeckel
Department of Neurobiology
Yale University School of Medicine
P.O. Box 208001
New Haven, CT 06520
USA

Jerry C.P. Yin
Department of Genetics
  and Department of Psychiatry
University of Wisconsin
425 Henry Mall
Madison, WI 53706
USA

Brian Yee Hong Lam
Department of Pharmacology
University of Cambridge
Tennis Court Road
Cambridge CB2 1PD
UK

Edward B. Ziff
Department of Biochemistry
New York University
School of Medicine
550 First Ave
New York, NY 10016
USA

Part I
Activity-regulated Transcription:
Getting to the Nucleus

Chapter 1
Transcriptional Regulation
of Activity-Dependent Genes by Birdsong

Tarciso A.F. Velho and Claudio V. Mello

Abstract Birdsong is a natural learned behavior used extensively for vocal communication and controlled by a well-characterized set of discrete brain areas in songbirds. The acts of hearing and producing birdsong lead to robust transcriptional regulation of expression of activity-dependent genes in auditory and vocal control areas respectively. Therefore, birdsong provides an ideal paradigm to study transcription regulation by a natural learned stimulus in the brain of awake behaving animals. In this chapter we first discuss briefly some basic aspects of birdsong neurobiology, focusing on the substrates for perceptual and motor aspects of vocal communication in songbirds. We then discuss our current knowledge of the influence of stimulus type, behavioral condition and context on transcriptional regulation by song; the mechanisms regulating induced gene expression in song-encoding neurons; and the possible functional significance of the transcriptional response to song in auditory and song control areas of the songbird brain.

1.1 The Neurobiology of Vocal Communication in Songbirds: the Substrates for Song Perception, Production and Learning

Birdsong is a complex learned behavior used extensively by songbirds for individual recognition, territorial defense and attracting potential reproductive partners (see Catchpole and Slater, 1995; Kroodsma and Miller, 1996 for reviews). Whereas many animals use a wide variety of signals to communicate, vocal learning, i.e. the ability to learn their vocal communication signals from exposure to a model, is a rare trait shared only by a few groups (humans, cetaceans, possibly bats, and three orders of birds: songbirds, parrots and hummingbirds) (Nottebohm, 1972; Brenowitz, 1997). The processing and memorization of complex auditory signals is essential for the learning and maintenance of vocalizations and for perceptual aspects of vocal communication in all vocal-learners, including humans (Doupe and Kuhl, 1999). Yet the underlying brain pathways and mechanisms are for the most part unknown. In contrast to other vocal learners, the brain circuits controlling song learning, production and perception in songbirds have been identified and are well characterized (see Brenowitz, Margoliash and Nordeen, 1997; Zeigler and

S. M. Dudek (ed.), *Transcriptional Regulation by Neuronal Activity.*
© Springer 2008

Marler, 2004 for reviews), placing birdsong as an excellent model for understanding the neuronal basis of vocal communication and vocal learning.

Vocal communication and learning in songbirds require two systems of brain pathways, the auditory system and the song control system. The avian ascending auditory pathways are quite similar, in their general organization, to the auditory pathways of other vertebrates (Butler and Hodos, 1996). Although the brainstem auditory pathway has not been studied in detail in songbirds, it is thought to consist of a series of cochlear, lemniscal, and midbrain nuclei that carry auditory input from the cochlea to the thalamic auditory nucleus ovoidalis (Ov; Fig. 1a, left panel). Auditory information then reaches the telencephalon through the projection from Ov to the primary thalamo-recipient zone field L, more specifically its L2a subdivision (Kelley and Nottebohm, 1979; Vates, Broome, Mello and Nottebohm, 1996). L2a sends projections to adjacent subfields L1 and L3, which in turn project to the main pallial targets of L, namely the caudomedial nidopallium (NCM), the caudomedial mesopallium (CMM), and the shelf and cup areas, the latter two in close proximity to nuclei of the song control system (Fig. 1a) (we follow here the revised avian brain nomenclature; see Reiner, Perkel, Bruce, Butler, Csillag, Kuenzel, Medina, Paxinos, Shimizu, Striedter, Wild, Ball, Durand, Gunturkun, Lee, Mello, Powers, White, Hough, Kubikova, Smulders, Wada, Dugas-Ford, Husband, Yamamoto, Yu, Siang and Jarvis, 2004). This auditory system is common to both males and females (Vates et al., 1996; Mello, Vates, Okuhata and Nottebohm, 1998c; Mello unpublished observations), and is necessary for the acquisition of song in juveniles and for its maintenance during adulthood (Konishi, 1965; Marler and Peters, 1977; Nordeen and Nordeen, 1992; Leonardo and Konishi, 1999; Woolley and Rubel, 2002). It is also necessary for the perceptual processing and memorization of birdsong in the context of identification and discrimination of other conspecifics (Catchpole et al., 1995; Kroodsma et al., 1996). Auditory pathways similar to those described in songbirds are thought to be present in other avian groups regardless of whether they learn their vocalizations, indicating that this is a highly conserved pathway among birds (Karten and Hodos, 1967; Karten, 1968; Bonke, Bonke and Scheich, 1979; Brauth, McHale, Brasher and Dooling, 1987; Brauth and McHale, 1988; Wild, 1993; Metzger, Jiang and Braun, 1998).

The song control system comprises two sets of interconnected nuclei, the direct vocal-motor and the anterior forebrain pathways (Fig. 1B, left panel). The direct vocal-motor pathway (Fig. 1B, black arrows) consists of projections from nucleus HVC (used as a proper name) to the robust nucleus of the arcopallium (RA), and from RA to midbrain and medullary centers involved in the control of vocal and respiratory function (Nottebohm, Kelley and Paton, 1982; see also review by Wild, 1997). This direct pathway is actively engaged during singing behavior (Yu and Margoliash, 1996; Jarvis and Nottebohm, 1997; Olveczky, Andalman and Fee, 2005) and is required for song production (Nottebohm, Stokes and Leonard, 1976). The anterior forebrain pathway (Fig. 1B, grey arrows) consists of projections from area X in the medial striatum to the medial part of the dorsolateral thalamic nucleus (DLM), from DLM to the lateral magnocellular nucleus of the anterior nidopallium (LMAN), and from LMAN back to area X (Bottjer, Halsema, Brown and Miesner, 1989; Johnson, Sablan and Bottjer, 1995; Vates, Vicario and

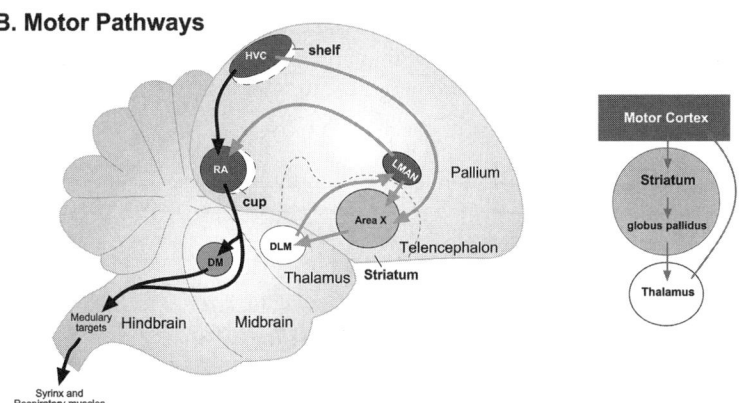

Fig. 1 Schematic representation of *auditory* (A) and *song control* (B) pathways and associated genomic responses to song in songbirds and the equivalent pathways in mammals. Dark grey indicates areas that show activity-dependent transcription in songbirds upon song-stimulation (Dark auditory pathway) and singing (motor pathway). A White indicates areas that lack this response. In B, black arrows indicate the direct vocal-motor pathway and grey arrows indicate the anterior forebrain pathway. CMM, caudomedial mesopallium; DLM, dorsal lateral nucleus of the medial thalamus; DM, dorsal medial nucleus; L1, L2 and L3, fields L1, L2, and L3; LMAN, lateral magnocellular nucleus of the anterior nidopallium; MLd, dorsal lateral nucleus of the anterior mesencephalon; NCM, caudal medial nidopallium; Ov, nucleus ovoidalis; RA, robustus nucleus of the arcopallium

Nottebohm, 1997; Luo, Ding and Perkel, 2001). This anterior pathway is connected with nuclei in the direct motor pathway via HVC-to-X and LMAN-to-RA projections (Fig. 1B) and is essential for vocal learning (Bottjer, Miesner and Arnold, 1984; Sohrabji, Nordeen and Nordeen, 1990; Scharff and Nottebohm, 1991) and vocal plasticity (Olveczky *et al.*, 2005). The song control system is typically more developed in males and is under the strong influence of sex steroids

(Nottebohm and Arnold, 1976; Arnold, 1980; Arnold, Bottjer, Brenowitz, Nordeen and Nordeen, 1986; Wade and Arnold, 2004). In some species, such as canaries, the song system shows marked seasonal fluctuations in neuronal cell numbers and morphology that are associated with vocal plasticity (Nottebohm, 1981; Nottebohm, Nottebohm and Crane, 1986; Brenowitz and Arnold, 1990; Brenowitz, 2004). Such observations led to the discovery of marked neuronal replacement during adulthood (Goldman and Nottebohm, 1983; Paton and Nottebohm, 1984).

It is important to note that the avian brain is currently considered to be more similar to the mammalian brain in its general organization than previously thought. Although birds lack the layered organization typical of the mammalian brain, it is clear that most of the avian telencephalon is pallial in its origin, and thus homologous as a field to the mammalian cortex (Jarvis, Gunturkun, Bruce, Csillag, Karten, Kuenzel, Medina, Paxinos, Perkel, Shimizu, Striedter, Wild, Ball, Dugas-Ford, Durand, Hough, Husband, Kubikova, Lee, Mello, Powers, Siang, Smulders, Wada, White, Yamamoto, Yu, Reiner and Butler, 2005). In addition, individual pallial and basal ganglia structures of the avian brain share many neurochemical markers and patterns of connectivity with their presumed mammalian counterparts (e.g. Fig. 1A and B, compare right and left panels) (also see (Jarvis *et al.*, 2005). This new view of the avian brain has been consolidated with a comprehensive nomenclature change in order to better reflect the known homologies between avian and mammalian brain structures (Reiner *et al.*, 2004).

1.2 Birdsong and activity-dependent transcription

1.2.1 Gene regulation by song auditory stimulation

When songbirds hear song, the expression of several activity-dependent genes is induced in discrete areas along the auditory pathway (dark grey areas in Fig. 1A) (Mello, Vicario and Clayton, 1992; Mello and Clayton, 1994; Mello and Ribeiro, 1998b; Kruse, Stripling and Clayton, 2000; Velho, Pinaud, Rodrigues and Mello, 2005). This phenomenon is most prominent in two major telencephalic areas, the caudomedial nidopallium (NCM) and the caudomedial mesopallium (CMM). Based on their connectivity, these areas are arguably analogous to supra-granular layers of the mammalian auditory cortex as they are targets of the primary thalamo-recipient zone field L (Fig. 1A, right panel). In addition, these areas are thought to participate in various aspects of song processing and discrimination (see reviews in Mello, 2002; Mello, Velho and Pinaud, 2004). The genomic response to song includes genes that encode early effectors (e.g., *Arc*), which are thought to exert a direct action on neuronal cell physiology, as well as inducible transcription factors (ITFs; e.g., *zenk*, *c-fos*, *c-jun*), which exert a more indirect and protracted effect through the regulation of the expression of downstream target genes (Fig. 2; we use italics for the genes and normal lettering for the protein, for example *zenk* and ZENK). In both cases, the expression levels of these activity-dependent genes,

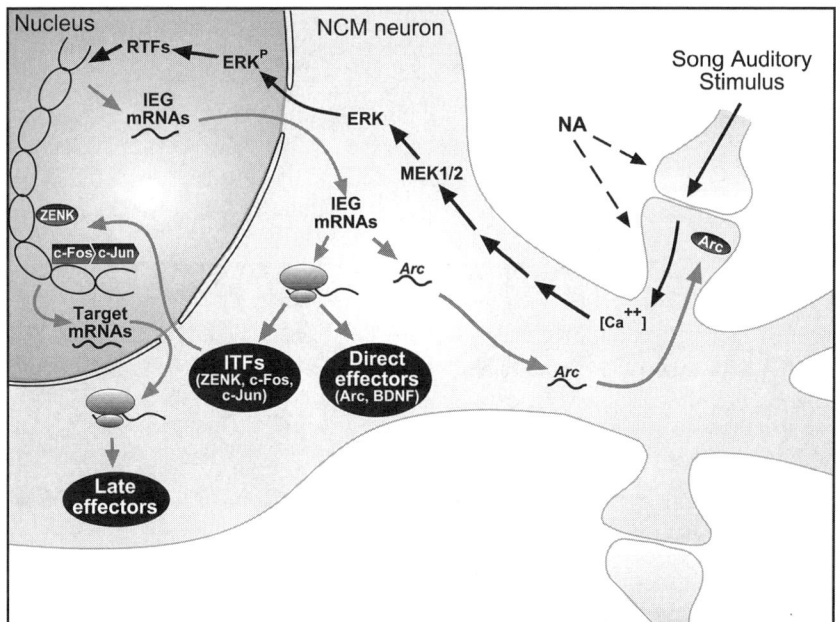

Fig. 2 Song-induced transcriptional response in auditory neurons. Black arrows indicate the second messenger signaling cascade triggering transcriptional activation; grey arrows indicate the genomic response to song stimulation; dashed arrows indicate possible modulatory mechanisms. ERK, extracellular-signal regulated kinase; ERK^P, phosphorylated ERK; MEK1/2, mitogen-activated extracellular-signal-regulated protein kinase kinase 1 and 2; IEG, immediate early genes; ITFs, inducible transcription factors; NA, noradrenaline; RTFs, regulatory transcription factors

as measured by either *in situ* hybridization or immunocytochemistry, is practically absent in auditory areas of unstimulated quiet controls (Figure 3A and B, control), indicating that the expression observed in song-stimulated birds (Fig 3A and B, stimulated) results from activation at the transcriptional level. The expression of several identified song-induced genes (*zenk, c-fos* and *Arc*) is induced in the same neuronal cells (Figure 3C), strongly suggesting that song-responsive neurons show a coordinated genomic response to song stimulation (Velho *et al.*, 2005). Most studies, however, have focused on song-induced regulation of *zenk* expression in NCM. Unless otherwise stated, this will also be the focus of our discussion below.

The initial song-induced response in auditory areas is rapid and transient (Mello *et al.*, 1992), appearing within minutes after stimulus onset, peaking at 30 minutes, and declining thereafter (Mello *et al.*, 1992; Mello *et al.*, 1994; Mello *et al.*, 1998b; Kruse *et al.*, 2000; Velho *et al.*, 2005). At the cellular level, activity-dependent gene expression is initially detected at the transcription sites within the nucleus, and the mRNA products progressively migrate into the cytoplasm of the activated cells (Velho, Pinaud, Jeong and Mello, 2003; Velho *et al.*, 2005). This transient nuclear accumulation confirms the transcriptional activation of song-inducible genes and is only detectable during a short interval that presumably comprises synthesis of the pre-mRNA, its processing into mature mRNA, and its export from the

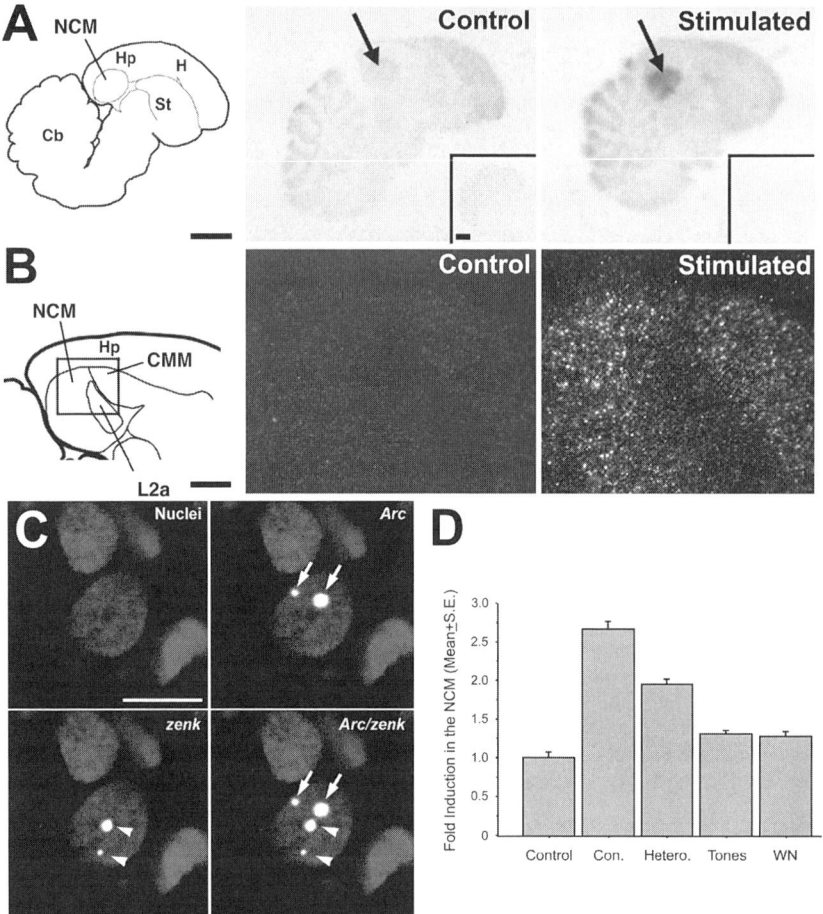

Fig. 3 Song-induced transcriptional response in the songbird telencephalon (adapted from Velho *et al.*, 2005). A) *Arc* expression in the NCM of silent control and song-stimulated zebra finches. Left panel, camera lucida drawing of a parasagittal section at the level of the NCM. Middle and right panels, autoradiograms of sections at the level depicted on the left panel, from control and song-stimulated birds, hybridized with a radioactively labeled *Arc* riboprobe. Insets, autoradiograms of sections hybridized with sense strand riboprobes. B) *Arc* expression in NCM, CMM and field L2a of silent control and song-stimulated birds. Left panel, camera lucida drawing of a parasagittal section at the level of the NCM, CMM and field L2a. Middle and right panels, dark field view of emulsion autoradiograms of sections at the level depicted on the left panel, from control and song-stimulated birds, hybridized with a radioactively labeled *Arc* riboprobe. C) Co-localization of song-induced *Arc* and *zenk* mRNAs in NCM neurons. Top left panel, Hoechst stained nuclei of NCM cells. Top right panel, NCM neuron showing two intranuclear foci of *Arc* mRNA. Bottom left panel, the same NCM neuron showing two intranuclear foci of *zenk* mRNA. Bottom right panel, overlay of the images shown in the other panels depicting the cellular co-localization of *Arc* and *zenk* (see Velho *et al.*, 2005 for a color version of this figure). D) Normalized *Arc* fold-induction levels in the NCM of birds stimulated for 30 min with conspecific song (Con.), heterospecific song (Hetero.), artificial tones (Tones), or white noise (WN) compared with unstimulated birds (Control). Cb, cerebellum; CMM, caudomedial mesopallium; H, hyperpallium; Hp, hippocampus; L2a, subfield L2a; NCM, caudomedial nidopallium; St, medial striatum. Scale bars: 2 mm for A and B; 10 μm for D

nucleus (~1-15 minutes) to the cytoplasm, where the newly synthesized mRNAs are translated into proteins before undergoing degradation (Fig. 2). The mRNA levels of two song-inducible genes encoding transcription factors, *zenk* and *c-fos*, return to levels close to controls by one hour after stimulus onset (Mello *et al.*, 1994; Velho *et al.*, 2005) whereas the direct effector *Arc* remains above unstimulated levels for several hours after the end of the stimulation (Velho *et al.*, 2005), indicating that distinct regulatory mechanisms determine the decay rates of these different mRNAs. Because the mRNA levels reflect a balance between synthesis and degradation, it is not clear whether the differences between *zenk* and *Arc* signal are at the transcriptional or post-transcriptional level (i.e. different rates of synthesis and/or degradation). The translated protein products of song-inducible ITFs have the expected nuclear distribution, as revealed by ICC (Kimpo and Doupe, 1997; Mello, 1998), and have been extensively used to map song-induced neuronal activation in several species and contexts (see Mello, 2002 for a review).

1.2.1.1 Stimulus Dependence

The strength of the genomic response observed in NCM is stimulus-dependent. Early studies showed that stimulation with conspecific song results in larger amounts of *zenk* mRNA compared to birds stimulated with heterospecific song or pure tones for similar durations and intensities (Mello *et al.*, 1992). This observation suggests that NCM neurons are more responsive (tuned) to acoustic (spectral and/or temporal) features present in conspecific song. Recent studies show that two other song-inducible genes, *Arc* and *c-fos*, present the same specificity, with stronger induction responses for natural sounds when compared to synthetic stimuli (Fig. 3D) (Velho *et al.*, 2005; Velho and Mello unpublished observations). It is not yet clear whether this observation, based on densitometric analysis of autoradiograms, reflects differences in the mRNA amounts per cell or in the total numbers of activated cells across stimuli.

The quality of the song, which directly influences the behavioral responses to song in females (mating choice), also modulates the song-induced genomic response in auditory areas. Females often show behavioral preferences for particular features present in conspecific songs. For instance, starling (*Sturnus vulgaris*) females prefer males singing longer bouts (Gentner and Hulse, 2000). Similarly, canary (*Serinus canaria*) females prefer songs containing more complex syllables, with rapid frequency modulations and higher repetition rates (Valet *et al.*, 1998). Stimulation of females with preferred songs induces higher levels of ZENK in NCM and CMM when compared with non-preferred songs, for canaries and starlings respectively (Gentner, Hulse, Duffy and Ball, 2001; Leitner, Voigt, Metzdorf and Catchpole, 2005). Thus, the strength of the *zenk* response to song in the NCM and CMM of females correlates with the quality of the male song.

The genomic response to song is tuned to specific acoustic components present in the stimulus, detecting subtle variations in spectral/temporal features. In canaries, auditory stimulation with whistles (i.e., syllabic elements from canary song that resemble pure tones) triggers gene expression in discrete subregions of NCM (Figure 4; Ribeiro, Cecchi, Magnasco and Mello, 1998). Specifically,

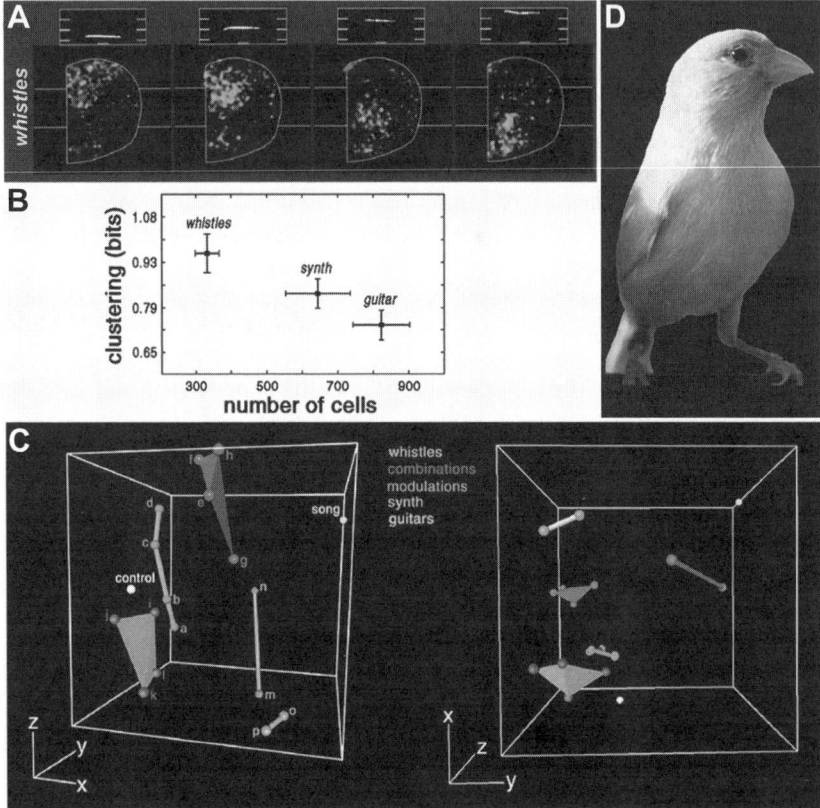

Fig. 4 Transcriptional activation analysis of syllabic representation in canary NCM (modified from Ribeiro et al., 1998). A) ZENK expression maps in the NCM of adult female canaries resulting from the presentation of natural canary whistles. The sonograms (top) show the respective frequencies (left to right), namely 1.4, 2.2, 2.8 and 3.5 kHz. B) Patterns elicited by natural whistles and artificial stimuli of the same frequencies can be clearly separated by quantifying total cell number and spatial clustering. C) Principal Component Analysis of ZENK expression maps in NCM. The first three components (plotted on the x, y and z axes) provide for a clear separation of the ZENK patterns that accords to the various groups of stimuli in this study, indicating that a syllabic representation of canary song is present in NCM. D) Female canary (*Serinus canaria*) (*See* Color Plate 1).

high-frequency whistles induce *zenk* expression in ventral portions and low-frequency whistles in dorsal portions of rostral NCM, in a graded and tonotopic fashion (Figure 4A; Ribeiro *et al.*, 1998). However, a comparison between the *zenk* response to naturally occurring canary whistles versus synthetic stimuli of the same frequencies but that lack subtle variations in amplitude and frequency reveals markedly different patterns of activation in terms of both intensity and spatial distribution of responsive cells (Figure 4B and C; Ribeiro *et al.*, 1998). Thus, the syllabic representation in canary NCM is tuned to specific acoustic elements present in canary song.

The amount of stimulation also influences the expression levels of song-induced genes. Contrary to stimulus type, the amount of stimulation has only a small effect on gene induction measured as the number of responding cells. For example, stimulation with a single song presentation ($\sim$2 seconds) is sufficient to recruit a large number of *zenk*-expressing cells in NCM. The number of responding cells peaks with a few further (4–10) song repetitions, but does not increase with even further stimulation (Kruse *et al.*, 2000). In contrast to the number of responding cells, the mRNA levels of song-induced genes increase continuously with increasing stimulation, peaking after about 30 minutes of song presentation (10–12 song repetitions per min), but declines thereafter, regardless of further stimulation (Mello, Nottebohm and Clayton, 1995a; Velho *et al.*, 2005). Thus, while the number of song-responsive neurons rises sharply after a brief stimulation, the amount of song-induced mRNA rises more gradually before reaching a plateau and declining. These observations likely reflect the very nature of the genomic response to neuronal excitation, where the neuronal population recruited depends primarily on properties intrinsic to the stimulus (such as spectral/temporal features of song), whereas the total amount of resulting transcription is determined by the recruitment of signal transduction pathways, which in turn is determined by the amount of stimulation.

1.2.1.2 Context Dependence

The strength of the genomic response observed in NCM and CMM is also modulated by the context in which the stimulus is presented. For instance, the associative pairing of conspecific song stimulation with a foot shock, in an active avoidance paradigm, enhances the *zenk* response in NCM and CMM, whereas shock alone or shock uncorrelated with song stimulation has no modulatory effect (Jarvis, Mello and Nottebohm, 1995). Similarly, pairing the song stimulus with salient visual stimulation, such as images of conspecifics or simply with an enriched light environment (multi-colored lights) results in higher levels of ZENK protein expression in these auditory areas (Kruse, Stripling and Clayton, 2004; Avey, Phillmore and MacDougall-Shackleton, 2005). Modulatory effects on transcriptional activation can also be observed by manipulating the distance to the stimulus source and/or intensity (i.e. sound pressure level) (Kruse *et al.*, 2004). Experimental conditions that alter the arousal state of the bird, such as restraint, can also affect the amount of *zenk* expression when compared to that seen in freely behaving birds (Park and Clayton, 2002). Such procedure-induced increases in gene expression can be reversed or attenuated by acclimation sessions consisting of brief and repeated (4–5) exposures to the experimental condition or apparatus (Jarvis *et al.*, 1995; Velho *et al.*, 2005). The mechanisms underlying the context dependence of song-induced gene expression most likely reflect changes in the behavioral state of the bird, particularly in attention. As discussed later in this chapter, this effect may involve catecholaminergic modulatory systems.

1.2.1.3 Experience Dependence

The expression of *zenk* in NCM and CMM is also heavily influenced by experience. Repeated presentations of the same stimulus lead to a progressive decline of the *zenk* response, a phenomenon that has been termed habituation (Mello *et al.*, 1995a). This reduction in *zenk* activation is accompanied by a similar reduction in firing rates of NCM neurons (Chew, Mello, Nottebohm, Jarvis and Vicario, 1995; Stripling, Volman and Clayton, 1997; Terleph, Mello and Vicario, 2006), and it can be relatively long-lasting, according to the amount of habituating stimulation used. Both the firing rates and the transcriptional induction of *zenk* in song-habituated NCM neurons can be reinstated by presentation of a novel song stimulus (Chew *et al.*, 1995; Mello *et al.*, 1995a; Stripling *et al.*, 1997). Importantly, habituation can be induced in response to multiple songs independently (Figure 5A). Therefore, the degree of habituation has been suggested as an index of the extent to which a specific song is remembered by song-responsive circuits (Chew *et al.*, 1995). In other words, habituation appears to be a physiological memory trace of a given song. Furthermore, habituation to song in NCM may play a significant contributing role to song discrimination.

Spaced brief pre-exposure to particular song or categories of song also regulates the strength of the transcriptional activation. In song sparrows, familiar songs, i.e. songs to which the birds have been previously exposed (30 min per day for seven days), induce ZENK expression in a smaller number of neurons in the ventral NCM when compared to novel songs (McKenzie, Hernandez and Macdougall-Shackleton, 2006), suggesting that some long-term habituation has occurred in this area. In starlings, pre-exposure to long-bout songs results in higher ZENK expression in both NCM and CMM in response to a novel long-bout song when compared to a novel short-bout song (Sockman, Gentner and Ball, 2002). Curiously, pre-exposure to short-bout songs increases Fos immunoreactivity in response to a novel short-bout song in these areas (Sockman, Gentner and Ball, 2004), suggesting in this case different modulatory mechanisms for ZENK and c-Fos.

In addition to recent experience (habituation), the birds' past experience also influences the magnitude of the genomic response to song. In zebra finches, the expression levels of *zenk* mRNA triggered by stimulation with the tutor's song (i.e. the song copied during the song learning period) are higher than the expression levels triggered by the bird's own song or a novel conspecific song, demonstrating a modulatory effect of the bird's previous experience (Bolhuis, Zijlstra, Den Boer-Visser and Van Der Zee, 2000; Terpstra, Bolhuis, Riebel, Van Der Burg and Den Boer-Visser, 2006). In addition, the number of ZENK-expressing cells in response to the tutor song correlates positively with how well that song was copied (Bolhuis *et al.*, 2000). This indicates that the genomic response in NCM is tuned to acoustic properties in the tutor's song and may be linked with an auditory representation, or memory, of that song. Consistent data have recently been obtained by recording the electrophysiological response of NCM to learned tutor songs (Phan, Pytte and Vicario, 2006).

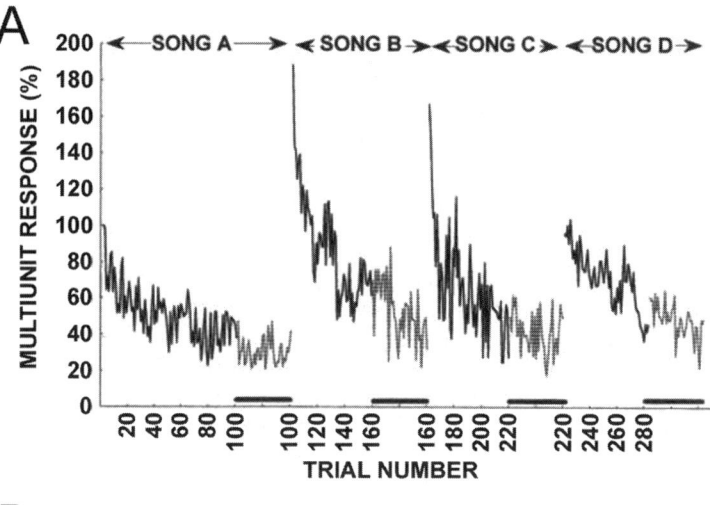

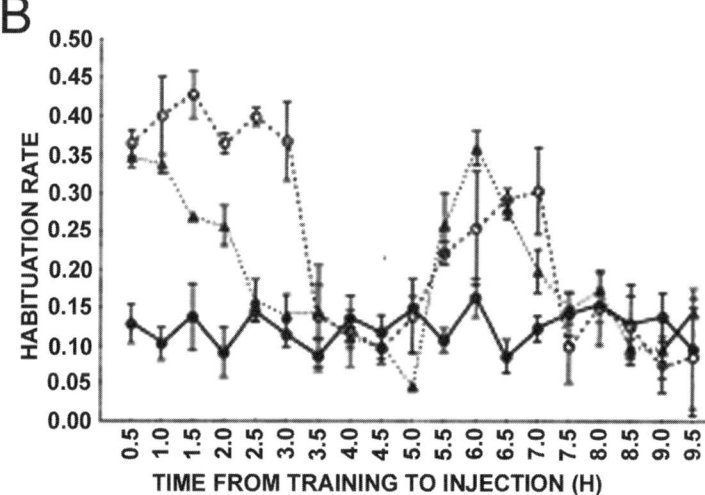

Fig. 5 Song-evoked electrophysiological responses of NCM neurons in zebra finches (adapted from Chew *et al.*, 1995). A) Electrophysiological habituation induced by repeated presentations of conspecific songs; the response amplitudes from multiunit recordings were normalized to the initial response to the first song presented. Dark bars on the x axis indicate the responses during test trials after all four songs were presented, showing that the habituated state is maintained even after training with different songs. The habituation rate represents the linear regression slope between normalized response magnitudes and iteration number. B) Effect of local injections of synthesis inhibitors on the long-term maintenance of the habituated state in NCM. Plotted are the habituation rates as a function of the interval from training onset to injection of the protein synthesis inhibitor cycloheximide (open circles), of the RNA synthesis inhibitor actinomycin (triangles), and of saline (solid circles)

1.2.1.4 Uncoupling of Auditory Activation and Gene Expression

Some auditory areas do not show song-induced transcription of *zenk* or other song-inducible genes (Mello *et al.*, 1992; Velho *et al.*, 2005), even though they are obligatory stations in the ascending auditory pathway (Fig. 1A, left panel) and are thought to be activated by song. These include the thalamic nucleus Ov and the telencephalic subfield L2a. The latter is the main entry site for auditory information into the telencephalon and precedes NCM and CMM in the auditory pathway (Vates *et al.*, 1996). The fact that L2a does not show song-induced transcription suggests that neuronal activation and activity-dependent gene expression are uncoupled in this area. In fact, even widespread depolarization triggered by drug-induced seizures does not lead to *zenk* induction in L2a (Mello and Clayton, 1995b). A possible function for the uncoupling would be the need to turn off activation-induced plasticity in L2a, preserving the fidelity of the auditory information that is relayed to higher-order auditory areas. Accordingly, in contrast to NCM and CMM, L2a neurons show strong phasic responses to auditory stimulation that do not habituate with repeated presentations (Terleph *et al.*, 2006), indicating the absence of experience-dependent plasticity. Although the mechanism of this uncoupling between electrical activation and transcriptional regulation is unknown, it could be due to the down-regulation or absence of critical components of the signal transduction pathway leading to induced-gene expression or, as occurs in the rodent hippocampus (Weaver, Cervoni, Champagne, D'Alessio, Sharma, Seckl, Dymov, Szyf and Meaney, 2004), by permanently altering the chromatin through methylation, which renders activity-dependent gene promoters unaccesible to activating transcription factors.

1.2.2 Gene Regulation by Singing Behavior

When birds sing, activity-dependent genes are markedly induced in the song control system (Jarvis *et al.*, 1997; Kimpo *et al.*, 1997; Velho *et al.*, 2005; Poopatanapong, Teramitsu, Byun, Vician, Herschman and White, 2006; Wada, Howard, McConnell, Whitney, Lints, Rivas, Horita, Patterson, White, Scharff, Haesler, Zhao, Sakaguchi, Hagiwara, Shiraki, Hirozane-Kishikawa, Skene, Hayashizaki, Carninci and Jarvis, 2006). Similar to the auditory system, the song control nuclei of birds that remained silent prior to sacrifice show little or no expression of known song-inducible genes (Jarvis *et al.*, 1997; Kimpo *et al.*, 1997; Velho *et al.*, 2005). In contrast, when birds sing, a genomic response is triggered in nuclei that control singing behavior, more specifically in song nuclei HVC and RA (Fig. 1B grey areas) (Jarvis *et al.*, 1997; Kimpo *et al.*, 1997; Mello, 1998; Velho *et al.*, 2005). This induction is also observed in the auditory processing areas discussed above (NCM, CMM, L1, L3, and the shelf and cup areas). In the latter regions, however, singing-induced expression is abolished by deafening, while it remains unaltered in song control nuclei. This demonstrates that the singing-induced transcriptional response in song control nuclei is motor-driven rather than auditorily triggered (see Fig. 2D in Jarvis *et al.*, 1997).

Interestingly, singing-induced gene regulation is also dependent on the context of the singing behavior, as clearly demonstrated in zebra finches (Jarvis, Scharff, Grossman, Ramos and Nottebohm, 1998). For example, when the bird sings in a solo context, or performs "undirected singing", i.e. song not directed to another male or female, *zenk* induction is observed in all telencephalic song control nuclei (HVC, RA, LMAN, and area X). In contrast, when the bird sings towards another male or female, i.e. performs "directed singing", as during courtship behavior, *zenk* induction is not observed in striatal area X, revealing a context-dependent regulation of its expression in the anterior forebrain pathway. This context-dependent regulation is also associated with differences in firing rates in area X, i.e. higher firing rates for undirected compared to directed singing (Hessler and Doupe, 1999a). This difference in neural activity is possibly the result of different neuromodulatory mechanisms acting under different behavioral conditions, as discussed later in this chapter.

1.3 Regulation of Song-induced Gene Expression in Auditory Neurons

As discussed above, the neuronal activation triggered by song perceptual processing and singing behavior induces an increase in activity-dependent gene expression in auditory and song control areas, respectively. This phenomenon is likely due to increased transcriptional activity because (*i*) the expression levels of song-induced genes are very low or absent in sound-isolated quiet controls; (*ii*) song presentation and singing behavior lead to a robust increase in the mRNA levels of several activity-dependent genes; and (*iii*) activity-dependent mRNAs are only detected at nuclear transcriptional sites of activated cells shortly after stimulation. Moreover, song stimulation appears to trigger a well-orchestrated gene regulatory program, because known song-inducible genes (*zenk*, *Arc* and *c-fos*) are co-expressed in activated neurons (Figure 3C).

The study of the transduction machinery that links neuronal activation by song to activity-dependent gene expression in songbirds is still incipient. Based mostly on studies in mammals, the induction of transcriptional regulation in neuronal cells is likely initiated by the activation of ionotropic and/or metabotropic membrane receptors leading to rises in intracellular Ca^{++} (Berridge, 1998). The precise nature of the synaptic input has not been determined in songbirds, but intracellular Ca^{++} elevation and second messenger recruitment through glutamate receptor activation is well-established in the mammalian brain (see Wang, Tang, Parelkar, Liu, Samdani, Choe and Yang, 2004 for a review; also see chapters 4, 5, 6 and 7 in this vollum). Although representatives of all classes of glutamate receptors are expressed in auditory and song control areas of zebra finches (Wada, Sakaguchi, Jarvis and Hagiwara, 2004; Velho and Mello unpublished observations), their functional role has not been directly tested. Alternatively, Ca^{++} could enter song-responsive neurons through activation of L-type calcium channels, also known to regulate the activation of transcriptional responses in mammals (Ghosh, Ginty, Bading and Greenberg, 1994; Finkbeiner and Greenberg, 1998).

It is important to point out that the inhibitory neurotransmitter GABA may also shape the genomic response to song. A very large percentage of NCM neurons (about 50%) are GABAergic and a significant portion of these inhibitory cells shows song-induced *zenk* transcription, indicating their direct participation in NCM's response to song (Pinaud, Velho, Jeong, Tremere, Leao, von Gersdorff and Mello, 2004). Furthermore, whole-cell patch-clamp recordings in NCM slices show very prominent spontaneous inhibitory post-synaptic potentials. This suggests that the GABAergic network tonically suppresses the activity of excitatory circuits in NCM at rest, and that song-induced activation of NCM likely involves the relief of such circuits from GABAergic suppression.

The initiation of activity-dependent transcriptional activation requires the recruitment of intracellular signal transduction pathways. In zebra finch auditory neurons, song-induced transcription is dependent on the activation of the mitogen-activated extracellular-signal-regulated protein kinase kinases 1 and 2 (MEK 1/2), whose main substrate is the extracellular-regulated kinase (ERK) (Fig. 2). Local blockade of MEK 1/2 prevents song-induced expression of *zenk*, *Arc* and *c-fos* in NCM (Cheng and Clayton, 2004; Velho *et al.*, 2005), suggesting that similar or converging signaling pathways regulate the expression of various song-induced genes and implicating ERK in song-induced transcription. Once phosphorylated, ERK is thought to translocate to the nucleus, where it activates constitutive regulatory transcription factors (RTFs) (Fig. 2) see review in (Grewal, York and Stork, 1999). Interestingly, the *zenk* gene promoter (in canaries) has binding sites for at least two RTFs that are known ERK targets (Cheng *et al.*, 2004), namely the cAMP response element-binding protein (CREB) and Elk-1, a member of the ETS oncogene family.

The kinetics of second-messenger activation may determine some fundamental aspects of song-induced transcriptional regulation. For instance, the phosphorylation of ERK by MEK1/2 in auditory structures (NCM, CMM and field L combined) shows very similar properties when compared to song-induced gene expression. ERK phosphorylation (*i*) is stimulus specific, i.e. it is higher for conspecific sounds compared to tones and white noise; (*ii*) is rapid and transient, returning to basal levels 5 min after the end of stimulation; (*iii*) habituates to repeated presentations of the same stimulus but is restored by presentation of a novel stimulus (Cheng *et al.*, 2004). Thus, even though neither the identity of ERK targets nor the duration of their activation is known, ERK phosphorylation mimics the genomic response to song and may determine some of its properties.

The fact that the transcriptional response to song is stimulus-dependent, regulated by experience, and influenced by the context in which the stimulus occurs, suggests that modulatory brain systems that alter the salience of the stimulus may regulate this response. A probable candidate is the noradrenergic system. In mammals, neuronal firing rates in the locus cerouleus (LC), the main source of noradrenergic projections, are proportional to the behavioral relevance of the stimulus, decrease with repeated stimulus presentations, and are restored by a novel stimulus (Aston-Jones and Bloom, 1981b; a). In addition, LC lesions disrupt the cortical induction of activity-dependent genes like *zenk* (aka *zif-268*, *egr-1*, *NGFI-A*, and *krox-24*) and *c-fos* (Cirelli, Pompeiano and Tononi, 1996; Cirelli and Tononi, 2000). Similar to mammals, the songbird telencephalon has a widespread noradrenergic innervation that includes pallial auditory processing and song control areas, but not the striatum, based on immunostaining for the noradrenaline-synthesizing enzyme dopamine-β-hydroxylase (DBH) (Mello, Pinaud and Ribeiro, 1998a). The blockade of adrenergic receptors by systemic injections of antagonists abolishes song-induced transcription of *zenk* in NCM, confirming a link between the noradrenergic system and song-induced gene expression (Ribeiro and Mello, 2000). While these systemic injections do not reveal the site of noradrenergic action, preliminary intracerebral injections confirm that this modulation occurs locally in NCM (Velho, Ribeiro, Pinaud and Mello, 2006).

Another catecholamine, dopamine, may regulate the higher gene expression induced by undirected vs. directed singing in striatal area X of the song control system (Jarvis *et al.*, 1998; Hessler and Doupe, 1999b). Similar to mammals, the songbird striatum, including area X, receives dense dopaminergic innervation from the ventral tegmental area (VTA) and the substantia nigra pars compacta (SNc) in the midbrain, and expresses high levels of dopamine receptors (Lewis, Ryan, Arnold and Butcher, 1981; Bottjer, 1993; Casto and Ball, 1996). Patch-clamp recordings in area X show that dopamine can modulate glutamatergic transmission in spiny neurons, reducing glutamatergic currents via activation of D1-like receptors (Ding, Perkel and Farries, 2003). In addition, dopamine levels in area X measured by *in vivo* microdialysis in awake behaving birds increase above baseline only during directed singing (Sasaki, Sotnikova, Gainetdinov and Jarvis, 2006). The increased dopaminergic input to area X in this context may decrease glutamate-dependent currents and therefore dampen the singing-induced transcriptional response in this nucleus. The singing context dependence of *zenk* expression can also be altered by pharmacological ablation of the noradrenergic system in the zebra finch (Castelino and Ball, 2005). However, because the VTA receives a dense noradrenergic projection (Mello *et al.*, 1998a), it is presently unclear whether the noradrenergic effect in area X is direct or mediated through the dopaminergic system (VTA-to-X projection).

1.4 Possible Functions of Song-induced Gene Expression in Auditory and Song Control Areas

Activity-dependent transcription has been postulated to link neuronal activation to long-term changes underlying neuronal plasticity, learning and memory (Goelet, Castellucci, Schacher and Kandel, 1986; Clayton, 2000). Indeed, the induction of activity-dependent genes by birdsong occurs in a wide variety of circumstances associated with learning (e.g., when birds hear the tutor song during the song learning period or an unfamiliar song in adulthood), as well as retrieval and/or possibly reconsolidation (e.g. when birds hear tutor song or their own song during adult singing, or the songs of familiar conspecifics during adulthood). Because the strength of the response is differentially regulated by novel (new learning) vs. familiar songs (retrieval and/or reconsolidation), it is likely that activity-dependent transcription exerts multiple functions in song-responsive circuits, depending on age and context. The fact that several components of the genomic response to song have been linked to neuronal plasticity, as well as learning and memory consolidation and reconsolidation in mammals (Morrow, Elsworth, Inglis and Roth, 1999; Guzowski, Lyford, Stevenson, Houston, McGaugh, Worley and Barnes, 2000; Jones, Errington, French, Fine, Bliss, Garel, Charnay, Bozon, Laroche and Davis, 2001; Lee, Everitt and Thomas, 2004) suggests that these components may play similar roles in song-responsive circuits in songbirds.

Importantly, gene expression is required for the stabilization of activity-dependent changes in the properties of song-responsive neurons in the songbird NCM (Chew *et al.*, 1995). More specifically, the long-term maintenance (several hours to days) of song-specific habituation to repeated presentations of a given conspecific song (Fig. 5A) requires local RNA and protein synthesis, as shown by injections of RNA and protein synthesis inhibitors in NCM (Fig. 5B). (Chew *et al.*, 1995). Two distinct *epochs* of *de novo* synthesis (i.e., the initial 2-3 hr period and the period between 5.5 and 7 hr after onset of song stimulation) are necessary for the stabilization of experience-dependent changes in NCM, as shown by varying the time of the pharmacological blockade relative to the training (song presentation) period (Fig 5B). In contrast, RNA and protein synthesis blockers do not affect the auditory responses to song or the initial acquisition of the habituated state in NCM neurons (Chew *et al.*, 1995).

The time-course of dependence on local RNA and protein synthesis suggests that song-induced genes might be involved in the long-term habituation of NCM responses to song. Indeed, song-induced genes like *Arc* and *zenk* have been linked to the long-term maintenance of synaptic plasticity (long-term potentiation, or LTP) in the mammalian hippocampus (Guzowski *et al.*, 2000; Jones *et al.*, 2001). Although the contribution of specific genes to long-term habituation to song has yet to be determined, the existence of distinct periods of *de novo* synthesis likely reflects a well-orchestrated cascade. For example, the two *epochs* could reflect sequential waves of gene expression in which the initially synthesized proteins influence and regulate late targets. More specifically, during the first wave, synthesis inhibitors likely prevent the induction of immediate-early genes such as *zenk*, *c-fos*, *c-jun* and *Arc*. During the second wave, synthesis inhibitors likely block the expression of targets of regulatory transcription factors induced during the first wave of gene expression. In addition, the second wave could include new synthesis of early genes induced by re-activation (reverberation) of song-responsive circuits. In favor of the latter hypothesis is the fact that the *Arc* response to song contains a second period of mRNA accumulation around 4 hours after stimulation onset (Velho *et al.*, 2005), a phenomenon that had been previously described for this gene during some forms of learning in rodents (Montag-Sallaz, Welzl, Kuhl, Montag and Schachner, 1999; Montag-Sallaz and Montag, 2003).

It is intriguing that while activity-dependent transcription in mammals has been linked to the maintenance of a potentiated state of activated synapses (hippocampal LTP), in songbirds it appears to be associated with a depression (habituation) of song-induced responses (Chew *et al.*, 1995). One possibility is that the mechanisms involved in long-lasting synaptic changes, and the functions of activity-dependent genes like *zenk*, *c-fos*, *c-jun*, *bdnf* and *Arc* are not conserved across birds and mammals. This possibility seems unlikely, given the high degree of conservation in sequence (e.g., the DNA-binding motif of ZENK is 100% identical between birds and mammals; Mello, unpublished observation), kinetics of activity-dependent regulation and cellular distribution patterns of activity-dependent gene expression between birds and mammals. A more likely possibility is that activity-dependent transcriptional regulation is associated with synaptic plasticity independently of the direction of the synaptic change. In this case, even though the early genomic responses to neuronal activation share several components, the end result is a circuit-specific activation of distinct target molecules leading to different synaptic modifications in birds and mammals. Of note, NCM is considered analogous to supragranular layers of the auditory cortex of mammals (see Fig. 1A); it remains to be determined whether stimulus-specific changes akin to habituation in NCM are also seen in comparable portions of the mammalian auditory cortex. At any rate, a better understanding of the significance of the transcriptional response to song requires an analysis of the function of individual song-induced genes.

The *zenk* gene (also known as *zif-268*, *egr-1*, *NGFI-A* and *Krox-24*) encodes a zinc-finger DNA-binding protein that can regulate the expression of downstream target genes with the appropriate motif in their promoters (Milbrandt, 1987; Christy and Nathans, 1989; Knapska and Kaczmarek, 2004). Two other known song-induced genes, *c-fos* and *c-jun*, together comprise the activator protein-1 (AP-1) complex (Sonnenberg, Rauscher, Morgan and Curran, 1989). Disruption of expression of one member of the AP-1 complex, *c-fos*, using antisense oligodeoxynucleotides (ODNs) prevents the long-term maintenance of some forms of hippocampal-dependent tasks (Grimm, Schicknick, Riede, Gundelfinger, Herdegen, Zuschratter and Tischmeyer, 1997; Guzowski and McGaugh, 1997). Since transcription factors such as *zenk* and *c-fos* exert their effects by regulating downstream targets, understanding the mechanisms of action of these genes in long-term plasticity and memory requires the identification of their targets. We identified putative ZENK targets through GenBank screenings for genes that have the conserved ZENK binding domain in their promoters (Mello and Velho, unpublished observations). Because an avian genome was not yet available, we relied on searches of genomic sequences of mammals. We then tested the potential song regulation of these candidates in NCM and obtained suggestive evidence that one candidate, synapsin II (*synII*), is upregulated by song stimulation in NCM (Velho and Mello, 2002), consistent with the fact that ZENK regulates *synI* and *II* in cultured cells (Thiel, Schoch and Petersohn, 1994; Petersohn, Schoch, Brinkmann and Thiel, 1995). More recently, with the eminent sequencing of the zebra finch genome, we are attempting chromatin immunoprecipitation (ChIP) screenings to directly identify ZENK targets in song-responsive neurons.

In contrast to transcription factors, other song-induced genes such as *Arc* and *bdnf* encode effector proteins thought to act by altering cellular properties independently of further transcription (Lanahan and Worley, 1998). Although the precise function of the ARC protein is not understood, its subcellular localization indicates that it acts specifically at recently activated synapses (Link, Konietzko, Kauselmann, Krug, Schwanke, Frey and Kuhl, 1995; Steward, Wallace, Lyford and Worley, 1998; Steward and Worley, 2001) through interactions with cytoskeletal (F-actin) and signal transduction (NMDA receptor complex and possibly calcium-dependent calmodulin kinase II) elements associated with the post-synaptic density (Husi,

Ward, Choudhary, Blackstock and Grant, 2000; Steward *et al.*, 2001; Guzowski, 2002). In songbirds, the cellular distribution we observed for *Arc* mRNA is consistent with a dendritic localization of this transcript in song-responsive neurons (Velho *et al.*, 2005). Interestingly, the properties of song habituation suggest that different songs activate the same NCM neurons differently, likely involving distinct portions of the dendritic arborization. Thus, ARC and associated molecules could serve as molecular tags that help differentiate activated vs. non-activated synapses, thus contributing to long-term song-specific habituation at the post-synaptic level in NCM. Another song-induced direct effector, *bdnf*, is thought to be involved in different aspects of synaptic plasticity, such as regulating dendritic and axonal morphology, synaptic efficacy and synaptogenesis (Lohof, Ip and Poo, 1993; Cabelli, Hohn and Shatz, 1995; Kang and Schuman, 1995; Levine, Dreyfus, Black and Plummer, 1995; McAllister, Lo and Katz, 1995; Patterson, Abel, Deuel, Martin, Rose and Kandel, 1996; Causing, Gloster, Aloyz, Bamji, Chang, Fawcett, Kuchel and Miller, 1997; Horch, Kruttgen, Portbury and Katz, 1999). It has also been associated with the consolidation of contextual fear memories in rats (Lee *et al.*, 2004), and could play a similar role in the formation of song-related memories in songbirds.

Singing-induced gene expression in song control nuclei of adult birds is likely associated with retrieval and perhaps reconsolidation of song memories. Species like zebra finches are called close-ended learners because once their song is learned and crystallized, it remains essentially unchanged throughout adulthood (Immelmann, 1969; Price, 1979; Eales, 1985). Since no new song-memories are thought to be formed after the song learning period in these birds, the transcriptional induction in song control nuclei triggered by singing behavior in adults is likely associated with the retrieval and/or reconsolidation of already well-established memory traces. Interestingly, *zenk* expression is required for the reconsolidation (i.e., protein synthesis-dependent re-stabilization of a memory trace that is being retrieved) of fear-conditioning memories in rodents (Lee *et al.*, 2004), suggesting that this could also be the case for singing-induced *zenk* expression. Singing-induced expression also occurs in open-ended learners, i.e. seasonal breeders that re-learn part of their songs during adulthood such as canaries. In open-ended learners, song-induced expression during singing could be associated with new-learning, retrieval and/or reconsolidation depending on the context. It would be interesting to investigate the extent to which singing-induced transcriptional responses during the learning period (new-learning and re-learning) and during adulthood are similar in open- and close-ended learners.

References

Arnold, A.P. (1980) Effects of androgens on volumes of sexually dimorphic brain regions in the zebra finch. Brain Research, **185**, 441–444.

Arnold, A.P., Bottjer, S.W., Brenowitz, E.A., Nordeen, E.J. & Nordeen, K.W. (1986) Sexual dimorphisms in the neural vocal control system in song birds: Ontogeny and phylogeny. Brain, Behavior and Evolution, **28**, 22–31.

Aston-Jones, G. & Bloom, F.E. (1981a) Activity of norepinephrine-containing locus coeruleus neurons in behaving rats anticipates fluctuations in the sleep-waking cycle. J Neurosci, **1**, 876–886.

Aston-Jones, G. & Bloom, F.E. (1981b) Nonrepinephrine-containing locus coeruleus neurons in behaving rats exhibit pronounced responses to non-noxious environmental stimuli. J Neurosci, **1**, 887–900.

Avey, M.T., Phillmore, L.S. & MacDougall-Shackleton, S.A. (2005) Immediate early gene expression following exposure to acoustic and visual components of courtship in zebra finches. Behav Brain Res, **165**, 247–253.

Berridge, M.J. (1998) Neuronal calcium signaling. Neuron, **21**, 13–26.

Bolhuis, J.J., Zijlstra, G.G., den Boer-Visser, A.M. & Van Der Zee, E.A. (2000) Localized neuronal activation in the zebra finch brain is related to the strength of song learning. Proc Natl Acad Sci U S A, **97**, 2282–2285.

Bonke, B.A., Bonke, D. & Scheich, H. (1979) Connectivity of the auditory forebrain nuclei in the guinea fowl (*numida meleagris*). Cell and Tissue Research, **200**, 101–121.

Bottjer, S.W. (1993) The distribution of tyrosine hydroxylase immunoreactivity in the brains of male and female zebra finches. Journal of Neurobiology, **24**, 51–69.

Bottjer, S.W., Halsema, K.A., Brown, S.A. & Miesner, E.A. (1989) Axonal connections of a forebrain nucleus involved with vocal learning in zebra finches. Journal of Comparative Neurology, **279**, 312–326.

Bottjer, S.W., Miesner, E.A. & Arnold, A.P. (1984) Forebrain lesions disrupt development but not maintenance of song in passerine birds. Science, **224**, 901–903.

Brauth, S.E. & McHale, C.M. (1988) Auditory pathways in the budgerigar. Ii. Intratelencephalic pathways. Brain, Behavior and Evolution, **32**, 193–207.

Brauth, S.E., McHale, C.M., Brasher, C.A. & Dooling, R.J. (1987) Auditory pathways in the budgerigar. I. Thalamo-telencephalic projections. Brain, Behavior and Evolution, **30**, 174–199.

Brenowitz, E.A. (1997) Comparative approaches to the avian song system. Journal of Neurobiology, **33**, 517–531.

Brenowitz, E.A. (2004) Plasticity of the adult avian song control system. Ann N Y Acad Sci, **1016**, 560–585.

Brenowitz, E.A. & Arnold, A.P. (1990) The effects of systemic androgen treatment on androgen accumulation in song control regions of the adult female canary brain. Journal of Neurobiology, **21**, 837–843.

Brenowitz, E.A., Margoliash, D. & Nordeen, K.W. (1997) The neurobiology of birdsong. Journal of Neurobiology, **33**.

Butler, A.B. & Hodos, W. (1996) *Comparative vertebrate neuroanatomy: Evolution and adaptation.* Wiley-Liss, New York, NY.

Cabelli, R.J., Hohn, A. & Shatz, C.J. (1995) Inhibition of ocular dominance column formation by infusion of nt-4/5 or bdnf. Science, **267**, 1662–1666.

Castelino, C.B. & Ball, G.F. (2005) A role for norepinephrine in the regulation of context-dependent zenk expression in male zebra finches (taeniopygia guttata). Eur J Neurosci, **21**, 1962–1972.

Casto, J.M. & Ball, G.F. (1996) Early administration of 17beta-estradiol partially masculinizes song control regions and alpha2-adrenergic receptor distribution in european starlings (sturnus vulgaris). Hormones and Behavior, **30**, 387–406.

Catchpole, C.K. & Slater, P.J.B. (1995) *Bird song: Biological themes and variations.* Cambridge University Press, U.K.

Causing, C.G., Gloster, A., Aloyz, R., Bamji, S.X., Chang, E., Fawcett, J., Kuchel, G. & Miller, F.D. (1997) Synaptic innervation density is regulated by neuron-derived bdnf. Neuron, **18**, 257–267.

Cheng, H.Y. & Clayton, D.F. (2004) Activation and habituation of extracellular signal-regulated kinase phosphorylation in zebra finch auditory forebrain during song presentation. J Neurosci, **24**, 7503–7513.

Chew, S.J., Mello, C., Nottebohm, F., Jarvis, E. & Vicario, D.S. (1995) Decrements in auditory responses to a repeated conspecific song are long-lasting and require two periods of protein synthesis in the songbird forebrain. Proc Natl Acad Sci U S A, **92**, 3406–3410.

Christy, B. & Nathans, D. (1989) DNA binding site of the growth factor-inducible protein zif268. Proceedings of the National Academy of Sciences U S A, **86**, 8737–8741.

Cirelli, C., Pompeiano, M. & Tononi, G. (1996) Neuronal gene expression in the waking state: A role for the locus coeruleus. Science, **274**, 1211–1215.

Cirelli, C. & Tononi, G. (2000) Differential expression of plasticity-related genes in waking and sleep and their regulation by the noradrenergic system. Journal of Neuroscience, **20**, 9187–9194.

Clayton, D.F. (2000) The genomic action potential. Neurobiol Learn Mem, **74**, 185–216.

Ding, L., Perkel, D.J. & Farries, M.A. (2003) Presynaptic depression of glutamatergic synaptic transmission by d1-like dopamine receptor activation in the avian basal ganglia. J Neurosci, **23**, 6086–6095.

Doupe, A.J. & Kuhl, P.K. (1999) Birdsong and human speech: Common themes and mechanisms. Annual Review of Neuroscience, **22**, 567–631.

Eales, L.A. (1985) Song learning in zebra finches: Some effects of song model availability on what is learnt and when. Animal Behavior, **33**, 1293–1300.

Finkbeiner, S. & Greenberg, M.E. (1998) Ca2+ channel-regulated neuronal gene expression. J Neurobiol, **37**, 171–189.

Gentner, T.Q. & Hulse, S.H. (2000) Female european starling preference and choice for variation in conspecific male song. Animal Behavior, **59**, 443–458.

Gentner, T.Q., Hulse, S.H., Duffy, D. & Ball, G.F. (2001) Response biases in auditory forebrain regions of female songbirds following exposure to sexually relevant variation in male song. Journal of Neurobiology, **46**, 48–58.

Ghosh, A., Ginty, D.D., Bading, H. & Greenberg, M.E. (1994) Calcium regulation of gene expression in neuronal cells. Journal of Neurobiology, **25**, 294–303.

Goelet, P., Castellucci, V.F., Schacher, S. & Kandel, E.R. (1986) The long and the short of long-term memory–a molecular framework. Nature, **322**, 419–422.

Goldman, S.A. & Nottebohm, F. (1983) Neuronal production, migration, and differentiation in a vocal control nucleus of the adult female canary brain. Proceedings of the National Academy of Sciences U S A, **80**, 2390–2394.

Grewal, S.S., York, R.D. & Stork, P.J.S. (1999) Extracellular-signal-regulated kinase signallin in neurons. Curr Opin Neurobiol, **9**, 544–553.

Grimm, R., Schicknick, H., Riede, I., Gundelfinger, E.D., Herdegen, T., Zuschratter, W. & Tischmeyer, W. (1997) Suppression of c-fos induction in rat brain impairs retention of a brightness discrimination reaction. Learn Mem, **3**, 402–413.

Guzowski, J.F. (2002) Insights into immediate-early gene function in hippocampal memory consolidation using antisense oligonucleotide and fluorescent imaging approaches. Hippocampus, **12**, 86–104.

Guzowski, J.F., Lyford, G.L., Stevenson, G.D., Houston, F.P., McGaugh, J.L., Worley, P.F. & Barnes, C.A. (2000) Inhibition of activity-dependent arc protein expression in the rat hippocampus impairs the maintenance of long-term potentiation and the consolidation of long-term memory. J Neurosci, **20**, 3993–4001.

Guzowski, J.F. & McGaugh, J.L. (1997) Antisense oligodeoxynucleotide-mediated disruption of hippocampal camp response element binding protein levels impairs consolidation of memory for water maze training. Proc Natl Acad Sci U S A, **94**, 2693–2698.

Hessler, N.A. & Doupe, A.J. (1999a) Singing-related neural activity in a dorsal forebrain-basal ganglia circuit of adult zebra finches. Journal of Neuroscience, **19**, 10461–10481.

Hessler, N.A. & Doupe, A.J. (1999b) Social context modulates singing-related neural activity in the songbird forebrain. Nature Neuroscience, **2**, 209–211.

Horch, H.W., Kruttgen, A., Portbury, S.D. & Katz, L.C. (1999) Destabilization of cortical dendrites and spines by bdnf. Neuron, **23**, 353–364.

Husi, H., Ward, M.A., Choudhary, J.S., Blackstock, W.P. & Grant, S.G. (2000) Proteomic analysis of nmda receptor-adhesion protein signaling complexes. Nat Neurosci, **3**, 661–669.

Immelmann, K. (1969) Song development in the zebra finch and other estrilid finches. In Hinde, R.A. (ed.) *Bird vocalizations*. Cambridge University Press, Cambridge, U.K., pp. 61–74.

Jarvis, E.D., Gunturkun, O., Bruce, L., Csillag, A., Karten, H., Kuenzel, W., Medina, L., Paxinos, G., Perkel, D.J., Shimizu, T., Striedter, G., Wild, J.M., Ball, G.F., Dugas-Ford, J., Durand, S.E., Hough, G.E., Husband, S., Kubikova, L., Lee, D.W., Mello, C.V., Powers, A.,

Siang, C., Smulders, T.V., Wada, K., White, S.A., Yamamoto, K., Yu, J., Reiner, A. & Butler, A.B. (2005) Avian brains and a new understanding of vertebrate brain evolution. Nat Rev Neurosci, **6**, 151–159.

Jarvis, E.D., Mello, C.V. & Nottebohm, F. (1995) Associative learning and stimulus novelty influence the song-induced expression of an immediate early gene in the canary forebrain. Learn Mem, **2**, 62–80.

Jarvis, E.D. & Nottebohm, F. (1997) Motor-driven gene expression. Proc Natl Acad Sci U S A, **94**, 4097–4102.

Jarvis, E.D., Scharff, C., Grossman, M.R., Ramos, J.A. & Nottebohm, F. (1998) For whom the bird sings: Context-dependent gene expression. Neuron, **21**, 775–788.

Johnson, F., Sablan, M.M. & Bottjer, S.W. (1995) Topographic organization of a forebrain pathway involved with vocal learning in zebra finches. Journal of Comparative Neurology, **358**, 260–278.

Jones, M.W., Errington, M.L., French, P.J., Fine, A., Bliss, T.V., Garel, S., Charnay, P., Bozon, B., Laroche, S. & Davis, S. (2001) A requirement for the immediate early gene zif268 in the expression of late ltp and long-term memories. Nat Neurosci, **4**, 289–296.

Kang, H. & Schuman, E.M. (1995) Long-lasting neurotrophin-induced enhancement of synaptic transmission in the adult hippocampus. Science, **267**, 1658–1662.

Karten, H.J. (1968) The ascending auditory pathway in the pigeon (*columba livia*). Ii. Telencephalic projections of the nucleus ovoidalis thalami. Brain Research, **11**, 134–153.

Karten, H.J. & Hodos, W. (1967) *A stereotaxic atlas of the brain of the pigeon (columba livia)*. Johns Hopkins Press, Baltimore, MD.

Kelley, D.B. & Nottebohm, F. (1979) Projections of a telencephalic auditory nucleus-field l-in the canary. J Comp Neurol, **183**, 455–469.

Kimpo, R.R. & Doupe, A.J. (1997) Fos is induced by singing in distinct neuronal populations in a motor network. Neuron, **18**, 315–325.

Knapska, E. & Kaczmarek, L. (2004) A gene for neuronal plasticity in the mammalian brain: Zif268/egr-1/ngfi-a/krox-24/tis8/zenk? Prog Neurobiol, **74**, 183–211.

Konishi, M. (1965) Effects of deafening on song development in american robins and black- headed grosbeaks. Z Tierpsychol, **22**, 584–599.

Kroodsma, D.E. & Miller, E.H. (1996) *Ecology and evolution of acoustic communication in birds.* Cornell University Press, Ithaca, NY.

Kruse, A.A., Stripling, R. & Clayton, D.F. (2000) Minimal experience required for immediate-early gene induction in zebra finch neostriatum. Neurobiology of Learning and Memory, **74**, 179–184.

Kruse, A.A., Stripling, R. & Clayton, D.F. (2004) Context-specific habituation of the zenk gene response to song in adult zebra finches. Neurobiol Learn Mem, **82**, 99–108.

Lanahan, A. & Worley, P. (1998) Immediate-early genes and synaptic function. Neurobiol Learn Mem, **70**, 37–43.

Lee, J.L., Everitt, B.J. & Thomas, K.L. (2004) Independent cellular processes for hippocampal memory consolidation and reconsolidation. Science, **304**, 839–843.

Leitner, S., Voigt, C., Metzdorf, R. & Catchpole, C.K. (2005) Immediate early gene (zenk, arc) expression in the auditory forebrain of female canaries varies in response to male song quality. J Neurobiol, **64**, 275–284.

Leonardo, A. & Konishi, M. (1999) Decrystallization of adult birdsong by perturbation of auditory feedback. Nature, **399**, 466–470.

Levine, E.S., Dreyfus, C.F., Black, I.B. & Plummer, M.R. (1995) Brain-derived neurotrophic factor rapidly enhances synaptic transmission in hippocampal neurons via postsynaptic tyrosine kinase receptors. Proc Natl Acad Sci U S A, **92**, 8074–8077.

Lewis, J.W., Ryan, S.M., Arnold, A.P. & Butcher, L.L. (1981) Evidence for a catecholaminergic projection to area x in the zebra finch. Journal of Comparative Neurology, **196**, 347–354.

Link, W., Konietzko, U., Kauselmann, G., Krug, M., Schwanke, B., Frey, U. & Kuhl, D. (1995) Somatodendritic expression of an immediate early gene is regulated by synaptic activity. Proc Natl Acad Sci U S A, **92**, 5734–5738.

Lohof, A.M., Ip, N.Y. & Poo, M.M. (1993) Potentiation of developing neuromuscular synapses by the neurotrophins nt-3 and bdnf. Nature, **363**, 350–353.

Luo, M., Ding, L. & Perkel, D.J. (2001) An avian basal ganglia pathway essential for vocal learning forms a closed topographic loop. J Neurosci, **21**, 6836–6845.

Marler, P. & Peters, S. (1977) Selective vocal learning in a sparrow. Science, **198**, 519–521.

McAllister, A.K., Lo, D.C. & Katz, L.C. (1995) Neurotrophins regulate dendritic growth in developing visual cortex. Neuron, **15**, 791–803.

McKenzie, T.L., Hernandez, A.M. & Macdougall-Shackleton, S.A. (2006) Experience with songs in adulthood reduces song-induced gene expression in songbird auditory forebrain. Neurobiol Learn Mem.

Mello, C., Nottebohm, F. & Clayton, D. (1995a) Repeated exposure to one song leads to a rapid and persistent decline in an immediate early gene's response to that song in zebra finch telencephalon. J Neurosci, **15**, 6919–6925.

Mello, C.V. (1998) Auditory experience, gene regulation and auditory memories in songbirds. Journal of the Brazilian Association for the Advancement of Science, **50**, 189–196.

Mello, C.V. (2002) Mapping vocal communication pathways in birds with inducible gene expression. J Comp Physiol A Neuroethol Sens Neural Behav Physiol, **188**, 943–959.

Mello, C.V. & Clayton, D.F. (1994) Song-induced zenk gene expression in auditory pathways of songbird brain and its relation to the song control system. J Neurosci, **14**, 6652–6666.

Mello, C.V. & Clayton, D.F. (1995b) Differential induction of the zenk gene in the avian forebrain and song control circuit after metrazole-induced depolarization. J Neurobiol, **26**, 145–161.

Mello, C.V., Pinaud, R. & Ribeiro, S. (1998a) Noradrenergic system of the zebra finch brain: Immunocytochemical study of dopamine-beta-hydroxylase. Journal of Comparative Neurology, **400**, 207–228.

Mello, C.V. & Ribeiro, S. (1998b) Zenk protein regulation by song in the brain of songbirds. Journal of Comparative Neurology, **393**, 426–438.

Mello, C.V., Vates, G.E., Okuhata, S. & Nottebohm, F. (1998c) Descending auditory pathways in the adult male zebra finch (*taeniopygia guttata*). Journal of Comparative Neurology, **395**, 137–160.

Mello, C.V., Velho, T.A. & Pinaud, R. (2004) Song-induced gene expression: A window on song auditory processing and perception. Ann N Y Acad Sci, **1016**, 263–281.

Mello, C.V., Vicario, D.S. & Clayton, D.F. (1992) Song presentation induces gene expression in the songbird forebrain. Proc Natl Acad Sci U S A, **89**, 6818–6822.

Metzger, M., Jiang, S. & Braun, K. (1998) Organization of the dorsocaudal neostriatal complex: A retrograde and anterograde tracing study in the domestic chick with special emphasis on pathways relevant to imprinting. Journal of Comparative Neurology, **395**, 380–404.

Milbrandt, J. (1987) A nerve growth factor-induced gene encodes a possible transcriptional regulatory factor. Science, **238**, 797–799.

Montag-Sallaz, M. & Montag, D. (2003) Learning-induced arg 3.1/arc mrna expression in the mouse brain. Learn Mem, **10**, 99–107.

Montag-Sallaz, M., Welzl, H., Kuhl, D., Montag, D. & Schachner, M. (1999) Novelty-induced increased expression of immediate-early genes c-fos and arg 3.1 in the mouse brain. J Neurobiol, **38**, 234–246.

Morrow, B.A., Elsworth, J.D., Inglis, F.M. & Roth, R.H. (1999) An antisense oligonucleotide reverses the footshock-induced expression of fos in the rat medial prefrontal cortex and the subsequent expression of conditioned fear-induced immobility. J Neurosci, **19**, 5666–5673.

Nordeen, K.W. & Nordeen, E.J. (1992) Auditory feedback is necessary for the maintenance of stereotyped song in adult zebra finches. Behavioral and Neural Biology, **57**, 58–66.

Nottebohm, F. (1972) The origins of vocal learning. American Naturalist, **106**, 116–140.

Nottebohm, F. (1981) A brain for all seasons: Cyclical anatomical changes in song control nuclei of the canary brain. Science, **214**, 1368–1370.

Nottebohm, F. & Arnold, A.P. (1976) Sexual dimorphism in vocal control areas of the songbird brain. Science, **194**, 211–213.

Nottebohm, F., Kelley, D.B. & Paton, J.A. (1982) Connections of vocal control nuclei in the canary telencephalon. Journal of Comparative Neurology, **207**, 344–357.

Nottebohm, F., Nottebohm, M.E. & Crane, L. (1986) Developmental and seasonal changes in canary song and their relation to changes in the anatomy of song-control nuclei. Behavioral and Neural Biology, **46**, 445–471.

Nottebohm, F., Stokes, T.M. & Leonard, C.M. (1976) Central control of song in the canary, serinus canarius. Journal of Comparative Neurology, **165**, 457–486.

Olveczky, B.P., Andalman, A.S. & Fee, M.S. (2005) Vocal experimentation in the juvenile songbird requires a basal ganglia circuit. PLoS Biol, **3**, e153.

Park, K.H. & Clayton, D.F. (2002) Influence of restraint and acute isolation on the selectivity of the adult zebra finch zenk gene response to acoustic stimuli. Behav Brain Res, **136**, 185–191.

Paton, J.A. & Nottebohm, F.N. (1984) Neurons generated in the adult brain are recruited into functional circuits. Science, **225**, 1046–1048.

Patterson, S.L., Abel, T., Deuel, T.A., Martin, K.C., Rose, J.C. & Kandel, E.R. (1996) Recombinant bdnf rescues deficits in basal synaptic transmission and hippocampal ltp in bdnf knockout mice. Neuron, **16**, 1137–1145.

Petersohn, D., Schoch, S., Brinkmann, D.R. & Thiel, G. (1995) The human synapsin ii gene promoter. Possible role for the transcription factor zif268/egr-1, polyoma enhancer activator 3, and ap2. J Biol Chem, **270**, 24361–24369.

Phan, M.L., Pytte, C.L. & Vicario, D.S. (2006) Early auditory experience generates long-lasting memories that may subserve vocal learning in songbirds. Proc Natl Acad Sci U S A, **103**, 1088–1093.

Pinaud, R., Velho, T.A., Jeong, J.K., Tremere, L.A., Leao, R.M., von Gersdorff, H. & Mello, C.V. (2004) Gabaergic neurons participate in the brain's response to birdsong auditory stimulation. Eur J Neurosci, **20**, 1318–1330.

Poopatanapong, A., Teramitsu, I., Byun, J.S., Vician, L.J., Herschman, H.R. & White, S.A. (2006) Singing, but not seizure, induces synaptotagmin iv in zebra finch song circuit nuclei. J Neurobiol, **66**, 1613–1629.

Price, P. (1979) Developmenteal determinants of structure in zebra finch song. Journal of Comparative Physiology and Psychology, **93**, 260–277.

Reiner, A., Perkel, D.J., Bruce, L.L., Butler, A.B., Csillag, A., Kuenzel, W., Medina, L., Paxinos, G., Shimizu, T., Striedter, G., Wild, M., Ball, G.F., Durand, S., Gunturkun, O., Lee, D.W., Mello, C.V., Powers, A., White, S.A., Hough, G., Kubikova, L., Smulders, T.V., Wada, K., Dugas-Ford, J., Husband, S., Yamamoto, K., Yu, J., Siang, C. & Jarvis, E.D. (2004) Revised nomenclature for avian telencephalon and some related brainstem nuclei. J Comp Neurol, **473**, 377–414.

Ribeiro, S., Cecchi, G.A., Magnasco, M.O. & Mello, C.V. (1998) Toward a song code: Evidence for a syllabic representation in the canary brain. Neuron, **21**, 359–371.

Ribeiro, S. & Mello, C.V. (2000) Gene expression and synaptic plasticity in the auditory forebrain of songbirds. Learn Mem, **7**, 235–243.

Sasaki, A., Sotnikova, T.D., Gainetdinov, R.R. & Jarvis, E.D. (2006) Social context-dependent singing-regulated dopamine. J Neurosci, **26**, 9010–9014.

Scharff, C. & Nottebohm, F. (1991) A comparative study of the behavioral deficits following lesions of various parts of the zebra finch song system: Implications for vocal learning. Journal of Neuroscience, **11**, 2896–2913.

Sockman, K.W., Gentner, T.Q. & Ball, G.F. (2002) Recent experience modulates forebrain gene-expression in response to mate-choice cues in european starlings. Proc R Soc Lond B Biol Sci, **269**, 2479–2485.

Sockman, K.W., Gentner, T.Q. & Ball, G.F. (2004) Complementary neural systems for the experience-dependent integration of mate-choice cues in european starlings. J Neurobiol, **62**, 72–81.

Sohrabji, F., Nordeen, E.J. & Nordeen, K.W. (1990) Selective impairment of song learning following lesions of a forebrain nucleus in the juvenile zebra finch. Behavioral and Neural Biology, **53**, 51–63.

Sonnenberg, J.L., Rauscher, F.J., Morgan, J.I. & Curran, T. (1989) Regulation of proenkephalin by fos and jun. Science, **246**, 1622–1625.

Steward, O., Wallace, C.S., Lyford, G.L. & Worley, P.F. (1998) Synaptic activation causes the mrna for the ieg arc to localize selectively near activated postsynaptic sites on dendrites. Neuron, **21**, 741–751.

Steward, O. & Worley, P.F. (2001) Selective targeting of newly synthesized arc mrna to active synapses requires nmda receptor activation. Neuron, **30**, 227–240.

Stripling, R., Volman, S.F. & Clayton, D.F. (1997) Response modulation in the zebra finch neostriatum: Relationship to nuclear gene regulation. J Neurosci, **17**, 3883–3893.

Terleph, T.A., Mello, C.V. & Vicario, D.S. (2006) Auditory topography and temporal response dynamics of canary caudal telencephalon. J Neurobiol, **66**, 281–292.

Terpstra, N.J., Bolhuis, J.J., Riebel, K., van der Burg, J.M. & den Boer-Visser, A.M. (2006) Localized brain activation specific to auditory memory in a female songbird. J Comp Neurol, **494**, 784–791.

Thiel, G., Schoch, S. & Petersohn, D. (1994) Regulation of synapsin i gene expression by the zinc finger transcription factor zif268/egr-1. J Biol Chem, **269**, 15294–15301.

Vates, G.E., Broome, B.M., Mello, C.V. & Nottebohm, F. (1996) Auditory pathways of caudal telencephalon and their relation to the song system of adult male zebra finches. J Comp Neurol, **366**, 613–642.

Vates, G.E., Vicario, D.S. & Nottebohm, F. (1997) Reafferent thalamo- "cortical" loops in the song system of oscine songbirds. Journal of Comparative Neurology, **380**, 275–290.

Velho, T.A., Pinaud, R., Rodrigues, P.V. & Mello, C.V. (2005) Co-induction of activity-dependent genes in songbirds. Eur J Neurosci, **22**, 1667–1678.

Velho, T.A.F. & Mello, C.V. (2002) Synapsin ii, a candidate zenk target, is regulated by song in the songbird ncm *Society for Neuroscience Annual Meeting*, Orlando, Fl.

Velho, T.A.F., Pinaud, R., Jeong, J. & Mello, C. (2003) Differential subcellular localization of zenk mrna reveals neuronal populations activated by two different songs in zebra finch ncm. *Society for Neuroscience Annual Meeting*, New Orleans.

Velho, T.A.F., Ribeiro, S., Pinaud, R. & Mello, C.V. (2006) Noradrenergic modulation of song-induced gene expression in the caudomedial nidopallium (ncm) of zebra finches *Society for Neuroscience Annual Meeting*. SFN, Atlanta.

Wada, K., Howard, J.T., McConnell, P., Whitney, O., Lints, T., Rivas, M.V., Horita, H., Patterson, M.A., White, S.A., Scharff, C., Haesler, S., Zhao, S., Sakaguchi, H., Hagiwara, M., Shiraki, T., Hirozane-Kishikawa, T., Skene, P., Hayashizaki, Y., Carninci, P. & Jarvis, E.D. (2006) A molecular neuroethological approach for identifying and characterizing a cascade of behaviorally regulated genes. Proc Natl Acad Sci U S A, **103**, 15212–15217.

Wada, K., Sakaguchi, H., Jarvis, E.D. & Hagiwara, M. (2004) Differential expression of glutamate receptors in avian neural pathways for learned vocalization. J Comp Neurol, **476**, 44–64.

Wade, J. & Arnold, A.P. (2004) Sexual differentiation of the zebra finch song system. Ann N Y Acad Sci, **1016**, 540–559.

Wang, J.Q., Tang, Q., Parelkar, N.K., Liu, Z., Samdani, S., Choe, E.S. & Yang, L.M., L. (2004) Glutamate signaling to ras-mapk in striatal neurons. Molecular Neurobiology, **29**, 1–14.

Weaver, I.C., Cervoni, N., Champagne, F.A., D'Alessio, A.C., Sharma, S., Seckl, J.R., Dymov, S., Szyf, M. & Meaney, M.J. (2004) Epigenetic programming by maternal behavior. Nat Neurosci, **7**, 847–854.

Wild, J.M. (1993) Descending projections of the songbird nucleus robustus archistriatalis. Journal of Comparative Neurology, **338**, 225–241.

Woolley, S.M. & Rubel, E.W. (2002) Vocal memory and learning in adult bengalese finches with regenerated hair cells. J Neurosci, **22**, 7774–7787.

Yu, A.C. & Margoliash, D. (1996) Temporal hierarchical control of singing in birds. Science, **273**, 1871–1875.

Zeigler, H.P. & Marler, P. (2004) *Behavioral neurology of birdsong*. The New York Academy of Sciences, New York.

Chapter 2
To the Nucleus with Proteomics

Bryen A. Jordan and Edward B. Ziff

Abstract In this chapter we review proteomic studies performed on purified postsynaptic density (PSD) fractions and analyze what their results reveal about nuclear signaling. Overall, these studies show that PSDs are similar to the cytoplasmic protein complexes found at other intercellular junctions. We find that PSDs contain numerous calmodulin (CAM)-associated transcriptional coactivators, nucleocytoplasmic shuttling proteins that bind RNA and potential splicing factors. Moreover, in conjunction with bioinformatics, we identify novel nucleocytoplasmic shuttling proteins at synapses. These results strongly suggest that control of the spatial distribution of synapse-associated nuclear factors regulates nuclear function.

2.1 Introduction

Proteomic studies of synaptic junctions have revealed an extensive and complex signaling network at synapses. Signaling pathways that rapidly adjust synaptic transmission co-exist with those that regulate long-term changes in interneuronal communication. The latter are mediated by activity-dependent signaling to the nucleus, a process required for neuronal survival, development and plasticity (Deisseroth et al., 2003; West et al., 2002). However, the nature of the components of these signaling pathways is widely debated. Neuronal excitation results in transient increases in nuclear Ca^{2+} levels (Hardingham et al., 2001; Lipscombe et al., 1988), suggesting that Ca^{2+} waves may regulate nuclear processes. This mechanism may underlie gene regulation by Ca^{2+}-binding transcription factors such as downstream regulatory element antagonistic modulator (DREAM) (Osawa et al., 2001) or activation of CREB (Hardingham et al., 2001), a transcription factor essential for learning and memory, fear conditioning and the regulation of circadian rhythms (Deisseroth et al., 2003; West et al., 2002). A complementary or competing hypothesis implicates the activity-dependent nucleocytoplasmic shuttling of transcription factors or their regulatory proteins in synapse-to-nucleus signaling. For example, neuronal activity results in the rapid nuclear accumulation of the NFATc4 and NF-kB transcription factors (Graef et al., 1999; Kaltschmidt et al., 1995), and the HDAC4 and HDAC5 histone deacetylases (Chawla et al., 2003). We have found numerous other nucleocytoplasmic shuttling factors at synaptic junctions

S. M. Dudek (ed.), *Transcriptional Regulation by Neuronal Activity.*   27
© Springer 2008

using mass-spectrometry based proteomics, bioinformatics and cell biology (Jordan et al., 2004). Indeed, most proteomic profiling studies show postsynaptic densities (PSDs) contain a variety of nucleocytoplasmic shuttling proteins such as β and γ catenin and associated cellular adhesion molecules (CAMs) such as N-cadherin and neurexin. Furthermore, research has also shown that synapses contain components of the nuclear import machinery. Most nuclear import events require the binding of importin subunits to nuclear localization sequences (NLSs) on proteins destined for the nucleus (Gorlich and Kutay, 1999). Both α and β importin subunits are present at synaptic junctions, and exhibit rapid nuclear translocation in response to NMDA receptor stimulation (Thompson et al., 2004). The presence of the nuclear import machinery, cytosolic anchors and associated cotranscriptional activators at synaptic junctions, strongly suggests that the spatial regulation of PSD-associated nuclear factors represents a mechanism in synapse-to-nucleus signaling.

In this chapter we briefly review the proteomic studies performed on purified PSD fractions and analyze what their results reveal about nuclear signaling. Overall, these studies show that PSDs are similar to the cytoplasmic protein complexes found at other intercellular junctions. This is important since the translocation of components of different junctional complexes to the nucleus is an increasingly noted mechanism via which extracellular signals are transferred to the nucleus. Using proteomics, we find that PSDs contain numerous CAM-associated transcriptional coactivators, nucleocytoplasmic shuttling proteins that bind RNA and potential splicing factors. Moreover, in conjunction with bioinformatics, we identify novel nucleocytoplasmic shuttling proteins at synapses. For example, we show that a novel PSD component, AIDA-1d, can shuttle to the nucleus in response to NMDA receptor stimulation and alter protein synthesis by regulating nucleolar assembly. The variety of nucleocytoplasmic shuttling proteins found at synaptic junctions suggests that neuronal activity not only regulates gene expression, but also alternative splicing, RNA trafficking and the protein biosynthetic machinery. At the end of this chapter, we outline experiments to identify large-scale changes in the nuclear and PSD proteome in response to neuronal activity using novel techniques in proteomics. Since only proteomic-based methods can reveal the multiple simultaneous effects of neuronal activity on the nuclear proteome, we believe they will be essential in developing a more rigorous model of synapse-to-nucleus signaling.

2.2 Proteomics of the PSD

The PSD is a large cytoplasmic protein complex associated with synaptic junctions in direct apposition to the presynaptic sites of vesicle release (Kennedy, 1993; Ziff, 1997). Within the PSD lie scaffolding proteins, cytoskeletal components, signaling molecules and neurotransmitter receptors that transform presynaptically released chemical messages into electrical and biochemical signals. The dynamic nature of PSD composition is critical for a synapse's ability to undergo long-term potentiation (LTP) or long-term depression (LTD), phenomena that alter synaptic strength and are thought to underlie learning and memory (Kandel, 2001). Given this central role in synaptic transmission many researchers have sought to elucidate its protein content.

Early studies to determine the composition of purified PSD fractions (Banker et al., 1974; Carlin et al., 1980; Cohen et al., 1977; Walikonis et al., 2000; Walsh and Kuruc, 1992; Ziff, 1997) have given way to a flurry of mass spectrometry-based proteomic studies (Jordan et al., 2004; Jordan et al., 2006; Li et al., 2004; Peng et al., 2004; Yoshimura et al., 2004). PSDs can be purified from crude brain tissue using a protocol developed in the lab of Phil Siekevitz (Carlin et al., 1980; Cohen et al., 1977), which relies on homogenization, sucrose step centrifugation and extraction with Triton X-100. This protocol was shown to yield a highly pure PSD fraction with relatively low contamination and has been the foundation for a majority of PSD profiling studies. Recent efforts have resulted in even cleaner fractions by immunopurying PSDs using beads containing antibodies to PSD-95 (an important PSD resident protein) (Vinade et al., 2003). Modern proteomic studies have relied on several prefractionation techniques to resolve the PSD complex prior to mass spectrometry, including one-dimensional SDS-PAGE (Jordan et al., 2004; Peng et al., 2004), two-dimensional electrophoresis (Li et al., 2004) or two-dimensional high performance liquid chromatography (Yoshimura et al., 2004). These prefractionation steps are essential for identifying low abundance proteins in the PSD. In most studies, tryptic peptides were then analyzed by tandem mass spectrometry (LC-MS/MS) followed by a statistical analysis to reveal high confidence protein precursors.

2.2.1 The PSD, CAMs and Nucleocytoplasmic Regulation

Together, these proteomic profiling studies have identified ∼1100 different proteins, comprising a richly diverse set of signaling components including proteins involved in cell polarity, protein translation and degradation, kinases/phosphatases, G-proteins and direct regulators of these groups as well as a large number of proteins of unidentified function (Collins et al., 2006). These studies confirm that PSDs associate with a variety of CAMs, including L-Afadin, neural cell adhesion molecule (NCAM), L1-CAM, integrins, SynCAM, N-cadherin, neurexin and add several novel CAMs such as cell adhesion (MOCA) and collapsin response-mediating protein (CRMP) (Jordan et al., 2004). Indeed N-cadherin is the most prominent glycoprotein in the PSD fraction (Beesley et al., 1995). Many of these CAMs have been shown to play an important role in synaptic function (Gerrow and El-Husseini, 2006). The similarities between synapses and other intercellular junctions may be an important clue in understanding activity-dependent nuclear signaling. While it was previously thought that CAMs solely regulate physical intercellular unions, it is now widely accepted that CAMs modulate, or directly activate downstream signaling pathways in response to extracellular signals (Aplin and Juliano, 2001). How this is accomplished is unclear, but an emerging theme is that CAMs function as cytosolic anchors for key signaling factors.

Specifically, there is increasing evidence that CAMs regulate nuclear signaling pathways by direct spatial control of proteins that control nuclear processes (Aplin and Juliano, 2001; Cyert, 2001). Thus abundant CAMs at the PSD suggest synaptic junctions could serve as repositories for proteins that control nuclear processes. There are several examples of CAMs and associated regulators of transcription at synaptic junctions. The synaptic CAMs neurexin and syndecan and binding partner CASK, a scaffolding protein with nucleocytoplasmic shuttling capabilities have all been identified in the PSD (Collins et al., 2006; Jordan et al., 2004). In the nucleus, CASK can associate with the Tbr-1 transcription factor and alter the expression of reelin, a secreted extracellular matrix glycoprotein whose gene is mutated in reeler mice (D'Arcangelo et al., 1995). Importantly, syndecan-3 acts as a cytosolic anchor as its overexpression prevents the nuclear accumulation of CASK (Hsueh and Sheng, 1999; Hsueh et al., 2000). A similar mechanism has been observed for the well-known cotranscriptional activator β-catenin, which is also present in the PSD (Murase et al., 2002). Like syndecan-3 for CASK, N-cadherin sequesters β-catenin in the cytoplasm. High levels of N-cadherin trap β-catenin at junctional plaques, which decreases its cotranscriptional activity (Fagotto et al., 1996). The level of cytosolic β-catenin, which is also regulated via the axin/GSK3b/APC degradation pathway, affects its nuclear accumulation where it acts on LEF-1/TCR transcription factors to affect cMyc expression (Gordon and Nusse, 2006). In addition to these known synaptic components, proteomics reveals PSDs contain several other proteins with identified nuclear functions.

2.2.2 Novel Transcriptional Regulators at the PSD

Using multiple proteomic techniques, several groups identified the armadillo-related proteins p120(ctn), armadillo repeat protein deleted in velo-cardio-facial syndrome (ARVCF), catenin p0071 and plakoglobin (γ-catenin) (Collins et al., 2006; Jordan et al., 2004; Li et al., 2004; Peng et al., 2004; Yoshimura et al., 2004) in the PSD fraction. ARVCF and p120(ctn) are catenins associated with cadherin-based junctions (Hatzfeld, 2005) that can shuttle to the nucleus (Kausalya et al., 2004; Roczniak-Ferguson and Reynolds, 2003). We found that in brain tissue, p120 and ARVCF were highly enriched in PSD fractions (Jordan et al., 2004). In the nucleus, p120 can bind to the BTB/POZ repressor kaiso and relieve its transcriptional repressing activity (Kelly et al., 2004). A Kaiso-p120(ctn) complex can regulate the expression of Rapsyn (Rodova et al., 2004), an important component of acetylcholine receptor clustering at neuromuscular junctions. ARVCF nuclear translocation requires an interaction with the scaffolding protein Zona Occludens-2 (ZO-2), where it may regulate transcriptional activity by affecting the Y-box transcriptional factor ZONAB (Kausalya et al., 2004). Plakoglobin is an important component of cadherin-based junctions that is involved in suppressing tumorigenicity in mouse and human cells (Simcha et al., 1996). Like β-catenin, inhibition of the ubiquitin proteasome pathway results in an increase in nuclear plakoglobin levels (Salomon et al., 1997). Interestingly plakoglobin and β-catenin binding to

E-cadherin/α-catenin/APC complexes is mutually exclusive (Butz and Kem-ler, 1994; Hulsken et al., 1994). Recent evidence has shown that plakoglobin can shuttle to the nucleolus in response to neuregulin/ErbB receptor stimulation (Li et al., 2003), where it can bind to and potentiate LEF-1 dependent transcription (Huber et al., 1996).

Our proteomic studies also provide evidence for integrin-based nuclear signaling. Integrins have been identified in PSDs by mass spectrometry (Collins et al., 2006; Rodriguez et al., 2000; Yoshimura et al., 2004) and have been shown to be impor-tant for synaptic plasticity and spatial memory (Chan et al., 2003). Using liquid-chromatography coupled to tandem mass spectrometry (LC-MS/MS), Collins et al., (Collins et al., 2006) identified a subunit of the JAB-1 signaling complex (subunit 3) in PSDs. JAB-1 is a well known c-Jun coactivator which localizes to both mem-branes and nuclei and has been shown to bind to the C-terminal tail of $\beta2$ integrin (Bianchi et al., 2000). Evidence of the JAB-1 signalosome at PSDs suggests that synapses posses the capacity to regulate c-Jun activity. Indeed c-Jun activity may be essential for learning and memory as mice lacking the N-terminal c-Jun kinase 2 (JNK2) exhibit impaired LTP (Chen et al., 2005).

By two-dimensional HPLC-coupled to MS/MS, Yoshimura et al., also identified the signal transducer and activator of transcription (STAT1) at PSDs (Yoshimura et al., 2004). STAT1 is an important transcriptional activator whose activity, along with Janus kinases (JAK) has been recently identified in brain and may help relay ciliary neurotrophic factor signaling (Rajan et al., 1996). How STAT1 may be asso-ciated with PSDs is unclear although it may be anchored at the cytoskeleton via described interactions with the PKC anchoring and PSD resident protein RACK1, or the neuregulin binding ErbB2 receptor (Jones et al., 2006; Usacheva et al., 2001). If verified, the presence of ARVCF, p120, p0071, plakoglobin, STAT1 and compo-nents of the JAB-1 signalosome complex at the synaptic junction may outline novel nuclear processes under synaptic control.

2.3 Scaffolding/Cytoskeletal Proteins and Nuclear Signaling

Proteomics studies have also revealed that PSDs contain a number of scaffolding/cytoskeletal proteins with a proposed link to nuclear signaling. While scaffold-ing proteins have traditionally been thought to link membrane bound proteins to signaling molecules, there is growing evidence that cytoskeletal and scaffolding proteins themselves link processes at the plasma membrane with the cellular nucleus (Hubner et al., 2001). Many scaffolding proteins, which typically are targeted to cellular junctions via PDZ domain-PDZ ligand interactions, also exhibit a nuclear distribution. Shuttling of cytoskeletal linkers could disrupt cytoskeletal anchoring and free CAMs or receptors for internalization or lateral diffusion. At the same time, the nuclear accumulation of these linkers could serve to relay activity-dependent signals to the nucleus. For example, there is evidence that the AMPA receptor scaffolding protein GRIP1 may participate in nuclear signaling. Ata-man et al., have shown that the drosophila GRIP orthologue dGRIP, is necessary for the nuclear translocation of the Wnt receptor Frizzled-2 following endocytosis

(Ataman et al., 2006). Drosophila with decreased dGRIP levels displayed a synaptic phenotype that was similar to wingless (Wnt drosophila orthologue) mutants, and also displayed reduced Frizzled-2 nuclear accumulation. A separate study showed that the GRIP1 splice variant GRIP1b can bind to the homeodomain proteins DLX2 and DLX5 and potentiate transcription (Yu et al., 2001). A second splice variant GRIPt can also function as a transcriptional coactivator when fused to a GAL4 DNA binding domain (Nakata et al., 2004). Studies of GRIP1 at synapses have shown that phosphorylation of the AMPA receptor GluR2 at S880 disrupts its interaction with GRIP (Chung et al., 2000; Matsuda et al., 1999), which may represent a mechanism to release GRIP1 for subsequent nucleocytoplasmic shuttling.

The GluR1 scaffolding protein SAP97, has also been observed in the nucleus (Kohu et al., 2002), although its nuclear function is unknown. Kohu et al., have shown that mutations that disrupt the SAP97 intramolecular interactions between the src Homology-3 (SH3) and guanylate kinase (GK) domains cause SAP97 to shuttle to the nucleus. Nuclear import is dependent on a functional NLS and is specific for SAP97. Similar mutations on a related protein, PSD95, do not cause it to become nuclear. The binding of factors to SAP97 that disrupt its SH3-GUK intracellular association may provide a mechanism for regulated nuclear import. Interestingly an additional GluR1 interacting protein, Band 4.1, which has been suggested to transiently link AMPA receptors to the cytoskeletal matrix (Shen et al., 2000) has also been identified in the nucleus. Nuclear band 4.1 associates with regions enriched in splicing factors. Indeed depletion of 4.1 inhibited splicing activity in HeLa nuclear extracts (Lallena et al., 1998). Using proteomics, we have identified additional novel scaffolding and cytoskeletal components at the PSD with reported nuclear functions (Table 1).

2.3.1 Novel Scaffolding/cytoskeletal Proteins with Potential Nuclear Functions at the PSD

Using one-dimensional SDS-PAGE and tandem MS, two groups have identified Zona occludens-1 (ZO-1) at the PSD (Collins et al., 2006; Peng et al., 2004). ZO-1 is an abundant component of tight junctions where it binds to occludin and is thought to act in signal transduction events. At neuronal synapses, ZO-1 may associate with the cadherin complex via a direct interaction between its PDZ domain and the C -terminal PDZ ligand on ARVCF (Kausalya et al., 2004). Cell-cell contact alters ZO-1 distribution; sparse cell cultures show ZO-1 to be principally nuclear, whereas high-density cultures show a primarily plasma membrane distribution (Gottardi et al., 1996). In the nucleus ZO-1 can form a functional interaction with the Y-box transcription factor ZONAB and the RNA processing factor symplekin (Kavanagh et al., 2006). Moreover, ZO-1 has been shown to regulate expression of the neuregulin receptor ErbB2 through ZONAB (Balda and Matter, 2000).

In neurons α-actinin has been proposed to link NMDARs to the cytoskeleton by binding to both NMDARs and actin (Wyszynski et al., 1997). α-actinin 1,2,3 and 4

Table 1 Novel nucleocytoplasmic shuttling proteins found in the postsynaptic density

Protein	Nuclear localization Sequence	Identified by	Potential role
Transcriptional regulators			
p120	arm repeats? [622] KKGKGGKP	1,5	Regulator of KAISO transcriptional activity
ARVCF	arm repeats? [606] RRRRDDASCFGGKKAKE	1	Regulator of ZONAB transcriptional activity via ZO-1?
p0071	arm repeats? [788] KKKKKKKR?	1,2,3	?
Plakoglobin	arm repeats?	1,4	Regulator of LEF-1 transcriptional activity
JAB-1 subunit-3	?	5	Regulator of c-Jun transcriptional activity
STAT 1	?	2	Transcription factor
ZO-1	[748] RKSARKLYERSHKLRKN?	3,5	Regulator of ZONAB transcriptional activity?
α-actinin-4	?	1,2,3,5	Regulator of Histone deactylase 7- Regulator of MEF2 transcriptional activity
Ezrin	?	5	?
RNA binding proteins			
hnRNP M,A1, H1,H2,A2/B1,L	Multiple	1,2,3,4,5	RNA Trafficking, transcription and splicing
Staufen	[194] KK(X14)KKR	1,3,5	mRNA trafficking
FMRP	N-terminal region	3	Supression of translation mRNA trafficking?
PSF	?	1	Splicing?
TLS/FUS	?	5	Splicing?
AIDA-1d	[161] HRKR	1,2,3,5	?

1. Jordan et al 2004, 2. Yoshimura et al 2004, 3. Peng et al 2004, 4. Li et al 2004,
5. Collins et al 2006

were identified in virtually every proteomic study of the PSD (Collins et al., 2006; Jordan et al., 2004; Li et al., 2004; Peng et al., 2004; Yoshimura et al., 2004). In human keratinocytes (HFK) cells, α-actinin-4 can shuttle to the nucleus in response to PI3K inhibition or actin depolymerization (Honda et al., 1998). Recently, it was shown that the C-terminal tail of α-actinin-4 binds to the histone deacetylase HDAC7. This interaction antagonizes HDAC7 activity and potentiates transcription via MEF2, a transcription factor that regulates expression of TAF 55, an important general transcription factor (Chakraborty et al., 2006). This signaling pathway may be important in neurons as MEF2 activity has been shown to control synaptogenesis by restricting excitatory synapse formation (Flavell et al., 2006).

Collins et al., also identified the cytoskeletal linker protein Ezrin (Collins et al., 2006). Ezrin is a widely studied cytoskeletal component with multiple roles ranging from polarity to migration to growth, all thought to result from its role in linking actin to the plasma membrane (Polesello and Payre, 2004). Interestingly, Ezrin was also shown to shuttle to the nucleolus in response to proteolytic cleavage by calpain I (Kaul et al., 1999). Calpain I is a major Ca^{2+}-dependent protease regulated by neuronal activity (Siman and Noszek, 1988), which suggests synaptic activity may result in Ezrin nuclear accumulation. Indeed proteolytic cleavage regulates the nuclear accumulation of several membrane bound and cytosolic proteins found at synaptic junctions, including N-cadherin, Amyloid precursor protein (APP), ErbB4 and huntingtin. The role of Ezrin in the nucleus, however, is unknown.

2.4 Nucleocytoplasmic Shuttling and RNA

Mass spectrometry of the PSD revealed a significant number of RNA binding proteins. Indeed we identified at least 31 different proteins containing RNA associated domains (such as RRM, RBD or DEAD/DEAH box helicase) (Jordan et al., 2004). The presence of RNA binding proteins at synapses is evidence for synapse-to-nucleus signaling. The largest RNA binding protein family identified in the PSD corresponds to the heterogeneous nuclear RNA binding proteins (hnRNPs), and included hnRNP M, A1, H1, H2, A2/B1 and L. hnRNPs are predominantly nuclear RNA binding proteins that participate in the trafficking of RNAs, but may also participate in transcription and splicing (Sossin and DesGroseillers, 2006). How RNAs are localized in neurons is only beginning to be understood. In general, translocation occurs when RNA is incorporated into transport particles via interactions with several RNA binding proteins (Sossin and DesGroseillers, 2006). Several nuclear proteins are enriched in these RNA granules and transport particles. Moreover, Kress et al., found that the recognition of localized RNAs by ribonucleoproteins (RNPs) in the nucleus is an early step in the transport of RNAs (Kress et al., 2004). Together this evidence suggests that mRNA transport particles originate in the nucleus. The presence of these nuclear hnRNPs at synaptic junctions outlines a mechanism for localizing mRNAs at dendritic spines. Activity-dependent nuclear signaling could initiate transport of mRNA granules to synapses, where interactions with PSD anchors could stabilize their subcellular distribution, or trigger activity-dependent translation. RNA processing and transport is a critical aspect of neuronal function. Misregulation of RNA distribution is the primary cause for several cognitive or neurodegenerative disorders including myotonic dystrophy type 2 (DM2), spinal cerebellar ataxias type 8, 10 and 12 and Fragile X Tremor Ataxia Syndrome (FXTAS). These disorders do not arise from a lack of gene function, but from the effects of expanded trinucleotide repeats at noncoding regions that cause a toxic accumulation of the RNA transcripts at regions of splicing activity in nuclei (Ranum and Cooper, 2006).

There is considerable evidence of activity-dependent trafficking of mRNAs along neuronal dendrites. NMDAR activity results in the synthesis and translocation of Arc mRNA to activated neuronal dendritic branches and spines (Steward and Worley, 2001a; Steward and Worley, 2001b). Arc protein is important for the maintenance of LTP and memory formation (Guzowski et al., 2000). CaMKIIα mRNA also shuttles along dendrites and accumulates at spines in response to stimuli leading to LTP (Rook et al., 2000; Thomas et al., 1994). Havik et al., found that high-frequency stimulation can result in the rapid delivery of existing CaMKIIα mRNA to synapses in vivo (Havik et al., 2003). In addition to these examples, neuronal stimulation using BDNF or depolarization has been shown to cause BDNF and TrkB mRNA to accumulate at dendrites, a process requiring PI3-kinase activity (Righi et al., 2000; Tongiorgi et al., 1997). How these transcripts shuttle to dendrites upon neuronal stimulation, however, is still unclear.

2.4.1 Novel RNA Binding Proteins Found at the PSD

Other than the hnRNP family members mentioned above, several groups (Collins et al., 2006; Jordan et al., 2004; Peng et al., 2004) identified the RNA binding mammalian staufen1 homolog at the PSD using SDS-PAGE coupled to LC-MS/MS. The mammalian homologue of the Drosophila RNA binding protein Staufen has been implicated in RNA transport into neuronal dendrites. Mutants, or overexpression of Staufen can significantly decrease or increase dendritic RNA pools, respectively (Tang et al., 2001). Staufen contains a functional bipartite NLS, exhibits nucleocytoplasmic shuttling capabilities and can accumulate in the nucleolus (Macchi et al., 2004; Martel et al., 2006). The nuclear and synaptic heterogeneous distribution of Staufen suggests it may help newly synthesized mRNA transcripts shuttle to synapses in response to stimulation.

Peng et al., (Peng et al., 2004) also identified the fragile-X related protein 1 (FMRP) at the PSD. FMRP is best known for its role in a form of mental retardation resulting from expanded CGG repeats in the FMR1 gene (reviewed in (Ranum and Cooper, 2006)). FMRP is a nucleocytoplasmic shuttling protein that associates with polyribosomes in neuronal dendrites (Feng et al., 1997). Its association with polyribosomes is thought to suppress the translation of certain associated mRNAs (Laggerbauer et al., 2001; Li et al., 2001). This process is important as FMR1 knockout mice and patients with fragile-X syndrome not only exhibit aberrant spine morphology (Irwin et al., 2001; Nimchinsky et al., 2001), but also exhibit enhanced mGLuR-induced LTP, which no longer requires protein synthesis (Huber et al., 2002; Pfeiffer and Huber, 2006).

2.4.2 Synapse-to-Nucleus Signaling and Alternative Splicing

In addition to RNA transport, there is growing evidence that synaptic activity regulates alternative splicing in neurons. Alternative splicing is a nuclear process in which a single gene transcript is processed to yield different mRNAs. This represents a powerful mechanism to increase the complexity of the human proteome, which stems from an estimated 20-25 thousand genes, less than the tiny Arabidopsis plant and only barely above the 19.5K found in C. Elegans (Consortium, 2004). Alternative splicing could also represent a mechanism for controlling protein abundance as ~30% of all splicing events can introduce premature termination codons, resulting in mRNA degradation by nonsense-mediated decay (Lewis et al., 2003).

Overall, alternative splicing has been shown to be essential for proper synaptic function. A mutation of the brain specific splicing factor NOVA, was shown to affect at least 49 different transcripts, many of which encode for proteins that are essential in synaptic function, such as CaMKII, NMDAR subunit R1 and Kainate receptor subunit 2 (Ule et al., 2005). Differential splicing of the reelin receptor ApoER, which binds to PSD-95 and NMDA receptors is important for plasticity and NMDA receptor subunit composition (Beffert et al., 2005). Neuroligins and neurexin splice variants affect transsynaptic interactions (Chih et al., 2006), a process which may be critical for selective synapse formation since neuroligins catalyze the formation of synaptic junctions (Scheiffele et al., 2000). Importantly, there is growing evidence that extracellular signaling regulates alternative splicing (Blaustein et al., 2005; Shin and Manley, 2004). In neurons, bicuculline-induced synaptic activity results in alternative splicing of the R1 subunit of ionotropic NMDA receptors (Mu et al., 2003). NMDAR activity or blockage regulates the insertion of a C2 or C2' cassette, respectively. The incorporation of a C2' cassette containing a newly identified ER export motif results in increased ER export and increased synaptic accumulation of this NMDAR splice variant. This represents an interesting novel mechanism to alter receptor distribution. Neuronal activity has also been shown to repress splicing of the STREX exon of BK potassium channel transcripts (Xie et al., 2005) and alter splicing of the POU family transcription factors Brn-3a and Brn-3b (Liu et al., 1996). The latter is an interesting novel mechanism to regulate gene expression.

It has been suggested that Ca^{2+} signaling activates nuclear Ca^{2+}-dependent kinases, which phosphorylate cis-acting pre-mRNA splicing enhancers and silencers (Xie et al., 2005). The best-characterized splicing activating factors are the serine-arginine rich SR proteins, which cooperate with hnRNPs to influence site-specific exon inclusion or exclusion. However several splicing factors, including a subset of hnRNPs, exhibit nucleocytoplasmic shuttling (Pinol-Roma and Dreyfuss, 1992). This suggests that control of the spatial distribution of these factors may regulate alternative splicing. For example, the yeast snu30 splicing component of the U1 snRNP (Black and Graveley, 2006) as well as the U2AF splice factor (Gama-Carvalho et al., 2001) exhibit nucleocytoplasmic shuttling and can affect splice-site choice. We have identified a few potential splicing factors at the PSD that exhibit nucleocytoplasmic shuttling.

2.4.3 Novel Splicing Factors at the PSD

In addition to hnRNP M (Gattoni et al., 1996), hnRNP A1 (Mayeda and Krainer, 1992) and hnRNP L (Rothrock et al., 2005), which have been shown to regulate alternative splicing, we found PSDs contain the polypyrimidine-tract-binding-protein (PTB)-associated splicing factor PSF (Jordan et al., 2004). PSF can bind to the nucleocytoplasmic shuttling cytoskeletal protein paxilin (Woods et al., 2002) and regulate alternative splicing (Dye and Patton, 2001; Patton et al., 1993). Using SDS-PAGE and LC-MS/MS, Collins et al., (Collins et al., 2006) identified TLS/FUS, a DNA/RNA binding protein that exhibits nucleocytoplasmic shuttling (Zinszner et al., 1997). TLS has been implicated in transport of an actin-stabilizing mRNAs to dendritic spines in response to mGluR5 activation (Fujii et al., 2005; Fujii and Takumi, 2005). Interestingly, TLS/FUS associates with NMDA receptors (Husi et al., 2000) and with components of the splicing machinery, including PTB, SRm160 and SR proteins, which influence splice site selection (Hallier et al., 1998; Kameoka et al., 2004; Meissner et al., 2003). Moreover, TLS participates in nucleocytoplasmic shuttling of bound RNA (Zinszner et al., 1997). This evidence suggests the possibility that TLS may coordinate alternative splicing with synaptic activity, a process that appears to be important for synaptic stability since TLS -/- mice exhibit defects in spine morphology (Fujii et al., 2005).

The identification of diverse RNA binding proteins at the PSD reveals an increasingly complex mechanism to regulate synaptic protein composition. There is a growing debate concerning the influence of local regulation of translation and transcription on synaptic function and plasticity (Schuman et al., 2006; Steward and Falk, 1986; Steward and Levy, 1982). Indeed studies suggest both translation of pre-existing mRNAs and activity-dependent transcription are important (Reymann and Frey, 2007). Evidence that certain transcripts, such as Arc mRNA are synthesized and transported to dendrites in response to synaptic activity represents a reasonable compromise between these two hypotheses (Steward and Worley, 2001a; Steward and Worley, 2001b). We have observed numerous RNA transport molecules at synaptic junctions, such as hnRNPs and Staufen, which could cause newly transcribed mRNAs to accumulate at dendritic spines (Table 1). However what signals catalyze the export of RNA transport particles or synaptic anchors that bind to these RNA transport molecules are currently unknown.

2.5 Uncovering Novel Synapse-to-Nucleus Signaling Pathways with Proteomics and Bioinformatics

In an effort to identify novel components of nuclear signaling pathways at synapses, we screened a PSD database for proteins with NLSs. We used SEEDTOP (BLAST program used to identify short exact matches on proteins) to identify NLSs (Nair et al., 2003) amongst proteins from a PSD database constructed from 8 different proteomic studies (Collins et al., 2006; Dosemeci et al., 2006; Jordan et al., 2004; Li et al., 2005; Li et al., 2004; Peng et al., 2004; Phillips et al., 2005; Yoshimura et al., 2004). We found that at least 166 proteins, of ~1100 PSD proteins, contained NLSs, a surprisingly high number. This included many of the proteins mentioned earlier, uncharacterized cDNAs, and proteins not previously associated with synaptic or neuronal function. However not all NLSs are functional, and biochemical contaminants are a significant factor in proteomic studies. Therefore we verified these results by selecting several proteins of unknown function (so as to not bias our selection) and cloned them from a rat brain RNA library. eGFP-tagged constructs were introduced into dissociated primary hippocampal neurons using Sindbis viral vectors or standard transfection techniques to assess their subcellular distribution (Jordan et al., 2004). Fig. 1 shows the heterogeneous distribution of three uncharacterized proteins (KIAA 1733, XP_225171 and mKIAA0417) in dendritic spines and in the nucleus, suggesting the possibility of a regulated subcellular localization. Out of 10 cloned proteins, 6 displayed both a nuclear and synaptic distribution and may thus represent novel components of synapse-to-nucleus signaling pathways (Jordan et al., 2004).

2.5.1 AIDA-1 and the Nucleolus

We identified AIDA-1 as a putative nuclear messenger using the methods described above. AIDA-1 is a brain specific protein, but overexpressed in certain immortalized cell lines, which is thought to participate in tyrosine kinase signaling pathways (Fu et al., 1999). AIDA-1 has been suggested to affect APP subcellular distribution and affect the stability of Cajal bodies (CBs) (Ghersi et al., 2004; Xu and Hebert, 2005). CBs are circular nuclear structures enriched in ribonucleoproteins (RNPs) involved in pre-mRNA and pre-ribosomal RNA (rRNA) processing. They

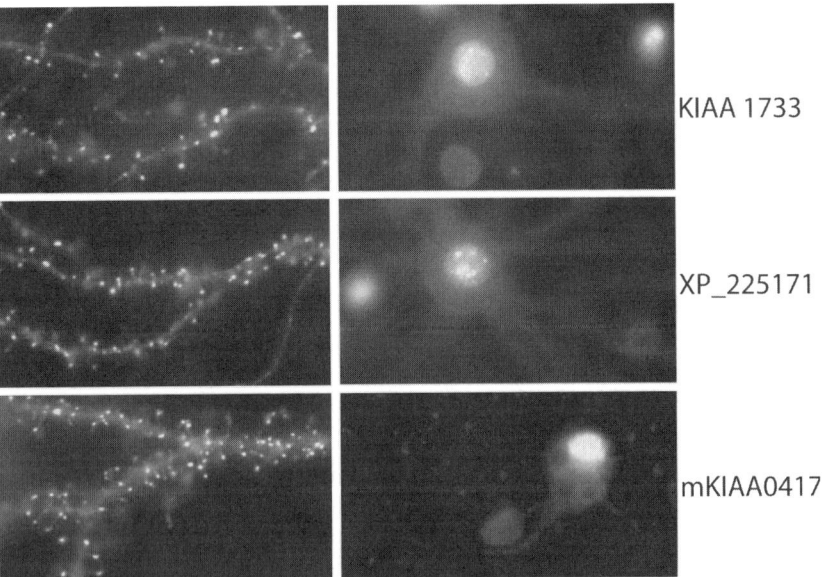

Fig. 1 KIAA 1733, XP_225171 and mKIAA0417 were cloned from a rat brain RNA library and tagged with eGFP and introduced into 21-DIV primary hippocampal neurons using lipofectamine 2000. All three proteins are present in both dendritic spines and nuclei. Note that XP_225171 is discretely expressed in subnuclear structures

have been suggested to participate in the maturation of RNPs and processing of RNA (Cioce and Lamond, 2005; Gall, 2003). We found that the identified AIDA-1 splice variant, AIDA-1d, was strikingly enriched in dendritic spines in dissociated primary hippocampal neurons. However, it was also nuclear in <20% of neurons (Figure 2). We found that KCl or glutamate-induced neuronal activity increased the number of cells displaying AIDA-1d in the nucleus (>85% of cells). Moreover, only the NMDA selective inhibitor APV could block KCl-mediated nuclear translocation. Treatment with NMDA alone, but not AMPA, caused rapid nuclear shuttling of AIDA-1d, confirming the requirement for NMDA receptor activation. We also found that AIDA-1d can bind directly to PSD-95 and is in a complex with NMDA receptors. While the provenance of nuclear AIDA-1d remains unclear, the enrichment of exogenous AIDA-1d at synaptic junctions and association with the NMDAR complex suggested that translocation originates at synapses (Jordan et al., 2007).

AIDA-1d was often confined to small circular bodies in the nucleus that were identified as CBs (Figure 3). Zatsepina et al., have recently shown that nucleolar formation require CBs (Zatsepina et al., 2003). Importantly, we found that AIDA-1 nuclear translocation regulates nucleolar stability. Prolonged bicuculine stimulation resulted in an increase in nucleolar numbers, an effect that could be blocked by AIDA-1 specific siRNAs. Given that the principle function of nucleoli is the synthesis and processing of rRNAs and ribosome generation, we speculated that more nucleoli translated to increased global protein synthesis. Indeed we found that the increase in nucleolar number was correlated to an increase in protein synthesis, an effect that was also blocked by AIDA-1 specific siRNAs. Therefore AIDA-1 is a component of an activity-dependent nuclear signaling pathway that regulates the neuronal protein biosynthetic capacity by altering nucleolar assembly.

How AIDA-1 regulates nucleologenesis is unclear, although it may do so by regulating rRNA synthesis or processing. A few observations support this hypothesis:

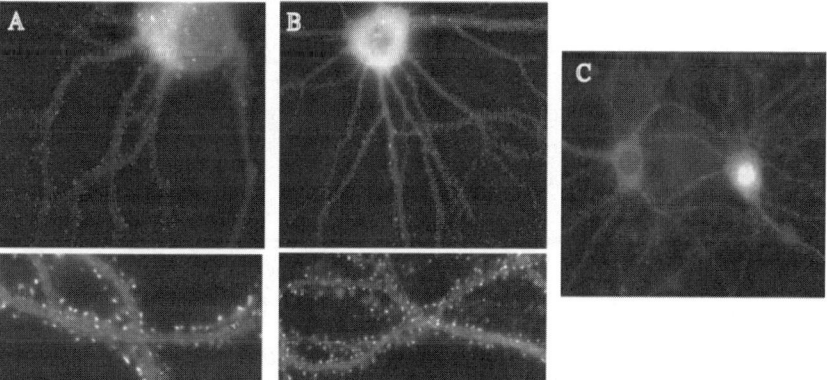

Fig. 2 A, B- AIDA-1d tagged with an N-terminal eGFP transfected into 21 DIV hippocampal is highly enriched in neuronal dendritic spines. C- AIDA-1d displays a nuclear localization in <20% of neurons. Shown here are two adjacent neurons showing nuclear and nonnuclear localizations of AIDA-1d

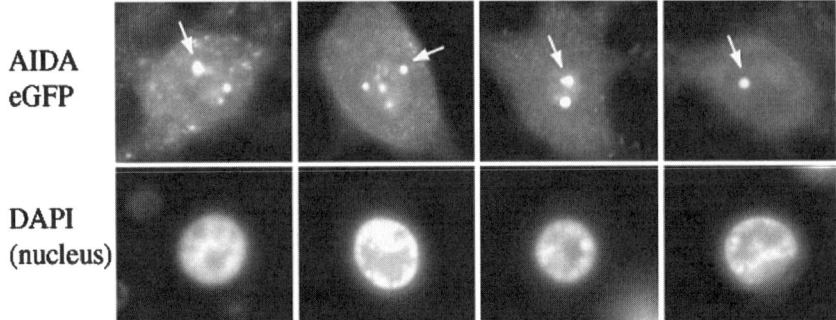

Fig. 3 AIDA-1D redistributes to Cajal bodies (CBs) in the nucleus in response to membrane depolarization by KCl. Neurons infected with AIDA-1D-eGFP were treated with 60mM KCl for 10 minutes. Arrowheads point out CBs. DAPI co-staining is shown to identify the region of the nucleus

rRNA synthesis and processing are the only established processes that regulate nucleolar assembly, as antibodies against RNA pol I can block the reformation of nucleoli during mitosis (Prieto and McStay, 2005). Moreover, neuronal activity has been shown to increase rRNA transcription as observed by an activity-dependent increase in ribosomal immunoreactivity using antibodies against rRNA (Hyson and Rubel, 1995). We show that AIDA-1 can associate with CBs and Xu et al., have shown AIDA-1 can directly bind to the CB marker p80-coilin (Xu and Hebert, 2005). Zatsepina et al (2003) have shown that Cajal bodies contribute to nucleolar assembly. Thus our work and that of others provides a possible mechanism for maturation of nucleoli by AIDA-1 involving AIDA-coilin interactions.

It will be important to determine if AIDA-1 participates in the long-term regulation of synaptic strength. Long-term storage of memories requires transcription and protein synthesis (McGaugh, 2000). LTP in particular, which underlies memory consolidation in rodents (Pastalkova et al., 2006), requires protein synthesis at later stages (Reymann and Frey, 2007). Several pathways contribute to activity-dependent increases in protein translation, including phosphorylation of general translation factors, and mRNA processing (Kelleher et al., 2004; Si et al., 2003; Wu et al., 1998). Ultrastructural studies have also shown that nucleolar numbers in neurons vary throughout development (Lafarga et al., 1995; Solovei et al., 2004), suggesting changes in the neuronal demands for protein synthesis are met by regulation of nucleolar assembly. However whether these processes and AIDA-1 contribute to long lasting changes in synaptic transmission remains to be determined.

2.6 Synapse-to-Nucleus Signaling and Future Steps

AIDA-1 is an example of how proteomics and bioinformatics can reveal novel synapse-to-nucleus signaling pathways. However searching for nucleocytoplasmic shuttling proteins by NLS identification is indirect and may yield false positives. To generate a more meaningful list of targets, we have started SILAC (stable isotope labeled amino acids in cell culture) studies to identify proteins that actually shuttle to the nucleus in response to neuronal stimulation. SILAC relies on the precise and quantitative ability of mass spectrometers to differentiate stable-isotope labeled proteins from their non-labeled counterparts. In this protocol, one population of neurons is grown in standard "light" media, while another is grown in "heavy" media containing ^{13}C-Arg and ^{13}C-Lys. Over time this label becomes incorporated into the neuronal proteome; over 90% incorporation occurs within 10 days in vitro (DIV) (Thomas A. Neubert, personal communication). We have initiated studies comparing bicuculine and 4-aminopyridine (4-AP) stimulated neurons to tetrodotoxin (TTX)-inhibited neurons. The neurons from both populations will be mixed and purified PSDs and nuclei will be analyzed by mass spectrometry. The mass spectrometer can then detect each identified protein as a paired peak (separated by the isotope mass) corresponding to the labeled and unlabeled protein. Thus the abundance of the proteins from both neuronal populations can be immediately compared. Changes in the nucleus could be correlated to changes in the protein composition of postsynaptic densities in response to the same stimulus to identify synaptic factors that traffic to the nucleus.

A second approach involves two-dimensional differential in-gel electrophoresis (2DE-DIGE). In this method, purified PSDs and nuclei from neurons treated with TTX are labeled with a cysteine-reactive fluorescent dye while PSDs and nuclei purified from bicuculine and 4-AP stimulated neurons are labeled with a second fluorescent dye. In this technique, both dyes are weight and pI matched so as to equally alter the migration of labeled proteins on 2D gels. The combined samples are then subjected to 2DE and visualized by fluorescence imagers to detect protein spots. Changes in the abundance of proteins can be measured as a shift in light emission towards one of the fluorophores. This technique is better suited for studying changes in nuclear protein composition as PSD proteins are generally hydrophobic and display poor resolution in 2DE gels (Santoni et al., 2000). Using SILAC and DIGE, we will obtain a global picture of changes in the nuclear and PSD composition in response to neuronal activity. As for all proteomics-based studies verification will be required to assess the quality of the data.

2.7 Conclusions

Proteomics-based profiling studies have shown that PSDs contain many of the proteins involved in nuclear signaling observed at other cellular junctions. The caveat to these studies is that mass spectrometry-based protein identification is very sensitive and able to detect low-level contaminants, most of which are introduced during the biochemical purification of fractions. Indeed most PSD profiling studies identified clear contaminants, including several mitochondrial resident proteins. Therefore verification of these results by conventional means will be essential. A second caveat is that most profiling studies were performed on PSDs isolated from whole brain and thus represent a composite "ideal" PSD. Since it is clear that significant heterogeneity exists between PSDs from different brain regions and neuronal types we do not suggest all proteins identified are present in each PSD.

Overall, we have identified at least 166 proteins with NLSs at synaptic junctions, which suggest PSDs have a large capacity for nuclear signaling. The presence of transcription factors, coactivators, putative splicing factors, RNA shuttling molecules and a whole host of transmembrane receptors found at the PSD whose

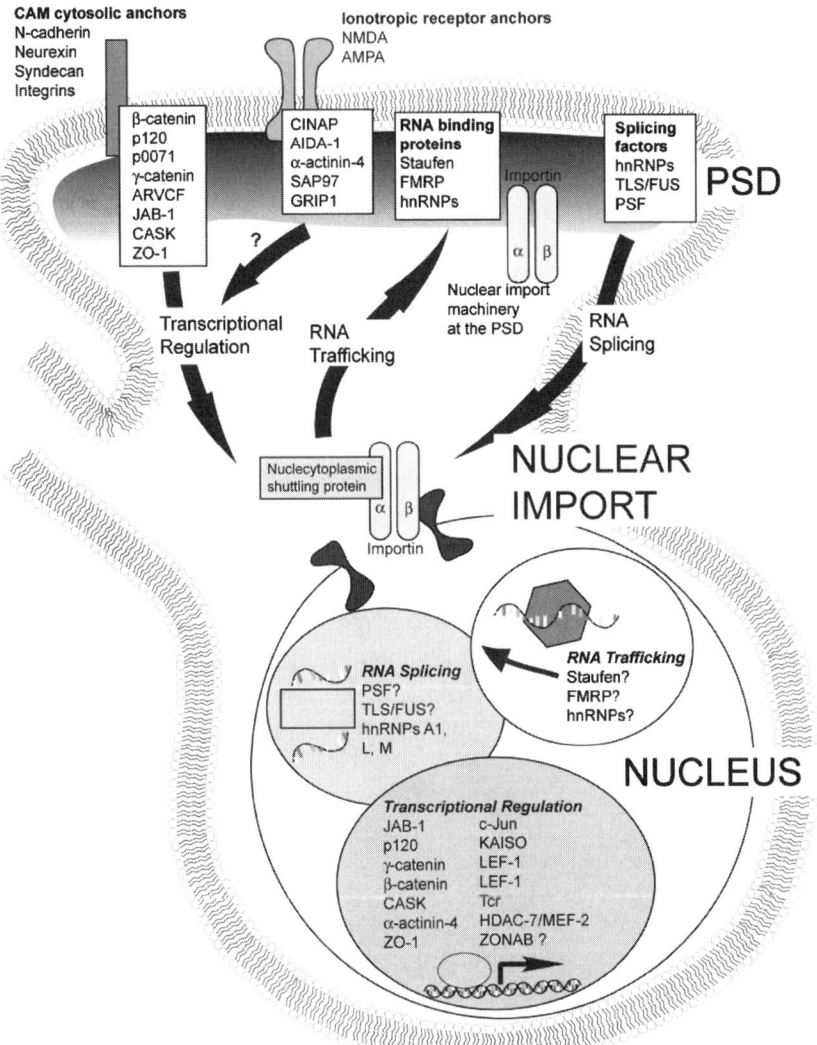

Fig. 4 Schematic representation of the major nucleocytoplasmic shuttling proteins found at the PSD. PSDs contain CAM-associated transcriptional regulators such as catenins α, β, γ and p120 as well as cytoskeletal/scaffolding proteins with newly identified roles in regulating transcription, such as ZO-1, GRIP1 and α-actinin-4. PSD-associated RNA binding proteins such as hnRNP A1, L and M, PSF and TLS/FUS may regulate activity-dependent alternative splicing. The RNA binding proteins Staufen, and other hnRNPs, which regulate RNA trafficking, may help localize newly synthesized mRNAs to synaptic junctions. (*See* Color Plate 2).

proteolytically cleaved intracellular domains shuttle to the nucleus (N-cadherin, ErbB4 and APP) (Cao and Sudhof, 2001; Marambaud et al., 2003; Williams et al., 2004), reveals a complex signaling pathway with the potential to regulate a variety of nuclear functions (Figure 4). We have shown that proteins that regulate nuclear processes, components of the nuclear import machinery and cytosolic anchors are all present at the PSD. These results strongly suggest that control of the spatial distribution of synapse-associated nuclear factors regulates nuclear function. There is increasing evidence that NMDA receptors themselves may also serve as activity-regulated cytoplasmic anchors. NMDARs associate with large signaling complexes (Husi et al., 2000). These complexes include several nuclear signaling factors such as CINAP, which exhibits nucleocytoplasmic shuttling and regulates nucleosome assembly (Wang et al., 2004), and the potential splicing factor TLS/FUS. We find AIDA-1 can also associate with NMDA receptors. The increasing complexity observed for synapse-to-nucleus signaling pathways is not surprising given that most extracellular signaling exists to regulate cellular behavior by altering gene expression. In this way and others, synaptic junctions appear as specialized cellular junctions whose widely publicized role in neuronal electrical computations is complemented by a firm and extensive capability to regulate the complex and critical nuclear functions necessary for neuronal function.

References

Aplin, A. E., and Juliano, R. L. (2001) Regulation of nucleocytoplasmic trafficking by cell adhesion receptors and the cytoskeleton. J. Cell Biol. 155, 187–191.

Ataman, B., Ashley, J., Gorczyca, D., Gorczyca, M., Mathew, D., Wichmann, C., Sigrist, S. J., and Budnik, V. (2006) Nuclear trafficking of Drosophila Frizzled-2 during synapse development requires the PDZ protein dGRIP. Proc. Natl. Acad. Sci. USA 103, 7841–7846.

Balda, M. S., and Matter, K. (2000) The tight junction protein ZO-1 and an interacting tran scription factor regulate ErbB-2 expression. EMBO J. 19, 2024–2033.

Banker, G., Churchill, L., and Cotman, C. W. (1974) Proteins of the postsynaptic density. J. Cell Biol. 63, 456–465.

Beesley, P. W., Mummery, R., and Tibaldi, J. (1995) N-cadherin is a major glycoprotein component of isolated rat forebrain postsynaptic densities. J. Neurochem. 64, 2288–2294.

Beffert, U., Weeber, E. J., Durudas, A., Qiu, S., Masiulis, I., Sweatt, J. D., Li, W. P., Adel mann, G., Frotscher, M., Hammer, R. E., and Herz, J. (2005) Modulation of synaptic plasticity and memory by Reelin involves differential splicing of the lipoprotein receptor ApoEr2. Neuron 47, 567–579.

Bianchi, E., Denti, S., Granata, A., Bossi, G., Geginat, J., Villa, A., Rogge, L., and Pardi, R. (2000) Integrin LFA-1 interacts with the transcriptional co-activator JAB1 to modulate AP-1 activity. Nature 404, 617–621.

Black, D. L., and Graveley, B. R. (2006) Splicing bioinformatics to biology. Genome Biol. 7, 317.

Blaustein, M., Pelisch, F., Tanos, T., Munoz, M. J., Wengier, D., Quadrana, L., Sanford, J. R., Muschietti, J. P., Kornblihtt, A. R., Caceres, J. F., et al. (2005) Concerted regulation of nuclear and cytoplasmic activities of SR proteins by AKT. Nat. Struct. Mol. Biol. 12, 1037–1044.

Butz, S., and Kemler, R. (1994) Distinct cadherin-catenin complexes in Ca(2+)-dependent cell-cell adhesion. FEBS Lett. 355, 195–200.

Cao, X., and Sudhof, T. C. (2001) A transcriptionally active complex of APP with Fe65 and histone acetyltransferase Tip60. Science 293, 115–120.

Carlin, R. K., Grab, D. J., Cohen, R. S., and Siekevitz, P. (1980) Isolation and characteriza tion of postsynaptic densities from various brain regions: enrichment of different types of postsynaptic densities. J. Cell Biol. 86, 831–845.

Chakraborty, S., Reineke, E. L., Lam, M., Li, X., Liu, Y., Gao, C., Khurana, S., and Kao, H. Y. (2006) α-Actinin 4 Potentiates Myocyte Enhancer Factor-2 Transcription Activ ity by Antagonizing Histone Deacetylase 7. J. Biol. Chem. 281, 35070–35080.

Chan, C. S., Weeber, E. J., Kurup, S., Sweatt, J. D., and Davis, R. L. (2003) Integrin require ment for hippocampal synaptic plasticity and spatial memory. J. Neurosci. 23, 7107–7116.

Chawla, S., Vanhoutte, P., Arnold, F. J., Huang, C. L., and Bading, H. (2003) Neuronal activ ity-dependent nucleocytoplasmic shuttling of HDAC4 and HDAC5. J. Neurochem. 85, 151–159.

Chen, J. T., Lu, D. H., Chia, C. P., Ruan, D. Y., Sabapathy, K., and Xiao, Z. C. (2005) Im paired long-term potentiation in c-Jun N-terminal kinase 2-deficient mice. J. Neurochem. 93, 463–473.

Chih, B., Gollan, L., and Scheiffele, P. (2006) Alternative splicing controls selective trans- synaptic interactions of the neuroligin-neurexin complex. Neuron 51, 171–178.

Chung, H. J., Xia, J., Scannevin, R. H., Zhang, X., and Huganir, R. L. (2000) Phosphorylation of the AMPA receptor subunit GluR2 differentially regulates its interaction with PDZ domain-containing proteins. J. Neurosci. 20, 7258–7267.

Cioce, M., and Lamond, A. I. (2005) CAJAL BODIES: A Long History of Discovery. Annu. Rev. Cell Dev. Biol. 21, 105–131.

Cohen, R. S., Blomberg, F., Berzins, K., and Siekevitz, P. (1977) The structure of postsynap tic densities isolated from dog cerebral cortex. I. Overall morphology and protein com position. J. Cell Biol. 74, 181–203.

Collins, M. O., Husi, H., Yu, L., Brandon, J. M., Anderson, C. N., Blackstock, W. P., Choudhary, J. S., and Grant, S. G. (2006) Molecular characterization and comparison of the components and multiprotein complexes in the postsynaptic proteome. J. Neurochem. 97 Suppl 1, 16–23.

Consortium, I. H. G. S. (2004) Finishing the euchromatic sequence of the human genome. Nature 431, 931–945.

Cyert, M. S. (2001) Regulation of nuclear localization during signaling. J. Biol. Chem. 276, 20805–20808.

D'Arcangelo, G., Miao, G. G., Chen, S. C., Soares, H. D., Morgan, J. I., and Curran, T. (1995) A rotein related to extracellular matrix proteins deleted in the mouse mutant reeler. Nature 374, 719–723.

Deisseroth, K., Mermelstein, P. G., Xia, H., and Tsien, R. W. (2003) Signaling from synapse to nucleus: the logic behind the mechanisms. Curr. Opin. Neurobiol. 13, 354–365.

Dosemeci, A., Tao-Cheng, J. H., Vinade, L., and Jaffe, H. (2006) Preparation of postsynaptic den sity fraction from hippocampal slices and proteomic analysis. Biochem. Biophys. Res. Commun. 339, 687–694.

Dye, B. T., and Patton, J. G. (2001) An RNA recognition motif (RRM) is required for the local ization of PTB-associated splicing factor (PSF) to subnuclear speckles. Exp. Cell Res. 263, 131–144.

Fagotto, F., Funayama, N., Gluck, U., and Gumbiner, B. M. (1996) Binding to cadherins antago nizes the signaling activity of beta-catenin during axis formation in Xenopus. J. Cell Biol. 132, 1105–1114.

Feng, Y., Gutekunst, C. A., Eberhart, D. E., Yi, H., Warren, S. T., and Hersch, S. M. (1997) Fragile X mental retardation protein: nucleocytoplasmic shuttling and association with somatodendritic ribosomes. J. Neurosci. 17, 1539–1547.

Flavell, S. W., Cowan, C. W., Kim, T. K., Greer, P. L., Lin, Y., Paradis, S., Griffith, E. C., Hu, L. S., Chen, C., and Greenberg, M. E. (2006) Activity-dependent regulation of MEF2 transcription factors suppresses excitatory synapse number. Science 311, 1008–1012.

Fu, X., McGrath, S., Pasillas, M., Nakazawa, S., and Kamps, M. P. (1999) EB-1, a tyrosine kinase signal transduction gene, is transcriptionally activated in the t(1;19) subset of pre-B ALL, which express oncoprotein E2a-Pbx1. Oncogene 18, 4920–4929.

Fujii, R., Okabe, S., Urushido, T., Inoue, K., Yoshimura, A., Tachibana, T., Nishikawa, T., Hicks, G. G., and Takumi, T. (2005) The RNA binding protein TLS is translocated to dendritic spines by mGluR5 activation and regulates spine morphology. Curr. Biol. 15, 587–593.

Fujii, R., and Takumi, T. (2005) TLS facilitates transport of mRNA encoding an actin-stabilizing protein to dendritic spines. J. Cell Sci. 118, 5755–5765.

Gall, J. G. (2003) The centennial of the Cajal body. Nat. Rev. Mol. Cell Biol. 4, 975–980.

Gama-Carvalho, M., Carvalho, M. P., Kehlenbach, A., Valcarcel, J., and Carmo-Fonseca, M. (2001) Nucleocytoplasmic shuttling of heterodimeric splicing factor U2AF. J. Biol. Chem. 276, 13104–13112.

Gattoni, R., Mahe, D., Mahl, P., Fischer, N., Mattei, M. G., Stevenin, J., and Fuchs, J. P. (1996) The human hnRNP-M proteins: structure and relation with early heat shock- induced splicing arrest and chromosome mapping. Nucleic Acids Res. 24, 2535–2542.

Gerrow, K., and El-Husseini, A. (2006) Cell adhesion molecules at the synapse. Front. Biosci. 11, 2400–2419.

Ghersi, E., Noviello, C., and D'Adamio, L. (2004) Amyloid-beta protein precursor (AbetaPP) intracellular domain-associated protein-1 proteins bind to AbetaPP and modulate its processing in an isoform-specific manner. J. Biol. Chem. 279, 49105–49112.

Gordon, M. D., and Nusse, R. (2006) Wnt signaling: multiple pathways, multiple receptors, and multiple transcription factors. J. Biol. Chem. 281, 22429–22433.

Gorlich, D., and Kutay, U. (1999) Transport between the cell nucleus and the cytoplasm. Annu. Rev. Cell Dev. Biol. 15, 607–660.

Gottardi, C. J., Arpin, M., Fanning, A. S., and Louvard, D. (1996) The junction-associated protein, zonula occludens-1, localizes to the nucleus before the maturation and during the remodeling of cell-cell contacts. Proc. Natl. Acad. Sci. USA 93, 10779–10784.

Graef, I. A., Mermelstein, P. G., Stankunas, K., Neilson, J. R., Deisseroth, K., Tsien, R. W., and Crabtree, G. R. (1999) L-type calcium channels and GSK-3 regulate the activity of NF-ATc4 in hippocampal neurons. Nature 401, 703–708.

Guzowski, J. F., Lyford, G. L., Stevenson, G. D., Houston, F. P., McGaugh, J. L., Worley, P. F., and Barnes, C. A. (2000) Inhibition of activity-dependent arc protein expression in the rat hippocampus impairs the maintenance of long-term potentiation and the consolidation of long-term memory. J. Neurosci. 20, 3993–4001.

Hallier, M., Lerga, A., Barnache, S., Tavitian, A., and Moreau-Gachelin, F. (1998) The transcription factor Spi-1/PU.1 interacts with the potential splicing factor TLS. J. Biol. Chem. 273, 4838–4842.

Hardingham, G. E., Arnold, F. J., and Bading, H. (2001) Nuclear calcium signaling controls CREB-mediated gene expression triggered by synaptic activity. Nat. Neurosci. 4, 261–267.

Hatzfeld, M. (2005) The p120 family of cell adhesion molecules. Eur J. Cell Biol. 84, 205–214.

Havik, B., Rokke, H., Bardsen, K., Davanger, S., and Bramham, C. R. (2003) Bursts of high-frequency stimulation trigger rapid delivery of pre-existing alpha-CaMKII mRNA to synapses: a mechanism in dendritic protein synthesis during long-term potentiation in adult awake rats. Eur J. Neurosci. 17, 2679–2689.

Honda, K., Yamada, T., Endo, R., Ino, Y., Gotoh, M., Tsuda, H., Yamada, Y., Chiba, H., and Hirohashi, S. (1998) Actinin-4, a novel actin-bundling protein associated with cell motility and cancer invasion. J. Cell Biol. 140, 1383–1393.

Hsueh, Y. P., and Sheng, M. (1999) Regulated expression and subcellular localization of syndecan heparan sulfate proteoglycans and the syndecan-binding protein CASK/LIN-2 during rat brain development. J. Neurosci. 19, 7415–7425.

Hsueh, Y. P., Wang, T. F., Yang, F. C., and Sheng, M. (2000) Nuclear translocation and tran scription regulation by the membrane-associated guanylate kinase CASK/LIN-2. Nature 404, 298–302.

Huber, K. M., Gallagher, S. M., Warren, S. T., and Bear, M. F. (2002) Altered synaptic plas ticity in a mouse model of fragile X mental retardation. Proc. Natl. Acad. Sci. USA 99, 7746–7750.

Huber, O., Korn, R., McLaughlin, J., Ohsugi, M., Herrmann, B. G., and Kemler, R. (1996) Nuclear localization of beta-catenin by interaction with transcription factor LEF-1. Mech. Dev. 59, 3–10.

Hubner, S., Jans, D. A., and Drenckhahn, D. (2001) Roles of cytoskeletal and junctional plaque proteins in nuclear signaling. Int. Rev. Cytol. 208, 207–265.

Hulsken, J., Birchmeier, W., and Behrens, J. (1994) E-cadherin and APC compete for the interaction with beta-catenin and the cytoskeleton. J. Cell Biol. 127, 2061–2069.

Husi, H., Ward, M. A., Choudhary, J. S., Blackstock, W. P., and Grant, S. G. (2000) Proteo mic analysis of NMDA receptor-adhesion protein signaling complexes. Nat. Neurosci. 3, 661–669.

Hyson, R. L., and Rubel, E. W. (1995) Activity-dependent regulation of a ribosomal RNA epitope in the chick cochlear nucleus. Brain Res. 672, 196–204.

Irwin, S. A., Patel, B., Idupulapati, M., Harris, J. B., Crisostomo, R. A., Larsen, B. P., Kooy, F., Willems, P. J., Cras, P., Kozlowski, P. B., et al. (2001) Abnormal dendritic spine characteristics in the temporal and visual cortices of patients with fragile-X syndrome: a quantitative examination. Am. J. Med. Genet. 98, 161–167.

Jones, R. B., Gordus, A., Krall, J. A., and MacBeath, G. (2006) A quantitative protein interaction network for the ErbB receptors using protein microarrays. Nature 439, 168–174.

Jordan, B. A., Fernholz, B. D., Boussac, M., Xu, C., Grigorean, G., Ziff, E. B., and Neubert, T. A. (2004) Identification and verification of novel rodent postsynaptic density proteins. Mol. Cell Proteomics. 3, 857–871.

Jordan, B. A., Fernholz, B. D., Neubert, T. A., and Ziff, E. B. (2006) *The Dynamic Synapse: Molecular Methods in Ionotropic Receptor Biology.* CRC/Taylor & Francis, Boc Raton.

Jordan, B. A., Fernholz, B. D., Khatri, L., Ziff, E. B. (2007) Activity-dependent AIDA-1 nuclear signaling regulates nucleolar numbers and protein synthesis in neurons. Nat. Neurosci. 10, 427–35.

Kaltschmidt, C., Kaltschmidt, B., and Baeuerle, P. A. (1995) Stimulation of ionotropic gluta mate receptors activates transcription factor NF-kappa B in primary neurons. Proc. Natl. Acad. Sci. USA 92, 9618–9622.

Kameoka, S., Duque, P., and Konarska, M. M. (2004) p54(nrb) associates with the 5' splice site within large transcription/splicing complexes. EMBO J. 23, 1782–1791.

Kandel, E. R. (2001) The molecular biology of memory storage: a dialogue between genes and synapses. Science 294, 1030–1038.

Kaul, S. C., Kawai, R., Nomura, H., Mitsui, Y., Reddel, R. R., and Wadhwa, R. (1999) Identi fication of a 55-kDa ezrin-related protein that induces cytoskeletal changes and localizes to the nucleolus. Exp. Cell Res. 250, 51–61.

Kausalya, P. J., Phua, D. C., and Hunziker, W. (2004) Association of ARVCF with zonula occludens (ZO)-1 and ZO-2: binding to PDZ-domain proteins and cell-cell adhesion regulate plasma membrane and nuclear localization of ARVCF. Mol. Biol. Cell 15, 5503–5515.

Kavanagh, E., Buchert, M., Tsapara, A., Choquet, A., Balda, M. S., Hollande, F., and Matter, K. (2006) Functional interaction between the ZO-1-interacting transcription factor ZONAB/DbpA and the RNA processing factor symplekin. J. Cell Sci. 119, 5098–5105.

Kelleher, R. J., 3rd, Govindarajan, A., Jung, H. Y., Kang, H., and Tonegawa, S. (2004) Trans lational control by MAPK signaling in long-term synaptic plasticity and memory. Cell 116, 467–479.

Kelly, K. F., Spring, C. M., Otchere, A. A., and Daniel, J. M. (2004) NLS-dependent nuclear localization of p120ctn is necessary to relieve Kaiso-mediated transcriptional repression. J. Cell Sci. 117, 2675–2686.

Kennedy, M. B. (1993) The postsynaptic density. Curr Opin Neurobiol 3, 732–737.

Kohu, K., Ogawa, F., and Akiyama, T. (2002) The SH3, HOOK and guanylate kinase-like domains of hDLG are important for its cytoplasmic localization. Genes Cells 7, 707–715.

Kress, T. L., Yoon, Y. J., and Mowry, K. L. (2004) Nuclear RNP complex assembly initiates cytoplasmic RNA localization. J. Cell Biol. 165, 203–211.

Lafarga, M., Andres, M. A., Fernandez-Viadero, C., Villegas, J., and Berciano, M. T. (1995) Number of nucleoli and coiled bodies and distribution of fibrillar centres in differentiating Purkinje neurons of chick and rat cerebellum. Anat. Embryol. (Berl) 191, 359–367.

Laggerbauer, B., Ostareck, D., Keidel, E. M., Ostareck-Lederer, A., and Fischer, U. (2001) Evidence that fragile X mental retardation protein is a negative regulator of translation. Hum. Mol. Genet. 329–338.

Lallena, M. J., Martinez, C., Valcarcel, J., and Correas, I. (1998) Functional association of nuclear protein 4.1 with pre-mRNA splicing factors. J. Cell Sci. 111 (Pt 14), 1963–1971.

Lewis, B. P., Green, R. E., and Brenner, S. E. (2003) Evidence for the widespread coupling of alternative splicing and nonsense-mediated mRNA decay in humans. Proc. Natl. Acad. Sci. USA 100, 189–192.

Li, K., Hornshaw, M. P., van Minnen, J., Smalla, K. H., Gundelfinger, E. D., and Smit, A. B. (2005) Organelle proteomics of rat synaptic proteins: correlation-profiling by isotope- coded affinity tagging in conjunction with liquid chromatography-tandem mass spectrometry to reveal post-synaptic density specific proteins. J. Proteome Res. 4, 725–733.

Li, K. W., Hornshaw, M. P., Van Der Schors, R. C., Watson, R., Tate, S., Casetta, B., Jimenez, C. R., Gouwenberg, Y., Gundelfinger, E. D., Smalla, K. H., and Smit, A. B. (2004) Proteomics analysis of rat brain postsynaptic density. Implications of the diverse protein functional groups for the integration of synaptic physiology. J. Biol. Chem. 279, 987–1002.

Li, Y., Yu, W. H., Ren, J., Chen, W., Huang, L., Kharbanda, S., Loda, M., and Kufe, D. (2003) Heregulin targets gamma-catenin to the nucleolus by a mechanism dependent on the DF3/MUC1 oncoprotein. Mol. Cancer Res. 1, 765–775.

Li, Z., Zhang, Y., Ku, L., Wilkinson, K. D., Warren, S. T., and Feng, Y. (2001) The fragile X mental retardation protein inhibits translation via interacting with mRNA. Nucleic Acids Res. 29, 2276–2283.

Lipscombe, D., Madison, D. V., Poenie, M., Reuter, H., Tsien, R. W., and Tsien, R. Y. (1988) Imaging of cytosolic Ca2+ transients arising from Ca2+ stores and Ca2+ channels in sympathetic neurons. Neuron 1, 355–365.

Liu, Y. Z., Dawson, S. J., and Latchman, D. S. (1996) Alternative splicing of the Brn-3a and Brn-3b transcription factor RNAs is regulated in neuronal cells. J. Mol. Neurosci. 7, 77- 85.

Macchi, P., Brownawell, A. M., Grunewald, B., DesGroseillers, L., Macara, I. G., and Kiebler, M. A. (2004) The brain-specific double-stranded RNA-binding protein Staufen2: nucleolar accumulation and isoform-specific exportin-5-dependent export. J. Biol. Chem. 279, 31440–31444.

Marambaud, P., Wen, P. H., Dutt, A., Shioi, J., Takashima, A., Siman, R., and Robakis, N. K. (2003) A CBP binding transcriptional repressor produced by the PS1/epsilon-cleavage of N-cadherin is inhibited by PS1 FAD mutations. Cell 114, 635–645.

Martel, C., Macchi, P., Furic, L., Kiebler, M. A., and Desgroseillers, L. (2006) Staufen1 is imported into the nucleolus via a bipartite nuclear localization signal and several modulatory determinants. Biochem. J. 393, 245–254.

Matsuda, S., Mikawa, S., and Hirai, H. (1999) Phosphorylation of serine-880 in GluR2 by protein kinase C prevents its C terminus from binding with glutamate receptor-interacting protein. J. Neurochem. 73, 1765–1768.

Mayeda, A., and Krainer, A. R. (1992) Regulation of alternative pre-mRNA splicing by hnRNP A1 and splicing factor SF2. Cell 68, 365–375.

McGaugh, J. L. (2000) Memory–a century of consolidation. Science 287, 248–251.

Meissner, M., Lopato, S., Gotzmann, J., Sauermann, G., and Barta, A. (2003) Proto- oncoprotein TLS/FUS is associated to the nuclear matrix and complexed with splicing factors PTB, SRm160, and SR proteins. Exp. Cell Res. 283, 184–195.

Mu, Y., Otsuka, T., Horton, A. C., Scott, D. B., and Ehlers, M. D. (2003) Activity-dependent mRNA splicing controls ER export and synaptic delivery of NMDA receptors. Neuron 40, 581–594.

Murase, S., Mosser, E., and Schuman, E. M. (2002) Depolarization drives beta-Catenin into neuronal spines promoting changes in synaptic structure and function. Neuron 35, 91–105.

Nair, R., Carter, P., and Rost, B. (2003) NLSdb: database of nuclear localization signals. Nucleic Acids Res 31, 397–399.

Nakata, A., Ito, T., Nagata, M., Hori, S., and Sekimizu, K. (2004) GRIP1tau, a novel PDZ domain-containing transcriptional activator, cooperates with the testis-specific transcription elongation factor SII-T1. Genes Cells 9, 1125–1135.

Nimchinsky, E. A., Oberlander, A. M., and Svoboda, K. (2001) Abnormal development of dendritic spines in FMR1 knock-out mice. J. Neurosci. 21, 5139–5146.

Osawa, M., Tong, K. I., Lilliehook, C., Wasco, W., Buxbaum, J. D., Cheng, H. Y., Penninger, J. M., Ikura, M., and Ames, J. B. (2001) Calcium-regulated DNA binding and oligomerization of the neuronal calcium-sensing protein, calsenilin/DREAM/KChIP3. J. Biol. Chem. 276, 41005–41013.

Pastalkova, E., Serrano, P., Pinkhasova, D., Wallace, E., Fenton, A. A., and Sacktor, T. C. (2006) Storage of spatial information by the maintenance mechanism of LTP. Science 313, 1141–1144.

Patton, J. G., Porro, E. B., Galceran, J., Tempst, P., and Nadal-Ginard, B. (1993) Cloning and characterization of PSF, a novel pre-mRNA splicing factor. Genes Dev. 7, 393–406.

Peng, J., Kim, M. J., Cheng, D., Duong, D. M., Gygi, S. P., and Sheng, M. (2004) Semi-quantitative proteomic analysis of rat forebrain postsynaptic density fractions by mass spec trometry. J. Biol. Chem. 14, 21003–21011.

Pfeiffer, B. E., and Huber, K. M. (2006) Current advances in local protein synthesis and synaptic plasticity. J. Neurosci. 26, 7147–7150.

Phillips, G. R., Florens, L., Tanaka, H., Khaing, Z. Z., Fidler, L., Yates, J. R., 3rd, and Colman, D. R. (2005) Proteomic comparison of two fractions derived from the transsynaptic scaffold. J. Neurosci. Res 81, 762–775.

Pinol-Roma, S., and Dreyfuss, G. (1992) Shuttling of pre-mRNA binding proteins between nucleus and cytoplasm. Nature 355, 730–732.

Polesello, C., and Payre, F. (2004) Small is beautiful: what flies tell us about ERM protein function in development. Trends Cell Biol. 14, 294–302.

Prieto, J. L., and McStay, B. (2005) Nucleolar biogenesis: the first small steps. Biochem. Soc. Trans. 33, 1441–1443.

Rajan, P., Symes, A. J., and Fink, J. S. (1996) STAT proteins are activated by ciliary neurotrophic factor in cells of central nervous system origin. J. Neurosci. Res 43, 403–411.

Ranum, L. P., and Cooper, T. A. (2006) RNA-mediated neuromuscular disorders. Annu. Rev. Neurosci. 29, 259–277.

Reymann, K. G., and Frey, J. U. (2007) The late maintenance of hippocampal LTP: Requirements, phases, 'synaptic tagging', 'late-associativity' and implications. Neuropharmacol. 52, 24–40.

Righi, M., Tongiorgi, E., and Cattaneo, A. (2000) Brain-derived neurotrophic factor (BDNF) induces dendritic targeting of BDNF and tyrosine kinase B mRNAs in hippocampal neurons through a phosphatidylinositol-3 kinase-dependent pathway. J. Neurosci. 20, 3165–3174.

Roczniak-Ferguson, A., and Reynolds, A. B. (2003) Regulation of p120-catenin nucleocyto plasmic shuttling activity. J. Cell Sci. 116, 4201–4212.

Rodova, M., Kelly, K. F., VanSaun, M., Daniel, J. M., and Werle, M. J. (2004) Regulation of the rapsyn promoter by kaiso and delta-catenin. Mol. Cell Biol. 24, 7188–7196.

Rodriguez, M. A., Pesold, C., Liu, W. S., Kriho, V., Guidotti, A., Pappas, G. D., and Costa, E. (2000) Colocalization of integrin receptors and reelin in dendritic spine postsynaptic densities of adult nonhuman primate cortex. Proc. Natl. Acad. Sci. USA 97, 3550–3555.

Rook, M. S., Lu, M., and Kosik, K. S. (2000) CaMKIIalpha 3' untranslated region-directed mRNA translocation in living neurons: visualization by GFP linkage. J. Neurosci. 20, 6385–6393.

Rothrock, C. R., House, A. E., and Lynch, K. W. (2005) HnRNP L represses exon splicing via a regulated exonic splicing silencer. EMBO J. 24, 2792–2802.

Salomon, D., Sacco, P. A., Roy, S. G., Simcha, I., Johnson, K. R., Wheelock, M. J., and Ben-Ze'ev, A. (1997) Regulation of beta-catenin levels and localization by overexpression of plakoglobin and inhibition of the ubiquitin-proteasome system. J. Cell Biol. 139, 1325–1335.

Santoni, V., Molloy, M., and Rabilloud, T. (2000) Membrane proteins and proteomics: unamour impossible? Electrophoresis 21, 1054–1070.

Scheiffele, P., Fan, J., Choih, J., Fetter, R., and Serafini, T. (2000) Neuroligin expressed in nonneuronal cells triggers presynaptic development in contacting axons. Cell 101, 657–669.

Schuman, E. M., Dynes, J. L., and Steward, O. (2006) Synaptic regulation of translation of dendritic mRNAs. J. Neurosci. 26, 7143–7146.

Shen, L., Liang, F., Walensky, L. D., and Huganir, R. L. (2000) Regulation of AMPA receptor GluR1 subunit surface expression by a 4. 1N-linked actin cytoskeletal association. J. Neurosci. 20, 7932–7940.

Shin, C., and Manley, J. L. (2004) Cell signalling and the control of pre-mRNA splicing. Nat Rev Mol Cell Biol 5, 727–738.

Si, K., Giustetto, M., Etkin, A., Hsu, R., Janisiewicz, A. M., Miniaci, M. C., Kim, J. H., Zhu, H., and Kandel, E. R. (2003) A neuronal isoform of CPEB regulates local protein synthesis and stabilizes synapse-specific long-term facilitation in aplysia. Cell 115, 893–904.

Siman, R., and Noszek, J. C. (1988) Excitatory amino acids activate calpain I and induce structural protein breakdown in vivo. Neuron 1, 279–287.

Simcha, I., Geiger, B., Yehuda-Levenberg, S., Salomon, D., and Ben-Ze'ev, A. (1996) Suppression of tumorigenicity by plakoglobin: an augmenting effect of N-cadherin. J. Cell Biol. 133, 199–209.

Solovei, I., Grandi, N., Knoth, R., Volk, B., and Cremer, T. (2004) Positional changes of pericentromeric heterochromatin and nucleoli in postmitotic Purkinje cells during murine cerebellum development. Cytogenetic and Genome Research 105, 302–310.

Sossin, W. S., and DesGroseillers, L. (2006) Intracellular trafficking of RNA in neurons. Traffic 7, 1581–1589.

Steward, O., and Falk, P. M. (1986) Protein-synthetic machinery at postsynaptic sites during synaptogenesis: a quantitative study of the association between polyribosomes and developing synapses. J. Neurosci. 6, 412–423.

Steward, O., and Levy, W. B. (1982) Preferential localization of polyribosomes under the base of dendritic spines in granule cells of the dentate gyrus. J. Neurosci. 2, 284–291.

Steward, O., and Worley, P. F. (2001a) A cellular mechanism for targeting newly synthesized mRNAs to synaptic sites on dendrites. Proc. Natl. Acad. Sci. USA 98, 7062–7068.

Steward, O., and Worley, P. F. (2001b) Selective targeting of newly synthesized Arc mRNA to active synapses requires NMDA receptor activation. Neuron 30, 227–240.

Tang, S. J., Meulemans, D., Vazquez, L., Colaco, N., and Schuman, E. (2001) A role for a rat homolog of staufen in the transport of RNA to neuronal dendrites. Neuron 32, 463–475.

Thomas, K. L., Laroche, S., Errington, M. L., Bliss, T. V., and Hunt, S. P. (1994) Spatial and temporal changes in signal transduction pathways during LTP. Neuron 13, 737–745.

Thompson, K. R., Otis, K. O., Chen, D. Y., Zhao, Y., O'Dell, T. J., and Martin, K. C. (2004) Synapse to nucleus signaling during long-term synaptic plasticity; a role for the classical active nuclear import pathway. Neuron 44, 997–1009.

Tongiorgi, E., Righi, M., and Cattaneo, A. (1997) Activity-dependent dendritic targeting of BDNF and TrkB mRNAs in hippocampal neurons. J. Neurosci. 17, 9492–9505.

Ule, J., Ule, A., Spencer, J., Williams, A., Hu, J. S., Cline, M., Wang, H., Clark, T., Fraser, C., Ruggiu, M., et al. (2005) Nova regulates brain-specific splicing to shape the synapse. Nat. Genet. 37, 844–852.

Usacheva, A., Smith, R., Minshall, R., Baida, G., Seng, S., Croze, E., and Colamonici, O. (2001) The WD motif-containing protein receptor for activated protein kinase C (RACK1) is required for recruitment and activation of signal transducer and activator of transcription 1 through the type I interferon receptor. J. Biol. Chem. 276, 22948–22953.

Vinade, L., Chang, M., Schlief, M. L., Petersen, J. D., Reese, T. S., Tao-Cheng, J. H., and Dosemeci, A. (2003) Affinity purification of PSD-95-containing postsynaptic complexes. J. Neurochem. 87, 1255–1261.

Walikonis, R. S., Jensen, O. N., Mann, M., Provance, D. W., Jr, Mercer, J. A., and Kennedy, M. B. (2000) Identification of Proteins in the Postsynaptic Density Fraction by Mass Spectrometry. J. Neurosci. 20, 4069–4080.

Walsh, M. J., and Kuruc, N. (1992) The postsynaptic density: constituent and associated proteins characterized by electrophoresis, immunoblotting, and peptide sequencing. J. Neurochem. 59, 667–678.

Wang, G. S., Hong, C. J., Yen, T. Y., Huang, H. Y., Ou, Y., Huang, T. N., Jung, W. G., Kuo, T. Y., Sheng, M., Wang, T. F., and Hsueh, Y. P. (2004) Transcriptional modification by a CASK-interacting nucleosome assembly protein. Neuron 42, 113–128.

West, A. E., Griffith, E. C., and Greenberg, M. E. (2002) Regulation of transcription factors by neuronal activity. Nat. Rev. Neurosci. 3, 921–931.

Williams, C. C., Allison, J. G., Vidal, G. A., Burow, M. E., Beckman, B. S., Marrero, L., and Jones, F. E. (2004) The ERBB4/HER4 receptor tyrosine kinase regulates gene expression by functioning as a STAT5A nuclear chaperone. J. Cell Biol. 167, 469–478.

Woods, A. J., Roberts, M. S., Choudhary, J., Barry, S. T., Mazaki, Y., Sabe, H., Morley, S. J., Critchley, D. R., and Norman, J. C. (2002) Paxillin associates with poly(A)-binding protein 1 at the dense endoplasmic reticulum and the leading edge of migrating cells. J. Biol. Chem. 277, 6428–6437.

Wu, L., Wells, D., Tay, J., Mendis, D., Abbott, M. A., Barnitt, A., Quinlan, E., Heynen, A., Fallon, J. R., and Richter, J. D. (1998) CPEB-mediated cytoplasmic polyadenylation and the regulation of experience-dependent translation of alpha-CaMKII mRNA at synapses. Neuron 21, 1129–1139.

Wyszynski, M., Lin, J., Rao, A., Nigh, E., Beggs, A. H., Craig, A. M., and Sheng, M. (1997) Competitive binding of alpha-actinin and calmodulin to the NMDA receptor. Nature 385, 439–442.

Xie, J., Jan, C., Stoilov, P., Park, J., and Black, D. L. (2005) A consensus CaMK IV-responsive RNA sequence mediates regulation of alternative exons in neurons. RNA 11, 1825–1834.

Xu, H., and Hebert, M. D. (2005) A novel EB-1/AIDA-1 isoform, AIDA-1c, interacts with the Cajal body protein coilin. BMC. Cell Biol. 6, 23.

Yoshimura, Y., Yamauchi, Y., Shinkawa, T., Taoka, M., Donai, H., Takahashi, N., Isobe, T., and Yamauchi, T. (2004) Molecular constituents of the postsynaptic density fraction revealed by proteomic analysis using multidimensional liquid chromatography-tandem mass spectrometry. J. Neurochem. 88, 759–768.

Yu, G., Zerucha, T., Ekker, M., and Rubenstein, J. L. (2001) Evidence that GRIP, a PDZ-domain protein which is expressed in the embryonic forebrain, co-activates transcription with DLX homeodomain proteins. Brain Res. Dev. Brain Res. 130, 217–230.

Zatsepina, O., Baly, C., Chebrout, M., and Debey, P. (2003) The step-wise assembly of a functional nucleolus in preimplantation mouse embryos involves the cajal (coiled) body. Dev. Biol. 253, 66–83.

Ziff, E. B. (1997) Enlightening the postsynaptic density. Neuron 19, 1163–1174.

Zinszner, H., Sok, J., Immanuel, D., Yin, Y., and Ron, D. (1997) TLS (FUS) binds RNA in vivo and engages in nucleo-cytoplasmic shuttling. J. Cell Sci. 110, 1741–1750.

Chapter 3
Transcriptional Regulation of the Tbr1-CASK-CINAP Protein Complex in Response to Neuronal Activity

Yi-Ping Hsueh

Abstract In neurons, postsynaptic density (PSD) refers to an electron dense structure underneath the plasma membrane at the postsynaptic site, containing various protein molecules essential for responding to the presynaptic signals, including ion channels, scaffold proteins, signaling molecules, and cytoskeletons. The scaffold proteins provide the linkage among ion channels, signaling molecules and the cytoskeleton. Accumulating data indicate that the membrane associated guanylate kinase (MAGUK) proteins are important postsynaptic scaffold proteins involved in synaptic targeting, clustering and signaling of ion channels (Sheng and Sala 2001; Kim and Sheng 2004; Montgomery, Zamorano and Garner 2004). Calcium/calmodulin-dependent serine protein kinase (CASK) belongs to MAGUK protein family. Although CASK contains a CaMK-like domain, it doesn't possess kinase activity. Instead, like other MAGUK proteins, CASK functions as multidomain scaffolding protein. Recent studies showed that CASK not only performs its function at the postsynaptic site but also enters the nuclei of neurons and regulates gene expression via the interaction with transcriptional factor T-brain-1 (Tbr-1) and nucleosome assembly protein CINAP (CASK-interacting nucleosome assembly protein). The Tbr-1-CASK-CINAP complex modulates expression of NMDA receptor subunit 2b (NR2b). More interestingly, CINAP protein levels in neurons are controlled by synaptic activity. The studies with the Tbr-1-CASK-CINAP complex provide a novel feedback mechanism that explains how synaptic activity regulates gene expression. In this chapter, the molecular characteristics of CASK, Tbr-1, and CINAP proteins will be described first, followed by a discussion of the regulation of NR2b expression by the CINAP-CASK-Tbr-1 protein complex.

3.1 Domain Structure and Protein-Protein Interactions of CASK

Members of the MAGUK family consist of several modular protein-protein interacting domains (Fig. 1). Based on sequence similarity and organization of the protein domains, the members are further classified into the PSD-95 and the p55 subfamilies. CASK, a member of the p55 subfamily (Fig. 1), is an evolutionally conserved protein. Rat CASK protein shares 99% amino acid sequence similarity with the mouse homolog, 98% with the human homolog, 97% with the

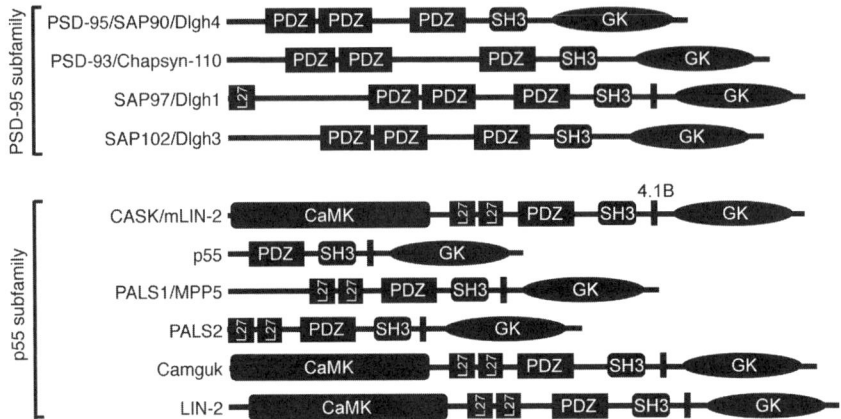

Fig. 1 Schematic of domain organization of MAGUK proteins

The members of the PSD-95 and p55 subfamilies are shown. PDZ, PSD-95, Dlg, ZO1; SH3, Src Homolog 3; GK, Guanylate Kinase; L27, LIN-2 and LIN-7 interacting domain; 4.1B, protein 4.1 binding site; CaMK, calcium/calmodulin-dependent kinase-like domain

Xenopus homolog, and 94% with the zebrafish homolog. In *C. elegans*, the corresponding CASK homolog is called LIN-2, which shares 63% similarity with mammalian CASK. In Drosophila, the CASK homolog is named camguk, which shares 74% similarity with mammalian CASK. Vertebrate CASK and invertebrate homologs share identical protein domain organization, which contains, from the N-terminal end, the CaMK-like domain, two L27 domains, the PDZ domain, the SH3 domain, protein 4.1 binding site and the guanylate kinase domain. The PDZ, SH3 and guanylate kinase domains are the characteristic domains shared in all MAGUK proteins (Fig. 1). The characteristics of these protein domains are described below.

3.1.1 The PDZ Domain

The PDZ domain was so named by the initial letters of the first three PDZ proteins to be identified which are postsynaptic density protein-95 (PSD-95) (Cho, Hunt and Kennedy 1992), Drosophila disc large tumor suppressor (Dlg) (Woods and Bryant 1991), and zonula occludens-1 (ZO-1) (Willott, Balda, Fanning, Jameson, Van Itallie and Anderson 1993). The PDZ domains, consisting of approximately 90 amino acid residues, are specialized for binding to short peptide motifs at the extreme C-terminal tail of their binding partners (reviewed by Sheng and Sala 2001), although other modes of interaction also occur (reviewed by Hung and Sheng 2002). Three types of PDZ domains have been classified based on their preference of binding sequence. The class I PDZ domains recognize the consensus X-S/T-X-V or X-S/T-X-L where X can represent any amino acid residue. The PDZ domains of the PSD-95 subfamily all belong to this class (Kim, Niethammer, Rothschild, Jan and Sheng 1995; Kornau, Schenker, Kennedy and Seeburg 1995). The consensus sequence of class II PDZ domains is X-F-X-V/A/I; F represents a hydrophobic amino acid, such as Tyr, Phe, Val, or Ile. The PDZ domains of the p55 subfamily belong to class II (Marfatia, Morais-Cabral, Kim, Byron and Chishti 1997; Songyang, Fanning, Fu, Xu, Marfatia, Chishti, Crompton, Chan, Anderson and Cantley 1997; Cohen, Woods, Marfatia, Walther, Chishti, Anderson and Wood 1998; Hsueh, Yang, Kharazia, Naisbitt, Cohen, Weinberg and Sheng 1998). The class III PDZ domain interacts with the sequence X-D-X-V. The PDZ domain of nNOS is a class III PDZ protein (Riefler and Firestein 2001). For CASK, several cellular proteins have been identified as binding partners via the PDZ domain, including neurexin (C-terminus: -E-Y-Y-V (Hata, Butz and Sudhof 1996)), syndecans (C-terminus: -E-F-Y-A (Hsueh et al., 1998), parkin (C-terminus: -W-F-D-V (Fallon,Moreau, Croft, Labib, Gu and Fon 2002), and SynCAM (C-terminus: -E-Y-F-I (Biederer, Sara, Mozhayeva, Atasoy, Liu, Kavalali and Sudhof 2002)). In fact, CASK was originally identified as a neurexin binding protein from yeast two-hybrid screening assay using the C-terminal cytoplasmic domain of neurexin β as bait (Hata et al., 1996). Neurexin β is a presynaptic cell adhesion molecule, which interacts with neuroligin at the postsynaptic site (Missler, Fernandez-Chacon and Sudhof 1998). The interaction between neurexin and neuroligin play a role in synaptic interaction (Irie, Hata, Takeuchi, Ichtchenko, Toyoda, Hirao, Takai, Rosahl and Sudhof 1997; Song, Ichtchenko, Sudhof and Brose 1999). Syndecans are the major transmembrane heparan sulfate proteoglycans. In mammals, there are four members in the syndecan family, syndecan-1, syndecan-2, syndecan-3 and syndecan-4. Syndecan-2 and syndecan-3 are the major syndecans in neurons (Hsueh and Sheng 1999), though the subcellular distributions of syndecan-2 and -3 are different: syndecan-2 is concentrated at synapse while syndecan-3 is distributed along the axon (Hsueh and Sheng 1999). All syndecans share the identical C-terminal tail and interact with CASK PDZ domain (Cohen et al., 1998; Hsueh et al., 1998). Since overexpression of syndecan-2 in young neurons accelerates dendritic spinogenesis and the syndecan-2 mutant lacking the C-terminal PDZ binding site loses this ability (Ethell and Yamaguchi 1999), the interaction between syndecan-2 and CASK has been suggested to play a role in synaptogenesis. SynCAM (Synaptic Cell Adhesion Molecule) was originally identified by a database screening with the criteria that the protein contains an extracellular immunoglobulin (Ig) domain and an intracellular PDZ binding motif (Biederer et al., 2002). SynCAM acts as a homophilic cell adhesion molecule at both pre- and post-synaptic sites. SynCAM is also involved in synaptic interaction and function (Biederer et al., 2002). The C-terminal CASK interacting site of SynCAM is also important for the function of SynCAM (Biederer et al., 2002). In conclusion, the interactions of CASK with neurexin, syndecan-2, and SynCAM are all involved in synaptic interaction and formation (Ethell and Yamaguchi 1999; Biederer et al., 2002; Levinson, Chery, Huang, Wong, Gerrow, Kang, Prange, Wang and El-Husseini 2005; Nam and Chen 2005). Thus far, the function of the interaction between CASK and parkin, a ubiquitin E3 ligase involved in Parkison's disease, is unclear.

3.1.2 The Src Homolog 3 (SH3) Domain

The SH3 domain originally identified from Src protein typically interacts with a proline-rich sequence. The CASK SH3 domain interacts with a strong class I SH3 domain-binding consensus sequence, R-Q-L-P-Q-T-P-L-T-P-R-P, present in the α_{1B1} subunit of the N-type Ca^{2+} channel (Maximov, Sudhof and Bezprozvanny 1999; Maximov and Bezprozvanny 2002). The CASK protein colocalizes with N-type Ca^{2+} channels at synapses. Moreover, mutation on the CASK binding site impairs the synaptic distribution of the N-type Ca^{2+} channel. These findings support the idea that interactions between the N-type Ca^{2+} channel and CASK are critical for synaptic targeting of the N-type channel (Maximov and Bezprozvanny 2002). In addition to interaction with the proline-rich motif, the CASK SH3 domain mediates interaction with its GK domain (Nix, Chishti, Anderson and Walther 2000). This interaction can occur intra- or inter-molecularly. CASK protein can therefore form a homodimer via the inter-molecular SH3–GK interaction. Via this SH3–GK intermolecular interaction, CASK also forms heterodimers with other MAGUK proteins, such as p55 (Nix et al., 2000). It has been hypothesized that MAGUK proteins may form their large scaffold complexes via their SH3–GK interactions (Nix et al., 2000).

3.1.3 The Guanylate Kinase (GK) Domain

CASK GK domain was named due to the similarity with guanylate kinase, which is able to catalyze the reversible phosphoryl transfer from ATP to GMP in the presence of Mg^{2+} (Agarwal, Miech and Parks 1978). Although the rat CASK GK domain shares 51% amino acid sequence similarity with rat guanylate kinase 1, thus far there is no evidence indicating that the CASK GK domain catalyzes the phosphoryl transfer activity. Instead, the GK domain of MAGUK proteins acts as a protein-protein interacting domain. For instance, the GK domain of CASK binds the SH3 domain of CASK and p55 as described above. In addition, CASK GK domain also interacts with the nuclear proteins Tbr-1 and CINAP. Details of the study of the interaction of CASK with Tbr-1 and CINAP will be described below.

3.1.4 The CaMK and L27 Domains

Although the CASK CaMK domain shares ~66% amino acid sequence similarity with CaMKII, it is not expected to have kinase activity because the important residues at the active site are mutated (Cohen et al., 1998). The CASK CaMK domain also functions as a protein-protein interacting domain. The first identified interacting protein for CASK CaMK domain was Mint1/X11. Actually, Mint1/X11 forms an evolutionarily conserved tripartite protein complex with CASK and Veli/MALS/mLIN-7, another CASK interacting protein, in mammalian brain and

C. elegans epithelial cells (Borg, Straight, Kaech, de Taddeo-Borg, Kroon, Karnak, Turner, Kim and Margolis 1998; Butz, Okamoto and Sudhof 1998; Kaech, Whitfield and Kim 1998). The L27 domain is a novel protein-protein interacting domain with approximately 50–60 amino acids originally identified from LIN-2/CASK and LIN-7/Veli/MALS (reviewed by Doerks, Bork, Kamberov, Makarova, Muecke and Margolis 2000). The second L27 domain of CASK interacts with the single L27 domain of Veli/MALS/mLIN-7 (Borg et al., 1998; Butz et al., 1998; Kaech et al., 1998; Lee, Fan, Makarova, Straight and Margolis 2002). In mammalian brain, the Mint1-CASK-Veli protein complex is involved in synaptic interaction (Butz et al., 1998) and in the trafficking and synaptic targeting of the NMDA receptor (Jo, Derin, Li and Bredt 1999; Setou, Nakagawa, Seog and Hirokawa 2000) and of the N-type calcium channel (Maximov et al., 1999; Maximov and Bezprozvanny 2002).

Table 1 Summary of CASK Interacting Proteins

CASK domain	Interacting protein	Function of the interaction	Reference
CaMK domain	Mint1/X11	synaptic protein trafficking and targeting	(1)
	Caskin1	competes CASK binding with Mint1	(2)
	CIP98	unknown	(3)
	Carom	unknown	(4)
	Liprin*	synaptic vesicle cycling	(5)
L27A domain	SAP97	links to Inward rectifier potassium channel	(6)
	Liprin*	synaptic vesicle cycling	(7)
L27B domain	Veli/mLIN-7/Mals	synaptic protein trafficking and targeting	(8)
PDZ domain	Neurexin	synaptic interaction	(9)
	syndecan-2	synaptic interaction and formation	(10)
	parkin	unknown	(11)
	SynCAM	synaptic interaction	(12)
SH3 domain	N-type Ca channel	synaptic targeting	(13)
	GK domain of MAGUKs	dimerization	(14)
4.1 binding site	Protein 4.1	Links to cytoskeleton	(15)
GK domain	Tbr-1	transcriptional regulation; brain development	(16)
	CINAP	transcriptional regulation; synaptic function	(17)

* Both CaMK and L27A domains of CASK are required for liprin interaction. (1) Maximov et al., 1999; Setou et al., 2000; Maximov and Bezprozvanny 2002; (2) Tabuchi, Biederer, Butz and Sudhof 2002; (3) Yap, Liang, Yamazaki, Muto, Kishida, Hayashida, Hashikawa and Yano 2003; (4) Ohno, Hirabayashi, Kansaku, Yao, Tajima, Nishimura, Ohnishi, Mashima, Fujita, Omata and Hata 2003; (5) Olsen, Moore, Fukata, Kazuta, Trinidad, Kauer, Streuli, Misawa, Burlingame, Nicoll and Bredt 2005; (6) Leonoudakis, Conti, Radeke, McGuire and Vandenberg 2004; Leonoudakis, Conti, Radeke, McGuire and Vandenberg 2004; (7) Olsen et al., 2005; (8) Setou et al., 2000; (9) Hata et al., 1996; (10) Hsueh et al., 1998; Ethell and Yamaguchi 1999; (11) Fallon et al., 2002; (12) Biederer et al., 2002 ; (13) Maximov et al., 1999; Maximov and Bezprozvanny 2002 ; (14) Nix et al., 2000; (15) Cohen et al., 1998; (16) Hsueh et al., 2000; Wang et al., 2004a; Wang et al., 2004b; (17) Wang et al., 2004a.

3.1.5 The Protein 4.1 Binding Motif

The protein 4.1 binding motif or so-called "hook motif" is a short lysine-rich stretch located between SH3 and GK domain. The in vitro pull down assay and overlay assay demonstrated that this short motif binds protein 4.1 (Cohen et al., 1998). Protein 4.1 interacts with actin molecules and promotes formation of actin/spectrin microfilaments (reviewed by Hoover and Bryant 2000; Bretscher, Edwards and Fehon 2002). Therefore, the interaction with protein 4.1 links CASK to actin cytoskeleton. Indeed, a biochemical reconstitution assay showed that CASK binds a brain-enriched isoform of protein 4.1 and nucleates local assembly of actin-spectrin filaments (Biederer and Sudhof 2001). More interestingly, this complex also recruits a CASK binding membrane protein, neurexin (Biederer and Sudhof 2001). The evidence supports the idea that CASK acts as an adaptor protein that links trans-membrane proteins to the actin cytoskeleton.

In conclusion, the major interactions of CASK include synaptic vesicle traf-ficking, synaptic targeting of ion channels, and synaptic interaction and formation (Table 1). The exception is the interaction with Tbr-1 and CINAP via the GK domain (Hsueh, Wang, Yang and Sheng 2000; Wang, Hong, Yen, Huang, Ou, Huang, Jung, Kuo, Sheng, Wang and Hsueh 2004a). Both Tbr-1 and CINAP are nuclear proteins. Tbr-1 is a sequence specific transcription factor (Bulfone, Smiga, Shimamura, Peter-son, Puelles and Rubenstein 1995; Hsueh et al., 2000), while CINAP functions as a nucleosome assembly protein (Wang et al., 2004a). The interactions with Tbr-1 and CINAP indicate a role of CASK in transcriptional regulation. (Hsueh et al., 2000; Wang et al., 2004a; Wang, Ding, Wang, Luo, Lin, Ruan, Hevner, Rubenstein and Hsueh 2004b).

3.2 Subcellular Distribution of CASK in Brain

CASK proteins are widely distributed at different subcellular regions of neurons. In adult rat brain, CASK proteins are mainly somatodendritic (Hsueh et al., 1998). Confocal analysis showed a punctate staining pattern of CASK (Hsueh et al., 1998), suggesting a synaptic distribution of CASK proteins. Immunogold EM further demonstrated that CASK is present at both pre- and post-synaptic sites (Hsueh et al., 1998). Biochemical fractionation further confirmed that CASK proteins are enriched at PSD (Hsueh et al., 1998). Synaptic distribution of CASK in mature neurons is consistent with the function of CASK in synaptic interaction and forma-tion, as described above. In addition to synaptic localization, a significant amount of CASK is also present in the cytoplasm (Hsueh et al., 1998), consistent with the idea that CASK is involved in protein trafficking.

At the embryonic stage, although the majority ($\sim$80%) of CASK proteins are still present in the cytoplasm, a portion ($\sim$20%) of CASK is detected in the nuclei of embryonic neurons (Hsueh et al., 2000), suggesting a nuclear function for CASK in embryonic neurons. Indeed, nuclear CASK in embryonic neurons interacts with

T-box transcription factor Tbr-1 and enhances the transcriptional activity of Tbr-1 (Hsueh et al., 2000), which is an important transcription factor in cerebral cortex development. The interaction of CASK with Tbr-1 and its enhancement of Tbr-1 transcriptional activity suggest a role for CASK in development during the embryonic period.

Presently, it is not clear whether there are any signals regulating subcellular distribution of CASK. When CASK alone was overexpressed in COS cells, the majority of CASK was present in the cytoplasm (Hsueh et al., 2000), similar to the observation in adult brain (Hsueh et al., 1998). When CASK was co-expressed with Tbr-1 in COS cells, CASK entered the nuclei with Tbr-1 (Hsueh et al., 2000). Interestingly, in the presence of both Tbr-1 and syndecan-3, CASK was retained in the cytoplasm and colocalized with syndecan-3. Even a trace amount of Tbr-1 is colocalized with CASK and syndecan-3 in the cytoplasm (Hsueh et al., 2000). These results suggest that subcellular distribution of CASK is regulated by the presence of its binding partners. For instance, Tbr-1 is expressed highly at the embryonic stage (Bulfone et al., 1995; Hsueh et al., 2000). In contrast, from the embryonic stage to the adult stage, the protein levels of CASK keep almost constant (Hsueh, Roberts, Volta, Sheng and Roberts 2001). High amounts of Tbr-1 at the embryonic stage bring a portion of CASK into the nuclei of neurons. When the expression levels of Tbr-1 decline during development, the nuclear signal of CASK is therefore reduced (Hsueh et al., 2000). Tbr-1 expression may be involved in nuclear distribution of CASK in neurons at the embryonic stage. On the other hand, the expression levels of the plasma membrane interacting proteins of CASK, such as syndecan-2 and syndecan-3, are increased after birth. Both syndecan-3 and syndecan-2 interact with the CASK PDZ domain (Hsueh et al., 1998; Hsueh and Sheng 1999). However, syndecan-3 is expressed highly at postnatal weeks 1–3 and is concentrated along the axon; syndecan-2 is a synaptic syndecan, which starts to express at the synapse after birth and gradually increases expression levels until adulthood (Hsueh and Sheng 1999). Therefore, CASK first interacts with syndecan-3 along the axon before 3 weeks after birth. Approaching neuronal maturation, syndecan-3 levels go down while syndecan-2 levels go up. CASK can then shift location to interact with syndecan-2 at the synapse in mature neurons. This speculation is consistent with the observation that CASK is present along the axon from postnatal days 1–14 in rat brain, and then its distribution shifts to the somatodendritic fraction in the adult brain. If subcellular distribution of CASK depends on the interaction with its binding partners, it will be interesting to explore in the future whether in addition to the expression levels of the binding partners, any other mechanism, such as protein phosphorylation, regulates the protein-protein interaction of CASK.

3.3 Tbr-1 Is Important for Cerebral Cortex Development

Tbr-1 belongs to the T-box transcription factor family which contains a conserved DNA binding domain T-box, namely the T-domain (Bulfone et al., 1995) (Fig. 2). Similarities in the T-box region have lead to the general belief that T-box

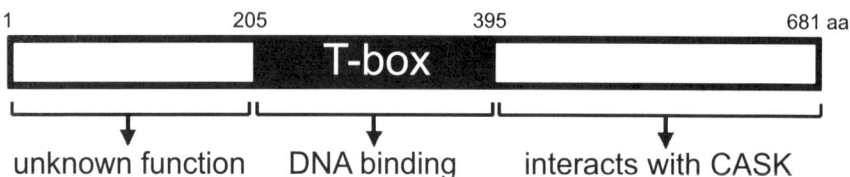

Fig. 2 Schematic of domain organization and function of murine Tbr-1

Tbr-1 is composed of the N-terminal region, central DNA binding T-box domain, and C-terminal region. The C-terminal region is the interaction site for CASK GK domain. The function of the N-terminal region is unknown

transcription factors recognize the same DNA target sequence. To date, approximately 20 T-box transcription factors have been identified in vertebrates (reviewed by Naiche, Harrelson, Kelly and Papaioannou 2005). The T-box transcription factors play important roles in development, including early embryonic cell fate determination, regulation of embryonic patterning, and many aspects of organogenesis. According to sequence similarity, the T-box transcription factors are further classified into five subfamilies, including T, Tbx1, Tbx2, Tbx6, and Tbr-1 subfamilies. There are three members in the Tbr-1 subfamily, which are Tbr-1, Tbr-2/Eomes, and Tbet/Tbx21. Tbr-1 and Tbr-2/Eomes are highly similar to each other in the T-box region. In the N- and C-terminal regions, these two proteins share approximately 28% and 51% similarity in the amino acid sequence, respectively. Both Tbr-1 and Tbr-2 are expressed in brain with differential expression patterns. At the embryonic stage, Tbr-1 is specifically expressed in postmitotic neurons in neocortex, olfactory bulb, and the nuclear transitory zone of hindbrain (Hevner, Shi, Justice, Hsueh, Sheng, Smiga, Bulfone, Goffinet, Campagnoni and Rubenstein 2001; Fink, Englund, Daza, Pham, Lau, Nivison, Kowalczyk and Hevner 2006). In addition, Tbr-1 is also detectable in thalamus of P1 and P7 mouse brain and Purkinje cells of the cerebellum in adult mouse brain (Bulfone et al., 1995). In contrast, Tbr-2/Eomes is expressed in neuronal progenitor cells in ventricular and subventricular zones of the telencephalon (Bulfone, Wang, Hevner, Anderson, Cutforth, Chen, Meneses, Pedersen, Axel and Rubenstein 1998; Englund, Fink, Lau, Pham, Daza, Bulfone, Kowalczyk and Hevner 2005). The biological significance of why Tbr-1 and Tbr-2/Eomes are expressed at different developmental stages of neuron is not clear. Tbet/Tbx21, the third member of Tbr-1 subfamily, is specifically expressed in T lymphocytes. This factor is critical for cell fate decision of Th1 cells (Szabo, Kim, Costa, Zhang, Fathman and Glimcher 2000; Szabo, Sullivan, Stemmann, Satoskar, Sleckman and Glimcher 2002). Tbet/Tbx21 is less similar to Tbr-1 and Tbr-2/Eomes, and only shares significant similarity with Tbr-1 and Tbr-2/Eomes in the T-box domain.

Tbr-1 was originally identified from a subtractive hybridization between cDNA libraries made from RNA expressed in the embryonic day14.5 telencephalon and the adult telencephalon (Porteus, Brice, Bulfone, Usdin, Ciaranello and Rubenstein 1992). Expression levels of Tbr-1 in neocortex are highest at the embryonic stage (Bulfone et al., 1995). After birth, its expression levels are gradually reduced, although Tbr-1 still remains at a significant level in the neocortex of adult brain (Hsueh et al., 2000). This temporal expression pattern suggests that Tbr-1 plays a role in neocortex development. Indeed, knocking out Tbr-1 in mice results in defective development of cerebral cortex and olfactory bulb (Bulfone et al., 1998; Hevner et al., 2001). Since mice with homozygous deficiency in the Tbr-1 gene lack most projecting neurons, mitral and tufted cells in the olfactory bulb (Bulfone et al., 1998), Tbr-1 is therefore suggested to control differentiation of projecting neurons in olfactory bulb.

In cerebral cortex, homozygous deficiency in the Tbr-1 gene leads to abnormal lamination (Hevner et al., 2001). Neurons in cerebral cortex are divided into six cell layers, with layers I to VI comprising the most superficial to the deepest positions, respectively. During development, most neocortical projection neurons arise from the ventricular zone. The newborn neurons migrate outwardly along the radial glia and deposit into the preplate, thereby splitting the preplate into the subplate and the marginal zone, forming the cortical plate. As cortical plate neurons continue to be generated, newborn neurons migrate past the older neurons and stop just beneath the marginal zone. This results in the five layers (layer II to VI) of the cortical plate having an inside-out pattern of birth dates, with the oldest neurons located in the deepest position, layer VI. In Tbr-1 -/- mice, this developmental process is disrupted. The newborn neurons cannot pass the older neurons and the subplate does not split efficiently from the marginal zone, resulting in the lack of normal layering in neocortex (Hevner et al., 2001). These results indicate that Tbr-1 is involved in neural migration.

In addition to migration phenotype, axonal projections emerging from the cortex are also defective. The corpus callosum is absent; the reciprocal connections between neocortex and the thalamus have also disappeared (Hevner et al., 2001). Since Tbr-1 expression is most prominent in glutamatergic neurons (Hevner et al., 2001), which are predominantly projection neurons, the effect of Tbr-1 deficiency on axonal projection from cerebral cortex is likely to be autonomous. The defective axonal projections from neocortex leave the neocortex isolated from other parts of brain, and the knockout mice die on the second postnatal day (Bulfone et al., 1998).

Because Tbr-1 deficient mice carry neural migration and axonal projection phenotypes and since Tbr-1 acts as a transcription factor, one can speculate that Tbr-1 regulates expression of genes involved migration and axon pathfinding. A couple of Tbr-1 down stream target genes has been identified. One of them is the reelin gene, which encodes a huge extracellular matrix protein important for neural migration. At the embryonic stage, reelin is expressed in Cajal-Retzius neurons. Tbr-1 is also expressed highly in Cajal-Retzius neurons. In Tbr-1 -/- mice, reelin expression, as examined by immunohistochemistry, is significantly lower than that in wild type or in heterozygous littermates (Hevner et al., 2001). In luciferase reporter assay, overexpression of Tbr-1 in cultured hippocampal neurons augments the activity of reelin enhancer (Hsueh et al., 2000), which contains a Tbr-1 binding site. These studies indicate that Tbr-1 regulates gene expression of reelin. As reeler mice, who carry a spontaneous mutation in the reelin gene, are also characterized by abnormal lamination of cerebral cortex (Ogawa, Miyata, Nakajima, Yagyu,

Seike, Ikenaka, Yamamoto and Mikoshiba 1995; Rice and Curran 2001; Tissir and Goffinet 2003), regulation of reelin expression accounts for, at least partially, the lamination phenotype of Tbr-1 knockout mice.

In addition to reelin, another Tbr-1 target gene is NMDA receptor subunit 2b (NR2b) gene (Wang et al., 2004b). NR2b is identified as a Tbr-1 target gene from a computational search using a conserved T-box binding sequence in the promoter of tyrosinase-related protein-1 (TRP-1) gene. NR2b promoter contains two T-box binding sites. Tbr-1 not only binds the two T-box binding sites in NR2b promoter in electrophoresis mobility shift assay (EMSA) but also activates NR2b promoter activity in luciferase reporter analysis in cultured hippocampal neurons. In Tbr-1 -/- mice, NR2b expression, as measured by immunoblotting and immunohistochemistry analysis, is significantly lower compared with wild type littermates (Wang et al., 2004b). Moreover, the result of chromatin immunoprecipitation studies also supports the idea that Tbr-1 binds the promoter of NR2b gene in vivo (Wang et al., 2004a).

NMDAR is a heteromeric ion channel composed of two NR1 subunits and two NR2 subunits. There are four NR2 subunits, NR2a, NR2b, NR2c, and NR2d, and their expression is regionally and developmentally regulated (Akazawa, Shigemoto, Bessho, Nakanishi and Mizuno 1994; Monyer, Burnashev, Laurie, Sakmann and Seeburg 1994; Portera-Cailliau, Price and Martin 1996; Wenzel, Fritschy, Mohler and Benke 1997). NR2a and NR2b are two major NR2 subunits in the forebrain: NR2b expression starts in the embryonic period, peaks around postnatal day 20, and gradually decreases; NR2a is first detected near birth and gradually increases toward maturation of brain development. Therefore, the NR2b expression profile shares a certain degree of similarity with Tbr-1. This is consistent with the finding that Tbr-1 regulates NR2b expression. However, NR2b is widely expressed in different regions of forebrain. In contrast, Tbr-1 is restricted in specific regions, suggesting that Tbr-1 controls NR2b expression in certain regions of brain, such as neocortex and olfactory bulb. In the regions (or neurons) without Tbr-1, other transcription factors, such as Tbr-2 or unidentified T-box transcription factors, might bind to the T-box binding sites of NR2b promoter and regulate NR2b expression.

The unique capabilities of NMDAR to gate Ca^{2+} ions and to link to Ca^{2+}-dependent intracellular signaling processes such as LTP have indicated important roles for NMDAR in synaptic plasticity and learning. In addition, NMADR also plays important roles in neural development, such as axonal guidance and neural circuit formation (Herkert, Rottger and Becker 1998; Behar, Scott, Greene, Wen, Smith, Maric, Liu Colton and Barker 1999; Dickson and Kind 2003), and pathophysiological conditions, such as schizophrenia, Parkinson's disease, Hungtington's disease, Alzheimer's disease, drug abuse, and excitotoxicity (reviewed by Loftis and Janowsky 2003). Because NR2b is expressed embryonically and involved in axonal outgrowth and neural map formation, it may mediate the function of Tbr-1 in axonal projections between the cerebral cortex and the thalamus. On the other hand, although Tbr-1 is important for cerebral cortex development, it may also play a role in the adult brain because it is still expressed in a significant amount in adult brain. Conditional Tbr-1 knockout mice will be helpful in addressing the function of Tbr-1 in adult brain and in investigating whether Tbr-1 also regulates NR2b expression in adults.

3.4 CASK Enhances Transcriptional Activity of Tbr-1

From a yeast two-hybrid screening assay using CASK GK domain as bait, Tbr-1 was identified as a CASK interacting protein (Hsueh et al., 2000). The interaction was confirmed by co-immunoprecipitation from rat brain extract. By deletion analysis, the C-terminal region of Tbr-1 was further mapped as the binding site of CASK GK domain (Hsueh et al., 2000). The result of immunofluorescence staining indicated that Tbr-1 is highly concentrated in the nuclei of embryonic neurons; a portion (~20%) of CASK enters the nuclei of neurons and colocalizes with Tbr-1 (Hsueh et al., 2000). The interaction of Tbr-1 with CASK in the nuclei of embryonic neurons may enhance transcriptional activity of Tbr-1 since co-expression of CASK enhances Tbr-1 transcriptional activity in luciferase assays (Hsueh et al., 2000). Because Tbr-1 is involved in brain development, CASK may play a role in development via its interaction with Tbr-1. This speculation is supported by the smaller brain phenotype observed in the transgenic mice carrying an insertional mutation in the CASK gene (Laverty and Wilson 1998) (Fig. 3).

In contrast with the nuclear distribution at the embryonic stage, Tbr-1 is no longer concentrated in the nuclei of neurons in the adult brain. It is also present in the cytoplasm of mature neurons (Hsueh et al., 2000). In cytoplasm, Tbr-1 is still colocalized with CASK. This observation introduces the intriguing possibility that in addition to acting as a transcriptional co-activator, CASK also plays a role in regulation of

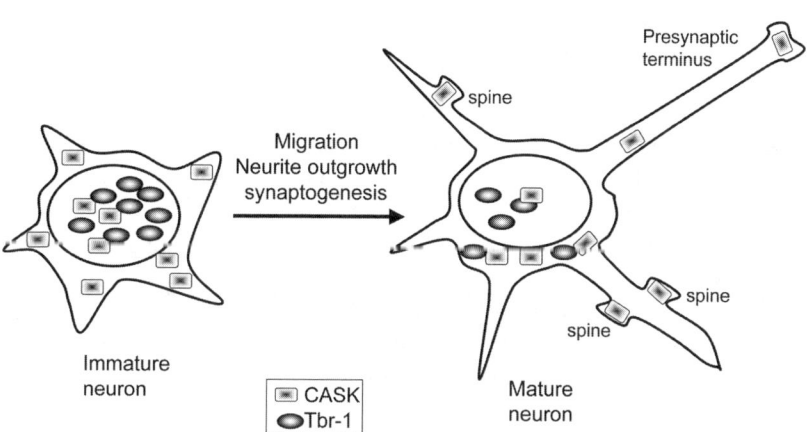

Fig. 3 The potential role of the Tbr-1-CASK complex in neural development

In immature neurons, a significant portion of CASK enters the nucleus of neuron, interacts with Tbr-1, and activates expression of genes involved in neural development. Several processes are involved in neuronal differentiation, including neural migration, neural outgrowth, and synaptogenesis. Based on the phenotype of Tbr-1 knockout mice, the Tbr-1-CASK complex may regulate expression of genes contributing to migration and axonal outgrowth. In mature neurons, the majority of CASK is present in the cytoplasm, dendrites, dendritic spines, and axonal termini. Some of Tbr-1 proteins are present in the cytoplasm and colocalized with CASK. Expression of genes involved in development controlled by the Tbr-1-CASK complex is therefore reduced in mature neurons nuclear translocation of Tbr-1. As described above, CASK synaptic interacting proteins continue to accumulate after birth. While CASK is retained in the cytoplasm through the interaction with the cytoplasmic or transmembrane interacting proteins, Tbr-1 may be also trapped in the cytoplasm via its interaction with CASK. If this is the case, it will be of interest to study the signal that regulates the interaction between CASK and Tbr-1 or nuclear translocation of CASK in neurons, which may contribute to control the activity of Tbr-1.

3.5 CINAP Regulates Expression of Tbr-1 Target Gene

Because the CASK protein itself does not possess any known transcriptional co-activation domain and functions mainly as an adaptor protein to link functionally related proteins together, it is very likely that CASK couples Tbr-1 and other transcriptional modulators in the nucleus, thus regulating transcriptional activity of Tbr-1. CINAP, the second CASK GK interacting protein identified from a yeast two-hybrid screening, may play a role in this aspect.

CINAP stands for CASK interacting nucleosome assembly protein because it contains a conserved nucleosome assembly protein (NAP) domain (Fig. 4). According to sequence comparisons, mouse CINAP shares a 77% identity with human CDA1/se20-4/DENTT (Ozbun, Martinez, Angdisen, Umphress, Kang, Wang, You and Jakowlew 2003); therefore, mouse CINAP has been classified as a homolog of human CDA1/se20-4/DENTT (Chai, Sarcevic, Mawson and Toh 2001; Eichmuller, Usener, Dummer, Stein, Thiel and Schadendorf 2001; Ozbun, You, Kiang, Angdisen, Martinez and Jakowlew 2001). More than a dozen NAPs have been identified. They act as histone chaperones, which deposit histones on the newly synthesized chromosomal DNA or remove histones from the chromosome during transcription (reviewed by Gruss and Sogo 1992; Adams and Kamakaka 1999). Therefore, NAPs play roles in both DNA replication and gene expression. The results of in vitro supercoiling assay indicates that CINAP indeed functions as a nucleosome

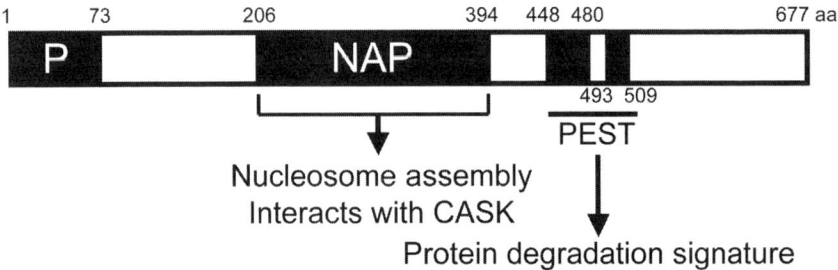

Fig. 4 Schematic of domain organization and function of murine CINAP

CINAP contains a proline-rich (P) region at the N-terminal end, a nucleosome assembly protein (NAP) domain in the central part, two PEST motifs in the C-terminal region. In addition to the nucleosome assembly activity, the NAP domain of CINAP is also the binding site of CASK. The PEST motif is generally recognized as a protein degradation signature

assembly protein (Wang et al., 2004a). It is likely that CINAP modulates chromatin structure in vivo, thereby regulating transcription. Indeed, in the presence of CASK, CINAP co-immunoprecipitated with Tbr-1 in heterologous cells. The results of immunofluorescence staining indicate that CINAP, CASK, and Tbr-1 colocalize in the nuclei of neurons in layer 6 of the cerebral cortex in postnatal day 1 rat brain (Wang et al., 2004a). More importantly, according to the results of chromatin immunoprecipitation, all Tbr-1, CASK, and CINAP associate with the Tbr-1 binding site of the NR2b promoter, indicating that CASK, CINAP, and Tbr-1 form a complex on the NR2b promoter (Wang et al., 2004a). The luciferase reporter assay further confirms that CINAP is required for NR2b expression. Decreasing endogenous CINAP reduced the activity of the NR2b promoter in cultured hippocampal neurons (Wang et al., 2004a). Taken together, these results support the idea that via the interaction with CASK, CINAP associates with Tbr-1, modifies the chromatin structure flanking the Tbr-1 binding sites, and thus regulates expression of Tbr-1 downstream target genes.

3.6 CINAP Proteins are Controlled by Synaptic Stimulation

CINAP is an unstable protein. When the cultured hippocampal neurons are treated with NMDA, CINAP protein levels are reduced to $\sim$30% relative to control neurons (Wang et al., 2004a), but this reduction can be prevented by pretreatment of neurons with proteasome inhibitor. Moreover, CINAP RNA levels are not significantly different in control and NMDA treated neurons (Wang et al., 2004a). These results indicate that CINAP protein levels in cultured hippocampal neurons are regulated by a proteasomal degradation pathway downstream from NMDAR. In the CINAP-CASK-Tbr-1 transcriptional complex, CINAP is the only component controlled by proteasomal degradation. The protein levels of both CASK and Tbr-1 do not change significantly upon NMDA stimulation (Wang et al., 2004a). By regulating the CINAP protein levels, NMDAR signaling controls transcriptional activity of the CINAP-CASK-Tbr-1 protein complex.

In addition to cerebral cortex and hippocampus, CINAP is also widely distributed in different regions of rodent brain, including striatum, Purkinje cell layer of cerebellum, and hypothalamus (Lin, Huang, Wang, Kuo, Yen and Hsueh 2006). We noticed prominent immunoreactivity of CINAP in the nuclei of hypothalamus, including the paraventricular nucleus (PVN), arcuate nucleus, suprachiasmatic nucleus, and supraoptic nucleus. The hypothamalus controls many physiological responses, such as osmoregulation, growth, metabolism, stress response, and circadian rhythm. To further explore the role of CINAP in neuronal activity in vivo, systematic saltwater administration was used to modulate neuronal activity in the PVN. Total immunoreactivity of CINAP in the PVN is around 2–3-fold higher in mice administered saltwater compared with control mice (Lin et al., 2006). In addition, the subcellular distribution of CINAP is also altered upon saltwater treatment. In control mice, CINAP immunoreactivity is present in both the nuclei and cytoplasm of neurons in PVN. In some neurons, CINAP is even absent from the nuclei. In contrast, nuclear distribution of CINAP is very prominent in PVN of mice treated with saltwater (Lin et al., 2006). The above study in PVN suggests that neuronal activity regulates protein stability of CINAP as well as subcellular distribution of CINAP. In addition to immunostaining, the result of biochemical fractionation also supports that a fraction of CINAP proteins is present in the light membrane, crude synaptosomal fractions as well as in the PSD fraction (Lin et al., 2006). In the cytoplasm, CINAP also colocalizes and cofractionates with CASK (Lin et al., 2006), suggesting that CINAP still interacts with CASK in the cytoplasm. If the nucleus of the neuron is the ultimate destination for CINAP to perform its biochemical function (e.g. nucleosome assembly), the CINAP proteins present in the cytoplasm need to be transported into the nuclei of neurons to perform their function. Although the signals controlling the protein levels and nuclear translocation of CINAP are still unclear, the studies have suggested that CINAP may be an important player in the control of neuronal activity. Further investigation to determine if CINAP is indeed required for synaptic response, if modulation of the protein levels or of nuclear translocation of CINAP modifies neuronal activity, and if the interaction with CASK is involved in the function of CINAP in neuronal activity will be intriguing.

3.7 Implication of CINAP-CASK-Tbr-1 Complex in Neurons

NR2b expression is regulated by neuronal activity. Blocking action potential-dependent synaptic activity with TTX enhances NR2b protein levels. On the other hand, increasing excitatory synaptic activity by blocking inhibitory GABAergic transmission with bicucullin can reduce NR2b protein expression (Ehlers, 2003). NR2b expression levels are also directly controlled by NMDAR signaling. Treatment with chronic NMDAR antagonists such as AP5 and CPP upregulates the production of NR2b polypeptide in cultured cortical and hippocampal neurons (Follesa and Ticku 1996; Rao and Craig 1997). Proteasomal degradation has been suggested to regulate NR2b expression (Ehlers 2003). However, since NR2b polypeptide itself is not ubiquitinated (Ehlers 2003), the blocking effect of proteasomal inhibitors on NR2b expression regulated by NMDA suggests that the NR2b polypeptide is indirectly controlled by the proteasome. Because NMDA stimulation down regulates promoter activity of NR2b gene (Wang et al., 2004a), it indicates that this kind of proteasome-dependent effect acts on transcriptional regulation. The regulation of NR2b expression by the CINAP-CASK-Tbr-1 protein complex provides an explanation for how neuronal activity controls NR2b gene expression in the nucleus. NMDA treatment triggers proteasome activation, down regulates CINAP protein levels, and thus controls transcription of the NR2b gene. Since the presence of the NR2b subunit in NMDAR results in longer duration and thus allows more calcium ion to enter the cytoplasm, down regulation of NR2b expression by synaptic stimulation via the CINAP-CASK-Tbr-1 complex can then change the composition and properties of NMDAR and thereby modulate synaptic response to glutamate (Fig. 5).

Because NMDAR is involved in excitotoxicity, learning/memory, neurological disorders, and neuronal development, the CINAP-CASK-Tbr-1 protein complex

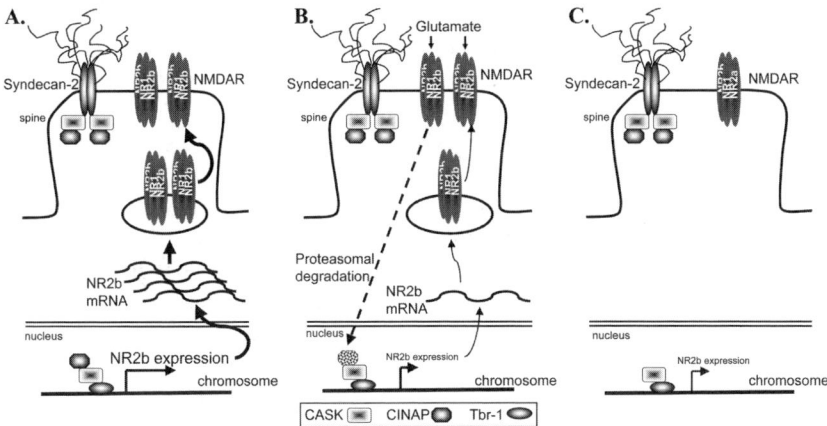

Fig. 5 Regulation of NR2b expression by the CINAP-CASK-Tbr-1 protein complex

(A) Through the interaction with CASK, Tbr-1 forms a complex with CINAP in the nucleus of neuron. The CINAP-CASK-Tbr-1 complex regulates expression of NR2b gene. In addition to the nucleus, a portion of CINAP is present in the cytoplasm and colocalizes with CASK at the synapse. Although a fraction of Tbr-1 is also present in the cytoplasm of the mature neuron, it is not clear if cytoplasmic Tbr-1 is colocalized with CIANP and CASK at the synapse. It is also unknown whether the synaptic CASK-CINAP (-Tbr-1?) complex can be translocated into the nucleus upon an unknown signal or whether this synaptic protein complex performs a function distinct from the regulation of gene expression. (B) CINAP proteins are unstable. In cultured hippocampal neurons, NMDAR activation by glutamate treatment results in protein degradation of CINAP via a proteasomal degradation pathway. Expression of NR2b gene is therefore reduced. (C) Consequently, reduction of NR2b mRNA expression by glutamate treatment will further reduce surface expression of NR2b subunit, leading to either a decrease of total number of surface NMDAR or to a shift of NMDAR subunit composition from NR1-NR2b to NR1-NR2a. The CINAP-CASK-Tbr-1 complex mediates a negative feedback mechanism by which NMDAR activation down regulates NR2b expression

may contribute to these processes through regulation of NR2b expression. For instance, high dose of glutamate induced cell death, which is a well documented effect of glutamate on neurons (Mattson, Dou and Kater 1988). Reduction of NR2b expression by down regulation of the CINAP-CASK-Tbr-1 complex may be a protection response that neurons have evolved to survive a challenge of high dose glutamate.

During neuronal development, extracellular glutamate can act as a chemoattractant for neurite outgrowth. It is evident that NMDAR is involved in filopodia protrusion from the growth cone and turning of the neural growth cone (Zheng, Wan and Poo 1996). In this process, extracellular calcium is required. NR2b is suggested to play a role in this process since NR2b is indeed present at the axonal growth cone (Herkert et al., 1998). When the extracellular glutamate is asymmetrically distributed across the growth cone, NMDARs are activated at only one side. This asymmetric activation of NMDAR leads to turning of the growth cone. Meanwhile, this signal also reduces CINAP protein levels and down regulates expression of the NR2b gene. The number of surface NR2b molecules at the growth cone is therefore reduced leading to shutdown or attenuation of the response to extracellular glutamate. The turned growth cone can then extend forward in a new direction. Therefore, it seems reasonable to hypothesize that the CINAP-CASK-Tbr-1 complex is involved in growth cone turning via controlling NR2b expression.

With respect to learning and memory, it has been proposed that NR2b is involved in long-term depression; NR2a contributes to long-term potentiation (Liu, Wong, Pozza Lingenhoehl, Wang, Sheng, Auberson and Wang 2004). Changing the subunit composition of NMDAR may result in a completely different neuronal response. Synaptic stimulation inhibiting NR2b expression via down regulation of CINAP protein levels can lead to a shift of NMDAR composition from NR1-NR2b to NR1-NR2a, and may therefore modulate neuronal response toward long-term potentiation. If this speculation is true, one can image that CINAP deficient mutant mice may have abnormal learning and memory behavior phenotypes. More experiments need to be performed to address this issue.

3.8 Conclusion

The formation of a CINAP–CASK–Tbr-1 protein complex in neurons provides a negative feedback mechanism with which NMDAR activation downregulates NR2b expression. This finding helps explain why synaptic activation inhibits NR2b expression and synaptic blockade increases NR2b expression in neurons. This complex may also contribute to neurite outgrowth and play a role in neuronal development before the synapse forms. However, several questions about the detailed mechanism still remain. First, how does NMDA or synaptic stimulation trigger proteasome activation? Since CINAP is present in both the nucleus and cytoplasm, it is not clear which population of CINAP is the target of the proteasome. Secondly, usually specific signals, such as protein phosphorylation, commit proteins to proteasomal degradation. Thus far, it is unknown what signal acts on CINAP and results in its protein degradation. Thirdly, because subcellular distribution of CINAP is also a mechanism controlling CINAP function, it will be interesting to explore what signal regulates nuclear translocation of CINAP and whether CINAP, Tbr-1, and CASK proteins transport together to the nucleus. Finally, although CINAP facilitates nucleosome assembly in vivo, it is not clear if CINAP regulates chromatin structure in vivo and whether nucleosome assembly is the only biochemical activity of CINAP required to regulate gene expression. More mechanisms of regulation for the CINAP-CASK-Tbr-1 protein complex in neurons remain to be discovered.

Acknowledgments This work was supported by grants from Academia Sinica, the National Science Council (NSC 95-2321-B-001-013), and the National Health Research Institute (NHRI-EX95-9403NI).

References

Adams, C.R. and Kamakaka, R.T. (1999) Chromatin assembly: biochemical identities and genetic redundancy. Curr. Opin. Genet. Dev. 9, 185–190.

Agarwal, K.C., Miech, R.P. and Parks, R.E., Jr. (1978) Guanylate kinases from human erythrocytes, hog brain, and rat liver. Methods Enzymol. 51, 483–490.

Akazawa, C., Shigemoto, R., Bessho, Y., Nakanishi, S. and Mizuno, N. (1994) Differential expression of five N-methyl-D-aspartate receptor subunit mRNAs in the cerebellum of developing and adult rats. J. Comp. Neurol. 347, 150–160.

Behar, T.N., Scott, C.A., Greene, C.L., Wen, X., Smith, S.V., Maric, D., Liu, Q.Y., Colton, C.A. and Barker, J.L. (1999) Glutamate acting at NMDA receptors stimulates embryonic cortical neuronal migration. J. Neurosci. 19, 4449–4461.

Biederer, T. and Sudhof, T.C. (2001) CASK and protein 4.1 support F-actin nucleation on neurexins. J. Biol. Chem. 276, 47869–47876.

Biederer, T., Sara, Y., Mozhayeva, M., Atasoy, D., Liu, X., Kavalali, E.T. and Sudhof, T.C. (2002) SynCAM, a synaptic adhesion molecule that drives synapse assembly. Science 297, 1525–1531.

Borg, J.P., Straight, S.W., Kaech, S.M., de Taddeo-Borg, M., Kroon, D.E., Karnak, D., Turner, R.S., Kim, S.K. and Margolis, B. (1998) Identification of an evolutionarily conserved heterotrimeric protein complex involved in protein targeting. J. Biol. Chem. 273, 31633–31636.

Bretscher, A., Edwards, K. and Fehon, R.G. (2002) ERM proteins and merlin: integrators at the cell cortex. Nat. Rev. Mol. Cell Biol. 3, 586–599.

Bulfone, A., Smiga, S.M., Shimamura, K., Peterson, A., Puelles, L. and Rubenstein, J.L. (1995) T-brain-1: a homolog of Brachyury whose expression defines molecularly distinct domains within the cerebral cortex. Neuron 15, 63–78.

Bulfone, A., Wang, F., Hevner, R., Anderson, S., Cutforth, T., Chen, S., Meneses, J., Pedersen, R., Axel, R. and Rubenstein J.L. (1998) An olfactory sensory map develops in the absence of normal projection neurons or GABAergic interneurons. Neuron 21, 1273–1282.

Butz, S., Okamoto, M.and Sudhof, T.C. (1998) A tripartite protein complex with the potential to couple synaptic vesicle exocytosis to cell adhesion in brain. Cell 94, 773–782.

Chai, Z., Sarcevic, B., Mawson, A. and Toh, B.H. (2001) SET-related cell division autoantigen-1 (CDA1) arrests cell growth. J. Biol. Chem. 276, 33665–33674.

Cho, K.O., Hunt, C.A. and Kennedy, M.B. (1992) The rat brain postsynaptic density fraction contains a homolog of the Drosophila discs-large tumor suppressor protein. Neuron 9, 929–942.

Cohen, A.R., Woods, D.F., Marfatia, S.M., Walther, Z., Chishti, A.H., Anderson, J.M. and Wood, D.F. (1998) Human CASK/LIN-2 binds syndecan-2 and protein 4.1 and localizes to the basolateral membrane of epithelial cells. J. Cell Biol. 142, 129–138.

Dickson, K.S. and Kind, P.C. (2003) NMDA receptors: neural map designers and refiners? Curr. Biol. 13, R920–922.

Doerks, T., Bork, P., Kamberov, E., Makarova, O., Muecke, S. and Margolis, B. (2000) L27, a novel heterodimerization domain in receptor targeting proteins Lin-2 and Lin-7. Trends Biochem. Sci. 25, 317–318.

Ehlers, M.D. (2003) Activity level controls postsynaptic composition and signaling via the ubiquitin-proteasome system. Nat. Neurosci. 6, 231–242.

Eichmuller, S., Usener, D., Dummer, R., Stein, A., Thiel, D. and Schadendorf, D. (2001) Serological detection of cutaneous T-cell lymphoma-associated antigens. Proc. Natl. Acad. Sci. USA 98, 629–634 Epub 2001 Jan 2009.

Englund, C., Fink, A., Lau, C., Pham, D., Daza, R.A., Bulfone, A., Kowalczyk, T. and Hevner, R.F. (2005) Pax6, Tbr2, and Tbr1 are expressed sequentially by radial glia, intermediate progenitor cells, and postmitotic neurons in developing neocortex. J. Neurosci. 25, 247–251.

Ethell, I.M. and Yamaguchi, Y. (1999) Cell surface heparan sulfate proteoglycan syndecan-2 induces the maturation of dendritic spines in rat hippocampal neurons. J. Cell Biol. 144, 575–586.

Fallon, L., Morcau, F., Croft, B.G., Labib, N., Gu, W.J. and Fon, E.A. (2002) Parkin and CASK/LIN-2 associate via a PDZ-mediated interaction and are co-localized in lipid rafts and postsynaptic densities in brain. J. Biol. Chem. 277, 486–491.

Fink, A.J., Englund, C., Daza, R.A., Pham, D., Lau, C., Nivison, M., Kowalczyk, T. and Hevner, R.F. (2006) Development of the deep cerebellar nuclei: transcription factors and cell migration from the rhombic lip. J. Neurosci. 26, 3066–3076.

Follesa, P. and Ticku, M.K. (1996) Chronic ethanol-mediated up-regulation of the N-methyl-D-aspartate receptor polypeptide subunits in mouse cortical neurons in culture. J. Biol. Chem. 271, 13297–13299.

Gruss, C. and Sogo, J.M. (1992) Chromatin replication. Bioessays 14, 1–8.

Hata, Y., Butz, S. and Sudhof, T.C. (1996) CASK: a novel dlg/PSD95 homolog with an N-terminal calmodulin-dependent protein kinase domain identified by interaction with neurexins. J. Neurosci. 16, 2488–2494.

Herkert, M., Rottger, S. and Becker, C.M. (1998) The NMDA receptor subunit NR2B of neonatal rat brain: complex formation and enrichment in axonal growth cones. Eur. J. Neurosci. 10, 1553–1562.

Hevner, R.F., Shi, L., Justice, N., Hsueh, Y., Sheng, M., Smiga, S., Bulfone, A., Goffinet, A.M., Campagnoni, A.T. and Rubenstein J.L. (2001) Tbr1 regulates differentiation of the preplate and layer 6. Neuron 29, 353–366.

Hoover, K.B. and Bryant, P.J. (2000) The genetics of the protein 4.1 family: organizers of the membrane and cytoskeleton. Curr. Opin. Cell Biol. 12, 229–234.

Hsueh, Y.P. and Sheng, M. (1999) Regulated expression and subcellular localization of syndecan heparan sulfate proteoglycans and the syndecan-binding protein CASK/LIN-2 during rat brain development. J. Neurosci. 19, 7415–7425.

Hsueh, Y.P., Wang, T.F., Yang, F.C. and Sheng, M. (2000) Nuclear translocation and transcription regulation by the membrane-associated guanylate kinase CASK/LIN-2. Nature 404, 298–302.

Hsueh, Y.P., Roberts, A.M., Volta, M., Sheng, M. and Roberts, R.G. (2001) Bipartite interaction between neurofibromatosis type I protein (neurofibromin) and syndecan transmembrane heparan sulfate proteoglycans. J. Neurosci. 21, 3764–3770.

Hsueh, Y.P., Yang, F.C., Kharazia, V., Naisbitt, S., Cohen, A.R., Weinberg, R.J. and Sheng, M. (1998) Direct interaction of CASK/LIN-2 and syndecan heparan sulfate proteoglycan and their overlapping distribution in neuronal synapses. J. Cell Biol. 142, 139–151.

Hung, A.Y. and Sheng, M. (2002) PDZ domains: structural modules for protein complex assembly. J. Biol. Chem. 277, 5699–5702.

Irie, M., Hata, Y., Takeuchi, M., Ichtchenko, K., Toyoda, A., Hirao, K., Takai, Y., Rosahl, T.W. and Sudhof, T.C. (1997) Binding of neuroligins to PSD-95. Science 277, 1511–1515.

Jo, K., Derin, R., Li, M. and Bredt, D.S. (1999) Characterization of MALS/Velis-1, -2, and -3: a family of mammalian LIN-7 homologs enriched at brain synapses in association with the postsynaptic density-95/NMDA receptor postsynaptic complex. J. Neurosci. 19, 4189–4199.

Kaech, S.M., Whitfield, C.W. and Kim, S.K. (1998) The LIN-2/LIN-7/LIN-10 complex mediates basolateral membrane localization of the C. elegans EGF receptor LET-23 in vulval epithelial cells. Cell 94, 761–771.

Kim, E. and Sheng, M. (2004) PDZ domain proteins of synapses. Nat. Rev. Neurosci. 5, 771–781.

Kim, E., Niethammer, M., Rothschild, A., Jan, Y.N. and Sheng, M. (1995) Clustering of Shaker-type K+ channels by interaction with a family of membrane-associated guanylate kinases. Nature 378, 85–88.

Kornau, H.C., Schenker, L.T., Kennedy, M.B. and Seeburg, P.H. (1995) Domain interaction between NMDA receptor subunits and the postsynaptic density protein PSD-95. Science 269, 1737–1740.

Laverty, H.G. and Wilson, J.B. (1998) Murine CASK is disrupted in a sex-linked cleft palate mouse mutant. Genomics 53, 29–41.

Lee, S., Fan, S., Makarova, O., Straight, S. and Margolis, B. (2002) A novel and conserved protein-protein interaction domain of mammalian Lin-2/CASK binds and recruits SAP97 to the lateral surface of epithelia. Mol. Cell Biol. 22, 1778–1791.

Leonoudakis, D., Conti, L.R., Radeke, C.M., McGuire, L.M. and Vandenberg, C.A. (2004) A multiprotein trafficking complex composed of SAP97, CASK, Veli, and Mint1 is associated with inward rectifier Kir2 potassium channels. J. Biol. Chem. 279, 19051–19063 Epub 12004 Feb 19011.

Leonoudakis, D., Conti, L.R., Anderson, S., Radeke, C.M., McGuire, L.M., Adams, M.E., Froehner, S.C., Yates, J.R., 3rd and Vandenberg, C.A. (2004) Protein trafficking and anchoring complexes revealed by proteomic analysis of inward rectifier potassium channel (Kir2.x)-associated proteins. J. Biol. Chem. 279, 22331–22346 Epub 22004 Mar 22315.

Levinson, J.N., Chery, N., Huang, K., Wong, T.P., Gerrow, K., Kang, R., Prange, O., Wang, Y.T. and El-Husseini, A. (2005) Neuroligins mediate excitatory and inhibitory synapse formation: involvement of PSD-95 and neurexin-1beta in neuroligin-induced synaptic specificity. J. Biol. Chem. 280, 17312–17319.

Lin, C.W., Huang, T.N., Wang, G.S., Kuo, T.Y., Yen, T.Y. and Hsueh, Y.P. (2006) Neural activity- and development-dependent expression and distribution of CASK interacting nucleosome assembly protein in mouse brain. J. Comp. Neurol. 494, 606–619.

Liu, L., Wong, T.P., Pozza, M.F., Lingenhoehl, K., Wang, Y., Sheng, M., Auberson, Y.P. and Wang, Y.T. (2004) Role of NMDA receptor subtypes in governing the direction of hippocampal synaptic plasticity. Science 304, 1021–1024.

Loftis, J.M. and Janowsky, A. (2003) The N-methyl-D-aspartate receptor subunit NR2B: localization, functional properties, regulation, and clinical implications. Pharmacol. Ther. 97, 55–85.

Marfatia, S.M., Morais-Cabral, J.H., Kim, A.C., Byron, O. and Chishti, A.H. (1997) The PDZ domain of human erythrocyte p55 mediates its binding to the cytoplasmic carboxyl terminus of glycophorin C. Analysis of the binding interface by in vitro mutagenesis. J. Biol. Chem. 272, 24191–24197.

Mattson, M.P., Dou, P. and Kater, S.B. (1988) Outgrowth-regulating actions of glutamate in isolated hippocampal pyramidal neurons. J. Neurosci. 8, 2087–2100.

Maximov, A. and Bezprozvanny, I. (2002) Synaptic targeting of N-type calcium channels in hippocampal neurons. J. Neurosci. 22, 6939–6952.

Maximov, A., Sudhof, T.C. and Bezprozvanny, I. (1999) Association of neuronal calcium channels with modular adaptor proteins. J. Biol. Chem. 274, 24453–24456.

Missler, M., Fernandez-Chacon, R. and Sudhof, T.C. (1998) The making of neurexins. J. Neurochem. 71, 1339–1347.

Montgomery, J.M., Zamorano, P.L. and Garner, C.C. (2004) MAGUKs in synapse assembly and function: an emerging view. Cell. Mol. Life Sci. 61, 911–929.

Monyer, H., Burnashev, N., Laurie, D.J., Sakmann, B. and Seeburg, P.H. (1994) Developmental and regional expression in the rat brain and functional properties of four NMDA receptors. Neuron 12, 529–540.

Naiche, L.A., Harrelson, Z., Kelly, R.G. and Papaioannou, V.E. (2005) T-box genes in vertebrate development. Annu. Rev. Genet. 39, 219–239.

Nam, C.I. and Chen, L. (2005) Postsynaptic assembly induced by neurexin-neuroligin interaction and neurotransmitter. Proc. Natl. Acad. Sci. USA 102, 6137–6142.

Nix, S.L., Chishti, A.H., Anderson, J.M. and Walther, Z. (2000) hCASK and hDlg associate in epithelia, and their src homology 3 and guanylate kinase domains participate in both intramolecular and intermolecular interactions. J. Biol. Chem. 275, 41192–41200.

Ogawa, M., Miyata, T., Nakajima, K., Yagyu, K., Seike, M., Ikenaka, K., Yamamoto, H. and Mikoshiba, K. (1995) The reeler gene-associated antigen on Cajal-Retzius neurons is a crucial molecule for laminar organization of cortical neurons. Neuron 14, 899–912.

Ohno, H., Hirabayashi, S., Kansaku, A., Yao, I., Tajima, M., Nishimura, W., Ohnishi, H., Mashima, H., Fujita, T., Omata, M. and Hata, Y. (2003) Carom: a novel membrane-associated guanylate kinase-interacting protein with two SH3 domains. Oncogene 22, 8422–8431.

Olsen, O., Moore, K.A., Fukata, M., Kazuta, T., Trinidad, J.C., Kauer, F.W., Streuli, M., Misawa, H., Burlingame, A.L., Nicoll, R.A. and Bredt, D.S. (2005) Neurotransmitter release regulated by a MALS-liprin-alpha presynaptic complex. J. Cell Biol. 170, 1127–1134.

Ozbun, L.L., You, L., Kiang, S., Angdisen, J., Martinez, A. and Jakowlew, S.B. (2001) Identification of differentially expressed nucleolar TGF-beta1 target (DENTT) in human lung cancer cells that is a new member of the TSPY/SET/NAP-1 superfamily. Genomics 73, 179–193.

Ozbun, L.L., Martinez, A., Angdisen, J., Umphress, S., Kang, Y., Wang, M., You, M. and Jakowlew, S.B. (2003) Differentially expressed nucleolar TGF-beta1 target (DENTT) in mouse development. Dev. Dyn. 226, 491–511.

Portera-Cailliau, C., Price, D.L. and Martin, L.J. (1996) N-methyl-D-aspartate receptor proteins NR2A and NR2B are differentially distributed in the developing rat central nervous system as revealed by subunit-specific antibodies. J. Neurochem. 66, 692–700.

Porteus, M.H., Brice, A.E., Bulfone, A., Usdin, T.B., Ciaranello, R.D. and Rubenstein, J.L. (1992) Isolation and characterization of a library of cDNA clones that are preferentially expressed in the embryonic telencephalon. Brain Res. Mol. Brain Res. 12, 7–22.

Rao, A. and Craig, A.M. (1997) Activity regulates the synaptic localization of the NMDA receptor in hippocampal neurons. Neuron 19, 801–812.

Rice, D.S. and Curran, T. (2001) Role of the reelin signaling pathway in central nervous system development. Annu. Rev. Neurosci. 24, 1005–1039.

Riefler, G.M. and Firestein, B.L. (2001) Binding of neuronal nitric-oxide synthase (nNOS) to carboxyl-terminal-binding protein (CtBP) changes the localization of CtBP from the nucleus to the cytosol: a novel function for targeting by the PDZ domain of nNOS. J. Biol. Chem. 276, 48262–48268.

Setou, M., Nakagawa, T., Seog, D.H. and Hirokawa, N. (2000) Kinesin superfamily motor protein KIF17 and mLin-10 in NMDA receptor-containing vesicle transport. Science 288, 1796–1802.

Sheng, M. and Sala, C. (2001) PDZ domains and the organization of supramolecular complexes. Annu. Rev. Neurosci. 24, 1–29.

Song, J.Y., Ichtchenko, K., Sudhof, T.C. and Brose, N. (1999) Neuroligin 1 is a postsynaptic cell-adhesion molecule of excitatory synapses. Proc. Natl. Acad. Sci. USA 96, 1100–1105.

Songyang, Z., Fanning, A.S., Fu, C., Xu, J., Marfatia, S.M., Chishti, A.H., Crompton, A., Chan, A.C., Anderson, J.M. and Cantley, L.C. (1997) Recognition of unique carboxyl-terminal motifs by distinct PDZ domains. Science 275, 73–77.

Szabo, S.J., Kim, S.T., Costa, G.L., Zhang, X., Fathman, C.G. and Glimcher, L.H. (2000) A novel transcription factor, T-bet, directs Th1 lineage commitment. Cell 100, 655–669.

Szabo, S.J., Sullivan, B.M., Stemmann, C., Satoskar, A.R., Sleckman, B.P. and Glimcher, L.H. (2002) Distinct effects of T-bet in TH1 lineage commitment and IFN-gamma production in CD4 and CD8 T cells. Science 295, 338–342.

Tabuchi, K., Biederer, T., Butz, S. and Sudhof, T.C. (2002) CASK participates in alternative tripartite complexes in which Mint 1 competes for binding with caskin 1, a novel CASK-binding protein. J. Neurosci. 22, 4264–4273.

Tissir, F. and Goffinet, A.M. (2003) Reelin and brain development. Nat. Rev. Neurosci. 4, 496–505.

Wang, G.S., Hong, C.J., Yen, T.Y., Huang, H.Y., Ou, Y., Huang, T.N., Jung, W.G., Kuo, T.Y., Sheng, M., Wang, T.F. and Hsueh, Y.P. (2004a) Transcriptional modification by a CASK-interacting nucleosome assembly protein. Neuron 42, 113–128.

Wang, T.F., Ding, C.N., Wang, G.S., Luo, S.C., Lin, Y.L., Ruan, Y., Hevner, R., Rubenstein, J.L. and Hsueh, Y.P. (2004b) Identification of Tbr-1/CASK complex target genes in neurons. J. Neurochem. 91, 1483–1492.

Wenzel, A., Fritschy, J.M., Mohler, H. and Benke, D. (1997) NMDA receptor heterogeneity during postnatal development of the rat brain: differential expression of the NR2A, NR2B, and NR2C subunit proteins. J. Neurochem. 68, 469–478.

Willott, E., Balda, M.S., Fanning, A.S., Jameson, B., Van Itallie, C. and Anderson, J.M. (1993) The tight junction protein ZO-1 is homologous to the Drosophila discs-large tumor suppressor protein of septate junctions. Proc. Natl. Acad. Sci. USA 90, 7834–7838.

Woods, D.F. and Bryant, P.J. (1991) The discs-large tumor suppressor gene of Drosophila encodes a guanylate kinase homolog localized at septate junctions. Cell 66, 451–464.

Yap, C.C., Liang, F., Yamazaki, Y., Muto, Y., Kishida, H., Hayashida, T., Hashikawa, T. and Yano, R. (2003) CIP98, a novel PDZ domain protein, is expressed in the central nervous system and interacts with calmodulin-dependent serine kinase. J. Neurochem. 85, 123–134.

Zheng, J.Q., Wan, J.J. and Poo, M.M. (1996) Essential role of filopodia in chemotropic turning of nerve growth cone induced by a glutamate gradient. J. Neurosci. 16, 1140–1149.

Chapter 4
Intracellular Calcium Waves Transmit Synaptic Information to the Nucleus in Hippocampal Pyramidal Neurons

Mark F. Yeckel, Amanda A. Sleeper, John S. Fitzpatrick, Daniel N. Hertle, Anna M. Hagenston, and Robin T. Garner

Abstract Ca^{2+} waves provide a spatially and temporally unique intracellular signal that carries information from one region of the neuron to another. Despite the computational potential of such a mechanism, relatively little is known about the consequences of Ca^{2+} waves on neuronal function. In this chapter we review the basic properties of internal Ca^{2+} release and Ca^{2+} waves in hippocampal CA1 pyramidal neurons and how synaptically elicited Ca^{2+} waves influence the transcription factor CREB in an age-dependent manner.

4.1 Complex Neuronal Structure Necessitates Complex Intracellular Signaling Strategies

The elaborate dendritic architecture of neurons, coupled with their role in transducing distributed synaptic inputs into electrical, biochemical, and molecular signals, places significant demands on intracellular signaling in this cell type. Based on the logistical complexity of such signaling, it has been proposed that neurons maintain coordinated local and long-distance signaling in order to process synaptic activation (Wells and Fallon 2000). Of the vast and growing number of important signaling entities, Ca^{2+}—by virtue of its fundamental role in a multitude of neuronal processes—is considered to be a preeminent signaling agent for regulating local signaling events and in triggering long-distance events. For example, Ca^{2+} is involved in neuronal differentiation and proliferation, dendritic and axon development, ion and ligand channel gating, synaptic transmission including short- and long-term synaptic plasticity, and programmed cell death. The regulation of these processes by Ca^{2+} depends in part on where in the neuron the Ca^{2+} signal originates and the extent to which it spreads. While some studies suggest that increases in intracellular Ca^{2+} concentration ($[Ca^{2+}]_i$) are restricted to the immediate region of the source (Koch and Zador 1993; Muller and Conner 1991; Sabatini et al., 2001; Svoboda and Mainen 1999; Wickens 1988; Yuste et al., 2000), others indicate that release of Ca^{2+} from intracellular stores can lead to the propagation of Ca^{2+} waves to distant regions of the neuron (Kapur et al., 2001; Nakamura et al., 1999; Watanabe et al., 2006). This raises the intriguing possibility that Ca^{2+} waves carry information from one region of the neuron to another and thus might contribute directly to long-distance signaling. In this chapter we review the basic properties of internal Ca^{2+} release and Ca^{2+} waves in hippocampal pyramidal neurons and how Ca^{2+} waves propagating to the soma might contribute to transcription events in the nucleus.

4.2 Sources of Ca^{2+}

The magnitude of Ca^{2+} signals and their spatial and temporal properties are likely to be important determinants of downstream signaling events. Rises in $[Ca^{2+}]_i$ in neurons are due to Ca^{2+} influx through voltage-gated Ca^{2+} channels (VGCCs) and/or ligand-gated channels (e.g., NMDA receptor-mediated channels), or by the release of Ca^{2+} from intracellular Ca^{2+} stores, located primarily on the endoplasmic reticulum (ER). The latter process is called internal Ca^{2+} release. In hippocampal pyramidal neurons, internal Ca^{2+} release is mediated by the inositol-1,4,5-trisphosphate (IP$_3$) signal transduction pathway that is triggered by activation of G-protein-coupled receptors (GPCRs) located on the plasma membrane. Each of these Ca^{2+} sources provides signals with unique properties determined by the location, magnitude, and kinetics of the Ca^{2+} signal. Rises in Ca^{2+} are further shaped by extrusion, sequestration into intracellular organelles (including the ER and mitochondria), buffering by Ca^{2+} binding proteins whose only known purpose is to buffer Ca^{2+}, and by binding to Ca^{2+} signaling proteins. While the processes regulating Ca^{2+} entry from extracellular sources and their subsequent functional consequences have been well characterized in neurons, relatively little is known about the characteristics and consequences of internal Ca^{2+} release and Ca^{2+} waves in neurons.

4.3 Internal Ca^{2+} Release in Neurons

4.3.1 The Mechanism of Internal Ca^{2+} Release

The release of Ca^{2+} from intracellular stores in cells is mediated by activation of IP$_3$ receptors and ryanodine receptors (RyRs). IP$_3$Rs and RyRs are two families of Ca^{2+} permeable channels located on membranes of cellular organelles that store Ca^{2+} and are ubiquitous across cell types (for review, see Berridge 1998). The basic mechanisms of internal Ca^{2+} release in non-neuronal cells have been thoroughly studied using a variety of reduced preparations and techniques (Berridge 1993; 1998). By comparison, relatively little is known about the basic properties of internal Ca^{2+} release and its function in neurons. Perhaps due to the profound structural and functional differences between non-neuronal cells and neurons, it is not surprising that the general mechanisms underlying internal Ca^{2+} release in non-neuronal cells differ in some respects from those of neurons. For example, in several non-neuronal cell types, concomitant or sequential activation of IP3Rs and RyRs shape Ca^{2+} release responses or lead to Ca^{2+} wave propagation (Johenning et al., 2002; Leite et al., 2002), whereas in hippocampal and neocortical pyramidal neurons there is little evidence, if any, that RyRs participate in release or waves (Kapur et al., 2001; Larkum et al., 2003; Nakamura et al., 1999), despite their widespread presence in these neurons (Sharp et al., 1993; Sharp et al., 1999). Based on a lack of evidence showing that RyRs participate in internal Ca^{2+} release in neurons, research has primarily focused on the properties of IP3-mediated release in neurons (Finch and Augustine 1998; Kapur et al., 2001; Morikawa et al., 2000; Nakamura et al., 1999; Nakamura et al., 2000; Pozzo Miller et al., 1996; Yeckel et al., 1999).

In hippocampal pyramidal neurons, internal Ca^{2+} release occurs downstream of the activation of G_q-coupled receptors such as Group I metabotropic glutamate receptors (mGluRs) or muscarinic acetylcholine receptors (mAChRs) (Nakamura et al., 1999; Power and Sah 2002; Pozzo Miller et al., 1996; Yeckel et al., 1999). G_q protein-coupled receptor-activation leads to the activation of phospholipase C (PLC), which cleaves the lipid precursor phosphatidylinositide 4,5-bisphosphate (PIP$_2$) to form IP3 and diacylglycerol (DAG) (Berridge 1993). IP3 diffuses in the cytosol (Allbritton et al., 1992; Dawson, 1997) and binds to type 1 IP3 receptors (IP3Rs) located on the ER. Binding of IP3 to IP3Rs opens the receptor's intrinsic ion channel, allowing Ca^{2+} to flow down its concentration gradient out of the ER and into the cytosol (Fig. 1).

An important feature of type 1 IP3Rs is that Ca^{2+} acts as an allosteric regulator when IP3 is present (Bezprozvanny et al., 1991; Finch et al., 1991; Iino, 1990; Marchant and Taylor 1997). In neurons, steady state $[Ca^{2+}]_i$ is $\sim$50–100 nM (Verkhratsky and Petersen 2002). Transient increases in $[Ca^{2+}]_i$ are primarily mediated by influx through VGCCs and/or NMDAR-mediated channels during depolarization and/or glutamatergic synaptic transmission. Facilitation of release by Ca^{2+} has been reported to result from increases in $[Ca^{2+}]_i$ up to $\sim$250 nM (Bezprozvanny et al., 1991). In pyramidal neurons, under relatively realistic physiological conditions, small rises in $[Ca^{2+}]_i$ via influx through VGCCs have been shown to be necessary for synaptically eliciting internal Ca^{2+} release (Kapur et al., 2001; Nakamura et al., 1999). This characteristic—the ability of IP3Rs to integrate glutamatergic synaptic transmission and postsynaptic neuronal activity—endows IP3Rs with the ability to act as coincidence detectors (Berridge 1998).

4.3.2 Propagating Ca^{2+} Waves

Internal Ca^{2+} release can lead to the propagation of Ca^{2+} waves in hippocampal CA1 and CA3 pyramidal cells (Kapur et al., 2001; Nakamura et al., 1999; Watanabe et al., 2006) and neocortical pyramidal cells (Larkum et al., 2003). This process is distinct from passive diffusion of Ca^{2+}, which results in a rapid decre-

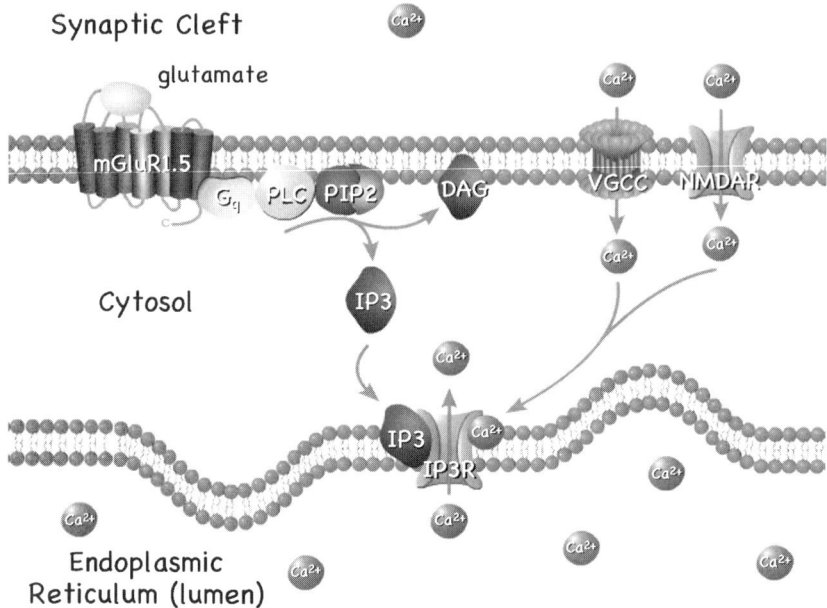

Fig. 1 Mechanism of internal Ca^{2+} release. Activation of group 1 mGluRs by glutamate or mGluR agonists such as ACPD leads to initiation of a signal transduction cascade that evokes internal Ca^{2+} release. Group 1 mGluRs are G_q protein-coupled receptors whose stimulation leads to activation of PLC. PLC cleaves membrane-bound PIP_2, forming IP_3 and DAG. IP_3 is freely mobile in the cytosol and binds to the IP_3Rs located on the ER. IP_3R activation opens an intrinsic ion channel within the receptor, allowing Ca^{2+} to flow out of the ER into the cytosol. Ca^{2+} acts as a coagonist on IP_3Rs. Modest elevations in cytosolic Ca^{2+} concentrations lead to sensitization on IP_3Rs to IP_3, lowering the threshold concentration of IP_3 necessary for initiation of internal Ca^{2+} release. This Ca^{2+} may enter through VGCC or NMDAR-mediated channels. Additionally, Ca^{2+} released from internal stores following initial IP_3R activation may serve to sensitize adjacent IP_3Rs, assisting in the wavelike propagation of internal Ca^{2+} release signals

ment of the Ca^{2+} signal (Allbritton et al., 1992; Kasai and Petersen 1994). Although the mechanism underlying the propagation of Ca^{2+} waves has not been identified in neurons, experimental data and computational models of wave propagation in Xenopus oocytes suggest that once a wave is initiated, its propagation is due to sequential activation of clusters of IP_3Rs by mobile IP_3 and released Ca^{2+} that diffuses to adjacent IP_3R clusters (Callamaras et al., 1998; Dargan et al., 2004; Dawson et al., 1999). The regenerative nature of internal Ca^{2+} release produces large Ca^{2+} signals capable of traveling long distances (Kapur et al., 2001; Nakamura et al., 1999; Watanabe et al., 2006).

IP_3-dependent internal Ca^{2+} release in CA1 neurons occurs predominantly in the primary apical dendrite. The waves typically originate at a point where an oblique dendrite branches from the main shaft (Nakamura et al., 1999; Nakamura et al., 2002). The frequency of synaptic stimulation, the number of synaptic pulses, the location of the stimulating electrode, and the recent history of the neuron's activity are important experimental factors in determining the reliability and the magnitude of IP3-mediated internal Ca^{2+} release and Ca^{2+} waves (Zhou and Ross 2002). Under optimal conditions, Ca^{2+} waves have been reported to travel greater than 100 μm (Nakamura et al., 1999), suggesting that these Ca^{2+} signals can provide a relatively far-reaching form of dendritic regulation.

4.3.3 Distinguishing Features of Internal Ca²⁺ Release

There are several characteristics of internal Ca^{2+} release that help to readily distinguish rises in $[Ca^{2+}]_i$ due to IP$_3$-mediated internal release from influx of extracellular Ca^{2+} through VGCCs and NMDAR-mediated channels (Figs 2 and 3): (1) Internal Ca^{2+} release typically has a delayed onset and regenerative kinetics. When neurons are clamped to rest and influx of Ca^{2+} through VGCCs is suppressed (but not blocked; see Kapur et al., 2001), brief trains of synaptic activation can lead to a delayed onset in the rise of Ca^{2+} due to internal Ca^{2+} release that

Ca^{2+} sources: Extracellular (VGCCs) and Intracellular (internal Ca^{2+} release)

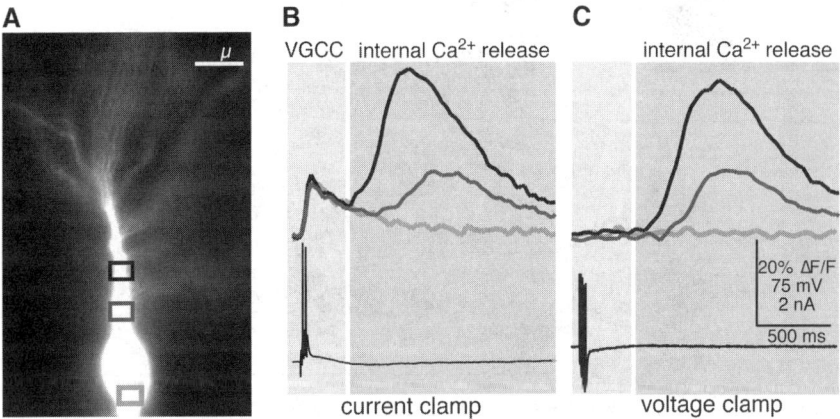

Fig. 2 Ca^{2+} Released from internal storess has unique spatial and temporal properties. (A) Fluorescence image of a CA1 pyramidal cell filled with the Ca^{2+} indicator bis-fura-2 (100 μM). Rectangles define three regions of interest. (B) A brief train of synaptic stimuli (5 impulses at 50 Hz) triggered Ca^{2+} influx into the cytosol from both the extracellular space and the internal stores. Influx from the extracellular space occurred earlier (top traces, gray levels used to indicate region) and resulted from action potentials (bottom trace) backpropagating into the primary apical dendrite and activating VGCCs. This earlier rise in $[Ca^{2+}]_i$ occurred simultaneously and equally in all three regions. On the other hand, Ca^{2+} was released from internal stores only later, and after the membrane had repolarized (bottom trace). Release occurred first in the black region and propagated into the dark gray region but not into the light gray region. Note that in the black region the Ca^{2+} signal due to internal release was nearly 3 times the amplitude of the Ca^{2+} signal due to activation of VGCCs. (C) When the same synaptic stimulation was applied under voltage clamp, it did not trigger any action potentials, and the earlier, VGCC-dependent component of the Ca^{2+} signal was absent. By contrast, the release of Ca^{2+} from internal stores was virtually unaffected

Synaptically or mGluR Agonist Elicited Ca²⁺ Waves

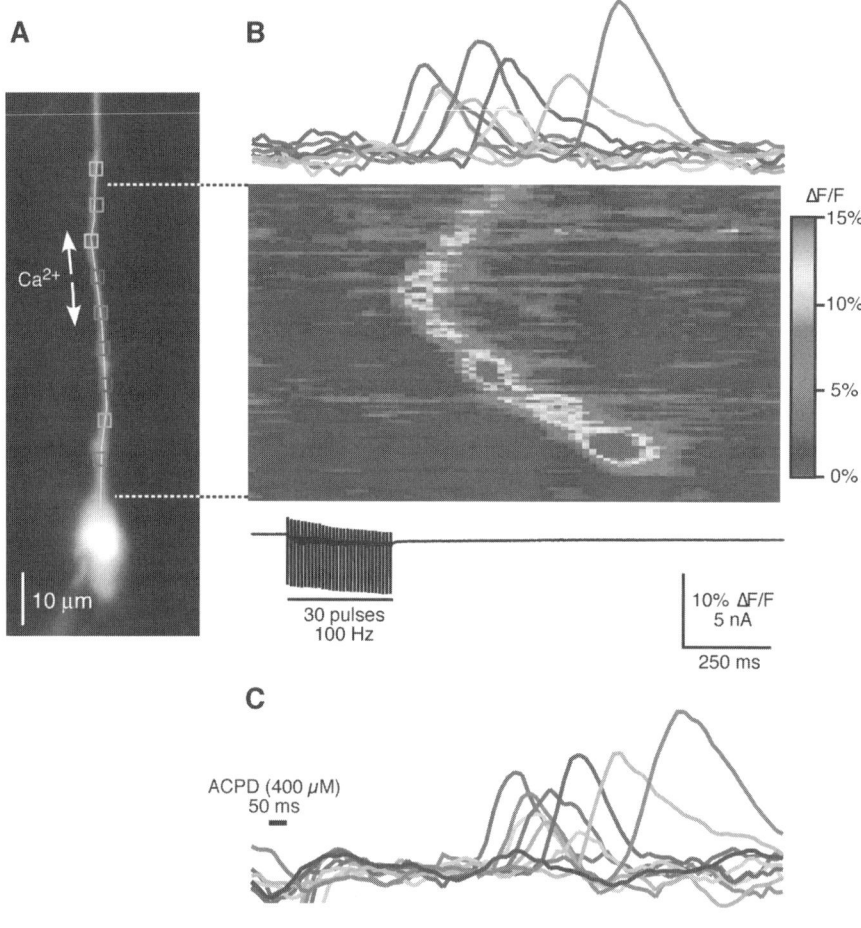

Fig. 3 A Ca²⁺ Wave Triggered By Synaptic Stimulation. (A) A CA1 pyramidal cell filled with bis-fura-2. Rectangles define regions of interest, and the line along the primary apical dendrite is used for the kymograph. (B) A train of synaptic stimulation (30 impulses at 100 Hz in voltage clamp, bottom) triggered a Ca²⁺ wave that propagated along the primary apical dendrite. The wave is represented in the top and middle panels using different methods. In the top panel, the wave is depicted by using regions of interest. $[Ca^{2+}]_i$ rises first in the black region and then sequentially in the other regions, indicating that the wave propagated toward the soma. In the middle panel the wave is depicted in a kymograph. In the kymograph the vertical dimension represents location along the line drawn in the image, and the horizontal dimension represents time. The wave appears as a diagonal band. The rise in $[Ca^{2+}]_i$ occurred earliest at the most distal location along the line and then propagated as a wave along the dendrite toward the soma, stopping where the dendrite meets the soma. (*See* Color Plate 3).

typically occurs when membrane potential has returned to rest. Frequently there is a gradual increase in $[Ca^{2+}]_i$ leading to a sharp, regenerative rise in $[Ca^{2+}]_i$ (Kapur et al., 2001; Nakamura et al., 1999; Yeckel et al., 1999). (2) Synaptically activated internal Ca^{2+} release in pyramidal neurons is almost always followed by propagation of Ca^{2+} waves. In contrast, action potential-evoked rises in $[Ca^{2+}]_i$ mediated by VGCCs is time-locked to the action potentials and occurs relatively simultaneously along the soma and dendrites. (3) Prior depolarization by current injection facilitates internal Ca^{2+} release for up to three minutes. This so-called *priming* or *charging* occurs after cytosolic $[Ca^{2+}]$ has returned to baseline, which typically takes less than a second. Although the mechanism of priming has not been identified, it is thought to result from filling of internal stores with Ca^{2+} that enters via VGCCs during the depolarization (Jaffe and Brown 1994; Watanabe et al., 2006; Yeckel et al., 1999). Priming has no effect on VGCC-mediated rises in Ca^{2+}. (4) Pharmacological manipulations that deplete internal stores by blocking SERCA pumps (CPA or thapsigargin) or block IP_3Rs (2-APB and heparin) confirm that rises in $[Ca^{2+}]_i$ with these characteristics are due to IP_3-mediated internal Ca^{2+} release (Kapur et al., 2001; Nakamura et al., 1999; Yeckel et al., 1999). (5) Lastly, direct pressure application of the wide-spectrum mGluR agonist, t-ACPD, elicits a similar rise in Ca^{2+} and Ca^{2+} waves in the absence of obvious changes in membrane potential (Kapur et al., 2001; Nakamura et al., 1999), consistent with previous findings showing that glutamatergic synaptic transmission leads to mGluR-mediated G_q protein activation and subsequent IP_3 mobilization.

4.3.4 Internal Ca^{2+} Release During Neuronal Maturation

To date, studies examining internal Ca^{2+} release in neurons have been performed almost exclusively on cultured neurons, slice cultures, and acute slices from developmentally immature rats (1–4 weeks old). We have examined internal Ca^{2+} release in neurons from animals ranging from 1–10 weeks of age. Under our experimental conditions we have shown that Ca^{2+} release is more permissive in CA1 pyramidal neurons from animals 1–3 weeks old (*immature*) compared to CA1 pyramidal neurons from animals > 4 weeks old (*mature*). More specifically, synaptic activation results in IP_3-mediated internal Ca^{2+} release in 100% of the cells tested from 1–3 week-old animals (n=16) and only in 40% of cells from animals > 4 weeks old (n=58). In addition, internal release-mediated Ca^{2+} waves observed in pyramidal neurons from 1–3 week-old animals travel on average twice as far and 45% faster than Ca^{2+} waves in cells from > 4 week-old animals (Fig. 4). Consistent with the findings of others, we didn't observe any obvious differences in their biophysical properties or in synaptic transmission across this age range.

The age-dependent profile of internal Ca^{2+} release suggests that this source of Ca^{2+} might play an important role in the maturation of CA1 pyramidal neurons. The early postnatal weeks are a dynamic period of development in the rat brain, during which many changes occur in the basic morphology of neurons and their dendritic processes. Dendrites elongate and branch (Bannister and Larkman 1995a; Pyapali

Age-Dependent Differences in Release and Wave Properties

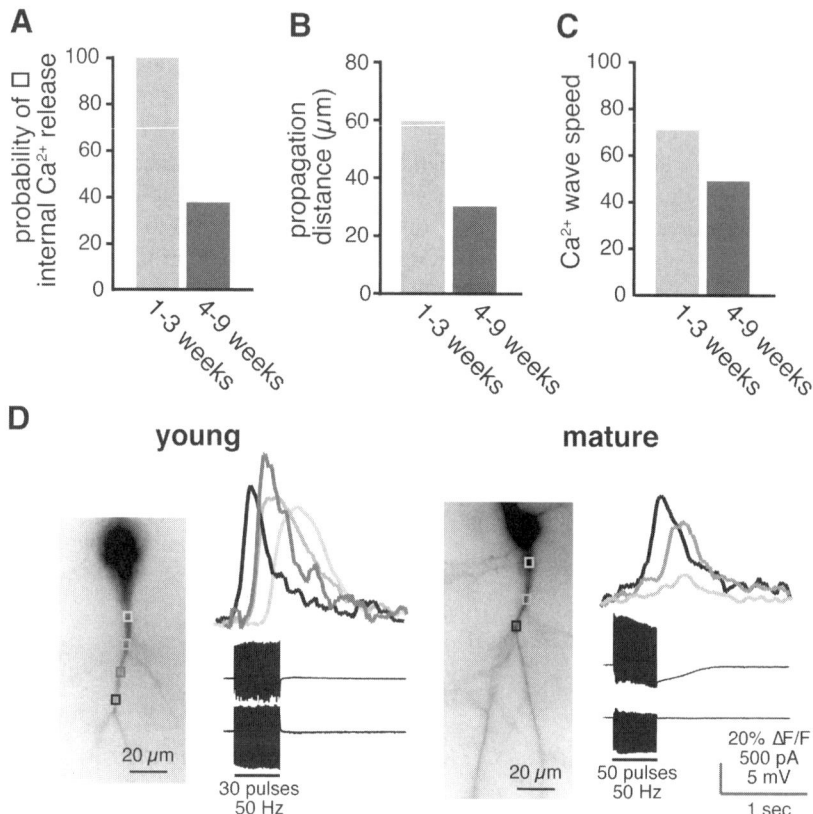

Fig. 4 Internal Ca^{2+} release and Ca^{2+} waves differ between immature and mature CA1 pyramidal neurons. (A) Synaptic stimulation of Schaffer collaterals reliably evokes internal Ca^{2+} release in neurons from 1–3 week-old rats. Stimulation intensity may be enhanced by increasing the number or frequency of stimuli, or by allowing the neuron to fire action potentials prior to/during stimulation. Even under maximal synaptic stimulation conditions, internal Ca^{2+} release is observed in less than 40% of neurons from animals older than 4 weeks (*left*). (B) When Ca^{2+} waves were observed in neurons from rats > 4 weeks old, they propagated for shorter distances and at slower speeds (C) than Ca^{2+} waves elicited in neurons from 1–3 week-old rats. (D) Representative data collected from the 1–3 week-old and > 4 week-old rats. Regions of interest are placed at 10 μm increments in each cell. Internal Ca^{2+} release is observed following synaptic stimulation. Ca^{2+} waves travel farther and faster in the neuron from the 1–3 week-old group than in the neuron from the >4 week-old group

et al., 1998; Turner et al., 1998), and spines are formed and refined (Bannister and Larkman 1995b). Hippocampal pyramidal neurons reach mature dendritic structure by approximately postnatal week 3 (Turner et al., 1998), which coincides with a diminution of internal Ca^{2+} release and Ca^{2+} waves. In addition to changes in cellular morphology, the expression of many signaling genes and proteins changes during development, as does the distribution of already expressed proteins. For example the expression and regulation of neurotransmitter receptors, ion channels, and ion transporters all change as neurons develop to their adult states (Costa 1996; Jones et al., 1997; Petralia et al., 2005; Rivera et al., 1999; Sala et al., 2000).

It is unclear how internal Ca^{2+} release might impact development, yet its prominence during this dynamic period suggests that this form of signaling may contribute to the developmental Ca^{2+}-dependent processes critical during this period. Due to the propagating nature of these signals, Ca^{2+} waves are particularly well poised to carry incoming information from activated synapses to the somatic region of neurons where they might participate in Ca^{2+}-dependent signal transduction cascades leading to gene transcription in the nucleus.

4.4 Ca^{2+}-dependent Gene Transcription

Ca^{2+} is a ubiquitous regulator of cellular function. In addition to regulating plasma membrane channels, triggering a vast array of signaling cascades in the cytosol, and regulating activity in subcellular structures such as the ER and mitochondria, Ca^{2+} plays an important role in gene transcription events occurring in the nucleus (Ghosh and Greenberg 1995). Reports suggest that Ca^{2+} can act directly in the nucleus or indirectly by triggering cytosolic signaling cascades in which the end product crosses into the nucleus (Deisseroth 1998; Hardingham et al., 2001; Hardingham et al., 1998; Mermelstein et al., 2000). These studies have largely focused on extracellular Ca^{2+} entering through VGCCs and NMDAR-mediated channels in cell culture models (Deisseroth 1998; Mermelstein et al., 2000). Synaptically elicited Ca^{2+} waves are also likely to provide an important—and efficient—means of transducing synaptic responses into Ca^{2+}-mediated events occurring in the nucleus. One way these Ca^{2+} signals may influence cellular function such as neuronal development is through the regulation of Ca^{2+}-dependent transcription factors that are necessary for the control of gene transcription.

4.5 The Nuclear Transcription Factor CREB

Several nuclear-mediated transcription events rely on elevated $[Ca^{2+}]_i$ in neurons, including activation of the transcription factor cyclic AMP response element binding protein (CREB) (Bonni et al., 1999; Nakagawa et al., 2002; Redmond et al., 2002; Riccio et al., 1999; Silva et al., 1998). CREB, when made active by phosphorylation, leads to the expression of immediate early genes that are important for dendritic development, synaptic plasticity, and neuronal survival (Bonni et al., 1999; Riccio et al., 1999). For example, CREB phosphorylation and transcription of the gene for brain-derived neurotrophic factor (BDNF) is important for dendritic elaboration and spine development (Chen and Ghosh 2005; Dijkhuizen and Ghosh 2005; McAllister et al., 1997; Redmond and Ghosh 2005; Redmond et al., 2002). Additionally, blocking CREB phosphorylation prevents the transcription of BDNF resulting in decreased cell survival (Ghosh et al., 1994). Finally, phosphorylation of CREB and BDNF have long been implicated in the expression of long-term potentiation

(Figurov et al., 1996; Messaoudi et al., 1998; Nguyen et al., 1994; Patterson et al., 2001).

4.5.1 CREB Activation

CREB is regulated through phosphorylation. Phosphorylation at serine 133 (S133) is obligatory for activation of gene transcription. S133 phosphorylation alone, though, is often insufficient for CREB-mediated gene transcription. Additional phosphorylation sites, transactivators and inhibitors participate in CREB regulation (Sheng et al., 1991; West et al., 2001).

Transactivators and inhibitors of CREB add an additional level of Ca^{2+}-dependent CREB regulation. CREB binding protein (CBP) is a transactivator of CREB. Phosphorylation of CREB at S133 enhances a protein-protein interaction between CREB and CPB. CPB binding to CREB augments CREB's gene transcription activity. This interaction may occur following Ca^{2+}-dependent or -independent S133 phosphorylation (Bading 2000; Bito et al., 1997; West et al., 2001). The repressor protein DREAM, named for its binding to downstream regulatory elements (DREs), competes with CBP for binding to CREB. In the absence of Ca^{2+}, DREAM may bind to CREB, independent of CREB's phosphorylation state at S133, and inhibits subsequent CBP binding. In this manner, DREAM represses CREB-mediated gene transcription (Ledo et al., 2002).

CREB's additional phosphorylation sites include S142 and S143. Phosphorylation at these residues, unlike S133, is thought to be exclusively dependent on Ca^{2+} elevation (Kornhauser et al., 2002). Early studies have shown that phosphorylation at these serine residues correlates with decreased binding of CREB to its co-activator CBP, which may provide a regulatory mechanism of CREB activity (Chrivia et al., 1993; Kornhauser et al., 2002; Kwok et al., 1994). In a neuronal culture model, however, phosphorylation at these sites correlates with robust transcription of a CREB-dependent reporter gene (Kornhauser et al., 2002; Mioduszewska et al., 2003), suggesting that S142/S143 phosphorylation may enhance neuronal CREB activity.

4.5.2 Ca^{2+}-dependent CREB Regulation

It is not entirely clear how Ca^{2+} triggers input-specific downstream responses like phosphorylation of CREB. It is generally thought, however, that the amount of Ca^{2+} entering the cytosol and the Ca^{2+} signal's temporal and spatial characteristics determine subsequent signaling events (Berridge 1998). Increases in $[Ca^{2+}]_i$ resulting from influx via L-type VGCCs or NMDAR-mediated channels has been shown to phosphorylate and activate CREB in cultured neurons (Deisseroth, 1998; Mermelstein et al., 2000; Sheng et al., 1991; West et al., 2001).

Ca^{2+} contributes to the regulation of CREB via the activation of Ca^{2+}-dependent kinases, which in turn phosphorylate and activate CREB (West et al., 2001). Multiple signal transduction cascades are capable of producing CREB phosphorylation. These cascades include PKC activation, adenylyl cyclase/PKA activation, CamKII and CamKIV activation, MAP kinase activation, and APV activation (Sanyal et al., 2002; Shaywitz and Greenberg 1999). The specific subset of signaling pathways that are activated, as well as the specificity and efficacy of subsequent gene transcription, may depend on the source of Ca^{2+} elevation. Such a mechanism could provide for input-specific Ca^{2+} signaling events.

The specificity and efficacy of any Ca^{2+} signal in driving gene transcription may depend on the proximity of the Ca^{2+} source to the necessary signal transduction molecules (Bading et al., 1995; Bito et al., 1997; Dolmetsch et al., 2001; Gallin and Greenberg 1995; Ginty, 1997; Levitan, 1999), and the likelihood of that Ca^{2+} elevation to invade the nucleus (Hardingham et al., 2001; Hardingham et al., 1998). Additionally, Ca^{2+}-dependent gene transcription may depend on the duration of the evoked phosphorylation. In neuronal cultures, the CaMK signaling pathway has been shown to mediate early S133 phosphorylation events ($\sim$ 0–10 min post stimulation) while the MAPK pathway is required for sustained S133 phosphorylation ($>$ 60 min) (Wu et al., 2001). Thus the cooperation of various signaling pathways and the time course of CREB phosphorylation may influence the transcription activity of this protein. Given the varied ability of distinct stimuli to produce specific patterns and durations of CREB phosphorylation, further studies will be needed to clarify the physiological consequences of stimuli leading to Ca^{2+}-dependent CREB phosphorylation via distinct Ca^{2+} sources.

4.6 Internal Ca^{2+} Release and CREB Phosphorylation

In an effort to determine whether internal Ca^{2+} release contributes to the regulation of Ca^{2+}-dependent transcription factors, we examined the ability of these signals to participate in phosphorylation of the transcription factor CREB, which is important for neuronal survival and dendritic development, and has been implicated in learning and memory. Specifically, we examined the ability of internal Ca^{2+} release to participate in phosphorylation of CREB at S133, the phosphorylation site obligatory for subsequent CREB activation.

4.6.1 Internal Ca^{2+} Waves Promote CREB Phosphorylation In Immature CA1 Pyramidal Neurons

We tested whether Ca^{2+} released from internal stores can trigger the phosphorylation of CREB at S133. Both pharmacological (mGluR agonist; 400 μM ACPD) and synaptic stimulation (100 pulses at 100 Hz) of internal Ca^{2+} release in acute hippocampal slices from 1–3 week-old rats produced significantly enhanced CREB phosphorylation in CA1 neurons (pharmacological stim., n = 19, p < 0.0001 ;

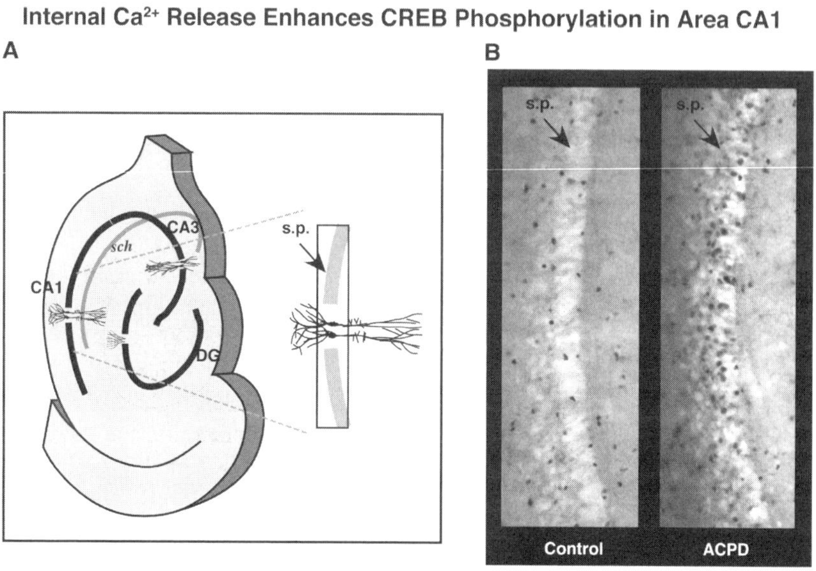

Internal Ca²⁺ Release Enhances CREB Phosphorylation in Area CA1

A   B

Fig. 5 Stimuli that evoke internal Ca^{2+} release enhance CREB phosphorylation in Area CA1. (A) Diagram of a hippocampal slice illustrates the location of area CA1 and the Schaffer collateral projections that synapse on the apical dendrites of CA1 neurons. Histochemistry data utilizing an antibody specific for CREB phosphorylated at S133 are presented with hippocampal slices in this orientation, magnified to illustrate the expanded region shown at the right. CA1 cell bodies lie in s.p. with punctate stain representing phosphorylated CREB in CA1 nuclei. (B) To the left is an example of a control hippocampal slice stained for phospho-CREB. A low amount of basal CREB phosphorylation is visible in s.p. (*left*). Slices treated with 400 μM ACPD exhibit enhanced phospho-CREB signal in s.p. (*right*). All slices were treated in the presence of antagonists of iGluRs (50 μM DL-APV, 20 μM MK-801, 20 μM DNQX) and VGCCs (200 μM $CdCl_2$), leaving internal Ca^{2+} release as the only significant Ca^{2+} source under these stimulation conditions. Inhibition of IP_3Rs or buffering of Ca^{2+} blocks the enhanced phospho-CREB signal (data not shown), suggesting that the increase in CREB phosphorylation is dependent on internal Ca^{2+} release mediated by IP_3Rs following mGluR activation

electrical stim, n $= 11$, $p < 0.05$). More specifically, western blot or immunohistochemistry analysis showed that CREB phosphorylation following stimulation was enhanced as much as 2.7-fold above baseline levels compared to non-stimulated control slices from the same animals (Fig. 5). The enhanced CREB phosphorylation was not accompanied by an increase in total CREB protein, indicating that the increase in phosphorylated CREB was due to a change in the phosphorylation state of CREB and not to an upregulation of CREB. To insure that extracellular Ca^{2+} was not contributing to the phosphorylation during pharmacologically elicited internal Ca^{2+} release both VGCCs and ionotropic glutamate receptors (iGluRs) were blocked with antagonists ($CdCl_2$ and APV/MK801/DNQX, respectively). Synaptically elicited postsynaptic rises in $[Ca^{2+}]_i$ were predominantly due to internal Ca^{2+} release because blockers of L-type Ca^{2+} channels and NMDARs (nimodipine and APV/MK801, respectively) were included in the media. Further evidence that enhancement of CREB phosphorylation under these conditions depended on IP_3-mediated rises in $[Ca^{2+}]_i$ was obtained by preventing phosphorylation with bath application of either a membrane permeant Ca^{2+} chelator (BAPTA-AM) or a membrane permanent antagonist of IP_3Rs (2-APB or xestospongin C). Consistent with the age-dependent differences we observed in the probability of synaptically eliciting Ca^{2+} waves and the properties wave propagation, we found that synaptic stimulation that reliably elicited internal Ca^{2+} release-mediated enhancement of CREB phosphorylation in immature neurons failed to enhance CREB phosphorylation in mature neurons. Mature neurons (7 week-old), however, exhibited an enhancement of CREB phosphorylation when either an mGluR agonist (ACPD) or a cAMP agonist (forskolin) was globally administered by including them in the bath solution. Similarly, and as shown by others (Hardingham et al., 2001; Moore et al., 1996), extensive electrical stimulation that triggered seizure-like activity characterized by spontaneous recurrent interictal activity, also resulted in a significant increase in CREB phosphorylation in mature neurons compared to control cells. The differences that we observed in synaptically elicited internal Ca^{2+} release-mediated phosphorylation of CREB during neuronal maturation, at least under our experimental conditions, suggest that Ca^{2+} waves might provide a developmentally important signal for CA1 neurons.

4.7 Conclusion

Ca^{2+} waves provide a spatially and temporally unique intracellular signal that carries information from one region of the neuron to another. Despite the computational potential of such a mechanism, relatively little is known about the consequences of Ca^{2+} waves on neuronal function. In this chapter we show how synaptically elicited Ca^{2+} waves that propagate to the somatic region might participate in transcription events occurring in the nucleus. As a first step, we characterized some of the basic properties of mGluR-mediated, IP_3-dependent internal Ca^{2+} release. We found that Ca^{2+} waves readily occur under physiologically realistic conditions in immature hippocampal CA1 pyramidal neurons. Although Ca^{2+} waves were also observed in neurons from maturing animals (> 4 weeks old), they occurred less frequently (40% vs. 100% in 1–3 week-old neurons) and the waves didn't propagate as fast or as far as waves observed in immature cells. Based on the ability of Ca^{2+} waves to propagate to the somatic region in immature neurons, we examined whether internal Ca^{2+} waves contributed to nuclear transcription events, and whether there was a similar age-dependence. Consistent with this prediction, we found that synaptically elicited internal release resulted in an enhancement of CREB phosphorylation in immature neurons, but not in mature neurons. Both immature and mature neurons exhibited enhanced phosphorylation of CREB when mGluRs were globally activated with pharmacological stimulation. Taken together, these findings suggest that Ca^{2+} waves might play an important role in CREB function in immature neurons.

Identifying a role for Ca^{2+} waves in CREB regulation is only a first step in unraveling the function of Ca^{2+} waves in gene transcription. Traveling Ca^{2+} waves undoubtedly participate in the regulation of many other Ca^{2+}-dependent transcription factors and in the regulation of other proteins involved in gene transcription.

More generally, our research examining the basic characteristics and consequences of internal Ca^{2+} release and Ca^{2+} waves provides compelling evidence that Ca^{2+} waves provide an efficient mechanism for transmitting synaptic information to the nucleus.

Acknowledgments Funded by the Whitehall Foundation, the Kavli Foundation, the Dart Foundation, The Hellman Family Fund, and NIMH (RO1-MH067830 and P50-MH068789).

References

Allbritton, N.L., Meyer, T., and Stryer, L. (1992) Range of messenger action of calcium ion and inositol 1,4,5-trisphosphate. Science 258, 1812–1815.

Bading, H. (2000) Transcription-dependent neuronal plasticity the nuclear calcium hypothesis. Eur. J. Biochem. 267, 5280–5283.

Bading, H., Segal, M.M., Sucher, N.J., Dudek, H., Lipton, S.A., and Greenberg, M.E. (1995) N-methyl-d-aspartate receptors are critical for mediating the effects of glutamate on intracellular calcium-concentration and immediate-early gene-expression in cultured hippocampal-neurons. Neuroscience 64, 653–664.

Bannister, N.J., and Larkman, A.U. (1995a) Dendritic morphology of CA1 pyramidal neurones from the rat hippocampus: I. Branching patterns. J. Comp. Neurol. 360, 150–160.

Bannister, N.J., and Larkman, A.U. (1995b) Dendritic morphology of CA1 pyramidal neurones from the rat hippocampus: Ii. Spine distributions. J. Comp. Neurol. 360, 161–171.

Berridge, M.J. (1993) Inositol trisphosphate and calcium signalling. Nature 361, 315–325.

Berridge, M.J. (1998) Neuronal calcium signaling. Neuron 21, 13–26.

Bezprozvanny, I., Watras, J., and Ehrlich, B.E. (1991) Bell-shaped calcium-response curves of ins(1,4,5)p3- and calcium-gated channels from endoplasmic reticulum of cerebellum. Nature 351, 751–754.

Bito, H., Deisseroth, K., and Tsien, R.W. (1997) Ca^{2+}-dependent regulation in neuronal gene expression. Curr. Opin. Neurobiol. 7, 419–429.

Bonni, A., Brunet, A., West, A.E., Datta, S.R., Takasu, M.A., and Greenberg, M.E. (1999) Cell survival promoted by the ras-mapk signaling pathway by transcription-dependent and -independent mechanisms. Science 286, 1358–1362.

Callamaras, N., Marchant, J.S., Sun, X.P., and Parker, I. (1998) Activation and co-ordination of insp3-mediated elementary Ca^{2+} events during global Ca^{2+} signals in Xenopus oocytes. J. Physiol. 509 (Pt 1), 81–91.

Chen, Y., and Ghosh, A. (2005) Regulation of dendritic development by neuronal activity. J. Neurobiol. 64, 4–10.

Chrivia, J.C., Kwok, R.P., Lamb, N., Hagiwara, M., Montminy, M.R., and Goodman, R.H. (1993) Phosphorylated CREB binds specifically to the nuclear protein CPB. Nature 365, 855–859.

Costa, P.F. (1996) The kinetic parameters of sodium currents in maturing acutely isolated rat hippocampal CA1 neurones. Brain Res. Dev. Brain Res. 91, 29–40.

Dargan, S.L., Schwaller, B., and Parker, I. (2004) Spatiotemporal patterning of IP3-mediated Ca^{2+} signals in Xenopus oocytes by Ca^{2+}-binding proteins. J. Physiol. 556, 447–461.

Dawson, A.P. (1997) Calcium signalling: How do IP3 receptors work? Curr. Biol. 7, R544–547.

Dawson, S.P., Keizer, J., and Pearson, J.E. (1999) Fire-diffuse-fire model of dynamics of intracellular calcium waves. Proc. Natl. Acad. Sci. USA 96, 6060–6063.

Deisseroth, K.H.E.K.T.R.W. (1998) Translocation of calmodulin to the nucleus supports CREB phosphorylation in hippocampal neurons. Nature 392, 198–202.

Dijkhuizen, P.A., and Ghosh, A. (2005) BDNF regulates primary dendrite formation in cortical neurons via the PI3-kinase and map kinase signaling pathways. J. Neurobiol. 62, 278–288.

Dolmetsch, R.E., Pajvani, U., Fife, K., Spotts, J.M., and Greenberg, M.E. (2001) Signaling to the nucleus by an L-type calcium channel-calmodulin complex through the MAP kinase pathway. Science 294, 333–339.

Figurov, A., Pozzo-miller, L.D., Olafsson, P., Wang, T., and Lu, B. (1996) Regulation of synaptic responses to high-frequency stimulation and LTP by neurotrophins in the hippocampus. Nature 381, 706–709.

Finch, E.A., and Augustine, G.J. (1998) Local calcium signalling by inositol-1,4,5-trisphosphate in purkinje cell dendrites. Nature 396, 753–756.

Finch, E.A., Turner, T.J., and Goldin, S.M. (1991) Calcium as a coagonist of inositol 1,4,5-trisphosphate-induced calcium release. Science 252, 443–446.

Gallin, W.J., and Greenberg, M.E. (1995) Calcium regulation of gene expression in neurons: The mode of entry matters. Curr. Opin. Neurobiol. 5, 367–374.

Ghosh, A., Carnahan, J., and Greenberg, M.E. (1994) Requirement for BDNF in activity-dependent survival of cortical neurons. Science 263, 1618–1623.

Ghosh, A., and Greenberg, M.E. (1995) Calcium signaling in neurons - molecular mechanisms and cellular consequences. Science 268, 239–247.

Ginty, D.D. (1997) Calcium regulation of gene expression: Isn't that spatial? Neuron 18, 183–186.

Hardingham, G.E., Arnold, F.J., and Bading, H. (2001) Nuclear calcium signaling controls CREB-mediated gene expression triggered by synaptic activity. Nat. Neurosci. 4, 261–267.

Hardingham, G.E., Cruzalegui, F.H., Chawla, S., and Bading, H. (1998) Mechanisms controlling gene expression by nuclear calcium signals. Cell Calcium 23, 131–134.

Iino, M. (1990) Biphasic Ca^{2+} dependence of inositol 1,4,5-trisphosphate-induced Ca^{2+} release in smooth muscle cells of the guinea pig taenia caeci. J. Gen. Physiol. 95, 1103–1122.

Jaffe, D.B., and Brown, T.H. (1994) Metabotropic glutamate-receptor activation induces calcium waves within hippocampal dendrites. J. Neurophys. 72, 471–474.

Johenning, F.W., Zochowski, M., Conway, S.J., Holmes, A.B., Koulen, P., and Ehrlich, B.E. (2002) Distinct intracellular calcium transients in neurites and somata integrate neuronal signals. J. Neurosci. 22, 5344–5353.

Jones, O.T., Bernstein, G.M., Jones, E.J., Jugloff, D.G., Law, M., Wong, W., and Mills, L.R. (1997) N-type calcium channels in the developing rat hippocampus: Subunit, complex, and regional expression. J. Neurosci. 17, 6152–6164.

Kapur, A., Yeckel, M., and Johnston, D. (2001) Hippocampal mossy fiber activity evokes Ca^{2+} release in CA3 pyramidal neurons via a metabotropic glutamate receptor pathway. Neuroscience 107, 59–69.

Kasai, H., and Petersen, O.H. (1994) Spatial dynamics of second messengers: IP3 and cAMP as long-range and associative messengers. Trends Neurosci. 17, 95–101.

Koch, C., and Zador, A. (1993) The function of dendritic spines: Devices subserving biochemical rather than electrical compartmentalization. J. Neurosci. 13, 413–422.

Kornhauser, J.M., Cowan, C.W., Shaywitz, A.J., Dolmetsch, R.E., Griffith, E.C., Hu, L.S., Haddad, C., Xia, Z., and Greenberg, M.E. (2002) CREB transcriptional activity in neurons is regulated by multiple, calcium-specific phosphorylation events. Neuron 34, 221–233.

Kwok, R.P., Lundblad, J.R., Chrivia, J.C., Richards, J.P., Bachinger, H.P., Brennan, R.G., Roberts, S.G., Green, M.R., and Goodman, R.H. (1994) Nuclear protein CBP is a coactivator for the transcription factor CREB. Nature 370, 223–226.

Larkum, M.E., Watanabe, S., Nakamura, T., Lasser-Ross, N., and Ross, W.N. (2003) Synaptically activated Ca^{2+} waves in layer 2/3 and layer 5 rat neocortical pyramidal neurons. J. Physiol. 549, 471–488.

Ledo, F., Kremer, L., Mellstrom, B., and Naranjo, J.R. (2002) Ca^{2+}-dependent block of CREB-cbp transcription by repressor dream. EMBO J. 21, 4583–4592.

Leite, M.F., Burgstahler, A.D., and Nathanson, M.H. (2002) Ca^{2+} waves require sequential activation of inositol trisphosphate receptors and ryanodine receptors in pancreatic acini. Gastroenterology 122, 415–427.

Levitan, I.B. (1999) It is calmodulin after all! Mediator of the calcium modulation of multiple ion channels. Neuron 22, 645–648.

Marchant, J.S., and Taylor, C.W. (1997) Cooperative activation of IP3 receptors by sequential binding of IP3 and Ca^{2+} safeguards against spontaneous activity. Curr. Biol. 7, 510–518.

McAllister, A.K., Katz, L.C., and Lo, D.C. (1997) Opposing roles for endogenous bdnf and nt-3 in regulating cortical dendritic growth. Neuron 18, 767–778.

Mermelstein, P.G., Bito, H., Deisseroth, K., and Tsien, R.W. (2000) Critical dependence of camp response element-binding protein phosphorylation on L-type calcium channels supports a selective response to EPSPs in preference to action potentials. J. Neurosci. 20, 266–273.

Messaoudi, E., Bardsen, K., Srebro, B., and Bramham, C.R. (1998) Acute intrahippocampal infusion of BDNF induces lasting potentiation of synaptic transmission in the rat dentate gyrus. J. Neurophys. 79, 496–499.

Mioduszewska, B., Jaworski, J., and Kaczmarek, L. (2003) Inducible cAMP early repressor (icer) in the nervous system–a transcriptional regulator of neuronal plasticity and programmed cell death. J. Neurochem. 87, 1313–1320.

Moore, A.N., Waxham, M.N., and Dash, P.K. (1996) Neuronal activity increases the phosphorylation of the transcription factor camp response element-binding protein (CREB) in rat hippocampus and cortex. J. Biol. Chem. 271, 14214–14220.

Morikawa, H., Imani, F., Khodakhah, K., and Williams, J.T. (2000) Inositol 1,4,5-triphosphate-evoked responses in midbrain dopamine neurons. J. Neurosci. 20, RC103.

Muller, W., and Conner, J.A. (1991) Dendritic spines as individual neuronal compartments for synaptic Ca^{2+} responses. Nature 354, 73–76.

Nakagawa, S., Kim, J.E., Lee, R., Chen, J., Fujioka, T., Malberg, J., Tsuji, S., and Duman, R.S. (2002) Localization of phosphorylated cAMP response element-binding protein in immature neurons of adult hippocampus. J. Neurosci. 22, 9868–9876.

Nakamura, T., Barbara, J.G., Nakamura, K., and Ross, W.N. (1999) Synergistic release of Ca^{2+} from IP3-sensitive stores evoked by synaptic activation of mGluRs paired with backpropagating action potentials. Neuron 24, 727–737.

Nakamura, T., Lasser-Ross, N., Nakamura, K., and Ross, W.N. (2002) Spatial segregation and interaction of calcium signalling mechanisms in rat hippocampal CA1 pyramidal neurons. J. Physiol. 543, 465–480.

Nakamura, T., Nakamura, K., Lasser-Ross, N., Barbara, J.G., Sandler, V.M., and Ross, W.N. (2000) Inositol 1,4,5-trisphosphate (IP3)-mediated Ca^{2+} release evoked by metabotropic agonists and backpropagating action potentials in hippocampal CA1 pyramidal neurons. J. Neurosci. 20, 8365–8376.

Nguyen, P.V., Abel, T., and Kandel, E.R. (1994) Requirement of a critical period of transcription for induction of a late phase of LTP. Science 265, 1104–1107.

Patterson, S.L., Pittenger, C., Morozov, A., Martin, K.C., Scanlin, H., Drake, C., and Kandel, E.R. (2001) Some forms of cAMP-mediated long-lasting potentiation are associated with release of BDNF and nuclear translocation of phospho-MAP kinase. Neuron 32, 123–140.

Petralia, R.S., Sans, N., Wang, Y.X., and Wenthold, R.J. (2005) Ontogeny of postsynaptic density proteins at glutamatergic synapses. Mol. Cell. Neurosci. 29, 436–452.

Power, J.M., and Sah, P. (2002) Nuclear calcium signaling evoked by cholinergic stimulation in hippocampal CA1 pyramidal neurons. J. Neurosci. 22, 3454–3462.

Pozzo Miller, L.D., Petrozzino, J.J., Golarai, G., and Connor, J.A. (1996) Ca^{2+} release from intracellular stores induced by afferent stimulation of CA3 pyramidal neurons in hippocampal slices. J. Neurophys. 76, 554–562.

Pyapali, G.K., Sik, A., Penttonen, M., Buzsaki, G., and Turner, D.A. (1998) Dendritic properties of hippocampal CA1 pyramidal neurons in the rat: Intracellular staining in vivo and in vitro. J. Comp. Neurol. 391, 335–352.

Redmond, L., and Ghosh, A. (2005) Regulation of dendritic development by calcium signaling. Cell Calcium 37, 411–416.

Redmond, L., Kashani, A.H., and Ghosh, A. (2002) Calcium regulation of dendritic growth via CAM kinase iv and CREB-mediated transcription. Neuron 34, 999–1010.

Riccio, A., Ahn, S., Davenport, C.M., Blendy, J.A., and Ginty, D.D. (1999) Mediation by a CREB family transcription factor of NGF-dependent survival of sympathetic neurons. Science 286, 2358–2361.

Rivera, C., Voipio, J., Payne, J.A., Ruusuvuori, E., Lahtinen, H., Lamsa, K., Pirvola, U., Saarma, M., and Kaila, K. (1999) The K^+/Cl^- co-transporter KCC2 renders GABA hyperpolarizing during neuronal maturation. Nature 397, 251–255.

Sabatini, B.L., Maravall, M., and Svoboda, K. (2001) Ca^{2+} signaling in dendritic spines. Curr. Opin. Neurobiol. 11, 349–356.

Sala, C., Rudolph-Correia, S., and Sheng, M. (2000) Developmentally regulated nmda receptor-dependent dephosphorylation of camp response element-binding protein (CREB) in hippocampal neurons. J. Neurosci. 20, 3529–3536.

Sanyal, S., Sandstrom, D.J., Hoeffer, C.A., and Ramaswami, M. (2002) Ap-1 functions upstream of CREB to control synaptic plasticity in drosophila. Nature 416, 870–874.

Sharp, A.H., McPherson, P.S., Dawson, T.M., Aoki, C., Campbell, K.P., and Snyder, S.H. (1993) Differential immunohistochemical localization of inositol 1,4,5- trisphosphate- and ryanodine-sensitive Ca^{2+} release channels in rat brain. J. Neurosci. 13, 3051–3063.

Sharp, A.H., Nucifora, F.C., Jr., Blondel, O., Sheppard, C.A., Zhang, C., Snyder, S.H., Russell, J.T., Ryugo, D.K., and Ross, C.A. (1999) Differential cellular expression of isoforms of inositol 1,4,5-triphosphate receptors in neurons and glia in brain. J. Comp. Neurol. 406, 207–220.

Shaywitz, A.J., and Greenberg, M.E. (1999) CREB: A stimulus-induced transcription factor activated by a diverse array of extracellular signals. Annu. Rev. Biochem. 68, 821–861.

Sheng, M., Thompson, M.A., and Greenberg, M.E. (1991) CREB: A Ca^{2+}-regulated transcription factor phosphorylated by calmodulin-dependent kinases. Science 252, 1427–1430.

Silva, A.J., Kogan, J.H., Frankland, P.W., and Kida, S. (1998) CREB and memory. Annu. Rev. Neurosci. 21, 127–148.

Svoboda, K., and Mainen, Z.F. (1999) Synaptic $[Ca^{2+}]$: Intracellular stores spill their guts. Neuron 22, 427–430.

Turner, D.A., Buhl, E.H., Hailer, N.P., and Nitsch, R. (1998) Morphological features of the entorhinal-hippocampal connection. Prog. Neurobiol. 55, 537–562.

Verkhratsky, A., and Petersen, O.H. (2002) The endoplasmic reticulum as an integrating signalling organelle: From neuronal signalling to neuronal death. Eur. J. Pharmacol. 447, 141–154.

Watanabe, S., Hong, M., Lasser-Ross, N., and Ross, W.N. (2006) Modulation of calcium wave propagation in the dendrites and to the soma of rat hippocampal pyramidal neurons. J. Physiol. 575, 455–468.

Wells, D.G., and Fallon, J.R. (2000) Dendritic mRNA translation: Deciphering the uncoded. Nat. Neurosci. 3, 1062–1064.

West, A.E., Chen, W.G., Dalva, M.B., Dolmetsch, R.E., Kornhauser, J.M., Shaywitz, A.J., Takasu, M.A., Tao, X., and Greenberg, M.E. (2001) Calcium regulation of neuronal gene expression. Proc. Natl. Acad. Sci. USA 98, 11024–11031.

Wickens, J. (1988) Electrically coupled but chemically isolated synapses: Dendritic spines and calcium in a rule for synaptic modification. Prof. Neurobiol. 31, 507–528.

Wu, G.Y., Deisseroth, K., and Tsien, R.W. (2001) Activity-dependent CREB phosphorylation: Convergence of a fast, sensitive calmodulin kinase pathway and a slow, less sensitive mitogen-activated protein kinase pathway. Proc. Natl. Acad. Sci. USA 98, 2808–2813.

Yeckel, M.F., Kapur, A., and Johnston, D. (1999) Multiple forms of LTP in hippocampal CA3 neurons use a common postsynaptic mechanism. Nat. Neurosci. 2, 625–633.

Yuste, R., Majewska, A., and Holthoff, K. (2000) From form to function: Calcium compartmentalization in dendritic spines. Nat. Neurosci. 3, 653–659.

Zhou, S., and Ross, W.N. (2002) Threshold conditions for synaptically evoking Ca^{2+} waves in hippocampal pyramidal neurons. J. Neurophys. 87, 1799–1804.

Chapter 5
Role of Action Potentials in Regulating Gene Transcription: Relevance to LTP

J. Paige Adams, Rachel A. Robinson, and Serena M. Dudek

Abstract The late phase of Long Term Potentiation (LTP) appears to require transcription, but how the nucleus is informed remains unknown. We propose that calcium elevation from multiple action potentials serves as the signal rather than an NMDA receptor-dependent signal transported from synapses. We find that NMDA receptor antagonists interfere with action potential generation and thus do not resolve the issue. Pharmacologic restoration of action potentials in the presence of NMDA receptor antagonists shows that ERK activation, transcription factor binding, and *arc* gene expression, previously all shown or thought to be NMDA receptor dependent, are maintained. These data demonstrate that types of signaling in the nucleus, previously attributed to NMDA-receptor dependent synapse-to-nucleus signals, can be initiated by action potentials. Action potential-mediated calcium increases can provide a fast and effective signal in the nucleus that may be an important factor in LTP consolidation.

5.1 Introduction

One proposed cellular basis of memory formation and consolidation is long-term potentiation (LTP). LTP is defined as a change in the efficacy of a synapse such that the connection between two neurons is stronger; it has been intensively studied in the hippocampus, a cortical structure located within the temporal lobes. Lesion and brain imaging studies have shown the hippocampus to be involved in memory formation and consolidation (Milner, Squire, and Kandel 1998; Shapiro and Eichenbaum 1999). Like memory, LTP can be divided temporally into short and long-term phases. The late phases of both memory and LTP (late-LTP) can be blocked with inhibitors of RNA transcription and protein translation (Davis and Squire 1984; Montarolo, Goelet, Castellucci, Morgan, Kandel, and Schacher 1986; Nguyen, Abel, and Kandel 1994), but see (Otani 1989). Perhaps the strongest evidence that maintenance of LTP depends on the nucleus, however, is that LTP induced in the dendrites will only be maintained for a few hours if the dendrites are severed from their cell bodies (Frey, Krug, Brodemann, Reymann, and Matthies 1989); LTP induced in the dendrites of an intact hippocampal slice preparation can be maintained for over 8 hours. While the pharmacological data for the dependence of

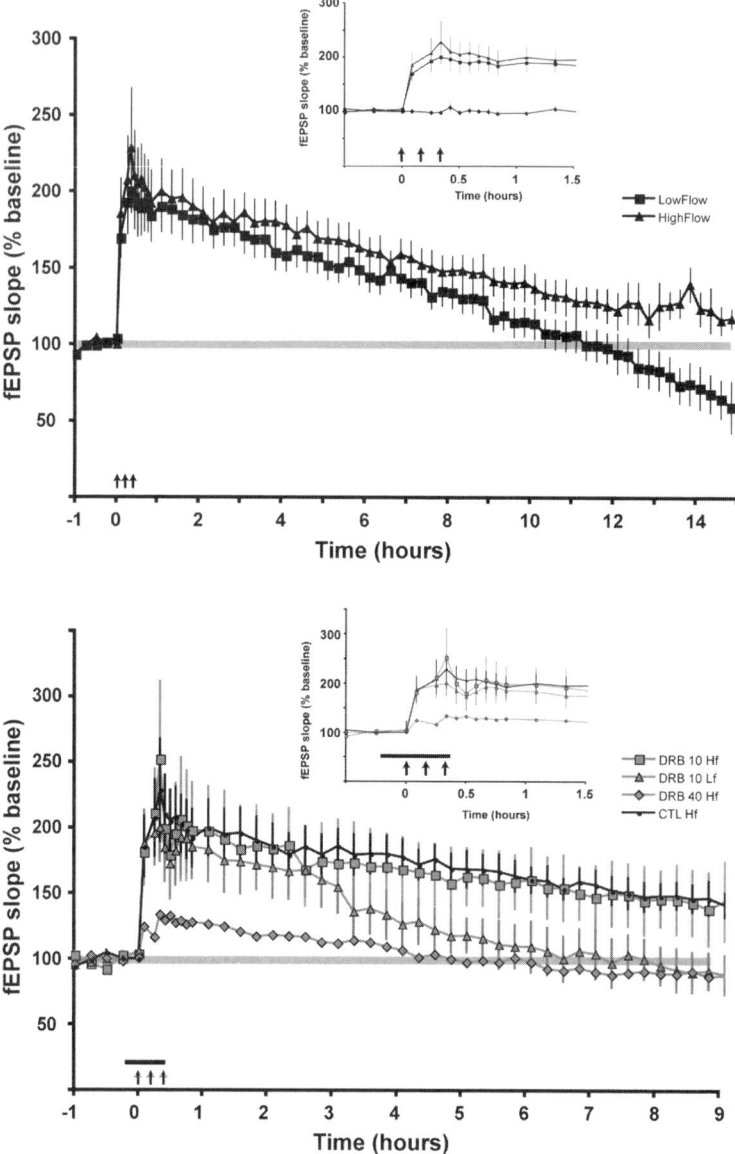

Fig. 1 Late-phase LTP is blocked by transcriptional inhibitors, depending on incubation conditions. A. Late-phase LTP is identical out to 8 hours in slices bathed with a lower flow rate (< 2.8 ml per minute) or higher flow rate (> 2.8 ml per minute). Only after 10 hours is the difference between the two significant (n= 12 and 5, respectively). B. Late-phase LTP is blocked by 10 µM DRB in low-flow conditions, but not in higher-flow conditions (n=6 and 3). LTP induction and early-LTP to 4 hours do not differ significantly between the two (see inset). LTP induced in the presence of 40 µM DRB was variable, and often the induction was impaired (average of 2 cases shown)

late-LTP on transcription do not entirely agree with each other on the duration of the early-LTP, work thus far is consistent on one point: if the drugs are applied even within a short time *after* induction, they are ineffective (Frey, Frey, Schollmeier, and Krug 1996; Nguyen et al., 1994). Therefore, the required transcriptional events induced with LTP must begin within a very short period around the time of LTP induction. Local dendritic protein synthesis most likely participates in some of the early maintenance of LTP, as the results of the new RNA synthesis are not seen for several hours and protein synthesis inhibitors can block LTP in severed dendrites (Sutton and Schuman 2006; Tsokas, Grace, Chan, Ma, Sealfon, Iyengar, Landau, and Blitzer 2005; Vickers, Dickson, and Wyllie 2005). For the purposes of this discussion, we define late-LTP as the portion of LTP lasting over eight hours *in vitro* that is sensitive to transcriptional inhibitors or severing dendrites from cell bodies (see Fig. 1). Our work indicates that the health of the hippocampal slices is an important factor in these types of experiments: apparently less healthy slices are more likely to be affected by transcriptional inhibitors, to the point where LTP induction is impaired by these drugs at high concentrations. Hence, the use of pharmacological experiments in which drugs are present during LTP induction is problematic due to the possibility that these drugs nonspecifically interfere with the induction mechanisms instead of specifically affecting transcription. We have therefore also considered the rapid induction of several genes after electrocon-vulsive shock and/or behavioral paradigms: both *arc/Arg3.1* and *zif268/EGR1* can be induced within two minutes *in vivo*, and additionally have been shown to be involved in LTP (Guzowski, Lyford, Stevenson, Houston, McGaugh, Worley, and Barnes 2000; Guzowski, McNaughton, Barnes, and Worley 1999; Jones, Errington, French, Fine, Bliss, Garel, Charnay, Bozon, Laroche, and Davis 2001; Steward and Worley 2001; Vazdarjanova, McNaughton, Barnes, Worley, and Guzowski 2002), but see (Rial Verde, Lee-Osbourne, Worley, Malinow, and Cline 2006; Shepherd, Rumbaugh, Wu, Chowdhury, Plath, Kuhl, Huganir, and Worley 2006). These data provide further evidence that transcription can occur within a short time of LTP induction, thus we have focused our efforts on signal transduction pathways leading to these rapid transcriptional events.

In summary, transcriptional mechanisms appear to be put into motion around the time of LTP induction. This consequently leads us to ask the question: how does the synapse tell the nucleus that LTP has taken place and new RNA and protein products need to be made? Because both of the models to be discussed require a synaptic tagging mechanism for use of the gene product (for a review, see Sajikumar and Frey 2004), we will leave aside the details of how the mRNA or protein products find their way to potentiated (or depressed) synapses. In this chapter we will describe evidence that action potentials are effective mediators of transcription related to LTP maintenance.

5.2 Getting to the Nucleus: Considerations and Constraints

In the absence of any knowledge on the exact gene(s) critical for the late phase of LTP, how do we determine the mechanisms employed by synapses to signal the nucleus that potentiation has occurred? Here we will discuss two largely theoretical mechanisms. The first is the widely accepted notion that a signal is physically transported from the synapse to the nucleus after LTP is induced. In this scenario a signal is created upon induction of LTP and is transported to the nucleus, where it enters and either directly affects transcription (e.g. NFκB, (Meffert, Chang, Wiltgen, Fanselow, and Baltimore 2003)) or triggers a biochemical cascade (e.g. calmodulin, for example, (Deisseroth, Heist, and Tsien 1998)), which would induce transcription. As LTP in hippocampal area CA1 is dependent upon the activation of NMDA-type ionotropic glutamate receptors (NMDARs) at the synapse, and blockade of these receptors inhibits transcription of several genes (Matsuo, Murayama, Saitoh, Sakaki, and Inokuchi 2000; Steward and Worley 2001; Worley, Bhat, Baraban, Erickson, McNaughton, and Barnes 1993), the logical conclusion has been that NMDARs induce a cascade of events that leads to the translocation of a physical entity from the synapse to the nucleus. Recently, however, we have shown that blockade of NMDA receptors has other effects in the cell, namely the inhibition of action potential generation from synaptic stimulation ((Zhao, Adams, and Dudek 2005), Fig. 2). These data indicate that a second model, that action potentials signal the nucleus directly through the resulting calcium influx, is an equally likely scenario. Thus one cannot infer that NMDA receptor blockade of transcription is due to signals transported from the synapse, unless action potentials are carefully controlled (Zhao et al., 2005, and see below). Not until the specific genes critical for LTP consolidation have been identified *and* those genes' promoters are thoroughly understood, however, will we know whether the signal to the nucleus is protein transport- or action potential-mediated. In the interim, however, we consider the following constraints.

5.2.1 Timing

As mentioned above, the time window during which critical LTP-maintenance genes are transcribed is noteworthy when determining the relevant mechanism for signaling the nucleus. Pharmacological data are found in published studies investigating the effect of transcriptional inhibitors on the late phase of LTP; all show that the transcriptional inhibitor must be applied at or before the time of LTP induction to have an effect on maintenance of L-LTP, and that the inhibitors are ineffective at later points. The important conclusions from this finding are either 1) lengthy cascades of gene expression are not required for LTP maintenance *in vitro*, or 2) transcription is *not* required for LTP maintenance at all, and the findings are secondary to non-specific drug effects on induction.

The first group to demonstrate the time dependence of transcriptional inhibitors was Nguyen, et al., (Nguyen et al., 1994), who showed that application of the transcriptional inhibitors actinomycin D (Act D) or 5,6-dichloro-1-β-D-ribofuranosyl benzimidazole (DRB) during tetanization led to a blockade of the late phase of LTP. In these experiments, LTP returned to baseline approximately 3 hours after tetanization, a time-course that was similar to the one observed by Kelleher, et al. (Kelleher, Govindarajan, Jung, Kang, and Tonegawa 2004). In a second study by

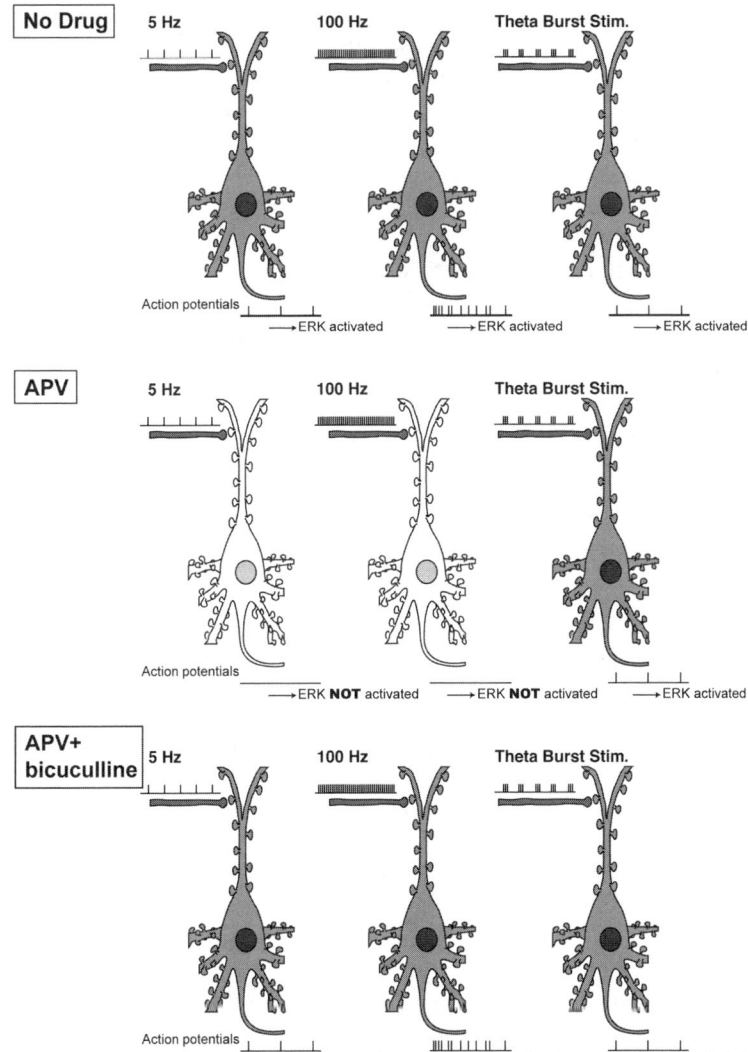

Fig. 2 Action potentials correlate with ERK activation. A. Synaptic stimulation at 5 Hz, 100 Hz, or in a theta burst pattern of stimulation (TBS), at intensities that induce action potentials, stimulate ERK phosphorylation/activation (Dudek and Fields 2001). B. Blockade of NMDA receptors with APV inhibits ERK activation by synaptic stimulation at 100 Hz (English and Sweatt 1996) and 5 Hz, but not when TBS is used (Dudek and Fields 2001). A similar pattern of sensitivity to NMDA receptor antagonists was observed when monitoring action potential generation: action potentials induced with 5 and 100 Hz were blocked by APV, while those induced with TBS were not blocked (Zhao et al., 2005). C. Bicuculline, which restores action potentials in APV, also restores ERK activation (Zhao et al., 2005). Note that when NMDARs were blocked as in (B), the entire neuron is free of stain for activated ERK, not just the dendrites

Frey et al. (Frey et al., 1996), concern about these drugs' effects on induction was circumvented with application before tetanization. Here also the transcriptional inhibitors caused a blockade of LTP's late phase, but the LTP did not show a significant decrease relative to control LTP until 6–8 hours after tetanization. This decay of LTP was similar to that observed in dendrites severed from cell bodies (Frey et al., 1989). Interestingly, similar findings were observed when long-term facilitation in *Aplysia* was tested (Montarolo et al., 1986). In our attempts to replicate the general results of this literature, we found that the narrow range of effectiveness can make these drugs difficult to work with, and that subtle differences in the functioning of the apparatus used to incubate the tissue can be important in the experimental outcomes. For instance, we found that minor deviations in the flow rate of ACSF across the tissue, despite not having a significant effect on apparent tissue health, level of LTP induction, or maintenance of late-LTP in untreated slices out to 8 hours, can lead to significant differences in levels of LTP that is maintained in slices treated with DRB (Fig. 1). Additionally, we found that small variations in drug concentration can translate to rather large differences in effect on LTP induction, which clearly precludes assessment of late LTP. One interpretation for our results is that the "healthier" slices contain slightly more of a protein signal critical for a maintaining an intermediate phase of LTP, but that ultimately new transcription is required. Evidence that such a factor could exist in limited supply has been shown under conditions of protein synthesis blockade (Fonseca, Nagerl, Morris, and Bonhoeffer 2004). Thus, the use of these drugs is complicated by the fact that sometimes the drug can interfere with the induction process and sometimes slices appear not to require new RNA synthesis. The difficulties of pharmacological studies notwithstanding, several genes are up-regulated within 2 minutes of neuronal activation (*arc*, for example (Guzowski et al., 1999)), and over 40 genes are up-regulated within 10 minutes (Matsuo et al., 2000). Therefore, if one were to accept the general premise that new RNA synthesis is required for the late phases of *in vitro* LTP, we conclude that that the critical RNA synthesis takes place shortly after induction, and not at any later point. While it is likely that protein synthesis is necessary for the maintenance of memory lasting days, it seems to require a delayed onset (after 12 hours, (Bekinschtein, Cammarota, Igaz, Bevilaqua, Izquierdo, and Medina 2007)), which cannot be assessed *in vitro*. Whether or not this later onset of protein synthesis requires transcription of new RNA is unknown.

If we proposed that a signal is transported from the synapse to the nucleus to regulate genes such as *arc* within two minutes, we would expect the protein to travel at nearly 75 microns per minute, assuming synapses 150 microns from the cell body; transport at over 30 microns per minute would be required to regulate a gene turned on within 5 minutes (Adams and Dudek 2005). While these may not be unreasonable rates for axonal transport, the fastest dendritic transport described is that of staufen, at 6 microns per minute (note that staufen moves away from the nucleus, not toward it (Kohrmann, Luo, Kaether, DesGroseillers, Dotti, and Kiebler 1999)), and a known transcription factor, NF-κB, which is able to move at only 2 microns per minute toward the nucleus in dendrites (Meffert et al., 2003). Translocation of calmodulin to the nucleus can occur rapidly within 1–2 minutes (Deisseroth et al., 1998, but see Hardingham, Arnold, and Bading 2001), but it is unknown whether the initial pool reaching the nucleus came from the cytosol in the soma or from the dendrites. Therefore we suggest that the time-consuming transport of a protein signal from a synapse in the distal dendrites to the nucleus cannot account for the rapid transcription of some genes and the finding that transcriptional inhibitors are ineffective if applied *after* LTP induction.

Which mechanisms could be fast enough to signal the nucleus? We propose that action potential-induced elevations in calcium (see below) or a wave of calcium-induced calcium release from intracellular stores (Yeckel, this volume) are sufficiently fast enough to regulate the rapid transcription of genes. Furthermore, large elevations of calcium levels, such as those produced after multiple action potentials, also avoid concerns regarding the stoichiometric constraints, described below.

5.2.2 Stoichiometric Constraints

A second consideration for rapidly regulating transcription in response to synaptic activity is the stoichiometry of the critical factors involved. That is, how many molecules of a transcription factor or other signal from a synapse are required to produce a transcriptional response in the nucleus? First, one must consider the number of molecules in a single dendritic spine that are activated/released/modified upon LTP induction. If we consider that dendritic spines' volumes range from 0.004–$0.6\,\mu m^3$ (we consider an average spine volume of $0.25\,\mu m^3$, or $0.25\,fL$) and compare that to the volume of an average-sized nucleus ($500\,\mu m^3$ or $0.5\,pL$), we calculate that the molecules would face a nearly 2000-fold dilution upon entering the nucleus, even if all of the protein makes it to the nucleus undiluted (Adams and Dudek 2005). Recent analyses of transcriptional regulation have posited that transcription factors spend a significant amount of time bound to proteins irrelevant to their target function, hence they embark on a three dimensional genome-wide search for their target (Phair, Scaffidi, Elbi, Vecerova, Dey, Ozato, Brown, Hager, Bustin, and Misteli 2004). High concentrations of the agent are required, therefore, for a transcription factor or other molecule to have a chance at regulating transcription within a reasonable amount of time. Consequently, many transcription factors and histone deacetylases are transported into or out of the nucleus when appropriate (Dolmetsch, Chawla, this volume). An exception to this large dilution faced by LTP-specific signals from synapses is when the signals localize into discrete, smaller nuclear organelles such as the Cajal bodies (see Jordan, this volume).

In order to illustrate our point, we have calculated that an average spine might contain $\sim$10 molecules of NFκB if the protein exists at $50\,nM$ ($5\times10^{-8}\,mol\,L^{-1}\times 0.25\times10^{-15}L=10^{-23}$ moles). Considering that the nuclear/cytoplasmic ratio typically increases from a ratio of 1:10 to 1:4 upon activation (Carlotti, Chapman, Dower, and Qwarnstrom 1999), we expect that hundreds of synapses must be recruited for transcription to occur; tens to hundreds of synapses would need to be recruited if the concentration within a spine were 10-fold higher ($500\,nM$). While such a number may not be unreasonable, it does suggest a requirement for cooperativity among synapses to evoke a nuclear response if one is relying on a transported synapse-to-nucleus signal. In the particular case of NFκB, one must also consider that IκB throughout the neuron will sequester free p65-p50 dimers *en route* to the nucleus if the inhibitor is not phosphorylated and subsequently degraded. It is likely, therefore, that a cell-wide regulation of IκB must occur in addition to a trigger at the synapse.

As an alternative to a signal transported from synapses, we propose a model in which action potentials generated during LTP induction lead to an increase in calcium in the cell body of the neuron. This calcium signal can then directly affect transcription by acting on transcription factors or their upstream activators. Synaptically generated action potentials would provide a large, almost *instantaneous* signal to the nucleus, with the only limiting factor being their number and frequency. In contrast to a transported synapse-to-nucleus signal, another advantage to this scenario, is that it is not necessary to postulate that a large number of synapses be potentiated: assuming the threshold number of action potentials was met to induce gene transcription, potentiation at a single "tagged" synapse could be maintained. Neuronal firing relies to some degree on the inhibitory circuitry and brain rhythms and so it is not always necessary to assume large numbers of synapses to fire post-synaptic cells or be potentiated. Thus the number and frequency of action potentials are the critical thresholds to be compared versus the number of synapses activated (cooperativity) and direction of change (i.e. LTP vs. LTD). While certain cellular processes such as kinase activation and gene transcription have been previously shown as sensitive to NMDAR blockade, we describe below evidence that NMDARs are unnecessary for the activation of kinases, increased transcription factor binding, and gene induction, provided that action potentials are maintained.

These time and stoichiometric constraints combined suggest that a transported synapse-to-nucleus signal is not an effective way for the synapse to rapidly signal the nucleus, unless completed in a cooperative manner. In our opinion, transported signals from synapses may be better suited for functions not requiring a particularly rapid transcriptional response, such as those related to the nucleus needing to sum up the number of activated synapses. Examples of this function might be found during development, when action potential generation by new synapses is relatively inefficient, and during seizures or periods of high neuronal activity. We propose that these transported signals might be ideal for inducing genes with homeostatic functions such as the following: chloride and potassium ion channels, phosphatases, and/or other function-blocking molecules like the 1a form of *homer*, which inhibits IP3-dependent calcium signaling (Yuan, Kiselyov, Shin, Chen, Shcheynikov, Kang, Dehoff, Schwarz, Seeburg, Muallem, and Worley 2003).

5.2.3 A Direct Test

As a result of the 1997 discovery of synaptic "tagging" (Frey and Morris 1997), we can directly test whether or not signals transported from synapses are required for the maintenance of LTP. Because LTP is known to be synapse-specific, meaning that non-stimulated synapses are not potentiated (Andersen, Sundberg, Sveen, and

Wigstrom 1977, but see Engert and Bonhoeffer 1997), it has been hypothesized by many that some type of "tag" temporarily marks potentiated synapses so that newly synthesized protein products are specifically directed to or captured by only those synapses where LTP was induced. Evidence for such synaptic tagging was first demonstrated in experiments where an early-LTP (that which decays after 3–4 hours, induced with a weak stimulation protocol) induced at one set of synapses was converted to late-LTP or "rescued" by a strong, late-LTP-inducing stimulation at an independent set of synapses. Both types of stimulation, in this experiment generated a tag at the respective synapses, but only the stronger stimulation was capable of inducing the gene product necessary for late-LTP. If the stimulations occurred within an hour or two of each other, the tag on the weakly-stimulated synapse was sufficient to ensure that new proteins were incorporated so as to maintain the potentiation in a manner similar to what occurred at the strongly-stimulated synapse. These data have been interpreted to mean that the gene product induced with late-LTP can be used only by the "tagged" synapses. If one were to slightly change the design of this experiment, to instead depolarize the soma without synaptic activation, action potentials alone can be assessed for their ability to substitute for synaptic late-LTP in inducing critical genes. Using stimulation of the alveus to antidromically back-fire pyramidal cell axons in a pattern that would induce late-LTP if delivered synaptically, Dudek and Fields (Dudek and Fields 2002) demonstrated that such non-synaptic activation could indeed rescue the early-LTP from decay (see also (Alarcon, Barco, and Kandel 2006)). The stimulation was in no way "stronger" than what was used to produce synaptic responses in that similar stimulation intensities were used, thus it does not likely represent an abnormally high level of postsynaptic activation. These experiments provided the first evidence that synaptic activity is not required for the protein/RNA synthesis-dependent portion of late-LTP; action potentials alone are sufficient. Action potentials delivered antidromically also induced a variety of cell signaling events that were similarly thought to require NMDA receptor activation ((Dudek and Fields 2002), discussed below). These data therefore suggest that events at the synapse can be considered independently from those in the nucleus.

5.3 ERK: A Cell-wide Player in Plasticity

The extracellular signal-regulated kinases 1 and 2 (ERK1/2) are members of the superfamily of mitogen-activated protein kinases (MAPKs) (for a review, see (Thomas and Huganir 2004). ERK is activated when the MAPK kinase MEK is phosphorylated, and consequently activated by a kinase, Raf; Raf itself can be activated by Ras, a small GTPase, through tyrosine kinase receptor activation or through calcium (Rosen, Ginty, Weber, and Greenberg 1994). ERK was first identified in mitotic cells, where it performs cellular differentiation, growth, and proliferation functions. ERK was found to be localized to post-mitotic cells (neurons) in the brain, and subsequently it was described to be identical to the tyrosine kinase regulated by various types of neuronal activity, including glutamate receptor activation, potassium-induced depolarization, and both LTP and long-term depression (LTD)-inducing stimulation (Bading and Greenberg 1991; Baron, Benes, Tan, Fagard, and Roisin 1996; English and Sweatt 1996; Thiels, Kanterewicz, Norman, Trzaskos, and Klann 2002). Additionally, ERK has been shown to be involved in both hippocampal and non-hippocampal-dependent memory formation in vertebrates and invertebrates (Adams and Sweatt 2002). ERK activity has been implicated in multiple neuronal processes including the phosphorylation and regulation of cytoskeletal elements (Quinlan and Halpain 1996), ion channels (Adams, Anderson, Varga, Dineley, Cook, Pfaffinger, and Sweatt 2000), and transcription factors, and has been shown to be required for induction of LTP (English and Sweatt 1997, but see Liu, Fukunaga, Yamamoto, Nishi, and Miyamoto 1999), and mGluR-LTD (Thiels et al., 2002). We have been particularly interested in ERK because of its exquisite sensitivity to neuronal activity and its ability to directly or indirectly impact transcription. Not so coincidentally, we find that the number of stimuli required for ERK activation is very similar to the requirement for inducing late-LTP (Dudek and Fields 2001).

5.3.1 Action Potentials Correlate with ERK Activation

Electrical stimulation of neurons induces ERK activation (Dudek and Fields 2001; English and Sweatt 1996; Thiels et al., 2002), and in most cases the ERK activation is sensitive to NMDAR antagonists. Interestingly, using immunohistochemistry on hippocampal slices with an antibody that recognizes dually phosphorylated, activated ERK1/2, Dudek and Fields (Dudek and Fields 2001) found that sensitivity of ERK activation to NMDAR antagonists depended on the *pattern* of synaptic stimulation; stimulation at 5 Hz and 100 Hz leads to ERK activation that is completely blocked with APV, but theta burst stimulation (TBS), which is based on a naturally occurring rhythm that induces LTP (Larson, Wong, and Lynch 1986), leads to ERK activation that is *independent* of NMDARs. An L-type calcium channel blocker such as nifedipine was needed in addition to block ERK activation. LTP in hippocampal area CA1 induced with TBS is dependent on post-synaptic activation of NMDAR. These data indicated that while various forms of stimulation patterns lead to ERK activation in hippocampal area CA1, some stimulation protocols such as 5 and 100 Hz induced ERK activation that is NMDAR-dependent, while others such as TBS induced ERK activation that is not dependent on NMDAR activation.

An initial interpretation of the data was that TBS is more likely to recruit voltage-dependent calcium channels than 5 or 100 Hz stimulation, as inhibition of NMDARs during TBS did not fully block postsynaptic calcium increases, and the ERK cascade was still activated. An alternative explanation was that NMDARs play a critical role in ERK activation through their effects on action potentials. Several lines of evidence led to this second hypothesis: first, Dudek and Fields had demonstrated that stimulation intensities that recruit action potentials must be used in order to activate ERK (Dudek and Fields 2001), as lower stimulation intensities were insufficient for ERK activation. Second, they presented evidence that prevention of action potential generation with a GABA receptor agonist (thus increasing overall inhibition in the circuit) also blocked ERK activation (Dudek and Fields 2001). Lastly, previous work suggested that APV, an NMDAR antagonist, inhibited action potential firing in the visual cortex and the hippocampus (Abraham and Mason 1988; Miller, Chapman, and Stryker 1989). In order to directly test if NMDAR's role in action potential generation could explain ERK's pattern of sensitivity to NMDAR antagonists, we used current clamp recordings from pyramidal neurons in hippocampal slices from juvenile rats (Zhao et al., 2005).

Consistent with a role for NMDARs in action potential generation, we found that bath application of the NMDAR antagonists APV, CPP, and MK801 all blocked action potentials induced with synaptic stimulation at 5 or 100 Hz. Postsynaptic NMDARs were found to be critical: MK801 in the patch pipette was equally as effective at blocking action potentials as bath application of the drug. The antagonists did not affect currents of another ionotropic glutamate receptor, the AMPA receptor type, thus antagonist-induced inhibition of action potentials is not merely due to shutting down the excitatory signaling in these circuits. These data demonstrate that postsynaptic NMDARs do play a critical role in action potential generation induced with 5 and 100 Hz synaptic stimulation independent of any non-specific effects on other glutamate receptors.

Will ERK activation similarly be restored if action potentials are re-established in the presence of NMDAR antagonists? Increasing the stimulus intensity overcame the blockade to only a limited degree, so we used a pharmacological approach. Using the GABA-A receptor antagonist bicuculline to block synaptic inhibition, we found that 5 and 100 Hz stimulation were able to induce action potentials even when NMDARs were blocked. Immunohistochemistry and western blotting methods allowed us to determine that *APV does not block ERK activation* when action potentials were preserved with bicuculline, which did not increase ERK activation on its own (Fig. 2). Our results indicate that action potentials induced with 5 and 100 Hz synaptic stimulation are sufficient for ERK activation, irrespective of NMDAR activation. This finding might have been predicted based on the observation that APV blocked ERK activation throughout the neurons, not just in the dendrites (Dudek and Fields 2001).

As previously stated, TBS-induced ERK activation is resistant to blockade by NMDAR antagonists (Dudek and Fields 2001). Consequently, for our idea to have any explanatory power, it was important that action potentials induced with TBS be *in*sensitive to the NMDAR antagonists. This certainly was the case: neither CPP nor APV had any effect on action potentials induced with TBS, just as the TBS-induced ERK activation was insensitive to the antagonists. As an aside, we found that if we increased the "within burst interval" (the time between each pulse in TBS), then the stimulation-induced action potentials became susceptible to the NMDAR antagonists. On the other hand, if we increased the "between-burst interval" (the time between each burst of pulses), the stimulation-induced action potentials remained resistant to blockade by NMDAR antagonists. From these data we conclude that temporal summation of AMPAR currents within the bursts of the TBS protocol allows for NMDAR-independent action potentials; without this temporal summation, as in the 5 and 100Hz stimulation protocols, action potentials require the

NMDARs to boost transmission above the threshold for action potential generation. In summary, when using synaptic stimulation, ERK is activated only when action potentials are present (stimulation with TBS, or with 5 or 100 Hz with bicuculline) (Zhao et al., 2005).

5.3.2 Action Potentials are Sufficient for ERK Activation

Action potentials appear to be required for ERK activation, but are they sufficient? To test this experimentally, Dudek and Fields used antidromically induced action potentials, which should depolarize the postsynaptic cell without need for activating synapses (Dudek and Fields 2002). These experiments showed that action potentials are clearly sufficient to signal new RNA or protein synthesis, the products of which could be used by a weakly stimulated synapse to convert an early LTP into a late LTP. Similar experiments by the authors demonstrated that action potentials are also sufficient for ERK activation throughout the cell bodies and in nuclei of the stimulated neurons. Thus, even though synaptic potentials appear to be more effective at activating L-channels (Mermelstein, Bito, Deisseroth, and Tsien 2000), action potentials are sufficient to activate ERK, as synaptic stimulation is unnecessary for activation. Further, this finding essentially rules out the requirement for neuromodulator or neurotrophin release to activate ERK upon stimulation. These data are also not surprising in light of the finding that potassium depolarization is also a very efficient way of inducing ERK phosphorylation (Baron et al., 1996).

In summary, our data illustrate that NMDARs are not required for ERK activation if action potentials are controlled by use of a GABA receptor antagonist or by TBS, a case where action potentials are NMDAR-independent due to temporal summation by AMPARs. Therefore, synapse-to-nucleus signals requiring NMDARs are not necessary for ERK activation in the nucleus. The most parsimonious explanation for these data is that ERK is activated throughout the neuron and in the nucleus by elevated calcium levels induced with action potentials. The role of L-type calcium channels has been discussed (Dolmetsch, Pajvani, Fife, Spotts, and Greenberg 2001).

5.4 Action Potential-generated Signals to the Nucleus

As mentioned above, some genes are transcribed rapidly after normal physiological learning behaviors and synaptic stimulation. For instance, Guzowski et al. (Guzowski et al., 1999; Vazdarjanova et al., 2002) demonstrated that the nuclear transcription of the immediate early genes *arc* and *zif-268* increased rapidly in hippocampal CA1 neurons after strong synaptic stimulation and exploration of a novel environment. These transcriptional increases were also observed immediately after contextual fear conditioning (Huff, Frank, Wright-Hardesty, Sprunger, Matus-Amat, Higgins, and Rudy 2006), and after hippocampal LTP induction (Link, Konietzko, Kauselmann, Krug, Schwanke, Frey, and Kuhl 1995; Rodriguez, Davies,

Silva, De Sousa, Peddie, Colyer, Lancashire, Fine, Errington, Bliss, and Stewart 2005; Steward, Wallace, Lyford, and Worley 1998). Blockade of Arc protein synthesis with antisense oligonucleotides interferes with *in vivo*-induced hippocampal LTP and learning (Guzowski et al., 2000). Moreover, knockout animals do not express late-phase LTP and have impaired learning (Plath, Ohana, Dammermann, Errington, Schmitz, Gross, Mao, Engelsberg, Mahlke, Welzl, Kobalz, Stawrakakis, Fernandes, Waltereit, Bick-Sander, Therstappen, Cooke, Blanquet, Wurst, Salmen, Bosl, Lipp, Grant, Bliss, Wolfer, and Kuhl 2006). What is relevant to this discussion is that *arc* expression has been shown to be NMDAR-dependent, as is the case for many other activity-regulated genes (Cole, Saffen, Baraban, and Worley 1989; Matsuo et al., 2000; Steward and Worley 2001). The role of *arc* in synaptic plasticity is becoming increasingly complex, however, in that the gene has now also been implicated in down-regulation of synaptic strength and/or occluding LTD (Flavell, Cowan, Kim, Greer, Lin, Paradis, Griffith, Hu, Chen, and Greenberg 2006; Rial Verde et al., 2006; Shepherd et al., 2006). Nevertheless it is clear from these studies that *arc* mRNA transcription can increase rapidly after plasticity-inducing events, and the resulting expression appears to be NMDA R-dependent.

Although ERK has been implicated in *arc* expression (Waltereit, Dammermann, Wulff, Scafidi, Staubli, Kauselmann, Bundman, and Kuhl 2001), and immunostaining for phosphorylated ERK appears in nuclei after stimulation, ERK's many functions do not necessarily allow for the conclusion that action potentials regulate transcription processes independently of NMDA Rs. Therefore, to directly investigate NMDA receptors' role in transcription, we tested for NMDA receptor antagonist inhibition of transcription factor binding to DNA and gene transcription when action potentials were controlled. In this section, we describe the results of experiments using stimulation of hippocampal CA1 mini-slices, providing evidence that neither transcription factor binding nor *arc* expression is sensitive to NMDAR antagonists when action potentials are preserved.

5.4.1 Effects on Transcription Factors

If activity-dependent genes such as *arc* are rapidly up-regulated due to LTP-inducing stimulation, then the transcription factors (TFs) controlling those genes can be studied using electrophoretic mobility shift assays (EMSAs) to aid in determining how neuronal activity modulates their binding. Earlier work in our lab used transcription factor arrays to identify transcription factors of interest from rat hippocampal nuclear extracts. Using the consensus sequences identified in the array and some of the known TF binding sites in the *arc* promoter region (Waltereit et al., 2001), we performed EMSAs on similar extracts to validate the TF array findings and test for the role of NMDARs. Experiments were conducted on slices that had been stimulated with or without an NMDAR antagonist to test if LTP-inducing stimulation could continue to activate TFs when synaptic inhibition was eliminated with biculculine. This treatment preserves action potentials during the continued NMDAR blockade (see above, section 5.3.1, Fig. 2).

When action potentials are maintained, our data show that NMDARs are not required for TF binding to AP-1, CBF, CREB, or NFκB consensus sequence oligonucleotides at five minutes post-stimulation (Fig. 3). Therefore, independent of NMDARs, synaptic stimulation can induce rapid modulation of transcription factor binding, possibly contributing to the maintenance and stability of LTP. Further studies are required to determine if either ERK or a calcium-dependent kinase (such as CAMKI, CAMKII, or CAMKIV) is critical for LTP-related nuclear signaling. Interestingly, CAMKIV is present in high concentrations in the nucleus and has been implicated in late-LTP and learning (Kang, Sun, Atkins, Soderling, Wilson, and Tonegawa 2001), while CAMKI reportedly translocates from the cytosol to the nucleus in less than one minute (Sakagami, Kamata, Nishimura, Kasahara, Owada, Takeuchi, Watanabe, Fukunaga, and Kondo 2005). Likewise, calcium-dependent production of cyclic AMP and subsequent activation of PKA could also possibly mediate transcription factor binding, as PKA activation has certainly been implicated in late-LTP and memory (Abel, Nguyen, Barad, Deuel, Kandel, and Bourtchuladze 1997; Impey, Mark, Villacres, Poser, Chavkin, and Storm 1996). PKA's speed entering the nucleus in response to neuronal activity, however, has not been assessed.

5.4.2 Effects on Gene Transcription: arc

Based on the data described in the preceding section, we sought to ascertain whether action potential-generated TF activation in fact leads to gene transcription. Using real-time PCR analysis, we determined that *arc*, an immediate early gene, could

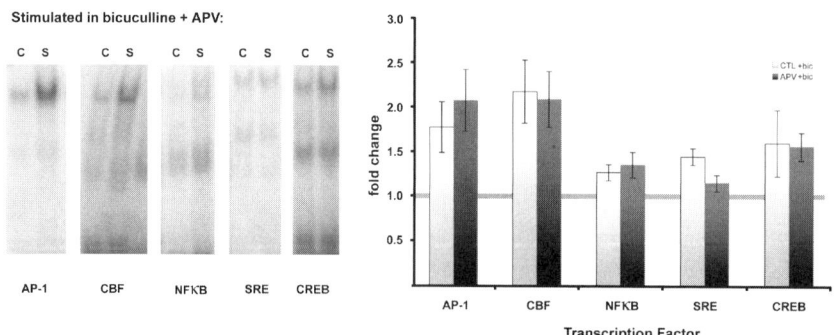

Fig. 3 Stimulation-induced transcription factor binding does not require NMDA receptors if action potentials are maintained. To insure that action potentials were maintained during NMDA receptor antagonist exposure (50 μM D-APV), synaptic stimulation was delivered in a theta burst pattern of stimulation (TBS) with bicuculline at intensities that induce action potentials and stimulate ERK phosphorylation/activation (Dudek and Fields 2001). Hippocampal CA1 mini-slices were snap-frozen 5 minutes after stimulation. The slices were pooled, homogenized, and spun to obtain a crude nuclear preparation. Nuclear proteins were extracted, incubated with ^{32}P-labled oligonucleotides, and run on a non-denaturing gel for electrophoretic mobility shift assays (EMSAs). Plotted are data from 8–11 (control) and 7–8 (APV) separate EMSAs. Binding to SRE may be affected by APV at this time point (p=0.04)

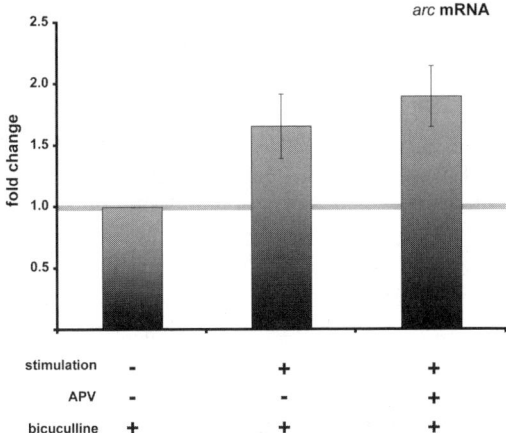

Fig. 4 Stimulation-induced *arc* expression does not require NMDA receptors if action potentials are maintained. To insure that action potentials were maintained during NMDA receptor antagonist exposure (50 µM D-APV), synaptic stimulation was delivered in a theta burst pattern of stimulation (TBS) with bicuculline (10 µM) at intensities that induce action potentials and stimulate ERK phosphorylation/activation (Dudek and Fields 2001). Hippocampal CA1 mini-slices were snap-frozen 15 minutes after stimulation. The slices were pooled, homogenized, and the mRNA purified. After cDNA was prepared from the mRNA, real-time quantitative PCR with SybrGreen was used to assess *arc* levels. Plotted are data from 4 separate determinations

be induced in the absence of NMDRs, while maintaining action potentials with the GABA-A antagonist bicuculline. We found that TBS in the presence of bicuculline led to an increase in *arc* transcription, and it was unaffected by the NMDAR antagonist APV (Fig. 4). As a result, we conclude that NMDARs are not necessary for stimulation-induced changes in *arc* transcription when action potentials are preserved. Hence, action potentials appear to be sufficient for increases in *arc* transcription. These data support our hypothesis that action potentials play a critical role in gene expression irrespective NMDAR activation for at least one immediate early gene.

5.5 Conclusions

Long-term potentiation has been a major focus of research for several decades, as we continue to search for the mechanisms of long-term memory. Our lab has focused on understanding the problem of maintenance during the later phases of LTP *in vitro*. Building upon the knowledge that new RNA is transcribed in the nucleus in response to LTP and learning, we have examined how synapses signal a need for the new RNA synthesis. For years assumptions concerning this synapse-to-nucleus signaling have been made based on experiments showing a role for NMDARs in transcription. While previously it was thought that signals such as enzymes or transcription factors must be transported from the synapse to the nucleus, we have provided evidence that action potential-generated calcium is necessary and sufficient to carry the signal in the transcription of an important plasticity-related gene. Research on late-LTP has given us information that imposes constraints on signaling models. For instance, experiments considering the effect of RNA synthesis inhibitors on the induction of some immediate early genes have defined certain time constraints for transcriptional events. In addition, the size differences between synapses and nuclei define the stoichiometric constraints on the type of signaling that is likely to occur. Our model of action potential-mediated signaling works well within both of these restrictions: calcium generated by action potentials provides a rapid, large signal that advantageously allows for the potentiation of a small number of synapses to be consolidated. In contrast, a transported signal from synapses would be slower and presumably requires the potentiation of hundreds of synapses. Transported signals offer no targeting advantage, as synaptic "tagging" after LTP is important no matter how a signal reaches the nucleus. While the threshold for nuclear response in the transported signal model necessitates that a large number of synapses be potentiated, a similar threshold using the action potential signal model is easily met with a minimum number or frequency of action potentials.

The mechanism proposed here, in which action potentials carry the signal from the synapse to the nucleus as a rise in calcium levels throughout the postsynaptic neuron, fulfills the requirements for LTP-related gene transcription. When combined with data examining how the new RNA/protein arrives at the synapse through synaptic tagging, this work presents a simplified picture about how LTP may be consolidated using gene transcription.

Acknowledgments This research was supported by the Intramural Research Program of the National Institutes of Health, National Institute of Environmental Health Sciences. Early EMSA and transcription factor array studies were performed by Eric Hudgins. Information on spine and nuclei sizes was found at http://synapse-web.org/atlas/contents.stm. We thank Dr. Ramendra Saha for his comments.

References

Abel, T., Nguyen, P.V., Barad, M., Deuel, T.A., Kandel, E.R. and Bourtchuladze, R. (1997) Genetic demonstration of a role for PKA in the late phase of LTP and in hippocampus-based long-term memory. Cell 88, 615–626.

Abraham, W.C. and Mason, S.E. (1988) Effects of the NMDA receptor/channel antagonists CPP and MK801 on hippocampal field potentials and long-term potentiation in anesthetized rats. Brain. Res. 462, 40–46.

Adams, J.P., Anderson, A.E., Varga, A.W., Dineley, K.T., Cook, R.G., Pfaffinger, P.J. and Sweatt, J.D. (2000) The A-type potassium channel Kv4.2 is a substrate for the mitogen-activated protein kinase ERK. J. Neurochem. 75, 2277–2287.

Adams, J.P. and Dudek, S.M. (2005) Late-phase long-term potentiation: Getting to the nucleus. Nat. Rev. Neurosci. 6, 737–743.

Adams, J.P. and Sweatt, J.D. (2002) Molecular psychology: Roles for the ERK MAP kinase cascade in memory. Ann. Rev. Pharm. and Tox. 42, 135–163.

Alarcon, J.M., Barco, A. and Kandel, E.R. (2006) Capture of the late phase of long-term potentiation within and across the apical and basilar dendritic compartments of CA1 pyramidal neurons: Synaptic tagging is compartment restricted. J. Neuro sci. 26, 256–264.

Andersen, P., Sundberg, S.H., Sveen, O. and Wigstrom, H. (1977) Specific long-lasting potentiation of synaptic transmission in hippocampal slices. Nature 266, 736–737.

Bading, H. and Greenberg, M.E. (1991) Stimulation of protein tyrosine phosphorylation by NMDA receptor activation. Science 253, 912–914.

Baron, C., Benes, C., Tan, H.V., Fagard, R. and Roisin, M.-P. (1996) Potassium chloride pulse enhances mitogen-activated protein kinase activity in rat hippocampal slices. J. Neurochem. 66, 1005–1010.

Bekinschtein, P., Cammarota, M., Igaz, L.M., Bevilaqua, L.R., Izquierdo, I. and Medina, J.H. (2007) Persistence of long-term memory storage requires a late protein synthesis- and BDNF-dependent phase in the hippocampus. Neuron 53, 261–277.

Carlotti, F., Chapman, R., Dower, S.K. and Qwarnstrom, E.E. (1999) Activation of nuclear factor kappaB in single living cells. Dependence of nuclear translocation and anti-apoptotic function on EGFPRELA concentration. J. Biol. Chem. 274, 37941–37949.

Cole, A.J., Saffen, D.W., Baraban, J.M. and Worley, P.F. (1989) Rapid increase of an immediate early gene messenger rna in hippocampal neurons by synaptic NMDA receptor activation. Nature 340, 474–476.

Davis, H.P. and Squire, L.R. (1984) Protein synthesis and memory: A review. Psy chol. Bull. 96, 518–559.

Deisseroth, K., Heist, E.K. and Tsien, R.W. (1998) Translocation of calmodulin to the nucleus supports CREB phosphorylation in hippocampal neurons. Nature 392, 198–202.

Dolmetsch, R.E., Pajvani, U., Fife, K., Spotts, J.M. and Greenberg, M.E. (2001) Signaling to the nucleus by an L-type calcium channel-calmodulin complex through the MAP kinase pathway. Science 294, 318–319.

Dudek, S.M. and Fields, R.D. (2001) Mitogen-activated protein kinase/extracellular signal-regulated kinase activation in somatodendritic compartments: Roles of action potentials, frequency, and mode of calcium entry. J. Neurosci. 21, RC122.

Dudek, S.M. and Fields, R.D. (2002) Somatic action potentials are sufficient for late-phase LTP-related cell signaling. Proc. Natl. Acad. Sci. USA 99, 3962–3967.

Engert, F. and Bonhoeffer, T. (1997) Synapse specificity of long-term potentiation breaks down at short distances. Nature 388, 279–284.

English, J.D. and Sweatt, J.D. (1996) Activation of p42 mitogen-activated protein kinase in hippocampal long term potentiation. J. Biol. Chem. 271, 24329–24332.

English, J.D. and Sweatt, J.D. (1997) A requirement for the mitogen-activated protein kinase cascade in hippocampal long term potentiation. J. Biol. Chem. 272, 19103–19106.

Flavell, S.W., Cowan, C.W., Kim, T.K., Greer, P.L., Lin, Y., Paradis, S., Griffith, E.C., Hu, L.S., Chen, C. and Greenberg, M.E. (2006) Activity-dependent regulation of MEF2 transcription factors suppresses excitatory synapse number. Science 311, 1008–1012.

Fonseca, R., Nagerl, U.V., Morris, R.G. and Bonhoeffer, T. (2004) Competing for memory: Hippocampal LTP under regimes of reduced protein synthesis. Neuron 44, 1011–1120.

Frey, J.U. and Morris, R.G. (1997) Synaptic tagging and long-term potentiation. Nature 385, 533–536.

Frey, U., Frey, S., Schollmeier, F. and Krug, M. (1996) Influence of actinomycin D, a RNA synthesis inhibitor, on long-term potentiation in rat hippocampal neurons in vivo and in vitro. J. Physiol. 490, 703–711.

Frey, U., Krug, M., Brodemann, R., Reymann, K. and Matthies, H. (1989) Long-term potentiation induced in dendrites separated from rat's CA1 pyramidal somata does not establish a late phase. Neurosci. Lett. 97, 135–139.

Guzowski, J.F., Lyford, G.L., Stevenson, G.D., Houston, F.P., McGaugh, J.L., Worley, P.F. and Barnes, C.A. (2000) Inhibition of activity-dependent arc protein expression in the rat hippocampus impairs the maintenance of long-term potentiation and the consolidation of long-term memory. J. Neurosci. 20, 3993–4001.

Guzowski, J.F., McNaughton, B.L., Barnes, C.A. and Worley, P.F. (1999) Environment-specific expression of the immediate-early gene arc in hippocampal neuronal ensembles. Nat. Neurosci. 2, 1120–1124.

Hardingham, G.E., Arnold, F.J.L. and Bading, H. (2001) Nuclear calcium signaling controls CREB-mediated gene expression triggered by synaptic activity. Nat. Neurosci. 4, 261–267.

Huff, N.C., Frank, M., Wright-Hardesty, K., Sprunger, D., Matus-Amat, P., Higgins, E. and Rudy, J.W. (2006) Amygdala regulation of immediate-early gene expression in the hippocampus induced by contextual fear conditioning. J. Neurosci. 26, 1616–1623.

Impey, S., Mark, M., Villacres, E.C., Poser, S., Chavkin, C. and Storm, D.R. (1996) Induction of CRE-mediated gene expression by stimuli that generate long-lasting LTP in area CA1 of the hippocampus. Neuron 16, 973–982.

Jones, M.W., Errington, M.L., French, P.J., Fine, A., Bliss, T.V., Garel, S., Charnay, P., Bozon, B., Laroche, S. and Davis, S. (2001) A requirement for the immediate early gene zif268 in the expression of late LTP and long-term memories. Nat. Neurosci. 4, 289–296.

Kang, H., Sun, L.D., Atkins, C.M., Soderling, T.R., Wilson, M.A. and Tonegawa, S. (2001) An important role of neural activity-dependent camkIV signaling in the consolidation of long-term memory. Cell 106, 771–783.

Kelleher, R.J., Govindarajan, A., Jung, H.Y., Kang, H. and Tonegawa, S. (2004) Translational control by MAPK signaling in long-term synaptic plasticity and memory. Cell 116, 467–479.

Kohrmann, M., Luo, M., Kaether, C., DesGroseillers, L., Dotti, C.G. and Kiebler, M.A. (1999) Microtubule-dependent recruitment of staufen-green fluorescent protein into large RNA-containing granules and subsequent dendritic transport in living hippocampal neurons. Mol. Biol. Cell 10, 2945–2953.

Larson, J., Wong, D. and Lynch, G. (1986) Patterned stimulation at the theta frequency is optimal for the induction of hippocampal long-term potentiation. Brain Res. 368, 347–350.

Link, W., Konietzko, U., Kauselmann, G., Krug, M., Schwanke, B., Frey, U. and Kuhl, D. (1995) Somatodendritic expression of an immediate early gene is regulated by synaptic activity. Proc. Natl. Acad. Sci. USA 92, 5734–5738.

Liu, J., Fukunaga, K., Yamamoto, H., Nishi, K. and Miyamoto, E. (1999) Differential roles of Ca(2+)/calmodulin-dependent protein kinase II and mitogen-activated protein kinase activation in hippocampal long-term potentiation. J. Neurosci. 19, 8292–8299.

Matsuo, R., Murayama, A., Saitoh, Y., Sakaki, Y. and Inokuchi, K. (2000) Identification and cataloging of genes induced by long-lasting long-term potentiation in awake rats. J. Neurochem. 74, 2239.

Meffert, M., Chang, J.M., Wiltgen, B.J., Fanselow, M.S. and Baltimore, D. (2003) NF-κB functions in synaptic signaling and behavior. Nat. Neurosci. 6, 1072–1078.

Mermelstein, P.G., Bito, H., Deisseroth, K. and Tsien, R.W. (2000) Critical dependence of camp response element-binding protein phosphorylation on L-type calcium channels supports a selective response to EPSPs in preference to action potentials. J. Neurosci. 20, 266–273.

Miller, K.D., Chapman, B. and Stryker, M.P. (1989) Visual responses in adult cat visual cortex depend on N-methyl D aspartate receptors. Proc. Natl. Acad. Sci. USA 86, 5183–5187.

Milner, B., Squire, L.R. and Kandel, E.R. (1998) Cognitive neuroscience and the study of memory. Neuron 20, 445–468.

Montarolo, P.G., Goelet, P., Castellucci, V.F., Morgan, J., Kandel, E.R. and Schacher, S. (1986) A critical period for macromolecular synthesis in long-term heterosynaptic facilitation in aplysia. Science 234, 1249–1254.

Nguyen, P.V., Abel, T. and Kandel, E.R. (1994) Requirement of a critical period of transcription for induction of a late phase of LTP. Science 265, 1104–1107.

Otani, S., Marshall, C.J, Tate, W.P, Goddard, G.V, Abraham, W.C (1989) Maintenance of long-term potentiation in rat dentate gyrus requires protein synthesis but not messenger RNA synthesis immediately post-tetanization. Neuroscience 28, 519–526.

Phair, R.D., Scaffidi, P., Elbi, C., Vecerova, J., Dey, A., Ozato, K., Brown, D.T., Hager, G., Bustin, M. and Misteli, T. (2004) Global nature of dynamic protein-chromatin interactions in vivo: Three-dimensional genome scanning and dynamic interaction networks of chromatin proteins. Mol. Cell. Biol. 24, 6393–6402.

Plath, N., Ohana, O., Dammermann, B., Errington, M.L., Schmitz, D., Gross, C., Mao, X., Engels-berg, A., Mahlke, C., Welzl, H., *et al.*, (2006) Arc/arg3.1 is essential for the consolidation of synaptic plasticity and memories. Neuron 52, 437–444.

Quinlan, E.M. and Halpain, S. (1996) Emergence of activity-dependent, bidirectional control of microtubule-associated protein MAP2 phosphorylation during postnatal development. J. Neurosci. 16, 7627–7637.

Rial Verde, E.M., Lee-Osbourne, J., Worley, P.F., Malinow, R. and Cline, H.T. (2006) Increased expression of the immediate-early gene arc/arg3.1 reduces AMPA receptor-mediated synaptic transmission. Neuron 52, 461–474.

Rodriguez, J.J., Davies, H.A., Silva, A.T., De Sousa, I.E., Peddie, C.J., Colyer, F.M., Lancashire, C.L., Fine, A., Errington, M.L., Bliss, T.V. and Stewart, M.G. (2005) Long-term potentiation in the rat dentate gyrus is associated with enhanced arc/arg3.1 protein expression in spines, dendrites and glia. Eur. J. Neurosci. 21, 2384–2396.

Rosen, L.B., Ginty, D.D., Weber, M.J. and Greenberg, M.E. (1994) Membrane depo larization and calcium influx stimulate MEK and MAP kinase via activation of Ras. Neuron 12, 1207–1221.

Sajikumar, S. and Frey, J.U. (2004) Late-associativity, synaptic tagging, and the role of dopamine during LTP and LTD. Neurobiol. Learn. Mem. 82, 12–25.

Sakagami, H., Kamata, A., Nishimura, H., Kasahara, J., Owada, Y., Takeuchi, Y., Watanabe, M., Fukunaga, K. and Kondo, H. (2005) Prominent expression and activity-dependent nuclear translocation of Ca2+/calmodulin-dependent protein kinase Idelta in hippocampal neurons. Eur. J. Neurosci. 22, 2697–2707.

Shapiro, M.L. and Eichenbaum, H. (1999) Hippocampus as a memory map: Synaptic plasticity and memory encoding by hippocampal neurons. Hippocampus 9, 365–384.

Shepherd, J.D., Rumbaugh, G., Wu, J., Chowdhury, S., Plath, N., Kuhl, D., Huganir, R.L. and Worley, P.F. (2006) Arc/arg3.1 mediates homeostatic synaptic scaling of AMPA receptors. Neuron 52, 475–484.

Steward, O., Wallace, C.S., Lyford, G.L. and Worley, P.F. (1998) Synaptic activation causes the mRNA for the IEG arc to localize selectively near activated postsynaptic sites on dendrites. Neuron 21, 741–751.

Steward, O. and Worley, P.F. (2001) Selective targeting of newly synthesized arc mRNA to active synapses requires NMDA receptor activation. Neuron 30, 227–240.

Sutton, M.A. and Schuman, E.M. (2006) Dendritic protein synthesis, synaptic plasticity, and memory. Cell 127, 49–58.

Thiels, E., Kanterewicz, B.I., Norman, E.D., Trzaskos, J.M. and Klann, E. (2002) Long-term depression in the adult hippocampus in vivo involves activation of extracellular signal-regulated kinase and phosphorylation of Elk-1. J. Neurosci. 22, 2054–2062.

Thomas, G.M. and Huganir, R.L. (2004) MAPK cascade signalling and synaptic plasticity. Nat. Rev. Neurosci. 5, 173–183.

Tsokas, P., Grace, E.A., Chan, P., Ma, T., Sealfon, S.C., Iyengar, R., Landau, E.M. and Blitzer, R.D. (2005) Local protein synthesis mediates a rapid increase in dendritic elongation factor 1a after induction of late long term potentiation. J. Neurosci. 25, 5833–5843.

Vazdarjanova, A., McNaughton, B.L., Barnes, C.A., Worley, P.F. and Guzowski, J.F. (2002) Experience-dependent coincident expression of the effector immediate-early genes *arc* and *Homer* 1a in hippocampal and neocortical neuronal networks. J. Neurosci. 22, 10067–10071.

Vickers, C.A., Dickson, K.S. and Wyllie, D.J.A. (2005) Induction and maintenance of late-phase long-term potentiation in isolated dendrites of rat hippocampal CA1 pyramidal neurones. J. Physiol. 568, 803–813.

Waltereit, R., Dammermann, B., Wulff, P., Scafidi, J., Staubli, U., Kauselmann, G., Bundman, M. and Kuhl, D. (2001) Arg3.1/arc mRNA induction by Ca2+ and cAMP requires protein kinase A and mitogen-activated protein kinase/extracellular regulated kinase activation. J. Neurosci. 21, 5484–5493.

Worley, P.F., Bhat, R.V., Baraban, J.M., Erickson, C.A., McNaughton, B.L. and Barnes, C.A. (1993) Thresholds for synaptic activation of transcription factors in hippocampus: Correlation with long-term enhancement. J. Neurosci. 13, 4776–4786.

Yuan, J.P., Kiselyov, K., Shin, D.M., Chen, J., Shcheynikov, N., Kang, S.H., Dehoff, M.H., Schwarz, M.K., Seeburg, P.H., Muallem, S. and Worley, P.F. (2003) Homer binds TRPC family channels and is required for gating of TRPC1 by IP3 receptors. Cell 114, 777–789.

Zhao, M., Adams, J.P. and Dudek, S.M. (2005) Pattern-dependent role of NMDA receptors in action potential generation: Consequences on extracellular signal-regulated kinase activation. J. Neurosci. 25, 7032–7039.

.

Chapter 6
L-type Channel Regulation of Gene Expression

Natalia Gomez-Ospina and Ricardo Dolmetsch

Abstract Calcium-regulated transcription plays a key role in converting electrical activity at the membrane into long-lasting structural and biochemical changes in excitable cells. Although several calcium influx pathways contribute to the intracellular calcium rise that follows membrane depolarization in neurons, calcium influx through L-type calcium channels (LTCs) and NMDA receptors is particularly effective at activating gene expression. In this chapter, we review some of the experiments implicating LTCs in the induction of gene expression in response to neuronal activity and discuss some of the mechanisms that explain the dependence of activity-induced transcription on LTCs. We will focus our discussion on studies that explore the features of LTCs that allow them to activate the transcription factor CREB, and we will discuss recent studies from our group that identify the C-terminus of the LTC as a protein that regulates transcription directly in the nucleus.

6.1 L-type Calcium Channels

Voltage-gated calcium channels (VGCC) are an important route of calcium entry into neurons and are essential for converting electrical activity into biochemical events in excitable cells (Catterall, Goldin and Waxman 2005). All VGCCs have a common ability to carry calcium in response to depolarization of the membrane but they differ in their subcellular localization, biophysical properties and in their ability to regulate specific biochemical processes. VGCCs are classified into L, N, P/Q, R and T types based on their pharmacological and biophysical properties, and are composed of four protein subunits: a pore forming $\alpha 1$ subunit, and β, $\alpha 2d$ and γ subunits that modulate gating and trafficking (Tsien and Tsien 1990). Neuronal L-type channels contain one of three $\alpha 1$ subunits: CaV1.2, CaV1.3 or CaV1.4. CaV1.2 and CaV1.3 form the predominant LTCs in the brain and have been implicated in a wide variety of neuronal functions including promoting survival, increasing dendritic arborization and regulating synaptic plasticity (Galli, Meucci, Scorziello, Werge, Calissano, and Schettini 1995; Moosmang, Haider, Klugbauer, Adelsberger, Langwieser, Muller, Stiess, Marais, Schulla, and Lacinova 2005; Redman, Kashani and Ghosh 2002).

LTCs have a number of features that set them apart from other types of VGCCs. They are blocked by dihydropyridines (DHPs), are activated at relatively depolarized potentials, and have slow rates of activation and inactivation (Tsien and Tsien 1990). LTCs are localized in the cell body, dendrites and postsynaptic membranes of adult neurons, making them ideally poised to control the signal transduction pathways that are activated post-synaptically (Hell, Westenbroek, Warner, Ahlijanian, Prystay, Gilbert, Snutch, and Catterall 1993; Westenbroek, Ahlijanian and Catterall 1990). Finally, LTCs are particularly effective at activating gene expression in response to electrical activity. A key question, however, is what features of LTCs allow them to activate the signaling pathways that lead to the nucleus.

6.2 L-Type Channels, CREB and c-*fos*

The first indication that LTCs were unusually effective at activating gene expression came from the experiments by Morgan and Curran who reported that depolarizing cells with high potassium provoked an influx of calcium ions via VGCCs that led to the transcription of the immediate-early gene c-*fos* (Morgan and Curran 1986). Inhibition of LTCs with DHPs blocked the induction of c-*fos*, suggesting a role for LTCs in regulating the transcription of this gene in response to neuronal activity. Murphy and colleagues then showed that blocking and activating LTCs respectively eliminated and increased basal c-*fos* expression in spontaneously active neuronal cultures (Murphy, Worley, and Baraban 1991). This implied that LTCs play a role in the induction of c-*fos* expression in response to endogenous electrical activity. Furthermore, they found that LTCs contributed less than 20% of the synaptically-induced calcium elevation, significantly less than NMDA or kainate receptors, suggesting that the route of calcium entry rather than the absolute amplitude of the calcium rise was important for the activation of *c-fos*. These early studies showed that calcium influx through LTCs specifically activates signaling pathways that lead to the induction of *c-fos* transcription.

Dissection of the *c-fos* promoter by a number of groups identified two main calcium-regulated response elements, the calcium response element (CRE) and the serum response element (SRE) (Miranti, Ginty, Huang, Chatila, and Greenberg 1995; Sheng, Dougan, McFadden, and Greenberg 1988). The CRE binds to the transcription factor CREB and the SRE binds to serum response factor (SRF) both of which are activated by calcium influx in neurons. CREB has emerged as a major regulator of calcium signaling in the brain and has been implicated in neuronal development, survival and plasticity (Lonze and Ginty 2002). Early studies by Sheng and Greenberg first demonstrated that calcium influx through LTCs is particularly effective at activating CREB-dependent transcription (Sheng, McFadden, and Greenberg 1990). Blockers of LTCs potently block the activation of CREB reporter genes, and calcium influx through LTCs in developing cortical neurons is substantially more effective at activating CREB than equivalent calcium elevations through NMDA receptors, suggesting that LTCs are specifically linked to CREB activity (Bading, Ginty, and Greenberg 1993). The most compelling illustration of the central role of LTCs in activating CREB is a study of LTC knockout mice. Eliminating CaV1.2 specifically in the hippocampus and cortex of mice using CRE recombinase-mediated recombination resulted in a loss of CREB phosphorylation in response to electrical activity, a reduction in an LTC-dependent form of long term potentiation, and in learning deficits (Moosmang et al., 2005). This result demonstrates the importance of LTCs in activating CREB and in regulating neuronal plasticity of neurons *in vivo*.

While the mechanisms that link calcium influx through LTCs to the activation of CREB are not completely understood, a great deal is known about how intracellular calcium elevations can activate CREB-dependent transcription. Activation of CREB is a multi-step process that involves both the recruitment of CREB to CRE elements and the phosphorylation of CREB. Recent studies suggest that CREB does not constitutively occupy CRE sites and that activation of CREB involves its recruitment to CREs via a nitric oxide-dependent cascade (Riccio 2006). At the same time as CREB is recruited to CRE elements, it is also phosphorylated at Ser[133]. This phosphorylation event is strongly calcium-dependent and is absolutely required for the activation of CREB-dependent transcription (Gonzalez and Montminy 1989). Phosphorylation of Ser[133] allows CREB to recruit transcriptional co-activators such as the CREB binding protein (CBP) which itself is subject to calcium regulation via calcium-dependent phosphorylation (Impey, Fong, Wang, Cardinaux, Fass, Obrietan, Wayman, Storm, Soderling, and Goodman 2002). Other CREB co-activators such as the Transducers of Regulated CREB activity (TORCs) translocate to the nucleus is response to intracellular calcium elevations (Conkright, Canettieri, Screaton, Guzman, Miraglia, Hogenesch, and Montminy 2003; Impey et al., 2002). In addition to phosphorylation at Ser[133], CREB is also phosphorylated at several other serines including Ser[142] and Ser[143] although how these phosphorylation events regulate transcription has not been elucidated yet (Kornhauser, Cowan, Shaywitz, Dolmetsch, Griffith, Hu, Haddad, Xia, and Greenberg 2002). Thus CREB is subject to calcium-dependent regulation at many different points during its activation.

Activation of several different signaling cascades leads to the phosphorylation of CREB or of CREB associated cofactors in neurons (Fig. 1, color plate). Two of these signaling systems, the calcium Calmodulin (CaM) activated kinases CaMKIV and CamKI and the mitogen activated kinases (MAPK), seem to be particularly important for linking CREB to calcium influx through LTCs. CaMKIV and CaMKI are activated both by calcium Calmodulin binding and by phosphorylation by CaMKK. The canonical MAP kinase cascade includes Ras, Raf, MEK, ERK, and the nuclear kinases RSK1, RSK2 and MSK1, which phosphorylate CREB on Ser[133]. Phosphorylation of CREB and CREB-dependent transcription are defective in mice lacking CaMKIV (Ho, Liauw, Blaeser, Wei, Hanissian, Muglia, Wozniak, Nardi, Arvin, and Holtzman 2000) or MAP kinase MSK1 (Arthur, Fong, Dwyer, Davare, Reese, Obrietan, and Impey 2004) (Wiggin, Soloaga, Foster, Murray-Tait, Cohen, and Arthur 2002) and in cells whose CaMKI levels have been reduced using siRNAs (Wayman, Impey, Marks, Saneyoshi, Grant, Derkach, and Soderling 2006), showing that activation of these signaling molecules is important for CREB-mediated transcription. Activation of LTCs therefore leads to the activation of several signaling cascades that result in phosphorylation of CREB.

Precisely what role each of these kinases plays in regulating the activation of CREB is still a subject of controversy. It has been proposed that the kinetics of activation of each of these kinases results in specific CREB phosphorylation profiles. The CaM kinases, for instance, are activated rapidly and transiently in response to calcium influx, whereas the MAP kinase cascade is activated more slowly and is more sustained. CaMK activation therefore leads to rapid, transient CREB phosphorylation, whereas activation of MAP kinase allows CREB to remain phosphorylated for a prolonged period of time (Wu, Deisseroth, and Tsien 2001).

6.3 What Mechanisms Link LTCs to CREB?

Despite a wealth of information on the biochemical signaling pathways that regulate CREB phosphorylation in response to depolarization, the mechanisms that specifically link calcium influx through LTCs to the activation of CREB are not completely understood. It is likely that multiple features of LTCs contribute to their ability to activate CREB. At least three features of LTCs seem to be important for their ability to activate transcription: their biophysical properties, their localization in the dendrites and cell bodies of neurons, and their association with signaling proteins that activate nuclear signaling cascades.

LTCs have distinct biophysical properties that make them particularly well suited to activate CREB. CREB-dependent transcription requires sustained CREB phosphorylation that persists for at least 30 minutes (Bito, Deisseroth, and Tsien 1996; Liu and Graybiel 1996). To maintain this sustained phosphorylation, calcium levels must be elevated for prolonged periods of time. LTCs open relatively slowly and thus require sustained bursts of action potentials or continuous depolarization for maximal activity. These are the conditions that normally lead to CREB-dependent transcription (Deisseroth, Bito, and Tsien 1996; Nakazawa and Murphy 1999). Furthermore, in contrast to other types of VGCCs, LTCs inactivate slowly and incompletely and so they contribute a disproportionate amount of the calcium current under conditions of tonic electrical stimulation (Liu, Ren, and Murphy 2003). Therefore, the biophysical properties of LTCs allow them to remain active under the stimulation conditions that lead to CREB-dependent transcription.

The biophysical properties of LTCs, however, do not account entirely for the ability of these channels to activate CREB-dependent transcription. In developing cortical and hippocampal neurons, sustained calcium elevations mediated by NMDA receptors or generated by the addition of calcium ionophores are significantly less effective at activating CREB-dependent transcription than calcium influx through LTCs (Bading et al., 1993). This suggests that there are additional features of LTCs that link them to the signaling pathways that activate transcription. Another feature of LTCs that may be involved in their ability to activate transcription is their subcellular localization. LTCs are found post-synaptically in proximal dendrites and in the cell body of neurons, and therefore they are well placed to generate calcium signals that invade the nucleus (Ahlijanian, Westenbroek, and Catterall 1990;

Hell, Westenbroek, Breeze, Wang, Chavkin, and Catterall 1996; Hell et al., 1993; Westenbroek et al., 1990). Experiments using nuclear injection of BAPTA-dextran have suggested that a nuclear calcium rise is required for CREB-dependent transcription (Hardingham, Arnold, and Bading 2001). This suggests that there are calcium-binding proteins in the nucleus that are required for transcriptional activation. In fact, both CaMKK and CaMKIV are localized in the nucleus, and therefore a nuclear calcium elevation would lead to activation of these signaling proteins and phosphorylation of CREB. In addition, the activity of CBP requires nuclear activation of CaMKIV, thus providing an additional set of targets for nuclear calcium. These studies suggest that a nuclear calcium rise is important for calcium-dependent transcriptional regulation, and that LTCs are well placed to generate these signals.

Nuclear calcium elevation is generally not sufficient for CREB-dependent transcription, since it has been observed that a number of other stimuli that elevate nuclear calcium are not effective at activating transcription (Bading et al., 1993; Deisseroth, Heist, and Tsien 1998; Dolmetsch, Pajvani, Fife, Spotts, and Greenberg 2001). This has led to the idea that LTCs activate an additional set of signaling molecules important for CREB-dependent transcription. Two sets of experiments support the hypothesis that LTCs are associated with calcium-binding proteins that detect calcium elevations close to the mouth of the channel. First, loading neurons with EGTA, a slow calcium buffer that prevents calcium elevation in the cell body and nucleus but allows calcium elevations close to the membrane, does not inhibit CREB phosphorylation in response to depolarization (Deisseroth et al., 1996). On the other hand loading neurons with BAPTA, a fast calcium buffer that chelates calcium close to the mouth of the channels, blocks CREB phosphorylation. This suggests that local calcium elevations near calcium channels are sufficient to induce CREB phosphorylation. Second, mutations of LTCs that do not alter their ability to carry calcium or to elevate nuclear calcium prevents LTCs from inducing CREB phosphorylation and CREB-dependent transcription (Dolmetsch et al., 2001). Using DHP-insensitive channels to distinguish ectopically expressed channels from endogenously expressed channels, it was found that point mutations that disrupt Calmodulin binding to the LTC prevent LTC activation of CREB. These mutations did not affect the ability of the LTC to activate the CaMK signaling pathway but prevented activation of MAP kinase, suggesting that local calcium elevations around LTCs activate MAP kinase signaling that is necessary for CREB-dependent transcription. This observation has led to the idea that global calcium elevations (including elevations of nuclear calcium) are required for activation of the CaMK signaling cascade, but that full activation of CREB-dependent transcription requires the activation of MAP kinase by signaling molecules close to the mouth of LTCs.

Other structural domains of LTCs have been found to alter the channels ability to signal to CREB. The C-termini of both CaV1.2 and CaV1.3 contain PDZ binding motifs and it has been suggested that these are necessary for the ability of the channels to activate CREB. Inhibition of the interaction of Cav1.2's PDZ motif with its endogenous binding proteins by overexpression of competitive PDZ-motif peptides and by removal of the sequence from the channel attenuated

CREB phosphorylation and CRE-dependent transcription following depolarization (Weick, Groth, Isaksen, and Mermelstein 2003). In the case of CaV1.3 an association with post-synaptic density protein Shank is necessary for CREB phosphorylation in response to Cav1.3 activation (zhang, Maximov, Fu, Xu, Tang, Tkatch, Surmeier, and Bezprozvanny 2005) (Zhang, Fu, Altier, Platzer, Surmeier, and Bezprozvanny 2006). Taken together these data further provides support for the idea that calcium responses at the mouth of calcium channels are centrally important for transcriptional regulation by LTCs.

The identity of the molecules that transmit signals from the LTCs to the nucleus has not been clearly established. Two candidates have been proposed, CaM itself and elements of the Ras/MAPK signaling cascade. Because CREB activation depends on CaM-binding proteins such as CaMKK and CaMKIV, and CaM is found enriched in the vicinity of the channels (Mori, Erickson, and Yue 2004), CaM has been proposed to be message that conveys local calcium signals to the nucleus. Under some conditions, CaM translocates to the nucleus of neurons in response to increases in intracellular calcium (Deisseroth et al., 1998). However CaM is also found in high levels in the nucleus of resting neurons so CaM is unlikely to be the only signal that conveys information from LTCs to the nucleus.

LTCs also activate the MAPK pathway and this signaling cascade seems to be important for the prolonged phosphorylation of CREB that is required for CREB-dependent transcription. The importance of the LTC induced activation of the Ras/MAPK pathway is highlighted by markedly reduced MAPK activation in response to strong LTC stimuli in the hippocampus and cortex of specific L-type channel knockouts (Moosmang et al., 2005). Surprisingly, little is known about how LTCs lead to sustained MAPK activation. In mammalian cells calcium can trigger Ras activity via PYK2, a calcium regulated tyrosine kinase/scaffolding protein (Lev, Moreno, Martinez, Canoll, Peles, Musacchio, Plowman, Rudy, and Schlessinger 1995), via calcium-sensitive K-Ras (Villalonga, Lopez-Alcala, Chiloeches, Gil, Marais, Bachs, and Agell 2002) or calcium-regulated Ras-guanine nucleotide exchange factors (GEFs) including RAS-GRF (Farnsworth, Freshney, Rosen, Ghosh, and Greenberg 1995), RAS-GRP (Ebinu, Bottorff, Chan, Stang, Dunn, and Stone 1998), CAPRI (Lockyer, Kupzig, and Cullen 2001), and RASL (Liu, Iqbal, Grundke-Iqbal, Rossie, and Gong 2005). However, which and how any of these proteins is preferentially activated by LTCs remains a question.

Despite more than 20 years of study, the signaling molecules that connect LTCs to the activation of CREB-dependent transcription have not been defined. It is likely that there is a complex of proteins around LTCs that senses calcium and converts calcium elevations into activation of the MAP kinase signaling cascade (Fig. 1). This complex includes calmodulin, which binds directly to the LTC. Calmodulin has multiple effects on LTCs, and mediates both calcium-dependent inactivation and calcium-dependent potentiation of the channel in addition to connecting the channels to activation of CREB. Calmodulin therefore alters the conformation of LTCs in response to local calcium elevations, and this conformational change might activate signaling proteins bound to the channel leading to CREB-dependent transcription. Understanding the molecular mechanisms that connect calmodulin to the activation of CREB-dependent transcription is a critical question channel signaling.

6.4 CaV1.2 and CCAT

Calcium influx through LTCs selectively activate signaling pathways leading to CREB, however recent studies have identified another mechanism by which LTCs are linked to transcription. Neurons produce a C-terminal fragment of the LTC CaV1.2, that can directly regulate transcription. This transcription factor has been named CCAT for calcium channel associated transcriptional regulator (Gomez-Ospina, Tsuruta, Barreto-Chang, Hu, and Dolmetsch 2006). CCAT contains transcriptional activation and nuclear localization domains. It is localized in the nucleus of a subset of neurons where it binds to nuclear transcription factors, associates with endogenous promoters and regulates the expression of endogenous genes. The nuclear localization of CCAT is regulated both developmentally and by changes in intracellular calcium, and low levels of calcium or reducing electrical activity causes import into the nucleus whereas membrane depolarization and calcium influx lead to export from the nucleus (Fig. 1). These findings revealed a novel and unexpected role for calcium channels but many important questions remain to be addressed.

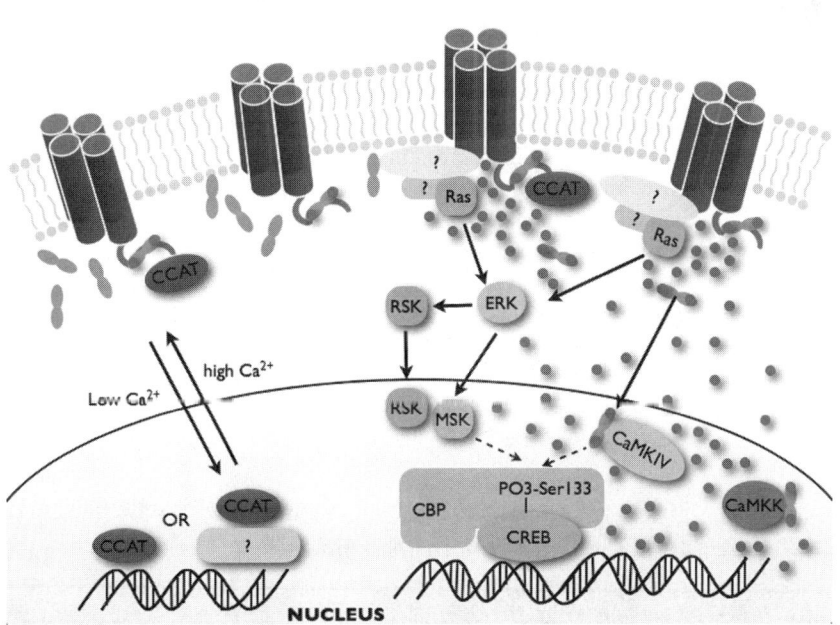

Fig. 1 LTC regulation of gene expression Calcium influx though LTCs activates signaling pathways that lead to the activation of CREB. Calcium close to the channels triggers the activation of the Ras/MAPK cascade while nuclear calcium activates nuclear Calmodulin dependent kinases. Both of these pathways are necessary for the transcriptional activation of CREB in response to neuronal depolarization. Another pathway by which LTCs regulate transcription is through CCAT, a C-terminal protein derived from LTCs. CCAT moves in and out of the nucleus in response to changes in intracellular calcium and can directly activate transcription of target genes (*See* Color Plate 4).

6.4.1 How is CCAT generated?

There are two possible mechanisms for the generation of CCAT in neurons. The first idea is that LTCs are cleaved at their C-terminus by an intracellular protease yielding a truncated channel and CCAT, the C-terminal peptide. In support of this hypothesis, it has been found that LTCs exist in two forms: a long full length form and a short form that lacks the C-terminal domain (De Jongh, Colvin, Wang, and Catterall 1994; Gerhardstein, Gao, Bunemann, Puri, Adair, Ma, and Hosey 2000; Hell et al., 1996). Treatment of hippocampal slices with NMDA causes a change in the mobility of LTCs that appears to correspond to cleavage of the channel and is attenuated by blockers of the calcium-regulated protease calpain (Hell et al., 1996). This suggests that CCAT might be generated by an activity-regulated process under some conditions. One difficulty with this idea is that in cultured neurons treatment with glutamate, NMDA or potassium chloride does not cause an increase in the amount of CCAT. In these cells, CCAT appears to be constitutively produced only in a small subpopulation of cells suggesting a separate pathway for CCAT synthesis. This has led to the idea that perhaps CCAT is produced by an alternative transcript from the CACNAIC gene that encodes only the C-terminus of CaV1.2 or by alternative translation initiation from the CaV1.2 mRNA. In fact, several putative mRNAs that encode only CCAT and not the full length CaV1.2 have been identified in genome-wide scans for mRNAs suggesting that this is a plausible mechanism for the generation of CCAT in the CNS. Other genes such as the p53 and p63 produce several alternative mRNAs driven by separate promoters (Murray-Zmijewski, Lane, and Bourdon 2006) that encode for proteins with different functions though this would be the first example of this occurring for proteins as dissimilar as a transcription factor and anion channel. Alternative translation initiation downstream of internal ribosome entry sequences (IRES) has also been described for several mammalian mRNAs and is also associated a variety of viral genes. While there is no data supporting either of these two possibilities, CCAT is found enriched in the nucleus of a very limited set of neurons in the developing nervous system and could thus be produced by a transcriptional or translational mechanism independent of channel cleavage.

6.4.2 How does CCAT activate transcription?

While depolarization does not increase the production of CCAT in neurons in culture it does have a profound effect on CCAT mediated transcription. Elevation of intracellular calcium triggers a decrease in nuclear CCAT and a decline in CCAT-dependent transcription. The activity-associated decline in nuclear CCAT arises from export and not from degradation of CCAT because there is a concomitant increase in cytoplasmic CCAT associated with depolarization and because the nuclear decrease in CCAT is not blocked by inhibitors of the proteosome. Interestingly microarray analysis of genes regulated by CCAT suggested that it suppresses at least as many genes as it activates. This suggests that export of CCAT from the nucleus of neurons could lead to both transcriptional activation and to transcriptional repression of different genes.

CCAT clearly regulates the transcription of genes when it is in the nucleus, however little is known about how this occurs. A region of 150 base pairs in the 3' region of the CX31.1 promoter is strongly responsive to nuclear CCAT levels, however a specific element with this region that binds to CCAT has not been identified. Chromatin immunoprecipitation reveals that CCAT binds to the Cx31.1 promoter region but it is not clear whether this involves a direct interaction with DNA or indirect binding to via other proteins. One protein that interacts with CCAT in the nucleus is the p54/NonO, a protein that contains both DNA and RNA binding motifs and that could link CCAT to transcriptional regulation (Wu, Yoo, Okuhama, Tucker, Liu, and Guan 2006). It is not clear if p54/NonO is absolutely required for transcription mediated by CCAT in neurons. Thus CCAT could bind to DNA directly or it might bind to a complex of proteins that recruits it to promoters in cells.

6.4.3 What is the physiological role of CCAT in neurons?

The activity of LTCs has been implicated in many neuronal processes including regulation survival, development, and dendritic morphology. All of these processes rely on transcriptional activity regulated by LTCs so CCAT expression could regulate multiple aspects of neuronal function. CCAT expression in cerebellar granule cells has a dramatic effect on neurite length. There are many possible mechanisms for this effect of CCAT on neuronal morphology. The observation that CCAT upregulates Cx31.1, formin, claudin 19, procolagen type XI, and a α-catenin-like protein suggests that it might promote the formation of adhesion complexes or junctional contacts between neurons and the extracellular matrix. Alternatively, since CCAT increases the production of Netrin4 and of two chemokines that regulate axonal and dendritic growth, it could lead to increases in neurite length via these mechanisms (Adler and Rogers 2005) (Barallobre, Pascual, Del Rio, and Soriano 2005). Finally, by down-regulating a potassium channel and a sodium-calcium exchanger, CCAT could increase the excitability of neurons and thus regulate their morphology indirectly. Understanding how CCAT modulates dendritic length might help uncover the mechanisms by which L-type calcium channels regulate neuronal morphology.

Our studies have identified many interesting genes regulated by CCAT, and these genes offer more clues to understanding CCAT's physiologic function. CCAT regulates the expression of several gap-junction proteins, a glutamate receptor, several potassium channels, a sodium-calcium exchanger, and signaling proteins such as RGS5, Formin, and Nitric oxide synthase. One of the main targets of CCAT in the nucleus is the gap-junction protein Cx31.1. Cx31.1 is expressed in the retina (Guldenagel, Sohl, Plum, Traub, Teubner, Weiler, and Willecke 2000), in developing embryos (Davies, Barr, Jones, Zhu, and Kidder 1996), and in GABAergic striatal-output neurons of the thalamus (Venance, Glowinski, and Giaume 2004). Connexins play a key role in forming electrical connections between developing neurons and form conduits for signaling molecules that can regulate a developing tissue. The expression of Cx31.1 during development in response to changes in CCAT could thus play an important role in regulating the development of inhibitory networks and the overall excitability of the brain.

6.5 Conclusions

LTCs regulate transcription in at least two important ways. First they generate intracellular calcium signals that lead to the activation of calcium-dependent signaling molecules in neurons. Calcium-regulated signaling pathways that control transcription are activated both in a microdomain of elevated calcium close to LTCs and in the nucleus of cells. The local complex of proteins that is necessary for LTCs-specific transcriptional regulation is not known but appears to link calcium channels to the activation of the Ras-MapK pathway in cells. Second, LTCs produce a transcription factor, CCAT, which is derived from their C-terminus. Precisely how this factor is produced is not clear, but it might be a consequence of channel cleavage or of a second transcriptional or translational product in derived from the CaV1.2 gene. CCAT localization and function are regulated by electrical activity, and CCAT has the potential to affect many aspects of neuronal function by controlling the expression of gap junctions and other proteins that regulate neuronal connectivity and morphology.

References

Adler, M.W. and Rogers, T.J. (2005) Are chemokines the third major system in the brain? J. Leukoc. Biol. 78, 1204–1209.

Ahlijanian, M.K., Westenbroek, R.E. and Catterall, W.A. (1990) Subunit structure and localization of dihydropyridine-sensitive calcium channels in mammalian brain, spinal cord, and retina. Neuron 4, 819–832.

Arthur, J.S., Fong, A.L., Dwyer, J.M., Davare, M., Reese, E., Obrietan, K. and Impey, S. (2004) Mitogen- and stress-activated protein kinase 1 mediates cAMP response element-binding protein phosphorylation and activation by neurotrophins. J. Neurosci. 24, 4324–4332.

Bading, H., Ginty, D.D. and Greenberg, M.E. (1993) Regulation of gene expression in hippocampal neurons by distinct calcium signaling pathways. Science 260, 181–186.

Barallobre, M.J., Pascual, M., Del Rio, J.A. and Soriano, E. (2005) The netrin family of guidance factors: Emphasis on netrin-1 signalling. Brain Res. Behav. Brain Res. 49, 22–47.

Bito, H., Deisseroth, K. and Tsien, R.W. (1996) CREB phosphorylation and dephosphorylation: A Ca2(+)- and stimulus duration-dependent switch for hippocampal gene expression. Cell 87, 1203–1214.

Catterall, W.A., Goldin, A.L. and Waxman, S.G. (2005) International union of pharmacology. XlVIII. Nomenclature and structure-function relationships of voltage-gated sodium channels. Pharmacol. Rev. 57, 397–409.

Conkright, M.D., Canettieri, G., Screaton, R., Guzman, E., Miraglia, L., Hogenesch, J.B. and Montminy, M. (2003) TORCS: Transducers of regulated CREB activity. Mol. Cell. 12, 413–423.

Davies, T.C., Barr, K.J., Jones, D.H., Zhu, D. and Kidder, G.M. (1996) Multiple members of the connexin gene family participate in preimplantation development of the mouse. Dev. Genet. 18, 234–243.

De Jongh, K.S., Colvin, A.A., Wang, K.K. and Catterall, W.A. (1994) Differential proteolysis of the full-length form of the L-type calcium channel alpha 1 subunit by calpain. J. Neurochem. 63, 1558–1564.

Deisseroth, K., Bito, H. and Tsien, R.W. (1996) Signaling from synapse to nucleus: postsynaptic CREB phosphorylation during multiple forms of hippocampal synaptic plasticity. Neuron 16, 89–101.

Deisseroth, K., Heist, E.K. and Tsien, R.W. (1998) Translocation of calmodulin to the nucleus supports CREB phosphorylation in hippocampal neurons. Nature 392, 198–202.

Dolmetsch, R.E., Pajvani, U., Fife, K., Spotts, J.M. and Greenberg, M.E. (2001) Signaling to the nucleus by an L-type calcium channel-calmodulin complex through the MAP kinase pathway. Science 294, 318–319.

Ebinu, J.O., Bottorff, D.A., Chan, E.Y., Stang, S.L., Dunn, R.J. and Stone, J.C. (1998) RasGRP, a Ras guanyl nucleotide- releasing protein with calcium- and diacylglycerol- binding motifs. Science 280, 1082–1086.

Farnsworth, C.L., Freshney, N.W., Rosen, L.B., Ghosh, A. and Greenberg, M.E. (1995) Calcium activation of Ras mediated by neuronal exchange factor Ras-GRF. Nature 376, 524–527.

Galli, C., Meucci, O., Scorziello, A., Werge, T.M., Calissano, P. and Schettini, G. (1995) Apoptosis in cerebellar granule cells is blocked by high KCl, forskolin, and IGF-1 through distinct mechanisms of action: The involvement of intracellular calcium and RNA synthesis. J. Neurosci. 15, 1172–1179.

Gerhardstein, B.L., Gao, T., Bunemann, M., Puri, T.S., Adair, A., Ma, H. and Hosey, M.M. (2000) Proteolytic processing of the C terminus of the alpha(1c) subunit of L-type calcium channels and the role of a proline-rich domain in membrane tethering of proteolytic fragments. J. Biol. Chem. 275, 8556–8563.

Gomez-Ospina, N., Tsuruta, F., Barreto-Chang, O., Hu, L. and Dolmetsch, R.E. (2006) The c terminus of the L-type voltage-gated calcium channel Ca(v)1.2 encodes a transcription factor. Cell 127, 591–606.

Gonzalez, G.A. and Montminy, M.R. (1989) Cyclic-amp stimulates somatostatin gene- transcription by phosphorylation of CREB at serine-133. Cell 59, 675–680.

Guldenagel, M., Sohl, G., Plum, A., Traub, O., Teubner, B., Weiler, R. and Willecke, K. (2000) Expression patterns of connexin genes in mouse retina. J. Comp. Neurol. 425, 193–201.

Hardingham, G.E., Arnold, F.J.L. and Bading, H. (2001) Nuclear calcium signaling controls CREB-mediated gene expression triggered by synaptic activity. Nat. Neurosci. 4, 261–267.

Hell, J.W., Westenbroek, R.E., Breeze, L.J., Wang, K.K., Chavkin, C. and Catterall, W.A. (1996) N-methyl-D-aspartate receptor-induced proteolytic conversion of postsynaptic class c L-type calcium channels in hippocampal neurons. Proc. Natl. Acad. Sci. USA 93, 3362–3367.

Hell, J.W., Westenbroek, R.E., Warner, C., Ahlijanian, M.K., Prystay, W., Gilbert, M.M., Snutch, T.P. and Catterall, W.A. (1993) Identification and differential subcellular localization of the neuronal class c and class d L-type calcium channel alpha 1 subunits. J. Cell Biol. 123, 949–962.

Ho, N., Liauw, J.A., Blaeser, F., Wei, F., Hanissian, S., Muglia, L.M., Wozniak, D.F., Nardi, A., Arvin, K.L., Holtzman, D.M., et al. (2000) Impaired synaptic plasticity and cAMP response element-binding protein activation in Ca2+/calmodulin-dependent protein kinase type IV/GR-deficient mice. J. Neurosci. 20, 6459–6472.

Impey, S., Fong, A.L., Wang, Y., Cardinaux, J.-R., Fass, D.M., Obrietan, K., Wayman, G., Storm, D.R., Soderling, T.R. and Goodman, R.H. (2002) Phosphorylation of CBP mediates transcriptional activation by neural activity and CaM kinase IV. Neuron 34, 235–244.

Kornhauser, J.M., Cowan, C.W., Shaywitz, A.J., Dolmetsch, R.E., Griffith, E.C., Hu, L.S., Haddad, C., Xia, Z.G. and Greenberg, M.E. (2002) CREB transcriptional activity in neurons is regulated by multiple, calcium-specific phosphorylation events. Neuron 34, 221–233.

Lev, S., Moreno, H., Martinez, R., Canoll, P., Peles, E., Musacchio, J.M., Plowman, G.D., Rudy, B. and Schlessinger, J. (1995) Protein tyrosine kinase PYK2 involved in Ca(2+)-induced regulation of ion channel and MAP kinase functions. Nature 376, 737–745.

Liu, F., Iqbal, K., Grundke-Iqbal, I., Rossie, S. and Gong, C.X. (2005) Dephosphorylation of tau by protein phosphatase 5: Impairment in Alzheimer's disease. J. Biol. Chem. 280, 1790–1796.

Liu, F.C. and Graybiel, A.M. (1996) Spatiotemporal dynamics of CREB phosphorylation: Transient versus sustained phosphorylation in the developing striatum. Neuron 17, 1133–1144.

Liu, Z., Ren, J. and Murphy, T.H. (2003) Decoding of synaptic voltage waveforms by specific classes of recombinant high-threshold Ca(2+) channels. J. Physiol. 553, 473–488.

Lockyer, P.J., Kupzig, S. and Cullen, P.J. (2001) CAPRI regulates Ca(2+)-dependent inactivation of the Ras-MAPK pathway. Curr. Biol. 11, 981–986.

Lonze, B.E. and Ginty, D.D. (2002) Function and regulation of CREB family transcription factors in the nervous system. Neuron 35, 605–623.

Miranti, C.K., Ginty, D.D., Huang, G., Chatila, T. and Greenberg, M.E. (1995) Calcium activates serum response factor-dependent transcription by a Ras- and Elk-1- independent mechanism that involves a Ca2+/calmodulin-dependent kinase. Mol. Cell. Biol. 15, 3672–3684.

Moosmang, S., Haider, N., Klugbauer, N., Adelsberger, H., Langwieser, N., Muller, J., Stiess, M., Marais, E., Schulla, V., Lacinova, L., et al. (2005) Role of hippocampal CaV1.2 Ca2+ channels in NMDA receptor-independent synaptic plasticity and spatial memory. J. Neurosci. 25, 9883–9892.

Morgan, J.I. and Curran, T. (1986) Role of ion flux in the control of c-fos expression. Nature 322, 552–555.

Mori, M.X., Erickson, M.G. and Yue, D.T. (2004) Functional stoichiometry and local enrich ment of calmodulin interacting with Ca2+ channels. Science 304, 432–435.

Murphy, T.H., Worley, P.F. and Baraban, J.M. (1991) L-type voltage-sensitive calcium channels mediate synaptic activation of immediate early genes. Neuron 7, 625–635.

Murray-Zmijewski, F., Lane, D.P. and Bourdon, J.C. (2006) P53/p63/p73 isoforms: An orchestra of isoforms to harmonize cell differentiation and response to stress. Cell Death Differ. 13, 962–972.

Nakazawa, H. and Murphy, T.H. (1999) Activation of nuclear calcium dynamics by synaptic stimulation in cultured cortical neurons. J. Neurochem. 73, 1075–1083.

Redman, L., Kashani, A.H. and Ghosh, A. (2002) Calcium regulation of dendritic growth via CaM kinase IV and CREB-mediated transcription. Neuron 34, 999–1010.

Riccio, A. (2006) A nitric oxide signaling pathway controls CREB-mediated gene expression in neurons. Mol. Cell. 21, 283–294.

Sheng, M., Dougan, S.T., McFadden, G. and Greenberg, M.E. (1988) Calcium and growth- factor pathways of c-fos transcriptional activation require distinct upstream regulatory sequences. Mol. Cell. Biol. 8, 2787–2796.

Sheng, M., McFadden, G. and Greenberg, M.E. (1990) Membrane depolarization and calcium induce c-fos transcription via phosphorylation of transcription factor CREB. Neuron 4, 571–582.

Tsien, R.W. and Tsien, R.Y. (1990) Calcium channels, stores, and oscillations. Annu. Rev. Cell Biol. 6, 715–760.

Venance, L., Glowinski, J. and Giaume, C. (2004) Electrical and chemical transmission between striatal GABAergic output neurones in rat brain slices. J. Physiol. 559, 215–230.

Villalonga, P., Lopez-Alcala, C., Chiloeches, A., Gil, J., Marais, R., Bachs, O. and Agell, N. (2002) Calmodulin prevents activation of Ras by PKC in 3T3 fibroblasts. J. Biol. Chem. 277, 37929–37935.

Wayman, G.A., Impey, S., Marks, D., Saneyoshi, T., Grant, W.F., Derkach, V. and Soderling, T.R. (2006) Activity-dependent dendritic arborization mediated by CaM-kinase I activation and enhanced CREB-dependent transcription of Wnt-2. Neuron 50, 897–909.

Weick, J.P., Groth, R.D., Isaksen, A.L. and Mermelstein, P.G. (2003) Interactions with PDZ proteins are required for L-type calcium channels to activate cAMP response element- binding protein-dependent gene expression. J. Neurosci. 23, 3446–3456.

Westenbroek, R.E., Ahlijanian, M.K. and Catterall, W.A. (1990) Clustering of L-type Ca2+ channels at the base of major dendrites in hippocampal pyramidal neurons. Nature 347, 281–284.

Wiggin, G.R., Soloaga, A., Foster, J.M., Murray-Tait, V., Cohen, P. and Arthur, J.S. (2002) MSK1 and MSK2 are required for the mitogen- and stress-induced phosphorylation of CREB and ATF1 in fibroblasts. Mol. Cell. Biol. 22, 2871–2881.

Wu, G.Y., Deisseroth, K. and Tsien, R.W. (2001) Activity-dependent CREB phosphorylation: convergence of a fast, sensitive calmodulin kinase pathway and a slow, less sensitive mitogen-activated protein kinase pathway. Proc. Natl. Acad. Sci. USA 98, 2808–2813.

Wu, X., Yoo, Y., Okuhama, N.N., Tucker, P.W., Liu, G. and Guan, J.L. (2006) Regulation of RNA-polymerase-II-dependent transcription by N-WASP and its nuclear-binding partners. Nat. Cell Biol. 8, 756–763.

Zhang, H., Fu, Y., Altier, C., Platzer, J., Surmeier, D.J. and Bezprozvanny, I. (2006) Ca1.2 and CaV1.3 neuronal L-type calcium channels: Differential targeting and signaling to pCREB. Eur. J. Neurosci. 23, 2297–2310.

Zhang, H., Maximov, A., Fu, Y., Xu, F., Tang, T.S., Tkatch, T., Surmeier, D.J. and Bezprozvanny, I. (2005) Association of CaV1.3 L-type calcium channels with shank. J. Neurosci. 25, 1037–1049.

Part II
Activity-regulated Transcription: Gene Regulation

Chapter 7
CREB-Dependent Transcription and Synaptic Plasticity

Angel Barco, Dragana Jancic, and Eric R. Kandel

Abstract The CREB family of transcription factors are involved in controlling the transcriptional responses to a wide range of extracellular signals in neurons. In this chapter we discuss the role of the CREB pathway in synaptic plasticity. We first describe how learning-related stimuli, of different nature and intensity, can activate signaling pathways that converge on the induction of CRE-driven gene expression and how the nuclear response orchestrated by CREB can alter future synaptic activity. Second, we will discuss how CREB's control of synaptic plasticity contributes to learning, memory and other complex brain function. Finally, we will briefly outline how dysfunction of this activation pathway can lead to psychiatric and neurological disorders.

7.1 The CREB Family of Transcription Factors

The CREB family of transcription factors refers to a group of highly homologous proteins encoded by the genes *creb, crem* and *atf-1*. This family is characterized by a highly conserved basic region/leucine zipper (bZIP) domain that bind to a specific DNA sequence called cAMP-responsive-element (CRE) found in one or several copies in the promoters of many genes (Fig. 1). Although both, the CRE sequence and CREB, the prototypic member of this family of regulatory molecules, were first identified through studies investigating the regulation of the expression of the hormone somatostatin (Montminy and Bilezikjian, 1987), it was later found that CRE sites were present in the promoter of many other genes and that CREB contributed to the regulation of a variety of cellular responses (Habener et al., 1995; Johannessen et al., 2004; Mayr and Montminy, 2001). In particular, CREB has been involved in many aspects of nervous system function (Carlezon et al., 2005; Lonze and Ginty, 2002), from activity-dependent synaptic plasticity during development (Pham et al., 1999) and in the adult brain (Barco et al., 2003; Dash et al., 1991; Pittenger and Kandel, 1998), to neuronal survival (Dawson and Ginty, 2002; Walton and Dragunow, 2000).

ATF1, CREB and CREM are not the only transcription factors that bind to CRE sites. The term ATF/CREB family of transcription factors is used to refer to a broader group of bZIP transcription factors, including ATF2, ATF3 and ATF4, that

---

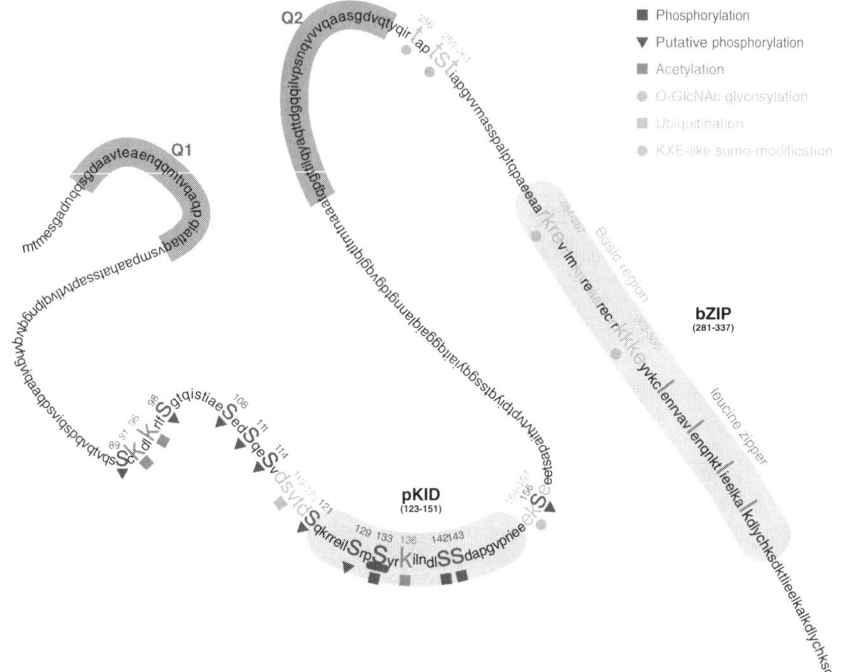

Fig. 1 CREB structure and relevant residues. CREB has a highly conserved leucine zipper and adjacent basic region responsible for DNA-binding, a regulatory kinase inducible domain (KID), and two glutamine-rich regions (Q1 and Q2). CREB is substrate of various enzymatic activities that affect its capability to bind CRE sites and activate transcription. Different symbols indicate the location of those residues relevant for CREB function discussed in the text (*See* Color Plate 5).

shares structural features with the CREB family and can also bind to CRE sites. Different ATF/CREB proteins can form selective heterodimers with each other and with other transcription factors such as AP-1 and C/EBP that do not belong to this family but share a bZIP DNA-binding domain.

Although homology in the bZIP domain is the main structural feature used to classify transcription factors belonging to the ATF/CREB family of transcription factors, other structural features are common to the several family members (Mayr and Montminy, 2001). Transcription activation in the CREB family is mediated by two types of transactivation domains: (a) The central kinase-inducible domain (KID), that contains several sites recognized by protein kinases and whose phosphorylation state determines the binding of the transcriptional co-activator CBP and that triggers the inducible transcriptional activity of CREB; and (b) the glutamine-rich domains that contribute to basal transactivation activity by interacting with the transcription machinery and stabilizing the interaction with CRE sites. In the case of CREB two glutamine-rich domains, designated Q1 and Q2, flank the KID domain (Fig. 1).

Whereas the *atf-1* gene encodes only one major protein product, the *creb* and *crem* genes have a complex structure, with multiple exons and introns (Habener et al., 1995; Hoeffler et al., 1990). The alternative splicing of CREB and CREM RNAs generates transcripts encoding both repressors and activators (Bartsch et al., 1998; Habener et al., 1995; Mayr and Montminy, 2001). The repressors are shorter variants that lack transactivation domains and form dimers with reduced or null transactivation capability, but still bind to DNA competing for CRE sites. In the case of CREM, the existence of different translation initiation codons, two alternative spliced bZIP domains and an alternative intronic CRE-driven promoter, which drive the expression of the inducible cAMP early repressor (ICER), provide additional diversity to the capability of dimerization with other bZIP proteins and binding to CRE sites (Habener et al., 1995).

CREB and ATF1 are ubiquitously expressed, whereas CREM is expressed primarily in testis and the neuroendocrine system. Since CREB is the most abundant CRE-binding protein expressed in the nervous system (Hummler et al., 1994) and a large number of studies support a critical role for CREB in controlling transcriptional responses in neurons, we will focus primarily in this transcription factor.

7.2 Regulation of CREB Activation by Synaptic Activity

CREB participates in numerous cell processes and stands at the crossroad of different signaling pathways. Synaptic activity, hormones, growth factors released during development, hypoxia and stress, among other stimuli, can trigger the phosphorylation of CREB, causing its activation and subsequent induction of CREB-dependent gene expression. Overall, more than 300 different stimuli have been reported to act through the CREB pathway (Johannessen et al., 2004).

7.2.1 Brief Overview of Signalling in the CREB Pathway

CREB binds constitutively to CRE sites present in the promoter of cAMP-responsive genes, but it is inactive. The activity of CREB-regulated promoters increases several folds when CREB is activated by phosphorylation. The activation of CREB is typically depicted as the consequence of an increase in the levels of the second messengers cAMP and Ca^{2+} (West et al., 2002). Both Ca^{2+} influx or cAMP production trigger the activation of protein kinase cascades that phosphorylates CREB (Fig. 2). The phosphorylated form of CREB (pCREB) then recruits the coactivator CREB-binding protein (CBP), which triggers transcription initiation by both bringing the RNA polymerase II complex to the promoter and acetylating histones in chromatin. This description provides a simplified view of a complex and highly regulated process. As we will discuss in the following sections, several kinase cascades can converge on CREB and a large number of proteins contribute to the activation of CREB-dependent gene expression.

7.2.2 Second Messenger Cascades Leading to CREB Phosphorylation at Ser133

The point of convergence of numerous signalling pathways is the phosphorylation of serine 133 (Ser133) on the KID domain of CREB. Similarly, Ser117 in CREM and Ser63 in ATF1 are the main phospho-acceptor sites in these proteins. Although dozens of kinases have been reported to phosphorylate these serine residues *in vitro*, only a few have been shown to contribute to the activation of CREB-dependent transcription *in vivo*. The phosphorylation of CREB *in vivo* is triggered by a wide variety of stimuli, such as an increase in cAMP after activation of G-coupled receptors, an increase of Ca^{2+} through activation of voltage- or ligand-gated channels, or the activation of receptor tyrosine kinases by growth factors (Lonze and Ginty, 2002) (Figure 2). Protein kinase A (PKA), Ca^{2+}/calmodulin-dependent (CaM) kinases and Ras/mitogen-activated protein kinase (MAPK) appear to be the most common kinase activities or cascades responsible for phosphorylating CREB at Ser133 after neuronal stimulation (Deisseroth and Tsien, 2002; Mayr and Montminy, 2001).

PKA was the first protein kinase found to target Ser133 of CREB (Gonzalez et al., 1989). PKA is a tetrameric enzyme that is regulated by intracellular levels of cAMP. This second messenger binds to the regulatory subunits of PKA causing its dissociation of the catalytic subunits, which can then catalyze the transfer of phosphate groups to serine or threonine residues of various proteins, including the

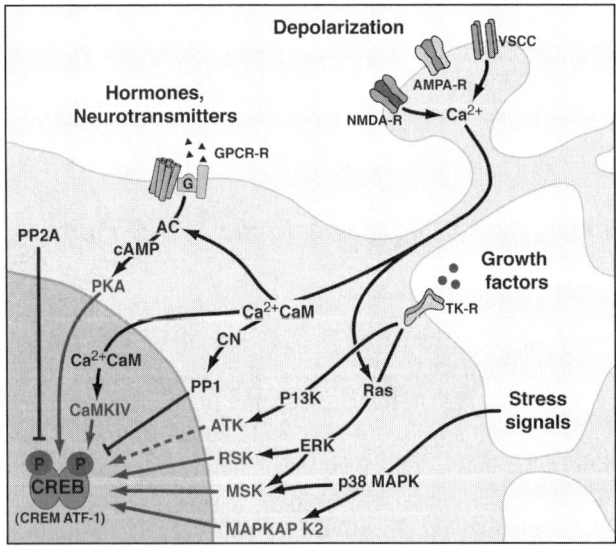

Fig. 2 Regulation of CREB activation by synaptic activity. Diverse external stimuli, such as stress, the activation of G protein-coupled receptors (GPC-R) by neurotransmitters, the activation of receptor tyrosine kinases (RTK) by neurotrophins, or the opening of Ca^{2+} channels (NMDA-R and VSCC, voltage-sensitive calcium channels) activate protein-kinase pathways that converge on the phosphorylation of CREB at Ser133 (modified from Lonze and Ginty, 2002) (*See* Color Plate 6).

major isoforms of CREB, CREM and ATF1. In neurons, the intracellular level of cAMP increases as a consequence of the activation G protein-coupled receptors (GPCRs), such as the receptors for serotonin, dopamine and other important neurotransmitters. The activation of these receptors stimulates the activity of adenylyl cyclases (AC) that converts ATP into cAMP. Some ACs also respond to increases in intracellular Ca^{2+}. Therfore, both Ca^{2+} influx and exposure to dopamine or other neurotransmitters can lead to CREB phosphorylation through PKA (Johannessen et al., 2004).

Membrane depolarization, for example during NMDA receptor mediated glutamatergic synaptic transmission, can increase the local concentration of Ca^{2+} in the postsynaptic cell by several orders of magnitude and trigger the phosphorylation of CREB (Deisseroth et al., 1996). Ca^{2+} influx occurs not only via ligand-gated cation channels, such as NMDA, but also through voltage-sensitive calcium channels (VSCC). Calmodulin (CaM), a small Ca^{2+} binding protein, transduces these changes in Ca^{2+} level to changes in CREB phosphorylation. In fact, CaM plays a critical role on both the activation and inactivation of the CREB pathway, first through its modulation of CaMKs, and second through interaction with the phosphatase calcineurin (Bito et al., 1996). The binding of Ca^{2+}/CaM causes the rapid activation of CaMKIV, a kinase highly expressed in neurons and located in the cell nucleus where it can directly and efficiently phosphorylate CREB (Deisseroth et al., 1998). Genetic deletion of CaMKIV or its inhibition by antisense oligonucleotides or specific drugs causes a significant reduction in Ser133 phosphorylation (Bito et al., 1996).

Although CaMKIV seems to be the most important CaMK regulating the phosphorylation of CREB by synaptic stimulation *in vivo*, CaMKI and CaMKII can also phosphorylate CREB (Dash et al., 1991; Deisseroth et al., 1996; Sheng et al., 1991).

The Ras/mitogen-activated protein kinase (MAPK) and the extracellular signal-regulated protein kinase (ERK) are part of a complex signalling cascade involving the activation of many other kinases, including different members of the ribosomal S6 kinase (RSK) and mitogen and stress-activated kinase (MSK) families. Both, MAPK and ERK, have been implicated in phosphorylation of CREB (Thomas and Huganir, 2004). In the case of MAPK, activation is typically triggered in response to various stressful stimuli or by the binding to receptor tyrosine kinases (RTK) of growth factors, such as the brain derived neurotrophic factor (BDNF) or the nerve growth factor (NGF). RTKs are also connected to the PI3-kinase/Akt pathway, still another signalling cascade that can control CREB phosphorylation under certain circumstances (Lin et al., 2001; Perkinton et al., 2002).

The number of protein kinases found to phosphorylate CREB *in vitro* is even larger, but the relevance of these activities in neuronal function is not clear (Johannessen et al., 2004; Lonze and Ginty, 2002). The convergence of all these signaling cascades, which respond to different stimuli and have different kinetics, allows a fine discrimination and integration of stimuli of different nature and enables the production of highly specific transcriptional responses. For example, the Ras/MAPK pathway is thought to be specifically effective in promoting a slow phase of CREB phosphorylation, whereas CaMKIV signalling is more rapid and transient

(Wu et al., 2001). Whereas stimuli that produce a modest Ca^{2+} influx trigger only the fast CaMKIV pathway, stronger stimuli can recruit, in addition, the MAPK pathway and cause a sustained increase in pCREB (Deisseroth et al., 1998; Deisseroth and Tsien, 2002). Furthermore, neurons are highly polarized and compartmentalized cells and not only the nature and intensity of the stimulus, but also the timing and cellular context of the stimulus are important. Thus, whereas synaptic activation of NMDA receptors induces CREB phosphorylation and triggers gene expression, its extrasynaptic activation, acting through divergent signaling, dephosphorylates CREB and promotes neuronal death (Hardingham et al., 2002).

7.2.3 CREB Can be Phosphorylated at Different Sites

The regulation of CREB activation is more complex than a single phosphorylation switch at Ser133. Several other serine residues are now known to be targets of kinase activities *in vitro*, including Ser89, Ser98, Ser108, Ser111, Ser114, Ser117, Ser121, Ser129, Ser142, Ser143 and Ser156 (Johannessen et al., 2004) (Fig. 1). The biological relevance of most of these variants is still unknown. The serine residues located in the KID domain contribute to the interaction with CBP and, therefore, are especially relevant. Although transfection studies in cell culture suggested that phosphorylation at Ser142 has a negative effect on CREB-dependent expression (Sun et al., 1994; Wu and McMurray, 2001) because this modification physically disrupts the interaction between CREB and CBP (Radhakrishnan et al., 1997), more recent studies have demonstrated positive regulation by phosphorylation at this site and suggest that some forms of CREB-dependent gene expression may require an alternative, CBP-independent, mechanism (Gau et al., 2002; Kornhauser et al., 2002). For effective stimulation of CREB-dependent transcription by Ca^{2+} influx is required the triple phosphorylation of CREB at Ser133, Ser142 and Ser143. The elevation of nuclear Ca^{2+} causes the phosphorylation by CaM kinases of both Ser 133 and Ser142, which, in turn, favors a the phosphorylation by casein kinase II (CKII) of Ser143 (Kornhauser et al., 2002). On the other hand, both glutamate and exposure to light induce the phosphorylation of CREB at Ser142 in the suprachiasmatic nucleus, a brain region known to control the circadian rhythm. Knock-in mice bearing a mutation that specifically blocks phosphorylation at Ser142 showed light-induced phase shifts of locomotor activity and an attenuated expression of c-Fos and mPer1, two genes known to participate in the control of circadian rhythms, suggesting that there is a positive role for phosphorylation of Ser142 in the circadian process (Gau et al., 2002).

7.2.4 CREB Can Be Dephosphorylated

To keep CREB activity precisely balanced is essential coordinated regulation of both kinases and phosphatases. The most important phosphatases known to directly dephosphorylate CREB are PP-1 and PP-2A, although PP-1 seems the most likely candidate to constraint CREB activation in neurons (Genoux et al., 2002; Jouvenceau et al., 2006). Calcineurin, also called PP-2B, is regulated by CaM and Ca^{2+} influx and functions as negative regulator of the CREB pathway by potentiating PP-1 activity through the inhibition of inhibitor-1 and DARPP-32, two negative regulators of PP-1 (Cohen, 1989). The balance of kinase and phosphatase activities modulates, among other parameters, the duration of CREB phosphorylation (Bito et al., 1996; Wu et al., 2001). Whereas weak synaptic stimulation leads to rapid dephosphorylation of pCREB via calcineurin-mediated PP-1 activation, strong stimuli provoke the inactivation of calcineurin thereby enabling an extended phosphorylation of CREB and, as a consequence, robust CREB-dependent transcription (Bito et al., 1996). Interestingly, histone deacetylase 1 (HDAC1) interacts directly with both CREB and PP1 and it is thought that the formation of HDAC1/PP1 complex may control the duration of CREB activation, as well as CREB-mediated gene expression (Canettieri et al., 2003). It has not been determined, however, whether this interaction, originally discovered in an immortalized cell line, takes also place in neurons. Furthermore, as we described for activation of CREB, there is cross-talk between different signalling cascades and, at least in non neuronal tissues, other phosphatases, such as the phosphatase and tensin homologue deleted on chromosome 10 phosphatase (PTEN) and the tyrosine phosphatase 1B (PTP1B), can play indirect roles in regulating the activation of the CREB pathway (Gum et al., 2003; Huang et al., 2001).

7.2.5 The CREB Pathway Can Be Modulated by Acetylation and Other Posttranslational Modifications

The phosphorylation of the KID domain, although necessary, is not sufficient to trigger CREB-dependent transcription. As mentioned above, the recruitment of the co-activators, CREB-binding protein (CBP) and p300, is critical for successful CREB activation (Chrivia et al., 1993). CBP and p300 are histone acetyltransferases (HATs), but, in addition, they also acetylate other proteins including CREB itself (Chan and La Thangue, 2001). Thus, CBP acetylates CREB at 3 lysine residues (at Lys-91, Lys-96, and Lys-136) within the activation domain (Lu et al., 2003). Interestingly, the location of Lys-136, proximal to Ser133, is conserved in other transcription factors of the CREB family. Although it has been postulated that acetylation may diminish phosphatase-dependent attenuation of CREB activity, the role of this posttranslational modification remains unclear (Lu et al., 2003).

CREB is also substrate of other enzymatic activities: ubiquitination and SUMOylation of CREB have been observed after hypoxia (Comerford et al., 2003) and O-GlcNAc glycosylation, at specific residues, has been found to be associated with reduced CREB-mediated transcription in cell culture (Lamarre-Vincent and Hsieh-Wilson, 2003). The relevance of these modifications in brain function is not known.

7.2.6 Regulation of the CREB Pathway Downstream of CREB

The induction of CRE-driven gene expression requires the interaction of the phosphorylated KID domain of CREB with the KIX domain of its co-activators CBP and p300, which are themselves targets of several posttranscriptional modifications. The ability of CBP to recruit specific transcription factors increases selectively after phosphorylation by growth factor–dependent signaling pathways at Ser436 (Zanger et al., 2001). CBP is also phosphorylated at Ser301 by CaM kinases (Impey et al., 2002) and is methylated at Arg residues by the methylase CARM1 (Xu et al., 2001). Whereas phosphorylation contributes to CBP-mediated transcription, methylation interferes with CREB interaction and prevents CREB-dependent gene expression. The activation of the PKA, CaMKIV and p42/44 MAPK pathways also increase the transactivation capability of CBP (Hu et al., 1999; Liu et al., 1999).

CBP and p300 are not exclusive coactivators of CREB. They interact with many other transcription factors and nuclear proteins (Chan and La Thangue, 2001). As a consequence, competition for occupancy of the KIX domain represents an additional mechanism of control in the CREB pathway. Our understanding of the significance of these regulatory steps in CREB-dependent gene expression *in vivo* is still limited, but regulation of CREB co-activators undoubtedly provides additional mechanisms of control for stimulus-induced gene expression.

7.2.7 CRE Availability and Competition for Binding Sites

In response to activation of CREB, different cell types express different sets of target genes. The availability of different CRE sites for binding is likely regulated by epigenetic changes in the chromatin, such as DNA methylation. Even with a open chromatin configuration, the availability of CRE sites also depends on the particular balance between inhibitor and activator forms of CRE-binding proteins expressed in that particular cell under those condition. Whereas some CRE sites can be occupied by CREB under basal conditions, as traditionally proposed, others may require the elimination of repressor constrains and would be occupied only after cAMP stimulation (Cha-Molstad et al., 2004; Guan et al., 2002).

The complex transcriptional regulation and alternative splicing of the *creb* and *crem* genes provide additional mechanisms for tissue or cell type-specific regulation of CREB-dependent gene expression. The expression of different repressor isoforms of CREM, such as ICER and the CREMα, β and γ variants, can be restricted to specific brain areas and play a pivotal role in the regulation of CREB-mediated gene expression (Mellstrom et al., 1993; Mioduszewska et al., 2003). Even non-phosphorylated CREB can contribute to the regulation of transcriptional responses by competing with other transcription factors for DNA binding sites. Similar balances of repressor and activators forms of CRE-binding proteins have been described in invertebrates. In *Aplysia*, the competition between the activator ApCREB1a and the repressors ApCREB2 and ApCREB1b controls the expression of CRE-driven genes and neuronal responses (Bartsch et al., 1998; Bartsch et al., 1995; Guan et al., 2002). In *Drosophila*, the nomenclature is different and the activator dCREB2 competes with the repressors dCREB1 to drive CRE-mediated gene expression (Yin et al., 1995; Yin et al., 1994).

There are, therefore, many mechanisms by which CRE-driven transcription is modulated. We have tried to present here a comprehensible although non-exhaustive overview of this highly regulated process. The complexity of the network of enzymatic activities and transcription factors that regulates the activation and shut-off of the CREB pathway emphasizes the physiological relevance of the transcriptional responses orchestrated by this family of transcription factors.

7.3 Regulation of Synaptic Function by CREB

Learning and memory storage are thought to depend on long-lasting changes in the strength of synaptic connections (Kandel, 2001). These stable changes may last hours, days or even years, and are known to depend on *de novo* gene expression. This is the case of the late phase of long-term potentiation (LTP) in the mammalian hippocampus or long-term facilitation (LTF) in *Aplysia* sensory-neurons, two processes that are blocked by inhibitors of transcription and that have provided so far the most compelling view of how the induction of gene expression can modify synaptic function (Barco et al., 2006). The CREB pathway has been identified as a major regulator of both processes. Although we will focus in hippocampal synaptic plasticity, CREB function has been related to the regulation of synaptic plasticity not only in the hippocampus, but also in the amygdala, basal ganglia and various regions of the neocortex and cerebellum.

7.3.1 Brief Overview of the Control of Synaptic Function by CREB

CREB is located in the cell nucleus where it regulates transcription. CREB is, therefore, not directly involved in synaptic function, although it can play a critical role in synapse formation and in synaptic transmission by driving the expression of synaptic proteins (Fig. 3). The electrical or chemical stimulation of a neuron can induce or enhance the expression of a number of genes, which based on their time-course of induction can be classified either as immediate early genes (IEGs), intermediate or delayed response genes. Immediate early genes represent the initial nuclear response to the activation of intracellular signaling cascades (Tischmeyer and Grimm, 1999) and have been extensively used for mapping neuronal activity (Guzowski et al., 2005; Kaczmarek and Chaudhuri, 1997). Many IEGs have one or more CRE sites in their promoters and are induced after activation of the CREB pathway. The expression of the mRNAS for IEGs is fast, transient, and does not depend on *de novo* protein synthesis indicating that the transcription factors controlling their induction, such as CREB, are already present in the basal state. Some

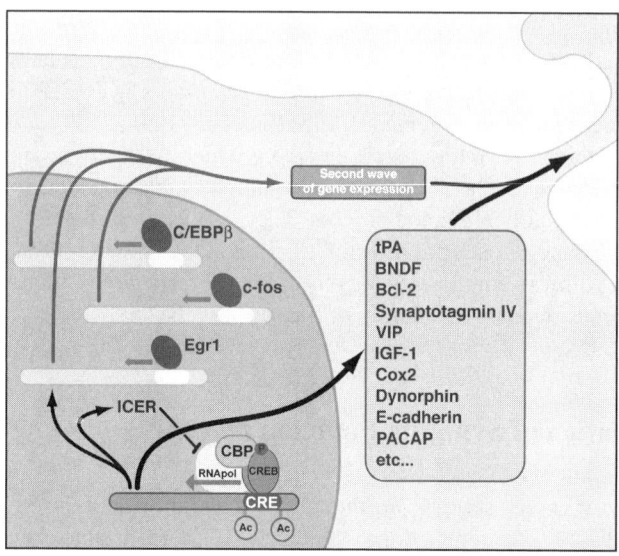

Fig. 3 Regulation of synaptic function by CREB-dependent gene expression. The phosphorylation of CREB at Ser133 promotes the recruitment of the co-activator CBP and initiates the transcription of targets genes. These gene products contribute to the stabilization of the short-term process (tPA, BDNF, only a few examples are listed etc...), or are themselves transcription factors that trigger a second wave of gene expression (zif268, c-fos, C/EBPβ). More extensive, but still non all-inclusive, lists of putative CREB target genes can be found in recent reviews (Barco and Kandel, 2005; Lonze and Ginty, 2002; Mayr and Montminy, 2001). In addition, the HAT activity of CBP can produce persistent changes in the acetylation state of the chromatin leading to long-term effects in the transcriptional activity of specific *loci* involved in synaptic plasticity (*See* Color Plate 7).

of the IEGs regulated by CREB, such as c-fos, egr1 or C/EBPβ, are themselves transcription factors, whose induction trigger a second wave of gene expression and may lead to the expression of intermediate and delayed response genes (Fig. 3). Overall, the gene expression cascade initiated by CREB activation seems to provide the building blocks required for the stabilization of the otherwise transient strengthening of synaptic connections. Both the formation of new synapses and the remodeling of pre-existing synapses are thought to play a critical role in this stabilization.

7.3.2 Evidence for a Role of CREB Controlling Synaptic Plasticity

Studies in the sea snail *Aplysia* (Brunelli et al., 1976) first established the critical role of the cAMP signaling pathway in long-term facilitation (LTF, a term equivalent to LTP as described in mammals) and have provided a refined view of the role of CREB in synaptic plasticity (Kandel, 2001). ApCREB-1 controls the transcription of several immediate-early response genes that contribute to the stabilization of the short-term process and its conversion to the long-term process. For example, two of the immediate reponse genes regulated by ApCREB-1 in *Aplysia* sensory neurons are a ubiquitin hydrolase, which regulates the degradation of the regulatory subunit of PKA that constrains long-term facilitation (Hegde et al., 1997), and the transcription factor C/EBP, which initiates a second wave of gene expression (Alberini et al., 1994). In agreement with this model, the injection of phosphorylated ApCREB-1 into cultured sensory neurons can by itself induce LTF (Bartsch et al., 1998; Casadio et al., 1999). Conversely, the inhibition of the repressor ApCREB-2 enhances long-term facilitation (Bartsch et al., 1995). This burst of gene expression stabilizes the strengthening of synaptic connections and leads to the growth of new synapses.

Many aspects of the role of CREB regulating synaptic plasticity seem to be conserved through evolution from mollusk to mammals. Thus, the enhanced expression of CRE-driven genes also favors the formation and stability of LTP in the mouse hippocampus (Barco et al., 2002; Marie et al., 2005). Conversely, inhibition of CRE-driven expression using a dominant negative form of CREB, which prevents the binding to DNA of different members of the CREB family, caused clear deficits in different forms of LTP (Huang et al., 2004; Pittenger et al., 2002). The results obtained with CREB hypomorphic mutants (mice homozygous for a deletion of the α and δ isoforms) and brain-restricted knockouts were, however, more controversial. The deficit in the late phase of LTP first detected in CREB hypomorphic mutant (Bourtchuladze et al., 1994) was later found to be sensitive to genetic background and gene dosage (Balschun et al., 2003; Gass et al., 1998) and no deficit was found in mice in which CREB was specifically depleted in forebrain neurons (Balschun et al., 2003). The overexpression of other CRE-binding proteins, such as CREM and the CREBβ isoform, observed in CREB hypomorphic and knockout mice suggests that the up regulation of these genes can compensate for the loss of CREB and attenuate the deficits (Balschun et al., 2003; Blendy et al., 1996; Hummler et al., 1994).

7.3.3 Identifying the CREB Transcriptome

CREB binds with high affinity to the palindromic consensus sequence for CRE, TGACGTCA, which is present on average once every 65 Kbp of random DNA sequence. This ratio represents 45,000 CRE sites in the human genome, the actual number is however much lower. A recent study identified only about 10,000 CRE palindromes in our genomic sequence (Zhang et al., 2005). The number of putative CREB binding sites is actually much higher because the promoters of many cAMP responsive genes often contain only a half-site TGACG sequence. The same study identified almost 750,000 half CREs in the human genome. However, the methylation of CRE sequences restricts CREB occupancy to functionally relevant sites and most of these sites do not bind CREB (Zhang et al., 2005). While classical biochemical and molecular studies, both *in vitro* and *in vivo*, demonstrated the participation of CREB in the regulation of the expression of more than one hundred genes, the availability of the complete sequence of the mouse and human genome, the constant refinement of bioinformatics tools for their analysis and the widespread application of genome-wide transcriptional profiling tools has allowed the recent identification of many more potential targets.

The widespread application of high-density cDNA and oligonucleotide microarrays, to both, studies on synaptic plasticity and behavior (Lee et al., 2005; Park et al., 2006) and the characterization of CREB mutant mice (Barco et al., 2005; McClung and Nestler, 2003), has confirmed many previously identified targets, and revealed hundred of candidate genes to be regulated directly or indirectly by CREB (Barco et al., 2005; Lee et al., 2005; McClung and Nestler, 2003; Park et al., 2006). Interestingly, a recent time-course microarray analysis of LTP-induced gene expression in the dentate gyrus revealed that activity-induced genes are frequently clustered on chromosomes. Many of the genes identified in this study located in chromosomal domains enriched with CREB-binding sites and displayed CREB-mediated transcription (Park et al., 2006).

An important recent development in the analysis of global transcription has been the application of chromatin immunoprecipitation (ChIP) techniques to genome-wide studies. In these techniques, chromatin is immunoprecipitated using a CREB antibody, and DNA bound sequences are amplified and identified. In contrast to expression arrays, this methodology can distinguish between direct and indirect targets of CREB. Using serial analysis of chromatin occupancy (SACO), a technique that combines ChIP assays with the sequencing of short tags similar to those used in serial analysis of gene expression (SAGE), Impey and colleagues identified a large number of genes in rat PC12 cells regulated by CREB (Impey et al., 2004). More than 60% of the CRE sites occupied by CREB in the basal condition were located in or near transcriptionally active regions, frequently within previously identified CREB-regulated genes. Strikingly, some of these CRE sites were located in bidirectional promoters and in promoters driving the expression of miRNAs and antisense transcripts (Impey et al., 2004).

An alternative approach for the analysis genome-wide transcription is the production of arrays of promoters suitable for what has been called "ChIP on chip" analysis. Here, immunoprecipitated DNA is amplified by PCR, labeled and hybridized to a microarray displaying promoter sequences. This technique was recently applied to screen for CREB target genes located in the chromosome 22, which represents about 1% of the human genome. This study revealed 215 binding sites corresponding to 192 loci. Most of the sites did not correspond to consensus CREs, but to shorter variants, and were located in regions outside known promoters. Only a subset of these candidate genes was affected by forskolin in cultured cells (Euskirchen et al., 2004). Another recent screen for CREB target genes, also based on ChIP-on-chip technology but with a genome-wide approach, demonstrated that CREB occupies at least 4000 promoter sites *in vivo*. Although the profiles of CREB occupancy were very similar in different human tissues, in a given cell type only a small proportion of CREB target genes were induced by forskolin (Zhang et al., 2005). Similar results were obtained using a different approach based on a hidden Markov model (HMM) trained on known CREB binding sites. This model, a bioinformatics tool, identified more than one 1,600 putative functional CRE sites in the human genome (Conkright et al., 2003), some of which were tested in cell culture. Only those promoters with a TATA box proximal to the CRE site exhibited a strong up regulation in response to forskolin. When the CRE site was moved farer from the TATA box the capability to drive cAMP responsive transcription was reduced. For technical reasons, most of these studies focused on culture cell lines, but it is only a matter of time that these unbiased, genome-wide screening approaches will be applied to the intact nervous system.

The current list of CREB target genes is heterogeneous and includes several hundred genes with very different functions, from transcription and metabolism regulation to cell structure or signaling. Some of the major classes are: transcriptional/nuclear factors (c-Fos, nurr77, zif268); growth factors (BDNF), signalling molecules (IGF-1, MKP-1, TGF-2, VGF), neuronal genes (synaptotagmin IV, presenilin-2), metabolic factors (cytochrome-c, SOD-2), molecules involved in the cell cycle, DNA repair and proliferation (DNA polymerase, cyclin 2), opioid receptors, neurotransmitter receptor subunits, molecules important for transport, structural proteins and factors of immune response (Barco and Kandel, 2005; Impey et al., 2004; Lonze and Ginty, 2002; Zhang et al., 2005). Despise the recent advances in the field, the complete set of CRE sites bound by CREB in a specific cell type or under an specific stimuli is still not known and it is still not clear how many of these downstream genes are really regulated in the brain under physiological conditions.

7.3.4 Transport of mRNAs, Local Protein Synthesis and Synaptic Capture

Upon transcription activation following high frequency stimulation, as occurs with LTP, or heterosynaptic stimulation, as described for LTF, the mRNAs produced *de novo* are translocated specifically to the synapses that received the stimulation (Steward and Worley, 2001). This specifity may reach the level of single synapses and is one of the most relevant properties of LTP and LTF as the cellular correlate of memory formation (Govindarajan et al., 2006; Martin et al., 2000). The participation of the cell nucleus and the requirement of *de novo* gene expression in long-lasting forms of synaptic plasticity impose a critical requirement for any model trying to explain learning-related plasticity: there must be mechanisms that restrict the action of the newly expressed gene products to active synapses but not to others. To address this problem, it has been suggested that the persistence of the changes in synaptic strenght is mediated by the generation of a transient local synaptic tag at recently activated synapses and by the production of plasticity-related proteins that can be used only at those synapses marked by a tag (Frey and Morris, 1997; Martin et al., 1997), an idea referred to us as the synaptic capture or synaptic tagging hypothesis. In the rat hippocampus, Frey and Morris found that once transcription-dependent LTP has been induced at one pathway, the long-term process can be "captured" at a second pathway receiving a stimulation that would normally produce only E-LTP, but can elicit the formation of the tag (Frey and Morris, 1997). Further studies revealed the precise time course of the tag and showed that synapses can be tagged shortly before, at the same time or after the stimulus that elicited the long-term process (Frey and Morris, 1998; Sajikumar and Frey, 2004).

In parallel, Martin and colleagues working in *Aplysia* cultured neurons independently described the phenomenon of synaptic tagging and synaptic capture in the nervous system of this organism (Martin et al., 1997). Indeed, studies on the gill-withdrawal reflex of *Aplysia* first revealed a direct role of CREB-dependent gene expression in synaptic capture. Kandel and co-workers demonstrated that repeated application of the neurotransmitter serotonin to one synapse branch caused CREB activation at the cell nucleus, and branch-specific LTF and the growth of new synaptic connections in the stimulated branch (Martin et al., 1997). Moreover, they found that the mere injection of phospho-CREB when paired with a single pulse of 5-HT in one of the branches increased varicosities formation. This single pulse of 5-HT marked the branch enabling the capture of the gene products produced by CREB activity and provided the building blocks necessary for the formation of new synaptic connections (Casadio et al., 1999). Similarly, in CA1 hippocampal neurons of transgenic mice, the expression of a constitutively active variant of CREB, VP16-CREB enhanced CRE-driven gene expression and reduced the threshold for eliciting a persistent late phase of LTP in the Schaffer collateral pathway (Barco et al., 2002). The pharmacological characterization of this form of facilitated L-LTP suggested that VP16-CREB activity can lead to a cell-wide priming for LTP by seeding the synaptic terminals with proteins and mRNAs required for the stabilization of L-LTP. As described in *Aplysia*, these gene products can then be used productively for L-LTP in synapses that have been tagged by stimulation of the sort normally needed for eliciting E-LTP (Barco et al., 2002). Transcription profiling analysis identified the neurotrophin BDNF as the most relevant effector molecule contributing to this enhanced LTP phenotype. Indeed, experiments in BDNF deficient mice suggested that presynapticaly released BDNF contributes to tagging the synapse in normal mice (Barco et al., 2005).

The presynaptic release of BDNF is not the only molecular process involved in synaptic tagging. Since the proposal that previously activated synapses are marked was introduced, the molecular nature of this mark has been intensively investigated both in mammals and *Aplysia* (Martin and Kosik, 2002; Morris, 2006). Several lines of evidence now point to a PKA-mediated phosphorylation event as being critical for the mark (Barco et al., 2002; Navakkode et al., 2004; Young et al., 2006) and that local protein synthesis is a required step that enable the functional capture of the transcripts produced at the nucleus (Casadio et al., 1999; Martin and Kosik, 2002; Si et al., 2003).

The presence of functional machinery for protein synthesis in neuronal processes suggests that local protein synthesis may play a major role in the control of synaptic strength (Schuman et al., 2006; Sutton and Schuman, 2005). Indeed, the induction of LTP in the Schaffer collateral pathway is accompanied by the transport of polysomes from dendritic shafts to active spines of CA1 neurons (Ostroff et al., 2002). In this way, the mRNAs synthesized as a consequence of nuclear activation can be translated exactly where the mRNAs are needed and lead to rapid changes in synaptic strength (Steward and Schuman, 2003).

7.3.5 Synaptic Growth and Remodeling

Long-lasting forms of LTF and LTP require *de novo* gene expression. The newly synthesized gene products are thought to participate in the formation of new synaptic connections (see recent reviews by (Hayashi and Majewska, 2005; Lamprecht and LeDoux, 2004; Segal, 2005)). Synaptic growth has been found to accompany various forms of learning-related plasticity, a phenomenon particularly well documented in *Aplysia* (Bailey and Kandel, 1993). In this organism, the injection of phosphorylated CREB-1 into cultured sensory neurons can causes long-lasting synaptic changes and leads to the formation of stable new synaptic connections when paired to weak synaptic stimulation (Casadio et al., 1999). However, in the mammalian brain, these structural changes are subtler and more difficult to study. The production of LTP has been associated with the generation and enlargement of dendritic spines in organotypic hippocampal slices (Matsuzaki et al., 2004; Nagerl et al., 2004) and acute slices of neonatal animals (Zhou et al., 2004), but the structural changes are much more discreet in the adult brain (Lang et al., 2004). In the adult brain, there is only a modest production of new spines (Zuo et al., 2005) and learning-related plasticity seems to rely more on subtle functional changes than in frank anatomical changes. However, as found in *Aplysia*, CREB-dependent gene expression when coupled to subthreshold synaptic activation seems to be sufficient to drive long lasting changes in synaptic function, including the conversion of silent synapses into active ones (Marie et al., 2005).

7.4 CREB and Brain Function

Given the important role of CREB regulating synaptic plasticity it is not surprising that this transcription factor had been involved in a variety of critical brain functions. Here will focus on the role of the CREB pathway in regulating different forms of learning and memory as an example of CREB's more general role controlling the persistence of changes in synaptic strength. However, it should be noted that the stabilization of the potentiation of synaptic connections is critical not only for learning but for many other aspects of brain function. Long-lasting changes in circuits controlling reward or emotional responses can also be regulated by activation of the CREB pathway, as we will discuss in the next section in reference to drug addiction and depression.

7.4.1 Learning and Memory

As described above for synaptic plasticity, studies in the sea snail *Aplysia* first identified the cAMP/CREB pathway as a core component of the molecular switch that converts short- to long-term memory for a simple form of learning called sensitization (Brunelli et al., 1976). This role was soon confirmed by the pioneer investigation by Benzer and coworkers on the genetic basis of memory in *Drosophila*. *Dunce* and *rutabaga*, two of the first memory mutants identified by genetic screenings in *Drosophila* (Dudai et al., 1976), were found to affect respectively a cAMP-dependent phosphodiesterase and a Ca^{2+}/CaM-regulated adenylyl cyclase, two important proteins gating the activation of CREB (Byers et al., 1981; Dudai et al., 1983; Waddell and Quinn, 2001). The demonstration of a direct role of CREB in memory formation in flies was provided few years later by the behavioral analysis of transgenic flies overexpressing opposing forms of CREB. Whereas the activator dCREB2 enhanced the formation of long-term memory, the repressor dCREB1suppressed it (Yin et al., 1995; Yin et al., 1994). Parallel studies on CREB hypomorphic mutant mice demonstrated that the reduced expression of CREB caused specific deficit in long-term memory that correlated with deficits in the late phase of LTP also in mammals (Bourtchuladze et al., 1994). These results were soon confirmed by experiments in rats, in which the intra-hippocampal infusion of CREB antisense oligos caused deficits in spatial learning (Guzowski and McGaugh, 1997). However, some of these earlier findings in flies and mammals are now appreciated to be controversial. Although there is agreement that over-expression of the dCREB1 repressor blocks long-term memory, the memory enhancing effect of overexpressing dCREB2 in flies could not be recently replicated (Perazzona et al., 2004). Similarly, further analyses of CREB hypomorphic and other CREB deficient mouse mutants demonstrated that some of the memory deficits associated to CREB were sensitive to gene dosage, genetic background or molecular mechanism of blocking CREB function (Balschun et al., 2003; Gass et al., 1998; Graves et al., 2002; Rammes et al., 2000), suggesting that CREB is not the only transcription factor controlling these changes in mammals. Further work with regulatable CREB deficient mutants has provided a more precise examination of the role of CREB on learning and memory. Thus, the inducible and transient repression of CREB function specifically blocked the consolidation of long-term fear memories (Kida et al., 2002) and caused reversible deficits in spatial navigation at the Morris water maze (Pittenger et al., 2002). In agreement with these results, the expression of a dominant negative CREB mutant in amygdala using a recombinant herpes virus inhibited the consolidation of fear conditioning memories (Josselyn et al., 2004), whereas the acute overexpression of CREB facilitated its formation (Josselyn et al., 2001).

The picture that emerges from these studies suggests that CREB itself may be dispensable for certain forms of explicit memory, likely because the lack of CREB can be compensated by the action of other CRE-binding factors. However, the induction of CRE-driven gene expression seems to be a general requirement for different types of long-lasting memory, both implicit and explicit. Genetic manipulation studies using recombinant viruses or mutant mice have shown that blocking the CREB pathway leads to deficits in a large number of memory processes, including spatial navigation, object recognition, social transmission of food preferences, conditional taste or odor aversion, contextual and cued fear conditioning (Balschun et al., 2003; Brightwell et al., 2005; Kogan et al., 1997; Pittenger et al., 2002; Zhang et al., 2003). Furthermore, the expression of diverse CREB downstream genes (Bito et al., 1996; Deisseroth et al., 1996; Impey et al., 1996; Lu et al., 1999) and the induction of a CRE-driven *lacZ* reporter construct (Impey et al., 1998) in specific brain regions correlate with the acquisition of long-term forms of memory dependent of activity in that region. By extension, these studies suggest that the CREB pathway is an attractive target for drugs aimed at improving or disrupting memory consolidation (Barco et al., 2003; Tully et al., 2003).

7.5 CREB and Brain Malfunction

The CREB pathway regulates cellular responses critical for the proper functioning of neurons and circuits. It is, therefore, not surprising that malfunction of this pathway has severe consequences for brain function and underlies several important brain disorders.

7.5.1 Pathologies of Synaptic Plasticity

Dysfunction of synaptic plasticity can have devastating consequences for the organism and can trigger pathological symptoms that rank from memory impairments and long-term alterations in behavior to neuronal cell loss. Interestingly, at least two human disorders characterized by cognitive impairments have been directly related to the CREB activation pathway (Trivier et al., 1996). Mutations in the gene encoding RSK-2, one of the kinases regulating CREB phosphorylation, cause the Coffin-Lowry syndrome a rare X-linked mental retardation disorder (Trivier et al., 1996). Notably, the cognitive performance of human patients with this syndrome correlates with the cellular capacity to activate RSK2 (Harum et al., 2001). On the other hand, mutations in the gene encoding for CREB co-activator CBP are associated with a second mental retardation syndrome, the Rubinstein-Taybi syndrome (RTS), a complex autosomal-dominant disorder. The recent characterization of several mouse models for RTS has revealed a direct role of both the histone acetyltransferase activity of CBP and its capability to activate the CREB pathway in the etiology of this condition (Alarcon et al., 2004; Bourtchouladze et al., 2003; Korzus et al., 2004; Wood et al., 2005). A new point of view emerging from these studies on mouse models is the thought that a component of the cognitive impairments associated to these syndromes could have a non-developmental origin and be primarily caused by the chronic reduction of CBP enzymatic activity in the postnatal or adult brain.

The dysregulation of the CREB pathway has also been associated with age-associated memory impairment (AAMI) (Brightwell et al., 2004; Chung et al., 2002). Drugs that enhance the activity of this pathway, such as inhibitor of phosphodiesterases, can prevent memory decline in old mice (Bach et al., 1999) and are currently evaluated in clinical trials in humans (Barco et al., 2003; Tully et al., 2003).

Alterations in synaptic plasticity may also underlie other pathological aspects of behavior. For example, much as is the case with long-term memory, addiction to drugs are life-long conditions responsible for permanent behavioral abnormalities and are considered nowadays a pathological manifestation of abnormal synaptic plasticity. Recent studies have highlighted the similarities between memory and addiction at the cellular and molecular level. Both processes depend on stimulus-induced long-lasting changes in neuronal function that correlate at the cellular level with changes in synaptic strenght and at the molecular level with the activation of CREB-dependent gene expression. After exposure to various drugs of abuse, such as ethanol or cocaine, the regions of the CNS known to be involved in addiction, such as the locus coeruleus and nucleus accumbens, show significant increases in CREB phosphorylation and CRE-mediated gene expression. Recent studies in mutant mice with an altered signaling to CREB or reduced CREB activity suggest that CREB-dependent gene expression may be universally involved in addiction, although its precise role may differ depending of the drug studied and its protocol of adminis-tration (see reviews by (Carlezon et al., 2005; Nestler, 2001; Pandey et al., 2005)).

Also, recent studies have revealed a critical role of the cAMP cascade in depres-sion (Conti and Blendy, 2004). Our understanding of the molecular bases of depres-sion and other mood disorders is still incomplete, but it is likely that long-term changes in synaptic plasticity underlie this persistent alteration in brain function. Both the expression and the activity of CREB are increased by chronic antidepres-sant treatment (Nibuya et al., 1996; Thome et al., 2000), suggesting that CRE-driven expression may be one of the targets for these treatments and, therefore, contribute to the etiology of this condition.

7.5.2 *Neurodegenerative Diseases*

There is evidence that dysfunction in the CREB pathway may be involved in early manifestations in Alzheimer's disease and that it may also play a central role in Huntington disease (HD) and other forms of polyglutamine pathogenesis. Although, we focused this review on the role of CREB in the signaling cascades that regulate synaptic plasticity, the CREB pathway has been also involved on the control of neuronal survival and response to neuronal stress (Dawson and Ginty, 2002; Lonze and Ginty, 2002; Walton and Dragunow, 2000). Whereas the pathological conditions discussed above are likely caused primarily by failures of synaptic plasticity, the par-ticipation of CREB signaling in the etiology of neurodegenerative disorders could be related to a different role for CREB, its role in controlling neuronal survival. However, given that abnormal synaptic plasticity can lead to atrophy or excitotox-icity and trigger neurodegenerative processes, and that neuronal loss by itself may have significant effects in synaptic plasticity, it is difficult to dissect what aspects of CREB malfunction are primarily associated to these pathologies.

CREB regulates the expression of a number of pro-survival factors, such as the neurotrophin BDNF and the anti-apoptotic protein bcl-2. Indeed, CREB-mediated gene expression is necessary and sufficient for the survival of several neuronal sub-types *in vitro* (Bonni et al., 1999; Riccio et al., 1999; Walton and Dragunow, 2000). Experiments with CREB-deficient mice supported this view and showed that dor-sal root ganglia (DRG) sensory neurons indeed require the expression of CREB for survival *in vivo* (Lonze et al., 2002). Further studies in knockout mice have demonstrated that different neuronal types differ in their requirement for transcription factors of the CREB family. Sensory neurons of the peripheral nervous system are much more dependent of CREB activity than neurons in the CNS. Experiments in $CREB^{-/-}/CREM^{-/-}$ double mutants demonstrated that when both factors are eliminated in the same cell, survival is compromised also for CNS neurons. (Mantamadiotis et al., 2002). These studies also revealed that different members of the CREB family can compensate for each other for the control of neuronal survival *in vivo*. The reduction on CREB expression causes an upregulation of CREM that likely ameliorates some of the deleterious consequences of the lack of CREB.

Based on this evidence, it is not surprising that transcriptional dysregulation in the CREB pathway had been proposed to play a central role in the pathogenesis of neurodegenerative disorders. For example, polyglutamine repeat disorders, a family of related neurological diseases whose more relevant member is Huntington's disease (HD), have been associated to deficient CREB-mediated gene expression. These diseases are caused by expansions of polyglutamine-encoding sequences that make the mutant protein toxic to neurons, possibly through abnormal interactions with polyglutamine tracts present in proteins important for neuronal survival (McMurray, 2001). CREB co-activator CBP exhibits a C-terminus very rich in glutamine residues that may interact with polyglutamine expansions (McCampbell et al., 2000; Steffan et al., 2000). As a consequence, CBP can be depleted from the cell nucleus by being sequestrated in the cytoplasmic aggregates, resulting in abnormal CREB-dependent transcriptional activity and cellular toxicity (Bates et al., 2006; Higgins et al., 1999; Obrietan and Hoyt, 2004; Sugars et al., 2004; Yu et al., 2002). Consistent with this view, the overexpression of CBP partially rescues the cell death accompanying expression of mutant huntingtin in neurons (Nucifora et al., 2001), whereas the expression of mutant huntingtin leads to a reduction on HAT activity *in vivo* (Igarashi et al., 2003). The administration of histone deacetylase (HADC) inhibitors ameliorates poly-Q dependent neurodegeneration in Drosophila and HD mouse models (Ferrante et al., 2003; Hockly et al., 2003; Steffan et al., 2001).

Also, the PKA/CREB pathway has been involved in early manifestations of Alzheimer's disease (AD), the most prevalent neurodegenerative disease in humans The treatment of cultured hippocampal neurons with Aβ peptide leads to the inactivation of PKA and reduces the activation of CREB in response to glutamate (Vitolo et al., 2002). Compounds that enhance the cAMP-signaling pathway, such as rolipram, could reverse these effects (Vitolo et al., 2002) and ameliorated synaptic plasticity and memory deficits in a mouse model for AD (Gong et al., 2006; Gong et al., 2004).

7.6 Conclusions

CREB plays a critical role translating specific patterns of synaptic stimulation, such as the temporal convergence of two different stimuli or the repetition of the same stimulus so that it has surpassed a given threshold, into long-term changes in synaptic plasticity. This is achieved through the induction of a cascade of gene expression that leads to the synthesis of proteins involved in synaptic function and structure that we are starting now to identify. The stabilization of, otherwise, transient alterations in synaptic plasticity has critical consequences for diverse aspects of brain function, from different forms of learning and memory in the hippocampus to experience-dependent plasticity in different areas of the cortex and addiction in the nucleus accumbens or the locus coeruleus. Although we do not have yet a complete picture of the complex network of protein interactions involved on the activation of CREB-dependent gene expression, the work during the last decade in this field has provided strong beginnings and a useful conceptual framework. Probably, the most important challenge ahead is the resolution of the CREB *transcriptome*, or better said, *transcriptomes*: the different sets of genes regulated by CREB in a given cell type in response to a given stimulus. Important efforts in this direction have been made, but the long lists of candidate genes provided by these studies needs to be refined and the targets need to be validated. The same techniques used to investigate CREB-dependent gene expression *in vitro* have to be applied *in vivo* and to different brain areas. The promised reward for these studies is likely to be well worth the effort. We will not only understand better what happens in a healthy brain during learning, but also the identification of specific CREB downstream genes may provide therapeutic targets for relevant neurodegenerative diseases and disorders of synaptic plasticity that are now still not accessible to treatment.

Acknowledgments Research in this review was supported in part by the Howard Hughes Medical Institute and the Kavli Institute for Brain Sciences (to E.R.K.), and the Marie Curie Excellence grant MEXT-CT-2003-509550 and the MEC grant BFU2005-00286 (A.B.). The authors thank S. Ingham for assistance in the elaboration of the figures.

References

Alarcon, J. M., Malleret, G., Touzani, K., Vronskaya, S., Ishii, S., Kandel, E. R., and Barco, A. (2004) Chromatin acetylation, memory, and LTP are impaired in CBP+/- mice: a model for the cognitive deficit in Rubinstein-Taybi syndrome and its amelioration. Neuron 42, 947–959.

Alberini, C. M., Ghirardi, M., Metz, R., and Kandel, E. R. (1994) C/EBP is an immediate- early gene required for the consolidation of long- term facilitation in Aplysia. Cell 76, 1099 1114.

Bach, M. E., Barad, M., Son, H., Zhuo, M., Lu, Y. F., Shih, R., Mansuy, I., Hawkins, R. D., and Kandel, E. R. (1999) Age-related defects in spatial memory are correlated with de fects in the late phase of hippocampal long-term potentiation in vitro and are attenuated by drugs that enhance the cAMP signaling pathway. Proc. Natl. Acad. Sci. USA 96, 5280–5285.

Bailey, C. H., and Kandel, E. R. (1993) Structural changes accompanying memory storage. Annu. Rev. Physiol. 55, 397–426.

Balschun, D., Wolfer, D. P., Gass, P., Mantamadiotis, T., Welzl, H., Schutz, G., Frey, J. U., and Lipp, H. P. (2003) Does cAMP response element-binding protein have a pivotal role in hippocampal synaptic plasticity and hippocampus-dependent memory? J. Neurosci. 23, 6304–6314.

Barco, A., Alarcon, J. M., and Kandel, E. R. (2002) Expression of constitutively active CREB protein facilitates the late phase of long-term potentiation by enhancing synaptic capture. Cell 108, 689–703.

Barco, A., Bailey, C. H., and Kandel, E. R. (2006) Common molecular mechanisms in ex plicit and implicit memory. J. Neurochem. 97, 1520–1533.

Barco, A., and Kandel, E. R. (2005) Role of CREB and CBP in brain function. In *Transcrip tion factors in the nervous system: Development, brain function and disease*, G. Thiel, ed. Wiley-VCH.

Barco, A., Patterson, S., Alarcon, J. M., Gromova, P., Mata-Roig, M., Morozov, A., and Kandel, E. R. (2005) Gene Expression Profiling of Facilitated L-LTP in VP16-CREB Mice Reveals that BDNF Is Critical for the Maintenance of LTP and Its Synaptic Capture. Neuron 48, 123–137.

Barco, A., Pittenger, C., and Kandel, E. R. (2003) CREB, memory enhancement and the treatment of memory disorders: promises, pitfalls and prospects. Expert Opin. Ther. Targets 7, 101–114.

Bartsch, D., Casadio, A., Karl, K. A., Serodio, P., and Kandel, E. R. (1998) CREB1 encodes a nuclear activator, a repressor, and a cytoplasmic modulator that form a regulatory unit critical for long-term facilitation. Cell 95, 211–223.

Bartsch, D., Ghirardi, M., Skehel, P. A., Karl, K. A., Herder, S. P., Chen, M., Bailey, C. H., and Kandel, E. R. (1995) Aplysia CREB2 represses long-term facilitation: relief of re pres sion converts transient facilitation into long-term functional and structural change. Cell 83, 979–992.

Bates, E. A., Victor, M., Jones, A. K., Shi, Y., and Hart, A. C. (2006) Differential contributions of Caenorhabditis elegans histone deacetylases to huntingtin polyglutamine toxicity. J. Neurosci. 26, 2830–2838.

Bito, H., Deisseroth, K., and Tsien, R. W. (1996) CREB phosphorylation and dephosphorylation: a Ca(2+)- and stimulus duration-dependent switch for hippocampal gene expression. Cell 87, 1203–1214.

Blendy, J. A., Kaestner, K. H., Schmid, W., Gass, P., and Schutz, G. (1996) Targeting of the CREB gene leads to up-regulation of a novel CREB mRNA isoform. EMBO J. 15, 1098–1106.

Bonni, A., Brunet, A., West, A. E., Datta, S. R., Takasu, M. A., and Greenberg, M. E. (1999) Cell survival promoted by the Ras-MAPK signaling pathway by transcription-dependent and -independent mechanisms. Science 286, 1358–1362.

Bourtchouladze, R., Lidge, R., Catapano, R., Stanley, J., Gossweiler, S., Romashko, D., Scott, R., and Tully, T. (2003) A mouse model of Rubinstein-Taybi syndrome: Defective long-term memory is ameliorated by inhibitors of phosphodiesterase 4. Proc. Natl. Acad. Sci. USA 100, 10518–10522.

Bourtchuladze, R., Frenguelli, B., Blęndy, J., Cioffi, D., Schutz, G., and Silva, A. J. (1994) Deficient long-term memory in mice with a targeted mutation of the cAMP-responsive element-binding protein. Cell 79, 59–68.

Brightwell, J. J., Gallagher, M., and Colombo, P. J. (2004) Hippocampal CREB1 but not CREB2 is decreased in aged rats with spatial memory impairments. Neurobiol Learn Mem 81, 19–26.

Brightwell, J. J., Smith, C. A., Countryman, R. A., Neve, R. L., and Colombo, P. J. (2005) Hippocampal overexpression of mutant CREB blocks long-term, but not short-term mem ory for a socially transmitted food preference. Learn. Mem. 12, 12–17.

Brunelli, M., Castellucci, V., and Kandel, E. R. (1976) Synaptic facilitation and behavioral sensitization in Aplysia: possible role of serotonin and cyclic AMP. Science 194, 1178–1181.

Byers, D., Davis, R. L., and Kiger, J. A., Jr. (1981) Defect in cyclic AMP phosphodiesterase due to the dunce mutation of learning in Drosophila melanogaster. Nature 289, 79–81.

Canettieri, G., Morantte, I., Guzman, E., Asahara, H., Herzig, S., Anderson, S. D., Yates, J. R., 3rd, and Montminy, M. (2003) Attenuation of a phosphorylation-dependent activator by an HDAC-PP1 complex. Nat. Struct. Biol. 10, 175–181.

Carlezon, W. A., Jr., Duman, R. S., and Nestler, E. J. (2005) The many faces of CREB. Trends Neurosci. 28, 436–445.

Casadio, A., Martin, K. C., Giustetto, M., Zhu, H., Chen, M., Bartsch, D., Bailey, C. H., and Kandel, E. R. (1999) A transient, neuron-wide form of CREB-mediated long-term facili tation can be stabilized at specific synapses by local protein synthesis. Cell 99, 221–237.

Cha-Molstad, H., Keller, D. M., Yochum, G. S., Impey, S., and Goodman, R. H. (2004) Cell-type-specific binding of the transcription factor CREB to the cAMP-response element. Proc. Natl. Acad. Sci. USA 101, 13572–13577.

Chan, H. M., and La Thangue, N. B. (2001) p300/CBP proteins: HATs for transcriptional bridges and scaffolds. J. Cell Sci. 114, 2363–2373.

Chrivia, J. C., Kwok, R. P., Lamb, N., Hagiwara, M., Montminy, M. R., and Goodman, R. H. (1993) Phosphorylated CREB binds specifically to the nuclear protein CBP. Nature 365, 855–859.

Chung, Y. H., Kim, E. J., Shin, C. M., Joo, K. M., Kim, M. J., Woo, H. W., and Cha, C. I. (2002) Age-related changes in CREB binding protein immunoreactivity in the cerebral cortex and hippocampus of rats. Brain Res. 956, 312–318.

Cohen, P. (1989) The structure and regulation of protein phosphatases. Annu. Rev. Biochem. 58, 453–508.

Comerford, K. M., Leonard, M. O., Karhausen, J., Carey, R., Colgan, S. P., and Taylor, C. T. (2003) Small ubiquitin-related modifier-1 modification mediates resolution of CREB-dependent responses to hypoxia. Proc. Natl. Acad. Sci. USA 100, 986–991.

Conkright, M. D., Guzman, E., Flechner, L., Su, A. I., Hogenesch, J. B., and Montminy, M. (2003) Genome-wide analysis of CREB target genes reveals a core promoter requirement for cAMP responsiveness. Mol. Cell 11, 1101–1108.

Conti, A. C., and Blendy, J. A. (2004) Regulation of antidepressant activity by cAMP re sponse element binding proteins. Mol. Neurobiol. 30, 143–155.

Dash, P. K., Karl, K. A., Colicos, M. A., Prywes, R., and Kandel, E. R. (1991) cAMP re sponse element-binding protein is activated by Ca2+/calmodulin- as well as cAMP- dependent protein kinase. Proc. Natl. Acad. Sci. USA 88, 5061–5065.

Dawson, T. M., and Ginty, D. D. (2002) CREB family transcription factors inhibit neuronal suicide. Nat. Med. 8, 450–451.

Deisseroth, K., Bito, H., and Tsien, R. W. (1996) Signaling from synapse to nucleus: postsyn aptic CREB phosphorylation during multiple forms of hippocampal synaptic plasticity. Neuron 16, 89–101.

Deisseroth, K., Heist, E. K., and Tsien, R. W. (1998) Translocation of calmodulin to the nucleus supports CREB phosphorylation in hippocampal neurons. Nature 392, 198–202.

Deisseroth, K., and Tsien, R. W. (2002) Dynamic multiphosphorylation passwords for activ ity-dependent gene expression. Neuron 34, 179–182.

Dudai, Y., Jan, Y. N., Byers, D., Quinn, W. G., and Benzer, S. (1976) dunce, a mutant of Drosophila deficient in learning. Proc. Natl. Acad. Sci. USA 73, 1684–1688.

Dudai, Y., Uzzan, A., and Zvi, S. (1983) Abnormal activity of adenylate cyclase in the Dro sophila memory mutant rutabaga. Neurosci. Lett. 42, 207–212.

Euskirchen, G., Royce, T. E., Bertone, P., Martone, R., Rinn, J. L., Nelson, F. K., Sayward, F., Luscombe, N. M., Miller, P., Gerstein, M., et al. (2004) CREB binds to multiple loci on human chromosome 22. Mol. Cell Biol. 24, 3804–3814.

Ferrante, R. J., Kubilus, J. K., Lee, J., Ryu, H., Beesen, A., Zucker, B., Smith, K., Kowall, N. W., Ratan, R. R., Luthi-Carter, R., and Hersch, S. M. (2003) Histone deacetylase inhibition by sodium butyrate chemotherapy ameliorates the neurodegenerative phenotype in Huntington's disease mice. J. Neurosci. 23, 9418–9427.

Frey, U., and Morris, R. G. (1997) Synaptic tagging and long-term potentiation. Nature 385, 533–536.

Frey, U., and Morris, R. G. (1998) Weak before strong: dissociating synaptic tagging and plasticity-factor accounts of late-LTP. Neuropharmacol. 37, 545–552.

Gass, P., Wolfer, D. P., Balschun, D., Rudolph, D., Frey, U., Lipp, H. P., and Schutz, G. (1998) Deficits in memory tasks of mice with CREB mutations depend on gene dosage. Learn. Mem. 5, 274–288.

Gau, D., Lemberger, T., von Gall, C., Kretz, O., Le Minh, N., Gass, P., Schmid, W., Schibler, U., Korf, H. W., and Schutz, G. (2002) Phosphorylation of CREB Ser142 regulates light-induced phase shifts of the circadian clock. Neuron 34, 245–253.

Genoux, D., Haditsch, U., Knobloch, M., Michalon, A., Storm, D., and Mansuy, I. M. (2002) Protein phosphatase 1 is a molecular constraint on learning and memory. Nature 418, 970–975.

Gong, B., Cao, Z., Zheng, P., Vitolo, O. V., Liu, S., Staniszewski, A., Moolman, D., Zhang, H., Shelanski, M., and Arancio, O. (2006) Ubiquitin Hydrolase Uch-L1 Rescues beta-Amyloid-Induced Decreases in Synaptic Function and Contextual Memory. Cell 126, 775–788.

Gong, B., Vitolo, O. V., Trinchese, F., Liu, S., Shelanski, M., and Arancio, O. (2004) Persis tent improvement in synaptic and cognitive functions in an Alzheimer mouse model after rolipram treatment. J. Clin. Invest. 114, 1624–1634.

Gonzalez, G. A., Yamamoto, K. K., Fischer, W. H., Karr, D., Menzel, P., Biggs, W., 3rd, Vale, W. W., and Montminy, M. R. (1989) A cluster of phosphorylation sites on the cyclic AMP-regulated nuclear factor CREB predicted by its sequence. Nature 337, 749–752.

Govindarajan, A., Kelleher, R. J., and Tonegawa, S. (2006) A clustered plasticity model of long-term memory engrams. Nat. Rev. Neurosci. 7, 575–583.

Graves, L., Dalvi, A., Lucki, I., Blendy, J. A., and Abel, T. (2002) Behavioral analysis of CREB alphadelta mutation on a B6/129 F1 hybrid background. Hippocampus 12, 18–26.

Guan, Z., Giustetto, M., Lomvardas, S., Kim, J. H., Miniaci, M. C., Schwartz, J. H., Thanos, D., and Kandel, E. R. (2002) Integration of long-term-memory-related synaptic plasticity involves bidirectional regulation of gene expression and chromatin structure. Cell 111, 483–493.

Gum, R. J., Gaede, L. L., Heindel, M. A., Waring, J. F., Trevillyan, J. M., Zinker, B. A., Stark, M. E., Wilcox, D., Jirousek, M. R., Rondinone, C. M., and Ulrich, R. G. (2003) Antisense protein tyrosine phosphatase 1B reverses activation of p38 mitogen-activated protein kinase in liver of ob/ob mice. Mol. Endocrinol. 17, 1131–1143.

Guzowski, J. F., and McGaugh, J. L. (1997) Antisense oligodeoxynucleotide-mediated disruption of hippocampal cAMP response element binding protein levels impairs consoli dation of memory for water maze training. Proc. Natl. Acad. Sci. USA 94, 2693–2698.

Guzowski, J. F., Timlin, J. A., Roysam, B., McNaughton, B. L., Worley, P. F., and Barnes, C. A. (2005) Mapping behaviorally relevant neural circuits with immediate-early gene expression. Curr. Opin. Neurobiol. 15, 599–606.

Habener, J. F., Miller, C. P., and Vallejo, M. (1995) cAMP-dependent regulation of gene tran scription by cAMP response element-binding protein and cAMP response element modulator. Vitam. Horm. 51, 1–57.

Hardingham, G. E., Fukunaga, Y., and Bading, H. (2002) Extrasynaptic NMDARs oppose synaptic NMDARs by triggering CREB shut-off and cell death pathways. Nat. Neurosci. 5, 405–414.

Harum, K. H., Alemi, L., and Johnston, M. V. (2001) Cognitive impairment in Coffin-Lowry syn drome correlates with reduced RSK2 activation. Neurology 56, 207–214.

Hayashi, Y., and Majewska, A. K. (2005) Dendritic spine geometry: functional implication and regulation. Neuron 46, 529–532.

Hegde, A. N., Inokuchi, K., Pei, W., Casadio, A., Ghirardi, M., Chain, D. G., Martin, K. C., Kandel, E. R., and Schwartz, J. H. (1997) Ubiquitin C-terminal hydrolase is an im mediate-early gene essential for long-term facilitation in *Aplysia*. Cell 89, 115–126.

Higgins, D. S., Hoyt, K. R., Baic, C., Vensel, J., and Sulka, M. (1999) Metabolic and gluta matergic disturbances in the Huntington's disease transgenic mouse. Ann. NY Acad. Sci. 893, 298–300.

Hockly, E., Richon, V. M., Woodman, B., Smith, D. L., Zhou, X., Rosa, E., Sathasivam, K., Ghazi-Noori, S., Mahal, A., Lowden, P. A., *et al.* (2003) Suberoylanilide hydroxamic acid, a histone deacetylase inhibitor, ameliorates motor deficits in a mouse model of Huntington's disease. Proc. Natl. Acad. Sci. USA 100, 2041–2046.

Hoeffler, J. P., Meyer, T. E., Waeber, G., and Habener, J. F. (1990) Multiple adenosine 3',5'-cyclic monophosphate response element DNA-binding proteins generated by gene diversification and alternative exon splicing. Mol. Endocrinol. 4, 920–930.

Hu, S. C., Chrivia, J., and Ghosh, A. (1999) Regulation of CBP-mediated transcription by neuronal calcium signaling. Neuron 22, 799–808.

Huang, H., Cheville, J. C., Pan, Y., Roche, P. C., Schmidt, L. J., and Tindall, D. J. (2001) PTEN induces chemosensitivity in PTEN-mutated prostate cancer cells by suppression of Bcl-2 expression. J. Biol. Chem. 276, 38830–38836.

Huang, Y. Y., Pittenger, C., and Kandel, E. R. (2004) A form of long-lasting, learning-related synaptic plasticity in the hippocampus induced by heterosynaptic low-frequency pairing. Proc. Natl. Acad. Sci. USA 101, 859–864.

Hummler, E., Cole, T. J., Blendy, J. A., Ganss, R., Aguzzi, A., Schmid, W., Beermann, F., and Schutz, G. (1994) Targeted mutation of the CREB gene: compensation within the CREB/ATF family of transcription factors. Proc. Natl. Acad. Sci. USA 91, 5647–5651.

Igarashi, S., Morita, H., Bennett, K. M., Tanaka, Y., Engelender, S., Peters, M. F., Cooper, J. K., Wood, J. D., Sawa, A., and Ross, C. A. (2003) Inducible PC12 cell model of Huntington's disease shows toxicity and decreased histone acetylation. Neuroreport 14, 565–568.

Impey, S., Fong, A. L., Wang, Y., Cardinaux, J. R., Fass, D. M., Obrietan, K., Wayman, G. A., Storm, D. R., Soderling, T. R., and Goodman, R. H. (2002) Phosphorylation of CBP mediates transcriptional activation by neural activity and CaM kinase IV. Neuron 34, 235–244.

Impey, S., Mark, M., Villacres, E. C., Poser, S., Chavkin, C., and Storm, D. R. (1996) Induction of CRE-mediated gene expression by stimuli that generate long-lasting LTP in area CA1 of the hippocampus. Neuron 16, 973–982.

Impey, S., McCorkle, S. R., Cha-Molstad, H., Dwyer, J. M., Yochum, G. S., Boss, J. M., McWeeney, S., Dunn, J. J., Mandel, G., and Goodman, R. H. (2004) Defining the CREB regulon: a genome-wide analysis of transcription factor regulatory regions. Cell 119, 1041–1054.

Impey, S., Smith, D. M., Obrietan, K., Donahue, R., Wade, C., and Storm, D. R. (1998) Stimulation of cAMP response element (CRE)-mediated transcription during contextual learning. Nat. Neurosci. 1, 595–601.

Johannessen, M., Delghandi, M. P., and Moens, U. (2004) What turns CREB on? Cell Signal 16, 1211–1227.

Josselyn, S. A., Kida, S., and Silva, A. J. (2004) Inducible repression of CREB function disrupts amygdala-dependent memory. Neurobiol. Learn. Mem. 82, 159–163.

Josselyn, S. A., Shi, C., Carlezon, W. A., Jr., Neve, R. L., Nestler, E. J., and Davis, M. (2001) Long-term memory is facilitated by cAMP response element-binding protein overexpression in the amygdala. J. Neurosci. 21, 2404–2412.

Jouvenceau, A., Hedou, G., Potier, B., Kollen, M., Dutar, P., and Mansuy, I. M. (2006) Partial inhibition of PP1 alters bidirectional synaptic plasticity in the hippocampus. Eur. J. Neuro sci. 24, 564–572.

Kaczmarek, L., and Chaudhuri, A. (1997) Sensory regulation of immediate-early gene expression in mammalian visual cortex: implications for functional mapping and neural plasticity. Brain Res. Brain Res. Rev. 23, 237–256.

Kandel, E. R. (2001) The molecular biology of memory storage: a dialogue between genes and synapses. Science 294, 1030–1038.

Kida, S., Josselyn, S. A., de Ortiz, S. P., Kogan, J. H., Chevere, I., Masushige, S., and Silva, A. J. (2002) CREB required for the stability of new and reactivated fear memories. Nat. Neurosci. 5, 348–355.

Kogan, J. H., Frankland, P. W., Blendy, J. A., Coblentz, J., Marowitz, Z., Schutz, G., and Silva, A. J. (1997) Spaced training induces normal long-term memory in CREB mutant mice. Curr. Biol. 7, 1–11.

Kornhauser, J. M., Cowan, C. W., Shaywitz, A. J., Dolmetsch, R. E., Griffith, E. C., Hu, L. S., Haddad, C., Xia, Z., and Greenberg, M. E. (2002) CREB transcriptional activity in neurons is regulated by multiple, calcium-specific phosphorylation events. Neuron 34, 221–233.

Korzus, E., Rosenfeld, M. G., and Mayford, M. (2004) CBP histone acetyltransferase activity is a critical component of memory consolidation. Neuron 42, 961–972.

Lamarre-Vincent, N., and Hsieh-Wilson, L. C. (2003) Dynamic glycosylation of the transcription factor CREB: a potential role in gene regulation. J. Am. Chem. Soc. 125, 6612–6613.

Lamprecht, R., and LeDoux, J. (2004) Structural plasticity and memory. Nat. Rev. Neurosci. 5, 45–54.

Lang, C., Barco, A., Zablow, L., Kandel, E. R., Siegelbaum, S. A., and Zakharenko, S. S. (2004) Transient expansion of synaptically connected dendritic spines upon induction of hippocampal long-term potentiation. Proc. Natl. Acad. Sci. USA 101, 16665–16670.

Lee, P. R., Cohen, J. E., Becker, K. G., and Fields, R. D. (2005) Gene expression in the conversion of early-phase to late-phase long-term potentiation. Ann. NY Acad. Sci. 1048, 259–271.

Lin, C. H., Yeh, S. H., Lu, K. T., Leu, T. H., Chang, W. C., and Gean, P. W. (2001) A role for the PI-3 kinase signaling pathway in fear conditioning and synaptic plasticity in the amygdala. Neuron 31, 841–851.

Liu, Y. Z., Thomas, N. S., and Latchman, D. S. (1999) CBP associates with the p42/p44 MAPK enzymes and is phosphorylated following NGF treatment. Neuroreport 10, 1239–1243.

Lonze, B. E., and Ginty, D. D. (2002) Function and regulation of CREB family transcription factors in the nervous system. Neuron 35, 605–623.

Lonze, B. E., Riccio, A., Cohen, S., and Ginty, D. D. (2002) Apoptosis, axonal growth de fects, and degeneration of peripheral neurons in mice lacking CREB. Neuron 34, 371–385.

Lu, Q., Hutchins, A. E., Doyle, C. M., Lundblad, J. R., and Kwok, R. P. (2003) Acetylation of cAMP-responsive element-binding protein (CREB) by CREB-binding protein enhances CREB-dependent transcription. J. Biol. Chem. 278, 15727–15734.

Lu, Y. F., Kandel, E. R., and Hawkins, R. D. (1999) Nitric oxide signaling contributes to late-phase LTP and CREB phosphorylation in the hippocampus. J. Neurosci. 19, 10250–10261.

Mantamadiotis, T., Lemberger, T., Bleckmann, S. C., Kern, H., Kretz, O., Martin Villalba, A., Tronche, F., Kellendonk, C., Gau, D., Kapfhammer, J., et al. (2002) Disruption of CREB function in brain leads to neurodegeneration. Nat. Genet. 31, 47–54.

Marie, H., Morishita, W., Yu, X., Calakos, N., and Malenka, R. C. (2005) Generation of silent synapses by acute in vivo expression of CaMKIV and CREB. Neuron 45, 741–752.

Martin, K. C., Casadio, A., Zhu, H., E, Y., Rose, J. C., Chen, M., Bailey, C. H., and Kandel, E. R. (1997) Synapse-specific, long-term facilitation of aplysia sensory to motor synapses: a function for local protein synthesis in memory storage. Cell 91, 927–938.

Martin, K. C., and Kosik, K. S. (2002) Synaptic tagging – who's it? Nat. Rev. Neurosci. 3, 813–820.

Martin, S. J., Grimwood, P. D., and Morris, R. G. (2000) Synaptic plasticity and memory: an evaluation of the hypothesis. Annu. Rev. Neurosci. 23, 649–711.

Matsuzaki, M., Honkura, N., Ellis-Davies, G. C., and Kasai, H. (2004) Structural basis of long-term potentiation in single dendritic spines. Nature 429, 761–766.

Mayr, B., and Montminy, M. (2001) Transcriptional regulation by the phosphorylation-dependent factor CREB. Nat. Rev. Mol. Cell Biol. 2, 599–609.

McCampbell, A., Taylor, J. P., Taye, A. A., Robitschek, J., Li, M., Walcott, J., Merry, D., Chai, Y., Paulson, H., Sobue, G., and Fischbeck, K. H. (2000) CREB-binding protein sequestration by expanded polyglutamine. Hum. Mol. Genet. 9, 2197–2202.

McClung, C. A., and Nestler, E. J. (2003) Regulation of gene expression and cocaine reward by CREB and DeltaFosB. Nat. Neurosci. 6, 1208–1215.

McMurray, C. T. (2001) Huntington's disease: new hope for therapeutics. Trends Neurosci. 24, S32–38.

Mellstrom, B., Naranjo, J. R., Foulkes, N. S., Lafarga, M., and Sassone-Corsi, P. (1993) Transcriptional response to cAMP in brain: specific distribution and induction of CREM antagonists. Neuron 10, 655–665.

Mioduszewska, B., Jaworski, J., and Kaczmarek, L. (2003) Inducible cAMP early repressor (ICER) in the nervous system–a transcriptional regulator of neuronal plasticity and pro grammed cell death. J. Neurochem. 87, 1313–1320.

Montminy, M. R., and Bilezikjian, L. M. (1987) Binding of a nuclear protein to the cyclic-AMP response element of the somatostatin gene. Nature 328, 175–178.

Morris, R. G. (2006) Elements of a neurobiological theory of hippocampal function: the role of synaptic plasticity, synaptic tagging and schemas. Eur. J. Neurosci. 23, 2829–2846.

Nagerl, U. V., Eberhorn, N., Cambridge, S. B., and Bonhoeffer, T. (2004) Bidirectional activity-dependent morphological plasticity in hippocampal neurons. Neuron 44, 759–767.

Navakkode, S., Sajikumar, S., and Frey, J. U. (2004) The type IV-specific phosphodiesterase inhibitor rolipram and its effect on hippocampal long-term potentiation and synaptic tagging. J. Neurosci. 24, 7740–7744.

Nestler, E. J. (2001) Molecular basis of long-term plasticity underlying addiction. Nat. Rev. Neurosci. 2, 119–128.

Nibuya, M., Nestler, E. J., and Duman, R. S. (1996) Chronic antidepressant administration increases the expression of cAMP response element binding protein (CREB) in rat hippocampus. J. Neurosci. 16, 2365–2372.

Nucifora, F. C., Jr., Sasaki, M., Peters, M. F., Huang, H., Cooper, J. K., Yamada, M., Takaha shi, H., Tsuji, S., Troncoso, J., Dawson, V. L., et al. (2001) Interference by huntingtin and atrophin-1 with CBP-mediated transcription leading to cellular toxicity. Science 291, 2423–2428.

Obrietan, K., and Hoyt, K. R. (2004) CRE-mediated transcription is increased in Huntington's disease transgenic mice. J. Neurosci. 24, 791–796.

Ostroff, L. E., Fiala, J. C., Allwardt, B., and Harris, K. M. (2002) Polyribosomes redistribute from dendritic shafts into spines with enlarged synapses during LTP in developing rat hippocampal slices. Neuron 35, 535–545.

Pandey, S. C., Chartoff, E. H., Carlezon, W. A., Jr., Zou, J., Zhang, H., Kreibich, A. S., Blendy, J. A., and Crews, F. T. (2005) CREB gene transcription factors: role in molecular mechanisms of alcohol and drug addiction. Alcohol Clin. Exp. Res. 29, 176–184.

Park, C. S., Gong, R., Stuart, J., and Tang, S. J. (2006) Molecular network and chromosomal clustering of genes involved in synaptic plasticity in the hippocampus. J. Biol. Chem. 281, 30195–30211.

Perazzona, B., Isabel, G., Preat, T., and Davis, R. L. (2004) The role of cAMP response element-binding protein in Drosophila long-term memory. J. Neurosci. 24, 8823–8828.

Perkinton, M. S., Ip, J. K., Wood, G. L., Crossthwaite, A. J., and Williams, R. J. (2002) Phosphatidylinositol 3-kinase is a central mediator of NMDA receptor signalling to MAP kinase (ERK1/2), Akt/PKB and CREB in striatal neurones. J. Neurochem. 80, 239–254.

Pham, T. A., Impey, S., Storm, D. R., and Stryker, M. P. (1999) CRE-mediated gene transcription in neocortical neuronal plasticity during the developmental critical period [pub lished erratum appears in Neuron 1999 Mar;22(3):635]. Neuron 22, 63–72.

Pittenger, C., Huang, Y. Y., Paletzki, R. F., Bourtchouladze, R., Scanlin, H., Vronskaya, S., and Kandel, E. R. (2002) Reversible inhibition of CREB/ATF transcription factors in region CA1 of the dorsal hippocampus disrupts hippocampus-dependent spatial memory. Neuron 34, 447–462.

Pittenger, C., and Kandel, E. (1998) A genetic switch for long-term memory. C. R. Acad. Sci. III 321, 91–96.

Radhakrishnan, I., Perez-Alvarado, G. C., Parker, D., Dyson, H. J., Montminy, M. R., and Wright, P. E. (1997) Solution structure of the KIX domain of CBP bound to the trans activation domain of CREB: a model for activator:coactivator interactions. Cell 91, 741–752.

Rammes, G., Steckler, T., Kresse, A., Schutz, G., Zieglgansberger, W., and Lutz, B. (2000) Synaptic plasticity in the basolateral amygdala in transgenic mice expressing dominant-negative cAMP response element-binding protein (CREB) in forebrain. Eur. J. Neurosci. 12, 2534–2546.

Riccio, A., Ahn, S., Davenport, C. M., Blendy, J. A., and Ginty, D. D. (1999) Mediation by a CREB family transcription factor of NGF-dependent survival of sympathetic neurons. Science 286, 2358–2361.

Sajikumar, S., and Frey, J. U. (2004) Resetting of 'synaptic tags' is time- and activity-dependent in rat hippocampal CA1 in vitro. Neuroscience 129, 503–507.

Schuman, E. M., Dynes, J. L., and Steward, O. (2006) Synaptic regulation of translation of dendritic mRNAs. J. Neurosci. 26, 7143–7146.

Segal, M. (2005) Dendritic spines and long-term plasticity. Nat. Rev. Neurosci.6, 277–284.

Sheng, M., Thompson, M. A., and Greenberg, M. E. (1991) CREB: a Ca(2+)-regulated transcription factor phosphorylated by calmodulin-dependent kinases. Science 252, 1427–1430.

Si, K., Giustetto, M., Etkin, A., Hsu, R., Janisiewicz, A. M., Miniaci, M. C., Kim, J. H., Zhu, H., and Kandel, E. R. (2003) A neuronal isoform of CPEB regulates local protein syn thesis and stabilizes synapse-specific long-term facilitation in aplysia. Cell 115, 893–904.

Steffan, J. S., Bodai, L., Pallos, J., Poelman, M., McCampbell, A., Apostol, B. L., Kazantsev, A., Schmidt, E., Zhu, Y. Z., Greenwald, M., et al. (2001) Histone deacetylase inhibitors arrest polyglutamine-dependent neurodegeneration in Drosophila. Nature 413, 739–743.

Steffan, J. S., Kazantsev, A., Spasic-Boskovic, O., Greenwald, M., Zhu, Y. Z., Gohler, H., Wanker, E. E., Bates, G. P., Housman, D. E., and Thompson, L. M. (2000) The Huntington's disease protein interacts with p53 and CREB-binding protein and represses transcription. Proc. Natl. Acad. Sci. USA 97, 6763–6768.

Steward, O., and Schuman, E. M. (2003) Compartmentalized synthesis and degradation of proteins in neurons. Neuron 40, 347–359.

Steward, O., and Worley, P. (2001) Localization of mRNAs at synaptic sites on dendrites. Results Probl. Cell Differ. 34, 1–26.

Sugars, K. L., Brown, R., Cook, L. J., Swartz, J., and Rubinsztein, D. C. (2004) Decreased cAMP response element-mediated transcription: an early event in exon 1 and full-length cell models of Huntington's disease that contributes to polyglutamine pathogenesis. J. Biol. Chem. 279, 4988–4999.

Sun, P., Enslen, H., Myung, P. S., and Maurer, R. A. (1994) Differential activation of CREB by Ca2+/calmodulin-dependent protein kinases type II and type IV involves phosphorylation of a site that negatively regulates activity. Genes Dev. 8, 2527–2539.

Sutton, M. A., and Schuman, E. M. (2005) Local translational control in dendrites and its role in long-term synaptic plasticity. J Neurobiol. 64, 116–131.

Thomas, G. M., and Huganir, R. L. (2004) MAPK cascade signalling and synaptic plasticity. Nat. Rev. Neurosci. 5, 173–183.

Thome, J., Sakai, N., Shin, K., Steffen, C., Zhang, Y. J., Impey, S., Storm, D., and Duman, R. S. (2000) cAMP response element-mediated gene transcription is upregulated by chronic antidepressant treatment. J. Neurosci. 20, 4030–4036.

Tischmeyer, W., and Grimm, R. (1999) Activation of immediate early genes and memory formation. Cell. Mol. Life Sci. 55, 564–574.

Trivier, E., De Cesare, D., Jacquot, S., Pannetier, S., Zackai, E., Young, I., Mandel, J. L., Sassone-Corsi, P., and Hanauer, A. (1996) Mutations in the kinase Rsk-2 associated with Coffin-Lowry syndrome. Nature 384, 567–570.

Tully, T., Bourtchouladze, R., Scott, R., and Tallman, J. (2003) Targeting the CREB pathway for memory enhancers. Nat. Rev. Drug Discov.2, 267–277.

Vitolo, O. V., Sant'Angelo, A., Costanzo, V., Battaglia, F., Arancio, O., and Shelanski, M. (2002) Amyloid beta -peptide inhibition of the PKA/CREB pathway and long-term potentiation: reversibility by drugs that enhance cAMP signaling. Proc. Natl. Acad. Sci. USA 99, 13217–13221.

Waddell, S., and Quinn, W. G. (2001) Flies, genes, and learning. Annu. Rev. Neurosci. 24, 1283–1309.

Walton, M. R., and Dragunow, I. (2000) Is CREB a key to neuronal survival? Trends Neuro sci. 23, 48–53.

West, A. E., Griffith, E. C., and Greenberg, M. E. (2002) Regulation of transcription factors by neuronal activity. Nat. Rev. Neurosci. 3, 921–931.

Wood, M. A., Kaplan, M. P., Park, A., Blanchard, E. J., Oliveira, A. M., Lombardi, T. L., and Abel, T. (2005) Transgenic mice expressing a truncated form of CREB-binding protein (CBP) exhibit deficits in hippocampal synaptic plasticity and memory storage. Learn. Mem. 12, 111–119.

Wu, G. Y., Deisseroth, K., and Tsien, R. W. (2001) Activity-dependent CREB phosphorylation: convergence of a fast, sensitive calmodulin kinase pathway and a slow, less sensitive mitogen-activated protein kinase pathway Proc. Natl. Acad. Sci. USA 98, 2808–2813.

Wu, X., and McMurray, C. T. (2001) Calmodulin kinase II attenuation of gene transcription by preventing cAMP response element-binding protein (CREB) dimerization and binding of the CREB-binding protein. J. Biol. Chem. 276, 1735–1741.

Xu, W., Chen, H., Du, K., Asahara, H., Tini, M., Emerson, B. M., Montminy, M., and Evans, R. M. (2001) A transcriptional switch mediated by cofactor methylation. Science 294, 2507–2511.

Yin, J. C., Del Vecchio, M., Zhou, H., and Tully, T. (1995) CREB as a memory modulator: induced expression of a dCREB2 activator isoform enhances long-term memory in Droso phila. Cell 81, 107–115.

Yin, J. C., Wallach, J. S., Del Vecchio, M., Wilder, E. L., Zhou, H., Quinn, W. G., and Tully, T. (1994) Induction of a dominant negative CREB transgene specifically blocks long-term memory in Drosophila. Cell 79, 49–58.

Young, J. Z., Isiegas, C., Abel, T., and Nguyen, P. V. (2006) Metaplasticity of the late-phase of long-term potentiation: a critical role for protein kinase A in synaptic tagging. Eur. J. Neurosci. 23, 1784–1794.

Yu, Z. X., Li, S. H., Nguyen, H. P., and Li, X. J. (2002) Huntingtin inclusions do not deplete polyglutamine-containing transcription factors in HD mice. Hum Mol Genet 11, 905–914.

Zanger, K., Radovick, S., and Wondisford, F. E. (2001) CREB binding protein recruitment to the transcription complex requires growth factor-dependent phosphorylation of its GF box. Mol. Cell 7, 551–558.

Zhang, J. J., Okutani, F., Inoue, S., and Kaba, H. (2003) Activation of the cyclic AMP re sponse element-binding protein signaling pathway in the olfactory bulb is required for the acquisition of olfactory aversive learning in young rats. Neuroscience 117, 707–713.

Zhang, X., Odom, D. T., Koo, S. H., Conkright, M. D., Canettieri, G., Best, J., Chen, H., Jenner, R., Herbolsheimer, E., Jacobsen, E., *et al.* (2005) Genome-wide analysis of cAMP-response element binding protein occupancy, phosphorylation, and target gene activation in human tissues. Proc. Natl. Acad. Sci. USA 102, 4459–4464.

Zhou, Q., Homma, K. J., and Poo, M. M. (2004) Shrinkage of dendritic spines associated with long-term depression of hippocampal synapses. Neuron 44, 749–757.

Zuo, Y., Lin, A., Chang, P., and Gan, W. B. (2005) Development of long-term dendritic spine stability in diverse regions of cerebral cortex. Neuron 46, 181–189.

Chapter 8
Activity-Dependent Regulation of *Brain-derived neurotrophic factor* Transcription

Anne E. West

Abstract Transcription of *Brain-derived neurotrophic factor (Bdnf)* is rapidly and robustly induced by neuronal activity, providing a powerful assay for the identification and characterization of neuronal activity-dependent transcriptional mechanisms. The *Bdnf* gene has a complex structure with multiple promoters that are differentially regulated by environmental stimuli through the activation of distinct complexes of activity-responsive transcription factors. Recent studies have revealed the importance of histone modifying enzyme complexes in the fine tuning of *Bdnf* transcription, including a role for the transcriptional repressor MeCP2, which is mutated in the human neurodevelopmental disorder Rett Syndrome. The cumulative data on transcriptional regulation of *Bdnf* paint a detailed portrait of this gene and define a number of key concepts for understanding the molecular mechanisms of long-lasting synaptic plasticity.

8.1 Introduction

Neuronal activity induces transcription of a large set of genes, many of which encode proteins that function at synapses. Transcriptional regulation of these gene products constitutes a mechanism for the use-dependent modification of synapses that is likely to contribute to synaptic plasticity both during development and in the adult brain.

Significant insights into the mechanisms and functions of activity-dependent transcription have come from studies of the gene encoding Brain-Derived Neurotrophic Factor (BDNF). BDNF is a secreted protein of the neurotrophin family that has numerous functions in the nervous system including the promotion of neuronal survival and the modulation of synaptic function (Lewin and Barde 1996; Poo 2001). Most importantly for this discussion, neuronal activity drives robust transcription at the *Bdnf* gene locus. Using the inducible expression of *Bdnf* mRNA as an assay, it has been possible to study mechanisms that mediate the transcriptional response to neuronal activity and subsequently to generate molecular tools that are being used to probe the functions of activity-dependent transcription in brain development and plasticity.

S. M. Dudek (ed.), *Transcriptional Regulation by Neuronal Activity.*   155
© Springer 2008

8.2 Neuronal Activity-Dependent Expression of *Bdnf* mRNA

BDNF was purified as a survival factor for CNS neurons, complementing the neurotrophic actions of NGF in the peripheral nervous system (Leibrock, et al., 1989).
However the expression pattern of BDNF – it is expressed primarily, though not
exclusively, in excitatory pyramidal neurons of the hippocampus and cerebral cortex – suggested that BDNF might also play a role in synaptic plasticity, and led
investigators to ask if expression of *Bdnf* mRNA might be induced by neuronal
activity.

Zafra and colleagues were the first to report that depolarization of cultured
embryonic rat hippocampal neurons leads to an increase in *Bdnf* mRNA that peaks
3-6 hours after stimulation and requires the elevation of intracellular calcium levels (Zafra, et al., 1990). *In vivo,* seizures were reported to drive induction of *Bdnf*
mRNA expression in both hippocampus and cortex (Ballarin, et al., 1991; Ernfors,
et al., 1991; Isackson, et al., 1991). Together these studies conclusively demonstrated that *Bdnf* mRNA expression is highly and reproducibly induced in CNS
neurons by neuronal activity. Subsequently, numerous groups have demonstrated
that a wide range of environmental stimuli upregulate expression of *Bdnf* mRNA
in corresponding activated brain regions (Table 1). The temporal and spatial correlation of *Bdnf* induction with neurons undergoing experience-dependent synaptic
plasticity strongly suggests that regulation of BDNF levels may be broadly used as
a mechanism for modulating behavioral responses to the environment.

8.3 Functional Organization of the *Bdnf* Gene

The *Bdnf* gene is comprised of multiple exons that span 52.3kB on chromosome 2 in
mouse, 50.2kB on chromosome 3 in rat, and 66.8kB on chromosome 11 in human
(Fig. 1; Genomic coordinates are from the February 2006 *Mus musculus* genome
assembly, the March 2006 *Homo sapiens* assembly, and the November 2004 *Rattus
norvegicus* assembly – http://genome.ucsc.edu). All three species appear to have at
least eight homologous exons that contribute to alternate 5' UTRs, while a ninth
exon contains the coding sequence and 3'UTR (Liu, et al., 2005; Liu, et al., 2006;
Aid, et al., 2007). The zebrafish *Bdnf* gene preserves a similar multi-exon organization suggesting that this genomic structure may have an conserved function through
evolution (Heinrich and Pagtakhan 2004). At the present time there is no evidence
for *Bdnf* orthologs in the genomes of non-vertebrate chordate or invertebrate species.

Extensive sequencing of both rodent and human cDNAs has revealed expression of a large number of distinct *Bdnf* transcripts. Differential usage of promoters,
transcription start sites, splice sites, and polyadenylation signals all contribute to
variation in these transcript sequences, and several steps in the production of *Bdnf*
mRNAs are subject to regulation by neuronal activity.

Please note, as many studies on the differential expression of *Bdnf* mRNAs were
done prior to identification of the full complement of *Bdnf* exons, the numbering of

Table 1 Differential induction of *Bdnf* transcript variants by environmental stimuli. References are as follows: 1. (Ernfors, et al., 1991), 2. (Ballarin, et al., 1991), 3. (Isackson, et al., 1991), 4. (Metsis, et al., 1993), 5. (Poulsen, et al., 2004), 6. (Castren, et al., 1992), 7. (Pattabiraman, et al., 2005), 8. (Lindvall, et al., 1992), 9. (Kokaia, et al., 1994), 10. (Patterson, et al., 1992), 11. (Lee, et al., 2005) 12. (Funakoshi, et al., 1993), 13. (Kobayashi, et al., 1996), 14. (Michael, et al., 1999), 15. (Nibuya, et al., 1995), 16. (Dias, et al., 2003), 17. (Smith, et al., 1995b), 18. (Smith, et al., 1995a), 19. (Marmigere, et al., 2003), 20. (Rocamora, et al., 1996), 21. (Nanda and Mack 1998), 22. (Nanda and Mack 2000), 23. (Neeper, et al., 1996), 24. (Russo-Neustadt, et al., 2000), 25. (Liang, et al., 1998), 26. (Berchtold, et al., 1999), 27. (Numan, et al., 1998), 28. (Grimm, et al., 2003), 29. (Liu, et al., 2006), 30. (Li, et al., 2000), 31. (Tokuyama, et al., 2000), 32. (Hall, et al., 2000), 33. (Aliaga, et al., 2002), 34. (Rattiner, et al., 2004), 35. (Pizarro, et al., 2004), 36. (Tsankova, et al., 2006), 37. (Katz and Meiri 2006)

Stimulus	Exons Induced	Brain Region	References
Seizure	I, II, IV, (VI)	Hippocampus, amygdala, piriform cortex	1–5
Light	I, II, IV, VI	Visual cortex	6, 7
Ischemia	IV	Dentate gyrus of hippocampus	8, 9
LTP	I	Hippocampus CA1	10, 11
Axotomy	VI	Motor neurons, denervated muscle, DRG neurons	12–14
Antidepressant drugs/ECS	I, II, IV, VI	Hippocampus	15, 16
Restraint stress	VI	Pituitary, hypothalamus, hippocampus	17–19
Whisker stimulation	I, IV, VI	Barrel cortex, hippocampus	20–22
Exercise	I, II	Hippocampus	23, 24
Circadian rhythms	IV	SCN, hippocampus	25, 26
Drugs of abuse	IV	Mesolimbic dopamine system (VTA, NAc, prefrontal cortex)	27–29
Bird singing		Song nucleus HVC	30
Long-term memory		Hippocampus, cortex	31, 32
Osmotic stress	I, II	Hypothalamus	33
Fear conditioning	I, IV	Basolateral amygdala	34
Social Defeat Stress	III, IV	Hippocampus	35, 36
Thermal exposure		Preoptic anterior hypothalamus	37

exons reported in the literature frequently differs from the current nine-exon nomenclature. Papers that describe a five-exon organization of the *Bdnf* gene reference 5' UTR exons I and II (which remain I and II), III and IV (now IV and VI) and the coding exon V (now IX). In this discussion I use the nine-exon rodent nomenclature described in Fig. 1B and by Aid, et al., (2007).

8.3.1 Alternative Promoters

The major source of *Bdnf* transcript variation stems from the alternate use of up to eight promoters within the *Bdnf* gene (Aoyama, et al., 2001; Bishop, et al., 1994; Liu, et al., 2005; Liu, et al., 2006; Ohara, et al., 1992; Timmusk, et al., 1993).

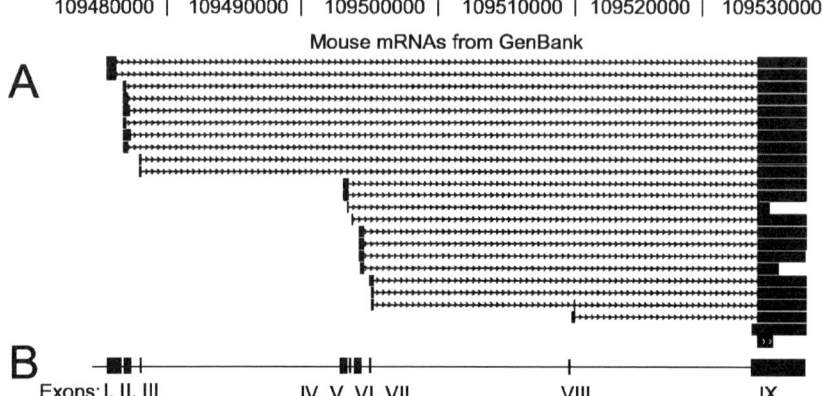

Fig. 1 Organization of the *Bdnf* gene. A) Mouse mRNAs mapped onto mouse chromosome 2. The numbers show genomic coordinates for the *Bdnf* exons from the February 2006 assembly of the mouse genome (http://genome.ucsc.edu). Black boxes represent exons and arrows show the direction of transcription. These clones represent the range of splicing patterns that occur among *Bdnf* transcripts. B) Summary of the position of all nine *Bdnf* exons along mouse chromosome 2

Transcription initiated at each of these promoters usually leads to expression of a single 5' UTR-containing exon that is spliced to form a two-part transcript with exon IX (Fig. 1B). A tripartite transcript variant consisting of exons VII, VIII, and IX has been reported by one group in rodents (Liu, et al., 2006, but see Aid, et al., 2007), however human exon VIII appears to be transcribed under its own promoter and forms a two-part transcript with human exon IX (Liu, et al., 2005). In situ hybridization, RNAse protection, and quantitative PCR studies have provided a detailed description of the patterns of differential 5' noncoding exon expression across brain regions and in non-neuronal cells (Liu, et al., 2006; Malkovska, et al., 2006; Timmusk, et al., 1993; Aid et al., 2007). In the *Molecular Mechanisms* section below I discuss how recruitment of transcription factor complexes to enhancer elements in the promoters upstream of the alternate 5' exons contributes to their extensive differential and activity-dependent regulation.

8.3.2 Transcription Start Sites

An additional layer of transcriptional complexity is implied by the observation that most if not all of the *Bdnf* promoters use multiple transcriptional start sites (TSSs). The position of the TSS is determined by the location of the core promoter elements to which RNA polymerase is recruited. Although canonically TBP is thought to recruit RNA pol II to a TATA box located 24bp upstream of the TSS, recent genome-level analyses suggest that as few as 10% of all mammalian transcripts are initiated by TATA-dependent transcription (Carninci, et al., 2006). TATA-like sequences can be found in some of the *Bdnf* promoters (Nakayama, et al., 1994;

A

                                AP-1   **USF/CREB**

...taagaagtttcctttttaccaag**ggagtcacagtgagt**tggtcacgtaactggctcagagaggctgcc

                                                    Exon I

ctggcccctcccctcgcccctccccgctgcgcttttctggtattcttat**TAAAGCAGTAGCCGGCTG**
        *   *

GTGCAGAAAAGCAACAAGTTCCCCAGCGGTCTTCCCGCCCTAGCTTGACAAGGCGAAGGGTTTCTTA...

B

                                                         Exon II

...tacgtggaaagggtctcatcaacatgtgatcaactattaacaggat**gGCTTTGGCAAAGCCATCCACA**

CGTGACAAAACGTAAGGAAGTGGAAGAAACCGTCTAGAGCAATATCAAGTACCACTTAATTAGAGAATA

                                                          RE-1

TTTTTTTAACCTTTTCCTCCTCCTGCGCCGGGTGTGTGATCCCGGAGAGCAGAGTCCATTCAGCACCTT

                                             IIA

GGACAGAGCCAGCGGATTTGTCCGAGGTGGTAGTACTTCATCCAGGTATTCTTTTCCTCGCTGTCAAGC

                  TATA box

CAACCCGGTGTCGCCCTTAAAAAGCGTCTTTTCCGAGGTTCGGCTCACACCGAGATCGGGGCTGGAGAG
   *

AGAGTCAGATTTTGGAGCGGAGCGTTTGGAGAGCGAGCCCCAGTTTGGTCCCCTCATTGAGCTCGCTGA

                                                  IIB

AGTTGGCTTCCTAGCGGTGTAGGCTGGAATAGACTCTTGGCAAGCTCCGGGTTGGTATACTGGGTTAAC

                                                  IIC

TTTGGGAAATGCAAGTGTTTATCACCAGGATCTAGCCACCGGGGTGGTGTAAGCCGCAAAGAAGgta...

C

                                                  Exon III

...tcctcctgcttccatccctccctcattttctctct**ACCTCTCCATCGCCCTCACGATCCTCGATGGA**
                  *   *

TAGTTCTTTATGTTTGGGTGATTTTTTTTTTTTTACCCCTTTCTATCATCCCTCCCCGAGAGTTCCGGG

                                              CpG island

TGCTGGCTTGGAGGGCTCCTGCTTCTCAAGGGAAGGGGGGCC*GCTGAGACTGCGCTCCACTCCCTGCC*

GGGCCGGATGCTTCATTGAGCCCAGgtccgagtcag...

Fig. 2 Transcriptional regulatory features of the mouse *Bdnf* promoters. Relevant sequence around the transcriptional start sites is shown. Lower case indicates introns and upper case indicates exons. Gray boxes mark the positions of key regulatory elements named above the sequence. The most upstream transcriptional start site for each exon is indicated by "Exon X", while asterisks (*) above the sequence show the position of alternative start sites. Genomic coordinates along mouse chromosome 2 are from the February 2006 assembly (http://genome.ucsc.edu). A) Exon I (starts at 109475538). A CpG island (not shown) is located upstream of the proximal promoter from 109474872-109475083. B) Exon II (starts at 109476735). IIA, IIB, and IIC indicate alternative splice donor sites. C) Exon III (starts at 109477870). A predicted CpG island overlaps the exon and is indicated with italics. D) Exon IV (starts at 109493274). E) Exon V (starts at109493865). F) Exon VI (starts at 109494401). *A indicates the TATA-dependent start site used in non-neuronal cells, *B indicates the activity-dependent start site. A predicted CpG island overlaps the promoter and exon and is indicated with italics. F) Exon VII (starts at 109493247)

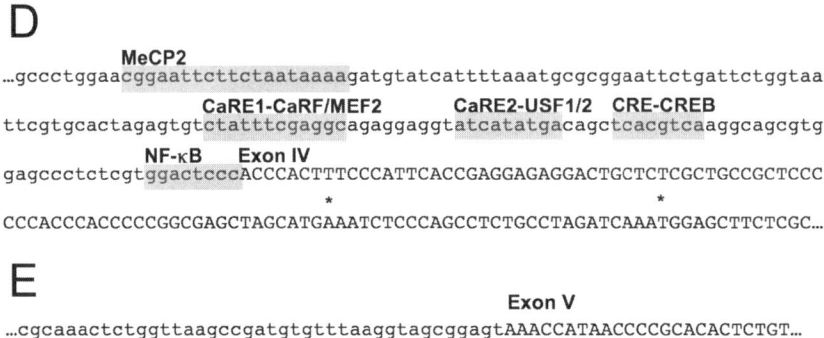

D

MeCP2
...gccctggaacggaattcttctaataaaagatgtgtatcattttaaatgcgcggaattctgattctggtaa

CaRE1-CaRF/MEF2 **CaRE2-USF1/2** **CRE-CREB**
ttcgtgcactagagtgtctatttcgaggcagaggaggtatcatatgacagctcacgtcaaggcagcgtg

NF-κB **Exon IV**
gagccctctcgtggactcccACCCACTTTCCCATTCACCGAGGAGAGGACTGCTCTCGCTGCCGCTCCC
                                    *   *
CCCACCCACCCCCGGCGAGCTAGCATGAAATCTCCCAGCCTCTGCCTAGATCAAATGGAGCTTCTCGC...

E

                                                            Exon V
...cgcaaactctggttaagccgatgtgtttaaggtagcggagtAAACCATAACCCCGCACACTCTGT...

F

      GRE
...gccctctcccacagaacttgggtgctgggatggaaaagatggggagagttgcaaaccaggggagaaag

atttgaacgtgtgtgagactcacactcgcttcctcctactctagagcacattgcggggagtagggtgccta

ggaaccgggattcgaccacccctcctagcaaagggcaccctgcgggagaccgcagagggcaggcacagag

ctttgggtttaagcagcgagctccggggaaatggagagaagccagtgcaaggcgatcagggataccccg

                                               CpG island
agggcttccgagggcttccgagggcttccgagggctgagccccc*gcaaggaaaaggcgcgtcgtcccct*

                                        GA/GC box-SP1
*ttaagcagccaccccaatgggtatcggtggcgcggcggagcagggaggggataggggcggcg*ctgtctg

Exon VI   *   **TATA box** *   *A
aCCAATCGAAGCTCAACCGAAGAGCTAAATAATGTCTGACCCCAGTGCCTGGCTCTGGCTGAGCTCTGG
                                                            *B
GTGCCCGTCGCTGCTGCCGTGCCGGGGCGCACCCGCTGGCTGGCTGTCGCACGGTTCCCAGTGCGC...

G

                                                            Exon VII
...tttggcgccttctgtcctcatctcttttcctgggtatcacaacaggctgttCACTGTCACCTGCTCTC
                                        *
TAGGGAGTACTACCAAAGCTTACTTACAGGTCCAAGGTCAACGTTTAA...

Fig. 2 (continued)

Timmusk, et al., 1993), however they lie at the correct distance from only a small number of the known TSSs, indicating that alternative core promoter elements mediate transcriptional initiation for the majority of *Bdnf* transcripts.

RNAse protection analyses have revealed that the TSS for *Bdnf* promoter VI is regulated by both cell type and neuronal activity. Exon VI-containing transcripts are the most common *Bdnf* mRNAs expressed in tissues outside of the nervous system, and these non-neuronal transcripts primarily use a TSS (Fig. 2E, *A) 24bp downstream of a consensus TATA box sequence (Timmusk, et al., 1993). By contrast in the nervous system, a much broader range of TSSs are used including several upstream of the TATA box, consistent with a TATA-independent mechanism of transcription (Timmusk, et al., 1993; Timmusk, et al., 1994b). In response to membrane depolarization of neurons, a new TSS is detected well downstream of the TATA box (Fig. 2E, *B) indicating that the location to which the core transcriptional machinery is recruited onto *Bdnf* promoter VI can be regulated in an activity-dependent manner (Timmusk, et al., 1994b). As this shorter transcript lacks a GC-rich region near the 5' end of Exon VI, it is predicted to be more easily translated, potentially enhancing the activity-dependent expression of BDNF protein. These data highlight the growing realization that RNA polymerase recruitment to core promoter elements is a target for regulation (Smale 2001) and suggest that further study of *Bdnf* promoter VI transcription may enhance understanding of activity-dependent core promoter choice mechanisms.

8.3.3 Transcript Splicing and Polyadenylation

Splice site choice and transcript termination also contribute to *Bdnf* mRNA variability, although there is no evidence that these variants are activity-regulated. Exon II has been well documented to use three alternative splice donor sites (Liu, et al., 2006; Nakayama, et al., 1994), while exon VIII-containing transcripts display variability in splice donor as well as splice acceptor sites (Liu, et al., 2005; Liu, et al., 2006). Transcript termination is signaled by recognition of two alternative polyadenylation signals, resulting in *Bdnf* transcripts with either short (0.3 kB) or long (2.9 kB) 3' UTRs (Timmusk, et al., 1993). Equimolar amounts of the short and long transcripts are associated with each of the 5' UTR exons (Maisonpierre, et al., 1990; Timmusk, et al., 1993) however the short transcripts are preferentially recovered in polysomal fractions, suggesting that they may be more easily translated (Timmusk, et al., 1994b).

8.4 Molecular Mechanisms of Activity-dependent *Bdnf* transcription

Most if not all are of the varied *Bdnf* transcripts are inducibly expressed in response to neuronal activity (Aid et al., 2007) However *in vivo* different stimuli regulate *Bdnf* expression via distinct promoters (Table 1), and between different brain regions there is a selective contribution of distinct calcium signaling pathways to *Bdnf* induction (Katoh-Semba, et al., 2001; Poulsen, et al., 2004). Although the functional significance of differential promoter regulation remains unknown, these observations point to the importance of characterizing the specific transcription factors that bind to and regulate transcription from each of the distinct *Bdnf* promoters.

A powerful first step toward identifying activity-dependent regulators of *Bdnf* transcription has been to map the calcium-response elements (CaREs) in activity-dependent promoters. By placing reporter genes such as choline acetyltransferase or luciferase under control of *Bdnf* promoter fragments, it is possible to conduct scanning mutagenesis to identify promoter regions required for activity-regulated induction of reporter gene expression. The sequence of the CaRE may match a known transcription factor binding site, suggesting candidate regulators, or it can serve as a template to screen for binding factors. Ideally the functional significance of the CaREs and their binding proteins can be confirmed through the generation of transgenic animals. Studies like these have lead to the characterization of a significant number of transcriptional regulators of *Bdnf*. One of the key findings is that distinct transcriptional regulatory mechanisms appear to be used by each of the *Bdnf* promoters – an observation that may help to explain the differential stimulus-dependent induction of *Bdnf* promoters.

8.4.1 Promoter I: CREB and USF

There are at least three interesting features of *Bdnf* promoter I activity. First, activity-dependent transcription from *Bdnf* promoter I is blocked by protein synthesis inhibitors, suggesting that it is likely to be a secondary target of immediate-early gene (IEG) transcription factors such as Fos and Jun (Lauterborn, et al., 1996). Promoter II also appears to be an IEG target, whereas promoters IV and VI are induced more rapidly in a protein-synthesis independent manner. Second, *Bdnf* promoter I is one of the later *Bdnf* promoters to become active during development, with levels of exon I continuing to rise into adulthood (Pattabiraman, et al., 2005; Timmusk, et al., 1994a). This developmental profile might suggest that exon I containing forms of *Bdnf* contribute selectively to adult rather than developmental plasticity, an idea that could be tested through the generation of *Bdnf* exon I knockout mice. Finally, promoter I is among the most highly activity-inducible promoters in the adult nervous system (Timmusk, et al., 1993). Deletion analysis of a *Bdnf* promoter I reporter gene has identified two CaREs, with a dominant contribution from an element proximal to the TSS (-173 to -154, Fig. 2A) (Tabuchi, et al., 2002). This element has overlapping consensus sites for two transcription factors – the well-known activity regulated factor CREB and the basic helix-loop-helix family members USF1/2. Though the USFs were first characterized as general transcription factors they have recently been shown to undergo calcium-dependent regulation (Chen, et al., 2003b) and both the USFs and CREB appear to contribute to the activity-dependence of the promoter I CaRE (Tabuchi, et al., 2002). It is interesting that several activity-dependent neuronal promoters including *Bdnf* promoter IV are regulated by both USFs and CREB (Chen, et al., 2003b), raising the possibility that these factors may cooperate to enhance the transcriptional response to neuronal activity.

8.4.2 Promoter II: REST-Dependent Repression

REST (NRSF) is a transcriptional repressor that plays a crucial role in restricting the expression of its target genes to neurons. REST binds to a well characterized element called the RE-1 (or NRSE), which has been identified in a number of neuronal genes including *Bdnf*. The *Bdnf* RE-1 site lies within exon II near its 5' end (Fig. 2B; Timmusk, et al., 1993). Consistent with a role for REST in regulating the timing of *Bdnf* expression during development, binding of REST to the *Bdnf* RE-1 is high in ES cells and neural progenitors which do not express *Bdnf*, while binding is lost in differentiated cortical neurons which upregulate *Bdnf* transcription (Ballas, et al., 2005).

A series of papers have argued that the RE-1 site and possibly REST may also regulate *Bdnf* in adult animals, however much remains to be determined. To test the function of the RE-1 site in *Bdnf* transcription, Timmusk et al., generated transgenic mice expressing *Bdnf* promoter I+II-CAT reporters with either an intact or mutant RE-1 site (Timmusk, et al., 1999). Loss of the RE-1 site did not induce transcription of the reporter gene in non-neuronal cells as was expected, but instead lead to elevation of *Bdnf* promoter activity in neurons within the adult brain. One possibility is that this effect of the RE-1 is mediated directly by REST itself. *Rest* mRNA is expressed in regions of the rat brain and has been reported to be induced in hippocampus by kainic acid (Palm, et al., 1998), however there is evidence against the expression of significant amounts of REST protein in adult brain (Ballas, et al., 2005). In addition to REST, it has been suggested that the huntingtin protein, which is mutated in Huntington's Disease, may regulate *Bdnf* transcription through the RE-1 site. Expression of mutant huntingtin is associated with reduced *Bdnf* expression in both mice and humans (Zuccato, et al., 2001). The authors propose that this defect is due to loss of the wild type function of huntingtin, which they show binds to REST and sequesters it in the cytoplasm, promoting the expression of RE-1 containing genes in the nucleus (Zuccato, et al., 2003). Alternatively, factors other than REST may bind to the RE-1, and huntingtin could exert its effects through interactions with these proteins. Future studies that address this pathway in targeted knockin and knockout mouse models will help to clarify the specific functions of the RE-1, REST, and huntingtin in transcriptional regulation of *Bdnf*.

8.4.3 Promoter IV: Cooperation Confers Calcium Selectivity

As the most highly transcribed and calcium inducible promoter in embryonic neurons, promoter IV has been extensively studied (West, et al., 2001). These studies reveal that a strikingly large number of calcium-responsive transcription factors contribute to regulation of this promoter. Assays using the upstream TSS for exon IV have uncovered roles for at least three calcium-response elements (CaREs) bound by at least four transcriptional activators- CaRF (Tao, et al., 2002), MEF2 (A.E. West and M.E. Greenberg, manuscript in preparation), USF1/2 (Chen, et al., 2003b) and CREB (Shieh, et al., 1998; Tao, et al., 1998) – each of which is independently responsive to calcium signaling (Fig. 2D). Assays using one of the downstream start sites have indicated a contribution from a additional complex of transcriptional regulators including the calcium-responsive activator NF-κB (Lipsky, et al., 2001), however whether this complex cooperates with the factors that regulate the upstream TSS or operates independently is not known. In reporter gene assays each of the three CaREs is required for activity-dependent transcription from *Bdnf* promoter IV suggesting that the factors that bind these CaREs need to cooperate in order to form an active transcriptional complex (Tao, et al., 2002). One consequence of this cooperative activation may be to confer stimulus specificity upon the induction of *Bdnf* promoter IV transcription (West, et al., 2002). Although the transcription of some CREB target genes such as *Fos* is equally well induced in neurons by both calcium and cAMP signaling, *Bdnf* exon IV expression shows a strong preference for calcium signaling pathways over cAMP. Calcium-selective regulatory events occur on CaRF (Tao, et al., 2002), MEF2D (Belfield, et al., 2006), CREB (Kornhauser, et al., 2002), and the transcriptional repressor MeCP2 (Zhou, et al., 2006) (see *Chromatin* section below) suggesting the existence of a complex multi-part mechanism for the calcium-selective activation of *Bdnf* exon IV transcription.

8.4.4 Promoter VI: An Expanding Repertoire of Regulated Factors

Reporter gene assays for promoter VI activity have identified three *cis* regulatory elements required for transcriptional activation by the MAP kinase, CaMKII, and PKA signaling pathways (Takeuchi, et al., 2002). The presence of a CCAAT-box in the most distal regulatory element and the presence of both a GA-rich sequence and a GC-box in the middle element suggested C/EBP/β and Sp1 respectively as candidate factors to mediate these responses (Fig. 2E). It is noteworthy that these promoter VI regulators have been more commonly studied outside of the nervous system, highlighting the fact that the transcriptional pathways regulated by neuronal activity may not necessarily be neuronal specific. In addition to neuronal activity, both NGF and corticosteroid hormones have been shown to regulate *Bdnf* promoter VI (Hansson, et al., 2006; Park, et al., 2006). NGF is likely to act through the MAP kinase pathway to induce transcription via the distal regulatory element. By contrast, steroid hormones lead to a reduction in *Bdnf* exon VI expression possibly by direct binding of a steroid hormone receptor repressor complex to promoter VI. A putative glucocorticoid response element-like sequence has been identified in promoter VI (Funakoshi, et al., 1993), however whether this sequence is required for the downregulation of *Bdnf* in response to steroid hormones remains to be determined.

8.5 Chromatin Regulation of *Bdnf* Transcription

In addition to modulating the function of CaRE-binding transcription factors, it is becoming increasingly clear that neuronal activity also regulates gene transcription by acting on enzymes that alter chromatin architecture, making it either more or less permissive for transcription. Strong evidence for activity-dependent chromatin remodeling at the *Bdnf* locus has come from chromatin immunoprecipitation (ChIP) studies that have examined the association of specific post-translationally modified histones with the various *Bdnf* promoters. Histones are the proteins that wrap genomic DNA into secondary structural units called nucleosomes. Posttranslational modifications (including acetylation, methylation, and phosphorylation) of specific amino acid residues of the histone tails are tightly correlated with either transcriptional activation or repression (Jenuwein and Allis 2001), suggesting that regulation of the enzymes that modify histones is a key step in transcriptional control.

8.5.1 Activity-Dependent Histone Modifications at **Bdnf** Promoters

Several stimuli that induce *Bdnf* transcription – including seizure, membrane depolarization, and cocaine exposure – have been demonstrated to drive increased acetylation of histones at *Bdnf* promoters (Chen, et al., 2003a; Huang, et al., 2002; Kumar, et al., 2005; Martinowich, et al., 2003; Tsankova, et al., 2004; Tsankova, et al., 2006). Active transcription is highly correlated with acetylation of histone H3 (at lysine 9 (K9) and K14) and histone H4 (at K5, K8, K12, and K16), and regulated acetylation of both of these histones has been detected on the *Bdnf* gene. Interestingly, some stimuli differentially drive acetylation of either histone H3 or histone H4 at different *Bdnf* promoters, or in response to acute versus chronic delivery of stimuli (Tsankova, et al., 2004). The significance of these differences remains unknown but they may be related to the identity of the specific histone acetyltransferases that mediate transcriptional activation of *Bdnf*.

Activity-dependent regulation of histone methylation has also been observed on *Bdnf* promoters, implicating an additional set of regulatory enzymes in transcriptional control. Histone methylation is more complex than acetylation as it has been associated with either transcriptional activation or repression depending on the particular lysine that is methylated, and because single amino acid residues may be either mono-, di-, or tri- methylated resulting in different effects on gene transcription (Lachner and Jenuwein 2002). On *Bdnf* promoter IV, membrane depolarization *in vitro* drives increased dimethylation of histone H3-K4, a modification associated with transcriptional activation (Martinowich, et al., 2003), while on the same promoter, a repressive methylation event – dimethylation of histone H3-K9 – is reduced by activity (Chen, et al., 2003a; Martinowich, et al., 2003).

However, the histone code can be complex. For example, *in vivo*, downregulation of *Bdnf* exons III and IV is seen in hippocampus in a paradigm of social defeat stress in mice (Tsankova, et al., 2006). This decrease in *Bdnf* transcription is correlated with an increase in repressive histone H3-K27 dimethylation on both promoters III and IV, without changes in H3-K9 dimethylation. Interestingly, acute treatment of defeated mice with the antidepressant imipramine restores *Bdnf* expression and induces the activating dimethylation of H3-K4 *without* diminishing the "repressive" dimethylation of H3-K27. One hypothesis is that this persistent histone methylation chronically alters the transcriptional state of *Bdnf* while still allowing it to be acutely induced by some stimuli. Further studies that address both the nature of the enzymes that mediate these histone modifications as well as the mechanisms of their effects on transcription will be essential for full understanding of this important process (Tsankova, et al., 2006).

8.5.2 MeCP2: A Link Between DNA Methylation, BDNF, and Disease

One intriguing candidate regulator of activity-modulated histone H3-K9 dimethylation on *Bdnf* promoter IV is the methyl-DNA binding transcriptional repressor MeCP2. MeCP2 is required for normal human brain development, as loss-of-function mutations in *MECP2* cause the severe neurodevelopmental disorder Rett Syndrome which is characterized by the lack of speech, loss of purposeful use of the hands, and an acquired microcephaly (Bienvenu and Chelly 2006). In resting neurons, in which *Bdnf* transcription is low, MeCP2 is bound to a methylated CpG dinucleotide in *Bdnf* promoter IV located at -148 bp 5' to the most upstream exon IV TSS (Chen, et al., 2003a; Martinowich, et al., 2003). Mice bearing mutations of either *Mecp2* or the maintenance methyltransferase *Dnmt1* which methylates *Bdnf* promoter IV, show an elevation of *Bdnf* mRNA expression in resting neurons (Chen, et al., 2003a; Martinowich, et al., 2003). MeCP2 has been shown to associate with both a histone deacetylase as well as a histone H3-K9 methyltransferase (Fuks, et al., 2003; Nan, et al., 1998), suggesting that it may recruit these enzymes to *Bdnf* promoter IV to repress transcription in the absence of neuronal activity.

In order for neuronal activity to drive transcription from *Bdnf* promoter IV, MeCP2-dependent repression of the promoter must first be relieved. In response to membrane depolarization, MeCP2 is rapidly phosphorylated at Ser421 in a manner dependent on CaMKII (Zhou, et al., 2006). This phosphorylation event is required for derepression of *Bdnf* exon IV transcription and has been suggested to act by releasing the complex of MeCP2 and its associated repressive histone modifying enzymes from their binding site in promoter IV (Ballas, et al., 2005; Chen, et al., 2003a; Zhou, et al., 2006).

Since MeCP2 requires DNA methylation to associate with its target gene promoters, it would be theoretically possible to induce transcription of its targets by regulating the state of DNA methylation. Although the identity of enzymes that demethylate DNA *in vivo* remains unknown, one study has suggested that membrane depolarization leads to demethylation of several CpG dinucleotides in *Bdnf* promoter IV (Martinowich, et al., 2003) although this effect was not observed by a second group (Chen, et al., 2003a). Several individual CpG dinucleotides in *Bdnf* promoter IV do show developmental and brain-region specific patterns of methylation that require neuronal expression of the maintenance DNA methyltransferase Dnmt1 and correlate with levels of *Bdnf* transcription (Dennis and Levitt 2005; Fan, et al., 2001; Martinowich, et al., 2003). The *Bdnf* gene also contains CpG islands that could contribute to *Bdnf* transcription (Figs. 1, 2; Levenson et al.,, 2006), however the methylation status of these regions is just beginning to be studied. Further study will be required to fully characterize functionally relevant sites of DNA methylation in the *Bdnf* gene and to understand whether and how neuronal activity may play a role in regulation of DNA methylation patterns.

8.6 Beyond Transcription – RNA Stability and Translation

Although the largest body of research on the activity-dependent induction of *Bdnf* expression has focused on transcriptional regulation, evidence for several posttranscriptional regulatory steps that affect the stability, localization, and translatability of *Bdnf* mRNA is beginning to emerge as well.

Neuronal activity prolongs the half-life of pre-existing *Bdnf* mRNA, potentially enhancing the effects of new transcription on total cellular *Bdnf* mRNA levels (Fukuchi, et al., 2005). Once synthesized, *Bdnf* mRNA is transported into apical dendrites, where it is hypothesized that synaptic activity could locally regulate its translation (Tongiorgi, et al., 1997; Tongiorgi, et al., 2004). Intriguingly, this transport appears to be selective for specific transcript variants of *Bdnf* as exon VI- but not exon IV-containing forms of *Bdnf* are selectively targeted to dendrites after stimulation of visual cortical neurons (Pattabiraman, et al., 2005). It remains unknown whether *Bdnf* mRNA undergoes regulated translation and processing in dendrites or what functional consequences dendritic mRNA targeting may have, however future studies that characterize the molecular mechanisms of *Bdnf* mRNA targeting should provide important tools for addressing these questions.

Stability and/or translatability of *Bdnf* mRNAs may also be influenced in part by the expression of regulatory non-coding RNAs. The *Bdnf* 3'UTR contains a putative binding site for the microRNA mIR-1 (Lewis, et al., 2003), although a functional role for miRNAs in the regulation of BDNF expression has not yet been established. Recently it has also been reported that the opposite strand of the human *Bdnf* gene encodes a variably spliced, apparently non-coding transcript spanning eleven exons transcribed in reverse orientation to *Bdnf* (Liu, et al., 2005). As one of the opposite strand exons overlaps in reverse orientation with the coding exon of *Bdnf,* and opposite strand splice variants containing this exon are expressed in brain, one possible function of these human-specific transcripts is to act as antisense regulators of *Bdnf* mRNA expression.

8.7 Conclusions

Given the important functions of BDNF in the regulation of neuron survival and synaptic plasticity, as well as its relevance to human diseases including depression and the neurodevelopmental disorder Rett Syndrome, intense effort has been devoted to identifying the molecular mechanisms of activity-dependent *Bdnf* transcription. These studies have revealed that the *Bdnf* gene is regulated by an intricate network of processes including the methylation of chromosomal DNA, the posttranslational modification of histones at *Bdnf* promoters, the association of distinct complexes of transcription factors with alternative promoters, and the selective regulation of these factors by calcium signaling pathways. That neurons expend such effort to obtain precise spatial and temporal regulation of *Bdnf* levels further suggests the functional importance of BDNF in the brain. Future studies that use information about these transcriptional regulatory mechanisms to experimentally manipulate the pattern and level of *Bdnf* expression will be likely to significantly advance our understanding of the mechanisms of neural plasticity.

Acknowledgments This work was supported in part by a grant from NARSAD: The Mental Health Research Association.

References

Aid, T., Kazantseva, A., Piirsoo, M., Palm, K., and Timmusk, T. (2007) Mouse and rat BDNF gene structure revisited. J Neurosci Res. 85: 525–535.

Aliaga, E., Arancibia, S., Givalois, L. and Tapia-Arancibia, L. (2002) Osmotic stress increases brain-derived neurotrophic factor messenger RNA expression in the hypothalamic supraoptic nucleus with differential regulation of its transcripts. Relation to arginine-vasopressin content. Neuroscience 112, 841–850.

Aoyama, M., Asai, K., Shishikura, T., Kawamoto, T., Miyachi, T., Yokoi, T., Togari, H., Wada, Y., Kato, T. and Nakagawara, A. (2001) Human neuroblastomas with unfavorable biologies express high levels of brain-derived neurotrophic factor mRNA and a variety of its variants. Cancer Lett. 164, 51–60.

Ballarin, M., Ernfors, P., Lindefors, N. and Persson, H. (1991) Hippocampal damage and kainic acid injection induce a rapid increase in mRNA for BDNF and NGF in the rat brain. Exp. Neurol. 114, 35–43.

Ballas, N., Grunseich, C., Lu, D. D., Speh, J. C. and Mandel, G. (2005) REST and its core-pressors mediate plasticity of neuronal gene chromatin throughout neurogenesis. Cell 121, 645–657.

Belfield, J. L., Whittaker, C., Cader, M. Z. and Chawla, S. (2006) Differential effects of Ca2+ and cAMP on transcription mediated by MEF2D and cAMP-response element-binding protein in hippocampal neurons. J. Biol. Chem. 281, 27724–27732.

Berchtold, N. C., Oliff, H. S., Isackson, P. and Cotman, C. W. (1999) Hippocampal BDNF mRNA shows a diurnal regulation, primarily in the exon III transcript. Brain Res. Mol. Brain Res. 71, 11–22.

Bienvenu, T. and Chelly, J. (2006) Molecular genetics of Rett syndrome: when DNA methylation goes unrecognized. Nat. Rev. Genet. 7, 415–426.

Bishop, J. F., Mueller, G. P. and Mouradian, M. M. (1994) Alternate 5' exons in the rat brain-derived neurotrophic factor gene: differential patterns of expression across brain regions. Brain Res. Mol. Brain Res. 26, 225–232.

Carninci, P., Sandelin, A., Lenhard, B., Katayama, S., Shimokawa, K., Ponjavic, J., Semple, C. A., Taylor, M. S., Engstrom, P. G., Frith, M. C., Forrest, A. R., Alkema, W. B., Tan, S. L., Plessy, C., Kodzius, R., Ravasi, T., Kasukawa, T., Fukuda, S., Kanamori-Katayama, M., Kitazume, Y., Kawaji, H., Kai, C., Nakamura, M., Konno, H., Nakano, K., Mottagui-Tabar, S., Arner, P., Chesi, A., Gustincich, S., Persichetti, F., Suzuki, H., Grimmond, S. M., Wells, C. A., Orlando, V., Wahlestedt, C., Liu, E. T., Harbers, M., Kawai, J., Bajic, V. B., Hume, D. A. and Hayashizaki, Y. (2006) Genome-wide analysis of mammalian promoter architecture and evolution. Nat. Genet. 38, 626–635.

Castren, E., Zafra, F., Thoenen, H. and Lindholm, D. (1992) Light regulates expression of brain-derived neurotrophic factor mRNA in rat visual cortex. Proc. Natl. Acad. Sci. USA 89, 9444–9448.

Chen, W. G., Chang, Q., Lin, Y., Meissner, A., West, A. E., Griffith, E. C., Jaenisch, R. and Greenberg, M. E. (2003a) Derepression of BDNF transcription involves calcium-dependent phosphorylation of MeCP2. Science 302, 885–889.

Chen, W. G., West, A. E., Tao, X., Corfas, G., Szentirmay, M., Sawadogo, M., Vinson, C. and Greenberg, M. E. (2003b) Upstream stimulatory factors are mediators of calcium-responsive transcription in neurons. J. Neurosci. 23, 2572–2581.

Dennis, K. E. and Levitt, P. (2005) Regional expression of brain derived neurotrophic factor (BDNF) is correlated with dynamic patterns of promoter methylation in the developing mouse forebrain. Brain Res. Mol. Brain Res. 140, 1–9.

Dias, B. G., Banerjee, S. B., Duman, R. S. and Vaidya, V. A. (2003) Differential regulation of brain derived neurotrophic factor transcripts by antidepressant treatments in the adult rat brain. Neuropharmacology 45, 553–563.

Ernfors, P., Bengzon, J., Kokaia, Z., Persson, H. and Lindvall, O. (1991) Increased levels of messenger RNAs for neurotrophic factors in the brain during kindling epileptogenesis. Neuron 7, 165–176.

Fan, G., Beard, C., Chen, R. Z., Csankovszki, G., Sun, Y., Siniaia, M., Biniszkiewicz, D., Bates, B., Lee, P. P., Kuhn, R., Trumpp, A., Poon, C., Wilson, C. B. and Jaenisch, R. (2001) DNA hypomethylation perturbs the function and survival of CNS neurons in postnatal animals. J. Neurosci. 21, 788–797.

Fuks, F., Hurd, P. J., Wolf, D., Nan, X., Bird, A. P. and Kourzarides, T. (2003) The Methyl-CpG-binding protein MeCP2 links DNA methylation to histone methylation. J. Biol. Chem. 278, 4035–4040.

Fukuchi, M., Tabuchi, A. and Tsuda, M. (2005) Transcriptional regulation of neuronal genes and its effect on neural functions: cumulative mRNA expression of PACAP and BDNF genes controlled by calcium and cAMP signals in neurons. J. Pharmacol. Sci. 98, 212–218.

Funakoshi, H., Frisen, J., Barbany, G., Timmusk, T., Zachrisson, O., Verge, V. M. and Persson, H. (1993) Differential expression of mRNAs for neurotrophins and their receptors after axotomy of the sciatic nerve. J. Cell Biol. 123, 455–465.

Grimm, J. W., Lu, L., Hayashi, T., Hope, B. T., Su, T. P. and Shaham, Y. (2003) Time-dependent increases in brain-derived neurotrophic factor protein levels within the mesolimbic dopamine system after withdrawal from cocaine: implications for incubation of cocaine craving. J. Neurosci. 23, 742–747.

Hall, J., Thomas, K. L. and Everitt, B. J. (2000) Rapid and selective induction of BDNF expression in the hippocampus during contextual learning. Nat. Neurosci. 3, 533–535.

Hansson, A. C., Sommer, W. H., Metsis, M., Stromberg, I., Agnati, L. F. and Fuxe, K. (2006) Corticosterone actions on the hippocampal brain-derived neurotrophic factor expression are mediated by exon IV promoter. J. Neuroendocrinol. 18, 104–114.

Heinrich, G. and Pagtakhan, C. J. (2004) Both 5' and 3' flanks regulate Zebrafish brain-derived neurotrophic factor gene expression. BMC Neurosci. 5, 19.

Huang, Y., Doherty, J. J. and Dingledine, R. (2002) Altered histone acetylation at glutamate receptor 2 and brain-derived neurotrophic factor genes is an early event triggered by status epilepticus. J. Neurosci. 22, 8422–8428.

Isackson, P. J., Huntsman, M. M., Murray, K. D. and Gall, C. M. (1991) BDNF mRNA expression is increased in adult rat forebrain after limbic seizures: temporal patterns of induction distinct from NGF. Neuron 6, 937–948.

Jenuwein, T. and Allis, C. D. (2001) Translating the histone code. Science 293, 1074–1080.

Katoh-Semba, R., Takeuchi, I. K., Inaguma, Y., Ichisaka, S., Hata, Y., Tsumoto, T., Iwai, M., Mikoshiba, K. and Kato, K. (2001) Induction of brain-derived neurotrophic factor by convulsant drugs in the rat brain: involvement of region-specific voltage-dependent calcium channels. J. Neurochem. 77, 71–83.

Katz, A. and Meiri, N. (2006) Brain-derived neurotrophic factor is critically involved in thermal-experience-dependent developmental plasticity. J Neurosci 26, 3899–3907.

Kobayashi, N. R., Bedard, A. M., Hincke, M. T. and Tetzlaff, W. (1996) Increased expression of BDNF and trkB mRNA in rat facial motoneurons after axotomy. Eur. J. Neurosci. 8, 1018–1029.

Kokaia, Z., Metsis, M., Kokaia, M., Bengzon, J., Elmer, E., Smith, M. L., Timmusk, T., Siesjo, B. K., Persson, H. and Lindvall, O. (1994) Brain insults in rats induce increased expression of the BDNF gene through differential use of multiple promoters. Eur. J. Neurosci. 6, 587–596.

Kornhauser, J. M., Cowan, C. W., Shaywitz, A. J., Dolmetsch, R. E., Griffith, E. C., Hu, L. S., Haddad, C., Xia, Z. and Greenberg, M. E. (2002) CREB transcriptional activity in neurons is regulated by multiple, calcium-specific phosphorylation events. Neuron 34, 221–233.

Kumar, A., Choi, K. H., Renthal, W., Tsankova, N. M., Theobald, D. E., Truong, H. T., Russo, S. J., Laplant, Q., Sasaki, T. S., Whistler, K. N., Neve, R. L., Self, D. W. and Nestler, E. J. (2005) Chromatin remodeling is a key mechanism underlying cocaine-induced plasticity in striatum. Neuron 48, 303–314.

Lachner, M. and Jenuwein, T. (2002) The many faces of histone lysine methylation. Curr. Opin. Cell Biol. 14, 286–298.

Lauterborn, J. C., Rivera, S., Stinis, C. T., Haynes, V. Y., Isackson, P. J. and Gall, C. M. (1996) Differential effects of protein synthesis inhibition on the activity-dependent expression of BDNF transcripts: evidence for immediate-early gene responses from specific promoters. J. Neurosci. 16, 7428–7436.

Lee, P. R., Cohen, J. E., Becker, K. G. and Fields, R. D. (2005) Gene expression in the conversion of early-phase to late-phase long-term potentiation. Ann. N. Y. Acad. Sci. 1048, 259–271.

Leibrock, J., Lottspeich, F., Hohn, A., Hofer, M., Hengerer, B., Masiakowski, P., Thoenen, H. and Barde, Y.-A. (1989) Molecular cloning and expression of brain-derived neurotrophic factor. Nature 341, 149–152.

Levenson, J. M., Roth, T. L., Lubin, F. D., Miller, C. A., Huang, I-C., Desai, P., Malone, L. M., and Sweatt, J. D. (2006) Evidence that DNA (Cytosine-5) methyltransferase regulates synaptic plasticity in the hippocampus. J. Biol. Chem. 23: 15763–15773.

Lewin, G. R. and Barde, Y. A. (1996) Physiology of the neurotrophins. Annu. Rev. Neurosci. 19, 289–317.

Lewis, B. P., Shih, I. H., Jones-Rhoades, M. W., Bartel, D. P. and Burge, C. B. (2003) Prediction of mammalian microRNA targets. Cell 115, 787–798.

Li, X. C., Jarvis, E. D., Alvarez-Borda, B., Lim, D. A. and Nottebohm, F. (2000) A relationship between behavior, neurotrophin expression, and new neuron survival. Proc. Natl. Acad. Sci. U S A 97, 8584–8589.

Liang, F. Q., Walline, R. and Earnest, D. J. (1998) Circadian rhythm of brain-derived neurotrophic factor in the rat suprachiasmatic nucleus. Neurosci. Lett. 242, 89–92.

Lindvall, O., Ernfors, P., Bengzon, J., Kokaia, Z., Smith, M. L., Siesjo, B. K. and Persson, H. (1992) Differential regulation of mRNAs for nerve growth factor, brain-derived neurotrophic factor, and neurotrophin 3 in the adult rat brain following cerebral ischemia and hypoglycemic coma. Proc. Natl. Acad. Sci. U S A 89, 648–652.

Lipsky, R. H., Xu, K., Zhu, D., Kelly, C., Terhakopian, A., Novelli, A. and Marini, A. M. (2001) Nuclear factor kappaB is a critical determininat in N-methyl-D-aspartate receptor-mediated neuroprotection. J. Neurochem. 78, 254–264.

Liu, Q. R., Walther, D., Drgon, T., Polesskaya, O., Lesnick, T. G., Strain, K. J., De Andrade, M., Bower, J. H., Maraganore, D. M. and Uhl, G. R. (2005) Human brain derived neurotrophic factor (BDNF) genes, splicing patterns, and assessments of associations with substance abuse and Parkinson's Disease. Am. J. Med. Genet. B Neuropsychiatr Genet 134, 93–103.

Liu, Q. R., Lu, L., Zhu, X. G., Gong, J. P., Shaham, Y. and Uhl, G. R. (2006) Rodent BDNF genes, novel promoters, novel splice variants, and regulation by cocaine. Brain Res. 1067, 1–12.

Maisonpierre, P. C., Belluscio, L., Friedman, B., Alderson, R. F., Wiegand, S. J., Furth, M. E., Lindsay, R. M. and Yancopoulos, G. D. (1990) NT-3, BDNF, and NGF in the developing rat nervous system: parallel as well as reciprocal patterns of expression. Neuron 5, 501–509.

Malkovska, I., Kernie, S. G. and Parada, L. F. (2006) Differential expression of the four untranslated BDNF exons in the adult mouse brain. J. Neurosci. Res. 83, 211–221.

Marmigere, F., Givalois, L., Rage, F., Arancibia, S. and Tapia-Arancibia, L. (2003) Rapid induction of BDNF expression in the hippocampus during immobilization stress challenge in adult rats. Hippocampus 13, 646–655.

Martinowich, K., Hattori, D., Wu, H., Fouse, S., He, F., Hu, Y., Fan, G. and Sun, Y. E. (2003) DNA methylation-related chromatin remodeling in activity-dependent BDNF gene regulation. Science 302, 890–893.

Metsis, M., Timmusk, T., Arenas, E. and Persson, H. (1993) Differential usage of multiple brain-derived neurotrophic factor promoters in the rat brain following neuronal activation. Proc. Natl. Acad. Sci. USA 90, 8802–8806.

Michael, G. J., Averill, S., Shortland, P. J., Yan, Q. and Priestley, J. V. (1999) Axotomy results in major changes in BDNF expression by dorsal root ganglion cells: BDNF expression in large trkB and trkC cells, in pericellular baskets, and in projections to deep dorsal horn and dorsal column nuclei. Eur. J. Neurosci. 11, 3539–3551.

Nakayama, M., Gahara, Y., Kitamura, T. and Ohara, O. (1994) Distictive four promoters collectively direct expression of brain-derived neurotrophic factor gene. Mol. Brain Res. 21, 206–218.

Nan, X., Ng, H. H., Johnson, C. A., Laherty, C. D., Turner, B. M., Eisenman, R. N. and Bird, A. (1998) Transcriptional repression by the methyl-CpG-binding protein MeCP2 involves a histone deacetylase complex. Nature 393, 386–389.

Nanda, S. and Mack, K. J. (1998) Multiple promoters direct stimulus and temporal specific expression of brain-derived neurotrophic factor in the somatosensory cortex. Mol. Brain Res. 62, 216–219.

Nanda, S. A. and Mack, K. J. (2000) Seizures and sensory stimulation result in different patterns of brain derived neurotrophic factor protein expression in the barrel cortex and hippocampus. Brain Res. Mol. Brain Res. 78, 1–14.

Neeper, S. A., Gomez-Pinilla, F., Choi, J. and Cotman, C. W. (1996) Physical activity increases mRNA for brain-derived neurotrophic factor and nerve growth factor in rat brain. Brain Res. 726, 49–56.

Nibuya, M., Morinobu, S. and Duman, R. S. (1995) Regulation of BDNF and trkB mRNA in rat brain by chronic electroconvulsive seizure and antidepressant drug treatments. J. Neurosci. 15, 7539–7547.

Numan, S., Lane-Ladd, S. B., Zhang, L., Lundgren, K. H., Russell, D. S., Seroogy, K. B. and Nestler, E. J. (1998) Differential regulation of neurotrophin and trk receptor mRNAs in catecholaminergic nuclei during chronic opiate treatment and withdrawal. J. Neurosci. 18, 10700–10708.

Ohara, O., Gahara, Y., Teraoka, H. and Kitamura, T. (1992) A rat brain neurotrophic factor-encoding gene generates multiple transcripts through alternative use of 5' exons and polyadenylation sites. Gene 121, 383–386.

Palm, K., Belluardo, N., Metsis, M. and Timmusk, T. (1998) Neuronal expression of zinc finger transcription factor REST/NRSF/XBR gene. J. Neurosci. 18, 1280–1296.

Park, S. Y., Lee, J. Y., Choi, J. Y., Park, M. J. and Kim, D. S. (2006) Nerve growth factor activates brain-derived neurotrophic factor promoter IV via extracellular signal-regulated protein kinase 1/2 in PC12 cells. Mol. Cells 21, 237–243.

Pattabiraman, P. P., Tropea, D., Chiaruttini, C., Tongiorgi, E., Cattaneo, A. and Domenici, L. (2005) Neuronal activity regulates the developmental expression and subcellular localization of cortical BDNF mRNA isoforms in vivo. Mol. Cell. Neurosci. 28, 556–570.

Patterson, S. L., Grover, L. M., Schwartzkroin, P. A. and Bothwell, M. (1992) Neurotrophin expression in rat hippocampal slices: a stimulus paradigm inducing LTP in CA1 evokes increases in BDNF and NT-3 mRNAs. Neuron 9, 1081–1088.

Pizarro, J. M., Lumley, L. A., Medina, W., Robison, C. L., Chang, W. E., Alagappan, A., Bah, M. J., Dawood, M. Y., Shah, J. D., Mark, B., Kendall, N., Smith, M. A., Saviolakis, G. A. and Meyerhoff, J. L. (2004) Acute social defeat reduces neurotrophin expression in brain cortical and subcortical areas in mice. Brain Res. 1025, 10–20.

Poo, M.-M. (2001) Neurotrophins as synaptic modulators. Nat. Rev. Neurosci. 2, 1–9.

Poulsen, F. R., Lauterborn, J., Zimmer, J. and Gall, C. M. (2004) Differential expression of brain-derived neurotrophic factor transcripts after pilocarpine-induced seizure-like activity is related to mode of Ca2+ entry. Neuroscience 126, 665–676.

Rattiner, L. M., Davis, M., French, C. T. and Ressler, K. J. (2004) Brain-derived neurotrophic factor and tyrosine kinase receptor B involvement in amygdala-dependent fear conditioning. J. Neurosci. 24, 4796–4806.

Rocamora, N., Welker, E., Pascual, M. and Soriano, E. (1996) Upregulation of BDNF mRNA expression in the barrel cortex of adult mice after sensory stimulation. J. Neurosci. 16, 4411–4419.

Russo-Neustadt, A. A., Beard, R. C., Huang, Y. M. and Cotman, C. W. (2000) Physical activity and antidepressant treatment potentiate the expression of specific brain-derived neurotrophic factor transcripts in the rat hippocampus. Neuroscience 101, 305–312.

Shieh, P. B., Hu, S.-C., Bobb, K., Timmusk, T. and Ghosh, A. (1998) Identification of a signaling pathway involved in calcium regulation of BDNF expression. Neuron 20, 727–740.

Smale, S. T. (2001) Core promoters: active contributors to combinatorial gene regulation. Genes Dev. 15, 2503–2508.

Smith, M. A., Makino, S., Kim, S. Y. and Kvetnansky, R. (1995a) Stress increases brain-derived neurotropic factor messenger ribonucleic acid in the hypothalamus and pituitary. Endocrinology 136, 3743–3750.

Smith, M. A., Makino, S., Kvetnansky, R. and Post, R. M. (1995b) Stress and glucocorticoids affect the expression of brain-derived neurotrophic factor and neurotrophin-3 mRNAs in the hippocampus. J. Neurosci. 15, 1768–1777.

Tabuchi, A., Sakaya, H., Kisukeda, T., Fushiki, H. and Tsuda, M. (2002) Involvement of an upstream stimulatory factor as well as cAMP-responsive element-binding protein in the activation of brain-derived neurotrophic factor gene promoter I. J. Biol. Chem. 277, 35920–35931.

Takeuchi, Y., Miyamoto, E. and Fukunaga, K. (2002) Analysis of the promoter region of exon IV brain-derived neurotrophic factor in NG108–15 cells. J. Neurochem. 83, 67–79.

Tao, X., Finkbeiner, S., Arnold, D. B., Shaywitz, A. J. and Greenberg, M. E. (1998) Ca2+ influx regulates BDNF transcription by a CREB family transcription factor-dependent mechanism. Neuron 20, 709–726.

Tao, X., West, A. E., Chen, W. G., Corfas, G. and Greenberg, M. E. (2002) A calcium-responsive transcription factor, CaRF, that regulates neuronal activity-dependent expression of BDNF. Neuron 33, 383–395.

Timmusk, T., Palm, K., Metsis, M., Reintam, T., Paalme, V., Saarma, M. and Persson, H. (1993) Multiple promoters direct tissue-specific expression of the rat BDNF gene. Neuron 10, 475–489.

Timmusk, T., Belluardo, N., Persson, H. and Metsis, M. (1994a) Developmental regulation of brain-derived neurotrophic factor messenger RNAs transcribed from different promoters in the rat brain. Neuroscience 60, 287–291.

Timmusk, T., Persson, H. and Metsis, M. (1994b) Analysis of transcriptional initiation and translatability of brain-derived neurotrophic factor mRNAs in the rat brain. Neurosci. Lett. 177, 27–31.

Timmusk, T., Palm, K., Lendahl, U. and Metsis, M. (1999) Brain-derived neurotrophic factor expression in vivo is under the control of neuron-restrictive silencer element. J. Biol. Chem. 274, 1078–1084.

Tokuyama, W., Okuno, H., Hashimoto, T., Li, Y. X. and Miyashita, Y. (2000) BDNF upregulation during declarative memory formation in monkey inferior temporal cortex. Nat. Neurosci. 3, 1134–1142.

Tongiorgi, E., Righi, M. and Cattaneo, A. (1997) Activity-dependent dendritic targeting of BDNF and TrkB mRNAs in hippocampal neurons. J. Neurosci. 17, 9492–9505.

Tongiorgi, E., Armellin, M., Giulianini, P. G., Bregola, G., Zucchini, S., Paradiso, B., Steward, O., Cattaneo, A. and Simonato, M. (2004) Brain-derived neurotrophic factor mRNA and protein are targeted to discrete dendritic laminas by events that trigger epileptogenesis. J. Neurosci. 24, 6842–6852.

Tsankova, N. M., Kumar, A. and Nestler, E. J. (2004) Histone modifications at gene promoter regions in rat hippocampus after acute and chronic electroconvulsive seizures. J. Neurosci. 24, 5603–5610.

Tsankova, N. M., Berton, O., Renthal, W., Kumar, A., Neve, R. L. and Nestler, E. J. (2006) Sustained hippocampal chromatin regulation in a mouse model of depression and antidepressant action. Nat. Neurosci. 9, 519–525.

West, A. E., Chen, W. G., Dalva, M. B., Dolmetsch, R. E., Kornhauser, J. M., Shaywitz, A. J., Takasu, M. A., Tao, X. and Greenberg, M. E. (2001) Calcium regulation of neuronal gene expression. Proc. Natl. Acad. Sci. USA 98, 11024–11031.

West, A. E., Griffith, E. C. and Greenberg, M. E. (2002) Regulation of transcription factors by neuronal activity. Nat. Rev. Neurosci. 3, 921–931.

Zafra, F., Hengerer, B., Leibrock, J., Thoenen, H. and Lindholm, D. (1990) Activity dependent regulation of BDNF and NGF mRNAs in the rat hippocampus is mediated by non-NMDA glutamate receptors. EMBO J. 9, 3545–3550.

Zhou, Z., Hong, E. J., Cohen, S., Zhao, W. N., Ho, H. Y., Schmidt, L., Chen, W. G., Lin, Y., Savner, E., Griffith, E. C., Hu, L., Steen, J. A., Weitz, C. J. and Greenberg, M. E. (2006) Brain-specific phosphorylation of MeCP2 regulates activity-dependent Bdnf transcription, dendritic growth, and spine maturation. Neuron 52, 255–269.

Zuccato, C., Ciammola, A., Rigamonti, D., Leavitt, B. R., Goffredo, D., Conti, L., Macdonald, M. E., Friedlander, R. M., Silani, V., Hayden, M. R., Timmusk, T., Sipione, S. and Cattaneo, E. (2001) Loss of huntingtin-mediated BDNF gene transcription in Huntington's disease. Science 293, 493–498.

Zuccato, C., Tartari, M., Crotti, A., Goffredo, D., Valenza, M., Conti, L., Cataudella, T., Leavitt, B. R., Hayden, M. R., Timmusk, T., Rigamonti, D. and Cattaneo, E. (2003) Huntingtin interacts with REST/NRSF to modulate the transcription of NRSE-controlled neuronal genes. Nat. Genet. 35, 76–83.

Chapter 9
Role of Signal-responsive Class IIa Histone Deacetylases in Regulating Neuronal Activity-dependent Gene Expression

Brian Yee Hong Lam and Sangeeta Chawla

Abstract Histone deacetylases (HDACs) act as transcriptional repressors by catalyzing the deacetylation of lysine residues on the N-terminal tails of histones and altering chromatin structure. Mammalian HDACs are grouped into 3 classes, Class I, II, and III based on their homology to known yeast enzymes. The Class II HDACs comprise of two subgroups of Class IIa and Class IIb enzymes. Class IIa histone deacetylases are subject to signal-dependent intracellular trafficking, which has emerged as an important regulatory mechanism for controlling gene expression through associated transcription factors. This chapter reviews our current understanding of the regulation and function of class IIa HDACs in neuronal cells.

9.1 Introduction

The categorization of mammalian histone deacetylases into 3 groups is based on their homology to three yeast enzymes. Class I histone deacetylases (HDAC1, -2, -3, -8 and −11) are similar to yeast Rpd3 (reduced potassium dependency 3) and display nuclear expression in a wide range of tissues. The Class II enzymes are homologous to the yeast protein Hda1 and are further sub-divided into class IIa (HDAC4, -5, -7, and -9) and class IIb (HDAC6 and -10) HDACs. Class IIb HDACs are distinguished from Class IIa enzymes by the presence of a duplicated HDAC domain. Class III HDACs (Sirt1, -2, -3, -4, -5, -6 and -7) are NAD^+ dependent deacetylases homologous to yeast Sir2 (silent information regulator 2). Unlike Class I HDACs, which are expressed in most tissues, the expression of class IIa enzymes is restricted. Class IIa HDACs are expressed predominantly in skeletal muscle, heart, brain and thymus where they act as transcriptional repressors to regulate gene expression during differentiation and development. In this chapter we focus on the regulation and function of Class IIa HDACs in neuronal cells.

Mammalian class IIa HDACs contain a C-terminal deacetylase domain, which shows some sequence homology to the catalytic domains of class I HDACs but not class III HDACs. In addition to the HDAC domain, Class IIa HDACs contain a conserved N-terminal region that binds transcription factors including the myocyte enhancer factor-2 (MEF2) family of transcription factors and can repress transcription independently of the C-terminal deacetylase domain (Miska, Karlsson, Langley,

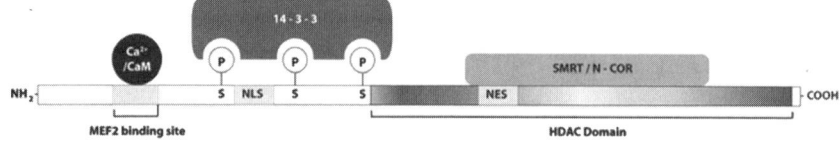

Fig. 1 Domain organization of class IIa HDACs

Nielsen, Pines and Kouzarides 1999; Wang, Bertos, Vezmar, Pelletier, Crosato, Heng, Th'ng, Han and Yang 1999; Lemercier, Verdel, Galloo, Curtet, Brocard and Khochbin 2000). Class IIa HDACs also contain nuclear localization and nuclear export sequences that play a key role in their regulation (McKinsey, Zhang and Olson 2001; Wang and Yang 2001) (Figure 1). Intriguingly, purified class IIa HDAC proteins do not show intrinsic enzymatic activity for histone substrates but function as histone deacetylases in the context of a large multi-protein complex containing silencing mediator for retinoic and thyroid receptors (SMRT), nuclear receptor co-repressor (N-CoR), and the class I deacetylase HDAC3 (Fischle, Dequiedt, Hendzel, Guenther, Lazar, Voelter and Verdin 2002). Similarly, HDAC3 is also enzymatically inactive alone and active when it binds SMRT and N-CoR.

The biological functions and regulation of class IIa HDACs have been studied most extensively in the heart, where they play a role in muscle differentiation and cardiac hypertrophy, and in the thymus, where they regulate T cell apoptosis. Here we review the role of class IIa HDACs in neurons where transcriptional repression mediated by class IIa HDACs is regulated by neuronal activity through changes in their subcellular localization and protein-protein interactions.

9.2 Interaction of Class IIa HDACs with Neuronal Transcription Factors

HDACs do not bind DNA directly but are recruited to gene promoters by sequence-specific transcription factors. In this section, we discuss activity-regulated neuronal transcription factors that interact with class IIa HDACs (Wang et al., 1999; Lemercier et al., 2000; Miska, Langley, Wolf, Karlsson, Pines and Kouzarides 2001). MEF2 transcription factors were one of the first class IIa HDAC partners to be identified. MEF2 proteins are expressed at high levels in the brain and interact with HDAC4, -5, -7 and -9 through a conserved MADS domain. The expression pattern of Class IIa HDACs and MEF2 proteins shows considerable overlap with both being expressed highly in muscle, immune cells and neurons. The MADS domain is also present in the related neuronal transcription factor, serum response factor (SRF) and mediates its interaction with HDAC4 (Davis, Gupta, Camoretti-Mercado, Schwartz and Gupta 2003). Another transcription factor expressed at high levels in neurons that has been shown to interact with HDAC5 is cAMP Response Element binding protein -2 (CREB2), also called activating transcription factor 4 (ATF4) (Guan, Giustetto, Lomvardas, Kim, Miniaci, Schwartz, Thanos and Kandel 2002). The regulation of MEF2, SRF and CREB2 by class IIa HDACs is reviewed in the following sections.

9.2.1 MEF2

The MEF2 family of transcription factors comprises of 4 members, MEF2A, -B, -C- and -D, each encoded by a different gene. The regulation of MEF2-mediated transcription by class IIa HDACs has been extensively studied in muscle during differentiation and cardiac hypertrophy and in T cells during apoptosis (reviewed in (McKinsey, Zhang and Olson 2002). In neurons, MEF2 mediates changes in gene expression in response to neuronal activity and in response to neurotrophins that are important for neuronal differentiation (Flavell, Cowan, Kim, Greer, Lin, Paradis, Griffith, Hu, Chen and Greenberg 2006; Shalizi, Gaudilliere, Yuan, Stegmuller, Shirogane, Ge, Tan, Schulman, Harper and Bonni 2006) and neuronal survival (Mao, Bonni, Xia, Nadal-Vicens and Greenberg 1999; Linseman, Bartley, Le, Laessig, Bouchard, Meintzer, Li and Heidenreich 2003). HDAC4, -5, -7 and -9 interact with MEF2 family members through a conserved motif in the N-terminal region and repress MEF2-mediated gene expression. To activate transcription in response to various stimuli MEF2 proteins are first released from class IIa HDAC-mediated repression. This involves phosphorylation of class IIa HDACs on conserved serine residues that triggers their CRM1-dependent nuclear export by creating docking sites for the 14-3-3 family of chaperone proteins (see section 9.3 below). Following their release from HDACs, MEF2 proteins are subject to further post-translational modifications such as phosphorylation/dephosphorylation, sumoylation and acetylation, which controls their activation, inhibition and degradation. In neurons, many studies have examined post-translational events occurring on MEF2 proteins themselves but the derepression of MEF2 transcription factors by class IIa HDACs has not been investigated in great detail although it is expected to be similar to that seen in muscle and immune cells (see section 9.3 below).

The neuronal activity-dependent activation of MEF2 is mediated by increases in the concentration of intracellular Ca^{2+} ions, which triggers signaling cascades that impinge on MEF2 proteins themselves as well as on class IIa HDACs. The MEF2 interacting motif of HDAC4 and -5 has been shown to interact with Ca^{2+}/calmodulin, the binding of which prevents their interaction with MEF2 (Youn, Grozinger and Liu 2000; Berger, Bieniossek, Schaffitzel, Hassler, Santelli and Richmond 2003) and provides a mechanism for the derepression of MEF2 proteins during neuronal activity. Ca^{2+} also triggers phosphorylation of the 3 conserved serine residues on class IIa HDACs, which induces their nuclear export, by activating Ca^{2+}/calmodulin-dependent protein kinases (CaMKs).

In addition to repression through modification of chromatin structure by their associated HDAC activity Class IIa HDACs can repress MEF2 factors by promoting their sumoylation (Gregoire and Yang 2005; Zhao, Sternsdorf, Bolger, Evans and Yao 2005; Gregoire, Tremblay, Xiao, Yang, Ma, Nie, Mao, Wu, Giguere and

Yang 2006). SUMO (small ubiquitin-related modifier) is a ubiquitin-like modification attached covalently to lysine residues. Sumoylation, which inhibits MEF2 transcriptional activity, is catalyzed by SUMO E3 ligases and has been observed on MEF2A, -C and -D on conserved lysine residues (Gregoire et al., 2006; Shalizi et al., 2006). Co-expression of HDAC4 promotes sumoylation of MEF2D and MEF2C and this effect requires the N-terminal HDAC domain and is independent of the C-terminal deacetylase domain (Gregoire and Yang 2005; Zhao et al., 2005). Similar to HDAC4, HDAC5, -7 and -9 also encourage sumoylation of MEF2 proteins (Gregoire and Yang 2005). Phosphorylation of MEF2 proteins at a particular conserved serine residue (serine 444 in MEF2D), which is dephosphorylated by the Ca^{2+}/calmodulin-dependent phosphatase calcineurin, promotes their sumoylation (Gregoire et al., 2006; Shalizi et al., 2006). The mechanism by which HDACs participate in sumoylation appears to involve the recruitment of kinases that phosphorylate this serine residue (Gregoire et al., 2006). Thus, in addition to histone deacetylation, the association of class IIa HDACs can repress MEF2 activity through sumoylation.

9.2.2 SRF

SRF belongs to the MADS box family of transcription factors and was initially identified as the transcription factor mediating serum induction of the immediate early gene *c-fos* (Treisman 1987). Subsequent studies revealed that SRF is also regulated by Ca^{2+} signals in neuronal cells (Misra, Bonni, Miranti, Rivera, Sheng and Greenberg 1994). In neurons, SRF directs gene expression that is important for neuronal survival (Chang, Poser and Xia 2004b) and neuronal plasticity, which underlies learning and memory (Ramanan, Shen, Sarsfield, Lemberger, Schutz, Linden9 and Ginty 2005; Etkin, Alarcon, Weisberg, Touzani, Huang, Nordheim and Kandel 2006). SRF has also recently emerged as a regulator of axonal outgrowth, guidance and synaptic targeting (Knoll, Kretz, Fiedler, Alberti, Schutz, Frotscher and Nordheim 2006).

The regulation of SRF by class IIa HDACs has been demonstrated in cardiac muscle where it plays a role in the genetic response during stress-induced cardiac hypertrophy (Davis et al., 2003). Davis et al. show that the association of SRF with HDAC4 through its MADS domain causes inhibition of SRF-mediated gene expression that is relieved by constitutively active CaMKIV through nuclear export of HDAC4.

9.2.3 CREB2

CREB2 (also called activating transcription factor 4 or ATF4) belongs to the basic region leucine zipper (bZIP) family of transcription factors that can activate or repress transcription. The interaction of CREB2 with class IIa HDACs has been observed in *Aplysia* neurons where the *Aplysia* homologue ApCREB2 has been shown to recruit HDAC5 to the promoter of the early response gene C/EBP and repress transcription (Guan et al., 2002). The same study also showed that application of the neurotransmitter FMRFa, which causes long-term depression of the synapses between sensory and motor neurons of the *Aplysia* gill withdrawal reflex, results in C/EBP gene repression. This repression is mediated by recruitment of CREB2 and HDAC5 to the C/EBP promoter and is accompanied by histone deacetylation at the promoter. Similar to *Aplysia*, CREB2 functions as a repressor in mammalian cells where it has been shown to negatively regulate transcription of the human enkephalin gene (Karpinski, Morle, Huggenvik, Uhler and Leiden 1992), although an interaction with class IIa HDACs in mammalian cells remains to be demonstrated.

9.3 Signal-dependent Intracellular Trafficking

The association of class IIa HDACs with target transcription factors results in repression of transcription. One mechanism for derepression of class IIa HDAC interacting transcription factors, such as MEF2, involves changes in the subcellular localization of class IIa HDACs by stimulus-induced signaling cascades (McKinsey, Zhang, Lu and Olson 2000a). This type of signal-dependent nucleocytoplasmic is exhibited by other transcriptional regulators including the transcription factor nuclear factor of activated T cells (NFAT) (Graef, Mermelstein, Stankunas, Neilson, Deisseroth, Tsien and Crabtree 1999).

In muscle cells, increases in intracellular Ca^{2+} through activation of CaMKs cause nuclear export of HDAC4, -5 and -7 by phosphorylation of the conserved serine residues. This phosphorylation results in their relocation to the cytoplasm by creating a docking site for the chaperone protein 14-3-3 (Grozinger and Schreiber 2000; McKinsey, Zhang and Olson 2000b; Wang, Kruhlak, Wu, Bertos, Vezmar, Posner, Bazett-Jones and Yang 2000; Kao, Verdel, Tsai, Simon, Juguilon Khochbin 2001; McKinsey et al., 2001; Zhao, Ito, Kane, Liao, Bolger, Lemrow, Means and Yao 2001) (Figure 2). The binding of 14-3-3 masks the nuclear localization signal of class IIa HDACs preventing nuclear import and unmasks a nuclear export sequence (McKinsey et al., 2001).

Early work in muscle showed that only constitutively CaMKs were capable of inducing HDAC5 nuclear export (McKinsey et al., 2000a). However, more recently other kinases have been identified that promote cytoplasmic accumulation of HDAC5 through phosphorylation of the 14-3-3 docking sites in cardiac cells. Using a pharmacological model of phenylephrine-induced cardiac hypertrophy, Vega et al. (Vega, Harrison, Meadows, Roberts, Papst, Olson and McKinsey 2004) showed that the inhibition of PKC blocked the hypertrophy-induced nuclear export of HDAC5 in cardiac myocytes. They showed that this PKC-dependent nuclear export of HDAC5 is mediated by direct phosphorylation of the protein by protein kinase D (PKD), a downstream effector of PKC, which promotes the binding of 14-3-3 to HDAC5. The PKC-induced HDAC5 nuclear export is important for

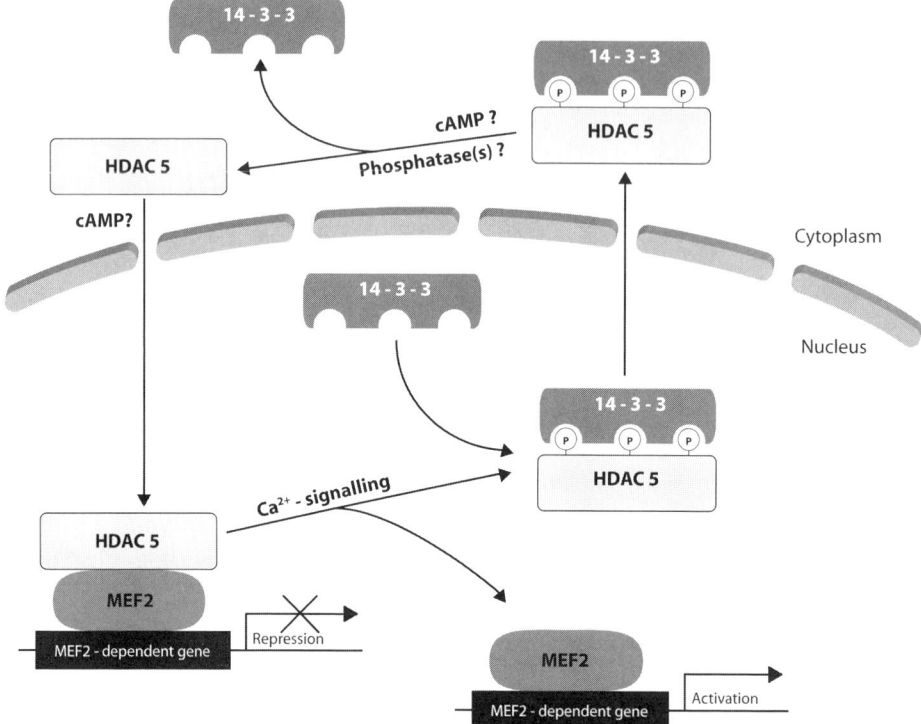

Fig. 2 Signal-responsive intracellular trafficking of HDAC5

the hypertrophic response as PKC inhibitors and a non-phosphorylatable mutant of HDAC5 abrogated cardiomyocyte hypertrophy mediated by phenylephrine (Vega et al., 2004).

Other kinases that regulate nucleocytoplasmic shuttling of class IIa HDACs include members of the microtubule affinity-regulating kinase (MARK)/Par-1 family (Chang, Bezprozvannaya, Li and Olson 2005; Dequiedt, Martin, Von Blume, Vertommen, Lecomte, Mari, Heinen, Bachmann, Twizere, Huang, Rider, Piwnica-Worms, Seufferlein and Kettmann 2006), which promote cytoplasmic localization of HDAC4, -5 and -7, and the extracellular signal-regulated kinases 1 and 2 (ERK1/2), which are unusual in promoting nuclear localization of HDAC4 (Zhou, Richon, Wang, Yang, Rifkind and Marks 2000). While the mechanisms of HDAC nuclear export clearly involve phosphorylation it is unclear how nuclear import of class IIa enzymes occurs. It has been presumed that dephosphorylation of the 14-3-3 binding sites by phosphatases in the cytoplasm returns class IIa HDACs to the nucleus in quiescent cells. However, to date, the phosphatase(s)-mediating this remains to be identified conclusively. The nuclear translocation of HDAC4 by activated ERKs raises the possibility that the nuclear import may also involve phosphorylation events on residues distinct from those involved in nuclear export. Related to this is the finding that increases in intracellular cAMP cause nuclear translocation of HDAC5 in hippocampal neurons (Belfield, Whittaker, Cader and Chawla 2006). Furthermore,

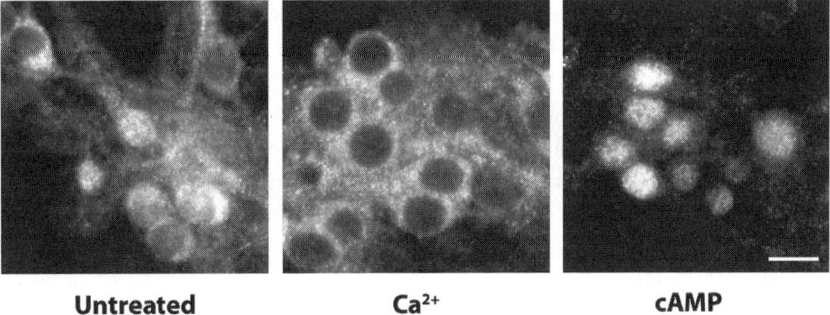

Untreated   **Ca²⁺**   **cAMP**

Fig. 3 HDAC5 immunoreactivity in primary hippocampal neurons left untreated (left); stimulated with KCl to increase intracellular Ca^{2+} (middle); and forskolin to increase intracellular cAMP (right). Scale bar = 10μm

HDAC5 nuclear export by Ca^{2+} signals is inhibited by elevations in intracellular cAMP (Belfield et al., 2006) (Figure 3). The cAMP-dependent nuclear localization of HDAC5 could be mediated either by phosphorylation of HDAC5 at residues that trigger its nuclear import or by the activation of phosphatases that target the 14-3-3 docking sites.

In neuronal systems, CaMKs have been implicated in regulating HDAC5 and HDAC4 localization (Linseman et al., 2003; Bolger and Yao 2005). Linseman et al. studied the neuronal activity-dependent shuttling of HDAC5 in cerebellar granule neurons where the CaMK inhibitor KN93 induced nuclear translocation of the depolarization-induced cytoplasmically localized HDAC5. They further implicated CaMKIIα in the nuclear export of HDAC5 by using specific antisense oligonucleotides to reduce CaMKIIα expression which resulted in nuclear accumulation of HDAC5 even under depolarizing conditions. Bolger and Yao showed that HDAC4, similarly exhibited nuclear localization in cerebellar granule cells treated with the CaMK inhibitor KN93. Consistent with the findings in cerebellar granule neurons, activity-induced HDAC4 and -5 nuclear export was inhibited by the pharmacologically related CaMK inhibitor KN-62 in hippocampal neurons (Chawla, Vanhoutte, Arnold, Huang and Bading 2003).

While there are many similarities in the regulation of different class IIa HDACs, significant differences have been reported. In hippocampal neurons, electrical activity-dependent nucleocytoplasmic shuttling of HDAC4 and HDAC5 exhibits different modes of regulation (Chawla et al., 2003). While nuclear export of both HDAC4 and HDAC5 can be triggered by Ca^{2+} influx through synaptic NMDA receptors or L-type Ca^{2+} channels there are differences in their threshold of activation. Chawla et al showed that spontaneous electrical activity is sufficient to induce HDAC4 nuclear translocation, whereas high frequency bursts of action potential firing are required to trigger HDAC5 nuclear export. These differences in the regulation of specific class IIa HDACs offers a mechanism by which they may regulate distinct subsets of genes in response to different stimuli. Specific class IIa family members may also show preferences for particular associated transcription factors under different conditions to regulate discrete neuronal processes.

9.4 Regulation through Degradation and other Posttranslational Events

In addition to phosphorylation dependent nucleocytoplasmic shuttling, class IIa HDACs undergo other post-translational modifications, which do not affect their subcellular localization but may impact on their function in other ways. For example, HDAC4 is subject to caspase-mediated cleavage (Liu, Dowling, Yang and Kao 2004; Paroni, Mizzau, Henderson, Del Sal, Schneider and Brancolini 2004). This cleavage of HDAC4 promotes nuclear localization of the N-terminal domain where it promotes cytochrome c release from mitochondria and represses MEF2-mediated gene expression resulting in cell death (Paroni et al., 2004). HDAC4 is also sumoylated by the SUMO E3 ligase RanBP2 but the functional significance of this modification has not been elucidated (Kirsh, Seeler, Pichler, Gast, Muller, Miska, Mathieu, Harel-Bellan, Kouzarides, Melchior and Dejean 2002).

9.5 Physiological Roles in the Nervous System

The biological functions of class IIa HDACs have been studied best in cardiac and skeletal muscle. Studies of HDAC deficient mutant mice and in cellular systems of cultured myocytes have shown that class IIa HDACs are critical regulators of skeletal and cardiac muscle differentiation and stress-induced cardiac remodelling. HDAC9 acts as a suppressor of cardiac hypertrophy in response to physiological stimuli such as exercise and in pathological conditions such as myocardial infarction (Zhang, McKinsey, Chang, Antos, Hill and Olson 2002). Mutant mice lacking HDAC9 protein exhibit a sensitization to stress signals, resulting in enhanced hypertrophy and also spontaneously develop hypertrophy with increasing age (Chang, McKinsey, Zhang, Richardson, Hill and Olson 2004a). Mice lacking HDAC5 display a similar phenotype to HDAC9 knockout animals and those lacking both HDAC5 and HDAC9 have lethal cardiac defects (Chang et al., 2004a). HDAC7 knockout has recently been reported to be embryonic lethal due to the repression of matrix metalloproteinase (MMP) 1 resulting in the rupture of blood vessels due to disruptions in cell-cell adhesion (Chang, Young, Li, Qi, Richardson and Olson 2006).

Neuronal phenotypes of class IIa HDAC mutant mice have not been reported, but work in cultured primary cells shows that in neurons class IIa HDACs are important for maintaining neuronal viability. A role for HDAC4 and -5 has been established in the survival of cerebellar granule neurons in vitro (Linseman et al., 2003; Bolger and Yao 2005). Cerebellar granule cells are cultured in medium containing depolarizing concentrations of extracellular potassium (25 mM) to mimic the in vivo neuronal activity that is needed to maintain cell viability. Exposure to low potassium (5 mM) containing medium causes cell death. This experimental system has been used extensively to study transcriptional mechanisms that influence neuronal survival. Linseman et al (2001) showed that HDAC5 is cytoplasmic in viable cerebellar granule cells and translocates to the nucleus when cells are transferred to low potassium medium. The CaMK inhibitor KN93 also stimulated HDAC5 nuclear import with a concomitant decrease in MEF2 transcriptional activity followed by apoptosis. Recent work on HDAC4 shows that similar to HDAC5 its subcellular localization is regulated by prosurvival and apoptotic agents (Bolger and Yao 2005). HDAC4, which is cytoplasmic in viable cerebellar granule neurons, redistributes to the nucleus when neurons are exposed to apoptotic stimuli such as low potassium or excitotoxic conditions. Furthermore, overexpression of a constitutively nuclear HDAC4 mutant causes cell death and repressess MEF2-mediated gene expression. Additionally, knock-down of HDAC4 protein in cerebellar neurons by RNA interference prevented their death in low potassium medium further supporting its role as a crucial component of survival mechanisms in neurons. (Bolger and Yao 2005). The cell viability-dependent changes in class IIa HDAC localization have also been observed in hippocampal neurons where HDAC4 and -5 translocate from the cytoplasmic to the nucleus when cells are exposed to excitotoxic concentrations of glutamate (Chawla et al., 2003).

In addition to their participation in neuronal survival, the interaction of class IIa HDACs with CREB family members and SRF points to a potential role in mammalian neuronal plasticity. Their interaction with MEF2 proteins also suggests that they may be involved in neuronal development. This is further supported by the observation that the expression of class II HDACs is upregulated during differentiation of neuronal progenitor cells (Ajamian, Salminen and Reeben 2004). Further work will doubtless reveal more details about their biological roles and regulation in neuronal cells.

9.6 Concluding Remarks

The interaction of Class IIa HDACs with key neuronal transcription factors suggests that they play important roles in many neuronal processes. Their signal-dependent nucleocytoplasmic shuttling allows them to respond to neuronal activity. It is conceivable that individual class IIa proteins have particular functions in the brain. Emerging differences in the regulation of individual HDACs implies a role in decoding patterns of neuronal activity.

Acknowledgments Work in the authors' laboratory is supported by the Biotechnology and Biological Sciences Research Council (SC) and the Cambridge Overseas Trust (BYHL).

References

Ajamian, F., Salminen, A. and Reeben, M. (2004) Selective regulation of class I and class II histone deacetylases expression by inhibitors of histone deacetylases in cultured mouse neural cells. Neurosci. Lett. 365, 64–8.

Belfield, J. L., Whittaker, C., Cader, M. Z. and Chawla, S. (2006) Differential effects of Ca^{2+} and cAMP on transcription mediated by MEF2D and CREB in hippocampal neurons. J. Biol. Chem. 281, 27724–27732.

Berger, I., Bieniossek, C., Schaffitzel, C., Hassler, M., Santelli, E. and Richmond, T. J. (2003) Direct interaction of Ca^{2+}/calmodulin inhibits histone deacetylase 5 repressor core binding to myocyte enhancer factor 2. J. Biol. Chem. 278, 17625–35.

Bolger, T. A. and Yao, T. P. (2005) Intracellular trafficking of histone deacetylase 4 regulates neuronal cell death. J. Neurosci. 25, 9544–53.

Chang, S., Bezprozvannaya, S., Li, S. and Olson, E. N. (2005) An expression screen reveals modulators of class II histone deacetylase phosphorylation. Proc. Natl. Acad. Sci. USA 102, 8120–5.

Chang, S., McKinsey, T. A., Zhang, C. L., Richardson, J. A., Hill, J. A. and Olson, E. N. (2004a) Histone deacetylases 5 and 9 govern responsiveness of the heart to a subset of stress signals and play redundant roles in heart development. Mol. Cell Biol. 24, 8467–76.

Chang, S., Young, B. D., Li, S., Qi, X., Richardson, J. A. and Olson, E. N. (2006) Histone deacetylase 7 maintains vascular integrity by repressing matrix metalloproteinase 10. Cell 126, 321–34.

Chang, S. H., Poser, S. and Xia, Z. (2004b) A novel role for serum response factor in neuronal survival. J. Neurosci. 24, 2277–85.

Chawla, S., Vanhoutte, P., Arnold, F. J., Huang, C. L. and Bading, H. (2003) Neuronal activity-dependent nucleocytoplasmic shuttling of HDAC4 and HDAC5. J. Neurochem. 85, 151–9.

Davis, F. J., Gupta, M., Camoretti-Mercado, B., Schwartz, R. J. and Gupta, M. P. (2003) Calcium/calmodulin-dependent protein kinase activates serum response factor transcription activity by its dissociation from histone deacetylase, HDAC4. Implications in cardiac muscle gene regulation during hypertrophy. J. Biol. Chem. 278, 20047–58.

Dequiedt, F., Martin, M., Von Blume, J., Vertommen, D., Lecomte, E., Mari, N., Heinen, M. F., Bachmann, M., Twizere, J. C., Huang, M. C., Rider, M. H., Piwnica-Worms, H., Seufferlein, T. and Kettmann, R. (2006) New Role for hPar-1 Kinases EMK and C-TAK1 in Regulating Localization and Activity of Class IIa Histone Deacetylases. Mol. Cell Biol. 26, 7086–102.

Etkin, A., Alarcon, J. M., Weisberg, S. P., Touzani, K., Huang, Y. Y., Nordheim, A. and Kandel, E. R. (2006) A role in learning for SRF: deletion in the adult forebrain disrupts LTD and the formation of an immediate memory of a novel context. Neuron 50, 127–43.

Fischle, W., Dequiedt, F., Hendzel, M. J., Guenther, M. G., Lazar, M. A., Voelter, W. and Verdin, E. (2002) Enzymatic activity associated with class II HDACs is dependent on a multiprotein complex containing HDAC3 and SMRT/N-CoR. Mol. Cell 9, 45–57.

Flavell, S. W., Cowan, C. W., Kim, T. K., Greer, P. L., Lin, Y., Paradis, S., Griffith, E. C., Hu, L. S., Chen, C. and Greenberg, M. E. (2006) Activity-dependent regulation of MEF2 transcription factors suppresses excitatory synapse number. Science 311, 1008–12.

Graef, I. A., Mermelstein, P. G., Stankunas, K., Neilson, J. R., Deisseroth, K., Tsien, R. W. and Crabtree, G. R. (1999) L-type calcium channels and GSK-3 regulate the activity of NF-ATc4 in hippocampal neurons. Nature 401, 703–8.

Gregoire, S., Tremblay, A. M., Xiao, L., Yang, Q., Ma, K., Nie, J., Mao, Z., Wu, Z., Giguere, V. and Yang, X. J. (2006) Control of MEF2 transcriptional activity by coordinated phosphorylation and sumoylation. J. Biol. Chem. 281, 4423–33.

Gregoire, S. and Yang, X. J. (2005) Association with class IIa histone deacetylases upregulates the sumoylation of MEF2 transcription factors. Mol. Cell Biol. 25, 2273–87.

Grozinger, C. M. and Schreiber, S. L. (2000) Regulation of histone deacetylase 4 and 5 and transcriptional activity by 14-3-3-dependent cellular localization. Proc. Natl. Acad. Sci. USA 97, 7835–40.

Guan, Z., Giustetto, M., Lomvardas, S., Kim, J. H., Miniaci, M. C., Schwartz, J. H., Thanos, D. and Kandel, E. R. (2002) Integration of long-term-memory-related synaptic plasticity involves bidirectional regulation of gene expression and chromatin structure. Cell 111, 483–93.

Kao, H. Y., Verdel, A., Tsai, C. C., Simon, C., Juguilon, H. and Khochbin, S. (2001) Mechanism for nucleocytoplasmic shuttling of histone deacetylase 7. J. Biol. Chem. 276, 47496–507.

Karpinski, B. A., Morle, G. D., Huggenvik, J., Uhler, M. D. and Leiden, J. M. (1992) Molecular cloning of human CREB-2: an ATF/CREB transcription factor that can negatively regulate transcription from the cAMP response element. Proc. Natl. Acad. Sci. USA 89, 4820–4.

Kirsh, O., Seeler, J. S., Pichler, A., Gast, A., Muller, S., Miska, E., Mathieu, M., Harel-Bellan, A., Kouzarides, T., Melchior, F. and Dejean, A. (2002) The SUMO E3 ligase RanBP2 promotes modification of the HDAC4 deacetylase. EMBO J. 21, 2682–91.

Knoll, B., Kretz, O., Fiedler, C., Alberti, S., Schutz, G., Frotscher, M. and Nordheim, A. (2006) Serum response factor controls neuronal circuit assembly in the hippocampus. Nat. Neurosci. 9, 195–204.

Lemercier, C., Verdel, A., Galloo, B., Curtet, S., Brocard, M. P. and Khochbin, S. (2000) mHDA1/HDAC5 histone deacetylase interacts with and represses MEF2A transcriptional activity. J. Biol. Chem. 275, 15594–9.

Linseman, D. A., Bartley, C. M., Le, S. S., Laessig, T. A., Bouchard, R. J., Meintzer, M. K., Li, M. and Heidenreich, K. A. (2003) Inactivation of the myocyte enhancer factor-2 repressor histone deacetylase-5 by endogenous Ca^{2+} //calmodulin-dependent kinase II promotes depolarization-mediated cerebellar granule neuron survival. J. Biol. Chem. 278, 41472–81.

Liu, F., Dowling, M., Yang, X. J. and Kao, G. D. (2004) Caspase-mediated specific cleavage of human histone deacetylase 4. J. Biol. Chem. 279, 34537–46.

Mao, Z., Bonni, A., Xia, F., Nadal-Vicens, M. and Greenberg, M. E. (1999) Neuronal activity-dependent cell survival mediated by transcription factor MEF2. Science 286, 785–90.

McKinsey, T. A., Zhang, C. L., Lu, J. and Olson, E. N. (2000a) Signal-dependent nuclear export of a histone deacetylase regulates muscle differentiation. Nature 408, 106–11.

McKinsey, T. A., Zhang, C. L. and Olson, E. N. (2000b) Activation of the myocyte enhancer factor-2 transcription factor by calcium/calmodulin-dependent protein kinase-stimulated binding of 14-3-3 to histone deacetylase 5. Proc. Natl. Acad. Sci. USA 97, 14400–5.

McKinsey, T. A., Zhang, C. L. and Olson, E. N. (2001) Identification of a signal-responsive nuclear export sequence in class II histone deacetylases. Mol. Cell Biol. 21, 6312–21.

McKinsey, T. A., Zhang, C. L. and Olson, E. N. (2002) MEF2: a calcium-dependent regulator of cell division, differentiation and death. Trends Biochem. Sci. 27, 40–7.

Miska, E. A., Karlsson, C., Langley, E., Nielsen, S. J., Pines, J. and Kouzarides, T. (1999) HDAC4 deacetylase associates with and represses the MEF2 transcription factor. EMBO J. 18, 5099–107.

Miska, E. A., Langley, E., Wolf, D., Karlsson, C., Pines, J. and Kouzarides, T. (2001) Differential localization of HDAC4 orchestrates muscle differentiation. Nucleic Acids Res. 29, 3439–47.

Misra, R. P., Bonni, A., Miranti, C. K., Rivera, V. M., Sheng, M. and Greenberg, M. E. (1994) L-type voltage-sensitive calcium channel activation stimulates gene expression by a serum response factor-dependent pathway. J. Biol. Chem. 269, 25483–93.

Paroni, G., Mizzau, M., Henderson, C., Del Sal, G., Schneider, C. and Brancolini, C. (2004) Caspase-dependent regulation of histone deacetylase 4 nuclear-cytoplasmic shuttling promotes apoptosis. Mol. Biol. Cell 15, 2804–18.

Ramanan, N., Shen, Y., Sarsfield, S., Lemberger, T., Schutz, G., Linden, D. J. and Ginty, D. D. (2005) SRF mediates activity-induced gene expression and synaptic plasticity but not neuronal viability. Nat. Neurosci. 8, 759–67.

Shalizi, A., Gaudilliere, B., Yuan, Z., Stegmuller, J., Shirogane, T., Ge, Q., Tan, Y., Schulman, B., Harper, J. W. and Bonni, A. (2006) A calcium-regulated MEF2 sumoylation switch controls postsynaptic differentiation. Science 311, 1012–7.

Treisman, R. (1987) Identification and purification of a polypeptide that binds to the c-fos serum response element. EMBO J. 6, 2711–7.

Vega, R. B., Harrison, B. C., Meadows, E., Roberts, C. R., Papst, P. J., Olson, E. N. and McKinsey, T. A. (2004) Protein kinases C and D mediate agonist-dependent cardiac hypertrophy through nuclear export of histone deacetylase 5. Mol. Cell Biol. 24, 8374–85.

Wang, A. H., Bertos, N. R., Vezmar, M., Pelletier, N., Crosato, M., Heng, H. H., Th'ng, J., Han, J. and Yang, X. J. (1999) HDAC4, a human histone deacetylase related to yeast HDA1, is a transcriptional corepressor. Mol. Cell Biol. 19, 7816–27.

Wang, A. H., Kruhlak, M. J., Wu, J., Bertos, N. R., Vezmar, M., Posner, B. I., Bazett-Jones, D. P. and Yang, X. J. (2000) Regulation of histone deacetylase 4 by binding of 14-3-3 proteins. Mol. Cell Biol. 20, 6904–12.

Wang, A. H. and Yang, X. J. (2001) Histone deacetylase 4 possesses intrinsic nuclear import and export signals. Mol. Cell Biol. 21, 5992–6005.

Youn, H. D., Grozinger, C. M. and Liu, J. O. (2000) Calcium regulates transcriptional repression of myocyte enhancer factor 2 by histone deacetylase 4. J. Biol. Chem. 275, 22563–7.

Zhang, C. L., McKinsey, T. A., Chang, S., Antos, C. L., Hill, J. A. and Olson, E. N. (2002) Class II histone deacetylases act as signal-responsive repressors of cardiac hypertrophy. Cell 110, 479–88.

Zhao, X., Ito, A., Kane, C. D., Liao, T. S., Bolger, T. A., Lemrow, S. M., Means, A. R. and Yao, T. P. (2001) The modular nature of histone deacetylase HDAC4 confers phosphorylation-dependent intracellular trafficking. J. Biol. Chem. 276, 35042–8.

Zhao, X., Sternsdorf, T., Bolger, T. A., Evans, R. M. and Yao, T. P. (2005) Regulation of MEF2 by histone deacetylase 4- and SIRT1 deacetylase-mediated lysine modifications. Mol. Cell Biol. 25, 8456–64.

Zhou, X., Richon, V. M., Wang, A. H., Yang, X. J., Rifkind, R. A. and Marks, P. A. (2000) Histone deacetylase 4 associates with extracellular signal-regulated kinases 1 and 2, and its cellular localization is regulated by oncogenic Ras. Proc. Natl. Acad. Sci. USA 97, 14329–33.

Chapter 10
NFAT-Dependent Gene Expression in the Nervous System: A Critical Mediator of Neurotrophin-Induced Plasticity

Rachel D. Groth and Paul G. Mermelstein

Abstract The modulation of synaptic transmission in response to particular spatial and temporal patterns of neuronal firing provides a means by which information can be stored within the nervous system. Activity-dependent synaptic plasticity is thought to underlie such diverse processes as the regulation and refinement of neuronal connections during development, learning and memory within the adult brain, and persistent pain states incurred following tissue injury and inflammation. While acute changes in synaptic strength are accomplished through modulation of existing proteins that influence cellular excitability, enduring modifications require the induction of gene expression and protein synthesis. Therefore, understanding the molecular mechanisms by which transcription factors bridge signaling at the synapse to gene expression in the nucleus is of particular importance. Recent evidence suggests that the NFATc family of transcription factors plays an important role in neuronal activity-dependent gene expression. This chapter reviews these data, focusing on activation of NFAT-dependent transcription by neurotrophins, signaling molecules involved in many aspects of neuronal development and plasticity.

10.1 The NFATc Transcription Factor

Adaptation to internal and/or external cues is an essential feature of cells throughout the body. A fundamental mechanism by which cells modify functioning in response to changes within their environment is through regulation of gene expression and protein synthesis. Activity-dependent transcription factors play a key role in this process by linking surface receptor stimulation to nuclear events that ultimately alter the fate of the cell.

One of the first and best-characterized classes of activity-dependent transcription factors is that termed nuclear factor of activated T-cells (NFATc). The NFATc family is comprised of four calcineurin-activated proteins, termed NFATc1-4, that are evolutionarily related to the Rel/NFκB family (Chytil and Verdine, 1996, Graef et al., 2001). The nomenclature of NFAT proteins has remained controversial and often confusing, with each NFAT protein having multiple names still in common use. While we will use NFATc1-4 in this review, their alternative names are as follows: NFATc1 (NFAT2, NFATc), NFATc2 (NFAT1, NFATp), NFATc3 (NFAT4,

S. M. Dudek (ed.), *Transcriptional Regulation by Neuronal Activity.*
© Springer 2008

NFATx) and NFATc4 (NFAT3). In addition, there is debate as to whether the transcription factor TonEBP (NFAT5) should be considered another NFAT isoform (Woo et al., 2002, Hogan et al., 2003). While TonEBP contains many structural similarities to NFATc1-4 within the DNA binding domain, it lacks the calcineurin regulatory domain found on the other NFAT transcription factors, rendering it insensitive to the phosphatase (Lopez-Rodriguez et al., 1999, Miyakawa et al., 1999). As such, this review will only consider the four calcineurin-sensitive, NFATc isoforms.

Under basal conditions, NFATc proteins are phosphorylated at multiple serine residues within their regulatory domain, resulting in their cytoplasmic localization. Following increases in intracellular calcium, calcineurin-mediated dephosphorylation of NFATc unmasks several nuclear localization signals, thereby triggering the rapid shuttling of the transcription factors into the nucleus (Clipstone and Crabtree, 1992, Emmel et al., 1989, Flanagan et al., 1991, Liu et al., 1991, Shaw et al., 1988). Once in the nucleus, NFATc proteins typically cannot bind DNA alone to trigger changes in gene expression, but instead require the cooperative binding of a nuclear partner (generically termed "NFATn") to initiate transcription. Because multiple transcription factors can play the role of NFATn, NFAT-dependent transcription provides a particularly versatile mechanism of gene expression as it can be initiated at response elements of varying sequences and spacing.

The activator protein-1 (AP-1) family members Fos and Jun commonly serve as NFATn (Chen et al., 1998, Macian et al., 2000; Macian et al., 2001). Other transcription factors that can act as NFATn include: Maf, ICER, and p21SNFT, of the AP-1 basic region-leucine zipper (bZIP) family (Ho et al., 1996, Bodor et al., 2000, Bower et al., 2002); the zinc finger proteins GATA and EGR (Molkentin et al., 1998; Decker et al., 1998; Decker et al., 2003); the helix–turn–helix domain proteins Oct, HNF3, and IRF-4 (Furstenau et al., 1999; Bert et al., 2000; Hu et al., 2002; Rengarajan et al., 2002); the MADS-box protein MEF2 (Olson and Williams, 2000; Crabtree and Olson, 2002; McKinsey et al., 2002); and the nuclear receptor PPAR-γ (Yang et al., 2000b). Interestingly, AP-1 and other NFATn proteins typically require PKC-mediated phosphorylation to become transcriptionally active. Thus, signaling mechanisms that lead to both kinase and calcineurin-based phosphatase activity are ideally suited to trigger assembly of NFATc/NFATn transcriptional complexes. The importance of this detail will be explored below.

Interestingly, particular NFATc isoforms may also bind to DNA as homodimers. For example, NFATc2 homodimer complexes have been characterized to bind within the IL-8 promoter and within the long-terminal repeat of HIV-1 (Giffin et al., 2003, Jin et al., 2003). The ability of NFATc proteins to form homodimers stems from their DNA-binding domains, also known as Rel homology regions (RHR), being evolutionarily conserved from NFκB transcription factors. The RHR contains two functionally distinct domains, an N-terminal specificity domain (RHR-N) that makes base-specific DNA contacts, and a C-terminal domain involved in dimer formation and IκB binding (RHR-C) (Huxford et al., 1998, Jacobs and Harrison, 1998). Many of the RHR-C interface residues identified in the NFATc2 homodimer are conserved in NFATc1 and NFATc3, suggesting that these isoforms may also be capable of binding DNA as a homodimer. Conversely, NFATc4 lacks the residues thought necessary for a homodimer interface and therefore may not be able to form a homodimer (Hogan et al., 2003).

Termination of NFAT-dependent transcription is mediated by any of several constitutive and/or inducible kinases that trigger the export of NFATc proteins from the nucleus through rephosphorylation. Casein kinase 1 (CK1) and glycogen synthase kinase-3β (GSK3β), unless inhibited, act as constitutive NFATc kinases (Beals et al., 1997, Zhu et al., 1998), while the MAP kinases p38 and c-Jun N-terminal kinase (JNK) appear to be inducible NFATc kinases (Chow et al., 1997; Gomez del Arco et al., 2000; Yang et al., 2002). Interestingly, JNK1 has been shown to phosphorylate NFATc1 and NFATc3, whereas p38 may specifically act on NFATc2 and NFATc4 (Chow et al., 1997; Gomez del Arco et al., 2000; Yang et al., 2002). Thus, despite their common reliance on calcineurin-mediated dephosphorylation for activation, NFATc1-4 possess several isoform-specific attributes that may impact both their binding to DNA and duration in the nucleus.

10.2 Discovery of NFATc in Brain

As their name implies, NFATc transcription factors were originally characterized for their critical role in coupling T-cell receptor activation to the expression of genes that underlie a coordinated immune response (Rao et al., 1997, Crabtree and Olson, 2002). As such, the calcineurin inhibitors FK506 and cyclosporin A (CsA) are routinely used clinically as immunosuppressants. More recently, NFAT-dependent transcription has been shown to regulate a variety of processes outside of the immune system, including adipogenesis (Ho et al., 1998), musculoskeletal development (Chin et al., 1998), cardiac valve formation (de la Pompa et al., 1998, Ranger et al., 1998), cardiac hypertrophy (Molkentin et al., 1998, van Rooij et al., 2002), chondrogenesis (Ranger et al., 2000, Tomita et al., 2002), and patterning of the vasculature (Graef et al., 2001). Given that proteins that serve a particular function in one cell type are often utilized across a variety of different systems, it stands to reason that NFAT-dependent transcription would also be utilized by the nervous system.

The notion that members of the NFATc family may function within the nervous system first emerged in 1994 when two independent laboratories detected NFATc2 within mouse brain by RNase protection assay (Northrop et al., 1994) and immunohistochemistry (Ho et al., 1994). A year later when NFATc4 was isolated, Northern blots of mouse brain extracts revealed that it too was present in nervous tissue (Hoey et al., 1995). Graef et al. (1999) later demonstrated that NFATc4 is expressed within the hippocampus and is activated by calcium entry through L-type calcium channels. Importantly, this paper was the first to provide functional relevance for NFATc transcription factors within the nervous system by implicating NFAT-mediated gene expression in the induction of hippocampal synaptic plasticity and memory formation. Since then, NFATc1 (Plyte et al., 2001) and NFATc3 (Asai et al., 2004, Jayanthi et al., 2005) have also been found within the brain, and the list of neuronal functions mediated by NFAT-dependent transcription has expanded substantially.

Much of the current work on NFATc proteins within the nervous system centers on their role in mediating neuronal survival. Within the striatum, NFATc3 and NFATc4 have been linked to methamphetamine-induced apoptosis

(Jayanthi et al., 2005). Similarly, activation of NFATc4 is correlated with stimuli that lead to the loss of pyramidal neurons in the hippocampus (Shioda et al., 2006). Taken together, these results suggest that NFAT-dependent transcription regulates the expression of pro-apoptotic genes to induce cell death. However, evidence from the developing cerebellum suggests that NFATc4 plays an important role in cell survival during development (Benedito et al., 2005). Whether this discrepancy is the result of a developmental switch in NFATc functioning or is based on cell autonomous differences remains to be determined, but this is certainly an intriguing area of research.

Other neuronal functions ascribed to NFATc family members have also recently been uncovered. For example, outgrowth of embryonic axons is dependent on activation of NFAT-dependent transcription (Graef et al., 2003). Further, NFATc2 regulates the expression of gonadal hormone-releasing hormone in the arcuate nucleus, suggesting that NFAT-dependent transcription may play an important role in neuroendocrinology (Asai et al., 2004). Within the zebrafish olfactory system, inhibition of NFATc disrupted processes of synaptic remodeling that normally occur during synaptogenesis (Yoshida and Mishina, 2005). Finally, NFAT-dependent gene expression has been implicated in hyperalgesia by virtue of its activation by the pro-nociceptive peptide substance P within spinal neurons (Seybold et al., 2006). Given this support for a diverse array of roles for NFATc proteins throughout nervous tissues, it is somewhat remarkable to recall that a little over ten years ago it was widely believed that these transcription factors were absent in neurons.

10.3 Characterization of NFATc Expression Throughout the Nervous System

With the accumulating evidence that NFAT-dependent transcription plays an important role in neuronal function, a systematic characterization of NFATc expression throughout the nervous system was warranted. The first such study was an immunohistochemical approach examining the expression of NFATc4, generally believed to be the predominant isoform expressed in brain. Bradley et al. (2005) found NFATc4 to be ubiquitously, but not uniformly, expressed throughout the adult mouse brain. Highest expression of NFATc4 was found within regions previously shown to express this transcription factor, including the hippocampus, striatum and cerebellum. However, pronounced expression was also found within the olfactory bulb, hypothalamus, globus palladium, and brainstem nuclei (Bradley et al., 2005).

But it is not just NFATc4 that appears expressed throughout the nervous system. Utilizing *in situ* hybridization techniques (lack of antibody specificity is a confounding factor when examining other NFATc isoforms with immunohistochemical procedures), mRNA for all four NFATc isoforms were detected in largely overlapping regions (Groth et al., 2006a). Sites that appeared to express at least three isoforms included the hippocampus, cerebellum and olfactory bulb. In comparison, regions that exhibited more isoform selectivity included the basal ganglia and individual layers of the cerebral cortex. Interestingly, NFATc expression was also found in neuronal precursor cells and glia (Groth et al., 2006a). Not surprisingly, the pattern of NFATc1-4 mRNA expression closely matched that of calcineurin expression (Kincaid et al., 1987, Polli et al., 1991, Takaishi et al., 1991).

It is quite interesting that the expression patterns of the four NFATc proteins largely coincide within the brain. This overlapping expression may be indicative of functional redundancy between the different isoforms, as has been suggested by the relatively mild phenotypes observed in mice lacking individual NFATc proteins. Indeed, pronounced functional abnormalities are generally not observed unless two or more of the NFATc isoforms have been removed (Crabtree and Olson, 2002, Hogan et al., 2003). Thus, expression of all or multiple NFATc isoforms within a cell may serve to ensure that NFAT-mediated gene expression persists even when the function of one or more NFATc proteins is lost.

That being said, NFATc1-4 in wild-type animals may still have differential effects within the same cell as a result of the aforementioned isoform-specific attributes. For example, given their ability to utilize different nuclear partners (or to form homodimers), NFATc isoforms likely regulate distinct complements of genes (Giffin et al., 2003, Jin et al., 2003). Additionally, NFATc1-4 may also have discrete nuclear export kinetics as a result of isoform-selective export kinases (Chow et al., 1997, Gomez del Arco et al., 2000). Finally, there is some evidence indicating that different calcium transients preferentially activate individual NFATc isoforms (Hogan et al., 2003). Thus, the functional significance behind multiple NFAT isoforms within an individual cell remains to be fully determined. While the rationale behind expression of multiple NFATc isoforms in individual neurons remains largely unknown, the activators of NFAT-mediated gene expression in neurons have been identified and characterized.

10.4 Neurotrophins: Initiators of Long-term Neuronal Changes in Both the Developing and Adult Nervous System

As previously described, long-lasting changes in synaptic efficacy are critically dependent on the induction of gene expression and protein synthesis. The transcription factors that mediate these changes are activated in response to particular patterns of activity within the cell. Although there are a number of neurotransmitter systems that participate in the initiation of plasticity, the signaling pathways activated by neurotrophins are particularly robust activators of the gene expression that mediate diverse forms of neural plasticity. Aside from their critical role in shaping neuronal connections during development (Huang and Reichardt, 2001), neurotrophins also modify synaptic transmission throughout the adult nervous system (Kang and Schuman, 1995, Kang and Schuman, 1996, Mendell et al., 1999). Yet, the mechanisms by which neurotrophins induce gene expression are largely unknown.

The neurotrophin family encompasses four proteins: nerve growth factor (NGF), brain derived neurotrophic factor (BDNF), neurotrophin 3 (NT-3), and neurotrophin

4/5 (NT-4/5). Neurotrophin-initiated signal transduction is mediated by a family of high-affinity tyrosine kinase receptors: TrkA is activated by NGF, TrkB is activated by BDNF or NT-4/5, and TrkC is activated by NT-3. Upon neurotrophin binding, dimerized Trk receptors cross-phosphorylate at specific tyrosine residues within their cytoplasmic domains (Lamballe et al., 1991; Kaplan et al., 1991; Soppet et al., 1991). The phosphorylated tyrosines serve as docking sites for three src homology domain 2 (SH2)-containing proteins: (1) phospholipase C (PLC), (2) phosphotidylinositol-3 kinase (PI3K), and (3) SH-2 containing protein (SHC), each of which initiate three distinct signaling pathways (Vetter et al., 1991, Ohmichi et al., 1992a, Ohmichi et al., 1992b, Obermeier et al., 1993, Stephens et al., 1994).

Following tyrosine phosphorylation, PLC cleaves phosphotidylinositol-4,5-bisphosphate to yield inositol trisphosphate (IP$_3$), which triggers the release of Ca^{2+} from intracellular stores, and diacylglycerol (DAG), which activates PKC. Phosphorylated PI3K catalyzes the synthesis of phosphoinositides that activate its putative effector, the serine and threonine kinase Akt. The active adaptor protein SHC induces interaction of the Grb2–SOS complex, which activates the small G-protein Ras by stimulating the exchange of guanosine diphosphate for guanosine triphosphate. This exchange elicits a conformational change in Ras, enabling it to bind to Raf and recruit it from the cytosol to the cell membrane, where Raf activation takes place. Activated Raf phosphorylates and activates MEK (MAPK/ERK kinase), which in turn phosphorylates and activates MAPK/ERK (Kaplan and Miller, 2000).

Given that neurotrophins trigger long-lasting changes in neurotransmission, it is not surprising that the signaling pathways activated by Trk receptors extend to the nucleus. For example, the SHC/Ras/ERK pathway leads to CREB phosphorylation and CRE-dependent transcription. CREB has been shown to regulate the expression of genes essential for normal differentiation and prolonged survival of neurons both *in vitro* and *in vivo* (Riccio et al., 1999, Lonze et al., 2002). Similarly, the PI3K/Akt pathway promotes neuronal survival through activation of NFκB-promoted gene transcription by phosphorylating and thus targeting IκB for degradation (Foehr et al., 2000, Wooten et al., 2001).

Yet, while several transcription factors have been identified that regulate neurotrophin-mediated cell survival programs, information regarding the transcription factors involved in other forms of neurotrophin-induced plasticity has been relatively scarce. Studies of the physiological functions of the TrkB-mediated PLC signaling pathway have revealed that this pathway, in particular, may be critical to synaptic plasticity. Mice engineered with mutations to prevent coupling of activated TrkB to PLC have significant deficiencies in both the early and late phases of hippocampal LTP (Minichiello et al., 2002). Although these mice had decreased levels of CaMKIV and CREB phosphorylation, it is becoming increasingly clear that diverse forms of synaptic plasticity require the coordinated activation of multiple transcription factors (West et al., 2002, Deisseroth et al., 2003; Dolmetsch, 2003). Thus, determining other mechanisms by which neurotrophin stimulation of PLC signaling activates gene expression will be critically important to understanding how neurotrophins trigger lasting changes in synaptic neurotransmission.

10.5 NFAT-dependent Transcription is Initiated by Neurotrophin Signaling

Given that assembly of NFATc/NFATn transcriptional complexes requires increases in intracellular calcium (to lead to calcineurin-mediated dephosphorylation of NFATc) concurrent with activation of protein kinases (to phosphorylate NFATn), it is not surprising that PLC signaling, which activates both components in parallel, is a potent stimulator of NFAT-dependent transcription in a variety of cell types. Thus, by extension, neurotrophin signaling through Trk receptors appears ideally suited to activate NFAT-dependent transcription. In support of this notion, we provide three examples of neurotrophin-induced plasticity involving gene expression mediated in part by NFAT-dependent transcription: (1) embryonic axonal outgrowth, (2) hippocampal synaptic plasticity, and (3) inflammation-induced persistent pain.

10.5.1 Embryonic Axonal Outgrowth

Neurotrophins, with the help of a variety of other extracellular cues, have the daunting task of shaping neuronal connections during embryonic development. This carefully orchestrated process relies both on local signaling at the growth cone to regulate the assembly of cytoskeletal proteins into axons, as well as activation of gene expression and protein synthesis to provide the supplies necessary for axon extension. The local actions of neurotrophins on peripheral axon outgrowth are mediated by the SHC/Ras/ERK pathway, which triggers microtubule assembly and axon expansion, and PI3K/Akt signaling, which is thought to regulate growth cone turning and branching (Markus et al., 2002). Determining how neurotrophins regulate the transcriptional/translational changes necessary for axon outgrowth, however, has proved more challenging, owing to the dependence of peripheral neurons on neurotrophins for survival as well as morphogenesis.

To determine whether the transcription factor CREB was involved in mediating neurotrophin effects on peripheral axonal outgrowth and survival, mice lacking CREB were generated. The result was a dramatic loss in the number of sensory and sympathetic neurons during the embryonic periods of neurotrophin dependence, which made discerning whether CREB is involved in axon outgrowth difficult (Lonze et al., 2002). To circumnavigate this problem, Lonze et al. (2002) crossed CREB null mice with mice lacking the pro-apoptotic gene Bax so that peripheral neurons would survive in the absence of CREB. Interestingly, these neurons exhibited attenuated axon growth, mimicking the effects observed in neurotrophin or Trk mutant mice (Patel et al., 2000, Tucker et al., 2001). Thus, CRE-dependent transcription appears to mediate neurotrophin-induced peripheral neuron outgrowth in addition to the studies supporting a role for survival (Finkbeiner, 2000). However, whether CREB mediates axon outgrowth directly or indirectly remains to be determined.

Conversely, it is possible that distinct signals control axonal elongation *independent* of neuronal survival. Graef et al. (2003) recently demonstrated that embryonic mice lacking NFATc2, c3, and c4 (and wild-type mice treated with calcineurin inhibitors) have attenuated sensory axon projections but do not display abnormalities in neuronal differentiation or survival. To determine whether this effect was cell autonomous, embryonic sensory ganglia explants were exposed to neurotrophins. In this classical model using embryonic sensory neurons, treatment with neurotrophins should produce a broad axon halo of neurite outgrowth after several hours. However, trigeminal explants from NFATc2, NFATc3, and NFATc4 mutant mice treated with neurons displayed little outgrowth. Importantly, explants from the mutant mice were able to grow when plated on Matrigel, a substrate that allows embryonic sensory neurons to extend in the absence of neurotrophins. Finally, neurotrophin stimulation elicited nuclear translocation of NFATc4 and activated NFAT-dependent gene expression in embryonic neurons. Collectively, these results suggest that NFATc transcription factors function downstream of neurotrophins to mediate the gene expression necessary for axon outgrowth, but not survival. Such independent transcriptional control of these distinct processes would explain how neurotrophins trigger axon outgrowth initially, and then primarily act as survival factors once the target cell has been reached. However, identification of the signal responsible for switching from one transcriptional program to another remains to be elucidated.

10.5.2 *Hippocampal Synaptic Plasticity*

The highest levels of BDNF mRNA are found within the hippocampal formation (Ayer-LeLievre et al., 1988, Ernfors et al., 1990, Hofer et al., 1990, Murer et al., 2001), a brain structure necessary for spatial learning and memory. As such, much of the current research involving BDNF has focused on its actions in mediating long-term modifications in both hippocampal synaptic structure and function. For example, BDNF increases the number of dendritic spines, translating to an increased number of hippocampal excitatory synapses (Tyler and Pozzo-Miller, 2001, Tolwani et al., 2002). Further, elevated levels of BDNF mRNA have been found in hippocampal neurons following stimulation leading to long-term potentiation (LTP) (Patterson et al., 1992; Castren et al., 1993; Dragunow et al., 1993), a paradigm used to study learning and memory in which repeated depolarization of neurons produces physical changes at the synapse that enhance the response to a subsequent stimulus (Barnes, 1979). In conjunction with this finding, BDNF exposure generates long-lasting enhancement of synaptic neurotransmission (Kang and Schuman, 1995; Kang and Schuman, 1996; Levine et al., 1995; Kang et al., 1997) while pretreatment of slices with the BDNF-scavenging protein TrkB-IgG prevents the induction of LTP (Figurov et al., 1996; Kang et al., 1997). Finally, two separate strains of BDNF knockout mice show no LTP following tetanic stimulation; the deficit is reversed by application of exogenous BDNF or transfection with an adenovirus that induces BDNF expression (Korte et al., 1995; Korte et al., 1996;

Patterson et al., 1996). Collectively, these data suggest that BDNF is necessary for LTP, and by extension learning and memory, within the adult brain.

Enduring modifications in synaptic strength, such as those thought to underlie learning and memory, require the induction of *de novo* gene expression and protein synthesis. We recently delineated a novel pathway whereby BDNF signaling, through activation of the transcription factor NFATc4, triggers the induction of gene expression within the hippocampus (Groth and Mermelstein, 2003). Specifically, BDNF was found to activate NFAT-dependent transcription in a concentration-dependent, calcineurin-sensitive manner. The primary mechanism underlying BDNF activation of the NFAT transcriptional complex involved TrkB-mediated PLC signaling to IP_3 and PKC. Increased expression of the type I 1,4,5-trisphosphate receptor (IP_3R1) and BDNF after neuronal exposure to BDNF was linked to NFAT-dependent transcription. These results support a model wherein activation of NFAT-dependent transcription underlies a significant component of BDNF-induced gene expression and long-term changes in neuroplasticity.

10.5.3 Inflammation-Induced Persistent Pain

While acute presentation of a noxious stimulus evokes a nociceptive sensation that disappears upon removal of the stimulus, tissue injury and inflammation induce a nociceptor-mediated increase in the excitability of dorsal horn neurons within the spinal cord that persists. This activity-dependent synaptic plasticity underlies hyperalgesia, an increase in pain sensitivity. Although hyperalgesia is beneficial in that it leads to the adaptive behavior of protecting the damaged tissue from further injury, the persistence of such changes is thought to contribute to the development of chronic pain (Millan, 1999).

Peripheral inflammation sets in motion a series of events mediated by the neurotrophins NGF and BDNF that lead to the development of hyperalgesia. This process begins when, following an inflammatory insult, levels of NGF are dramatically increased in peripheral tissues (Weskamp and Otten, 1987; Aloe et al., 1992; Donnerer et al., 1992). Increases in NGF lead to the acute sensitization of nociceptors to noxious heat both indirectly, by triggering mast cell degranulation (Mazurek et al., 1986; Horigome et al., 1994), and directly, by sensitizing TRPV1 channels within the subset of primary afferent neurons that express the NGF receptor TrkA (Shu and Mendell, 1999; Bonnington and McNaughton, 2003; Zhuang et al., 2004). Collectively, these mechanisms of peripheral sensitization underlie the rapid induction of hyperalgesia elicited by subcutaneous application of NGF (Woolf et al., 1994; Andreev et al., 1995; Bennett et al., 1998; Shu and Mendell, 1999).

In addition to its relatively acute peripheral actions, NGF also triggers more lasting effects by increasing the expression of pro-nociceptive genes in nociceptive afferents, including that of fellow neurotrophin BDNF (Apfel et al., 1996; Cho et al., 1997; Michael et al., 1997). Following its upregulation, BDNF is anterogradely transported from the DRG to terminals within the dorsal horn of the spinal cord, where it is packaged in dense-core vesicles (Michael et al., 1997). The release of BDNF in the dorsal horn is encoded by specific patterns of primary afferent neuron stimulation, such as that elicited following sensitization of nociceptors (Lever et al., 2001). The cognate receptor for BDNF, TrkB, is also upregulated in the dorsal horn after peripheral inflammation (Lever et al., 2001; Groth and Aanonsen, 2002), paralleling the change in BDNF levels within the DRG. Further, BDNF elicits central sensitization by increasing the excitability of sensory processing in the dorsal horn of the spinal cord (Kerr et al., 1999). Consistent with these findings, sequestration or impaired translation of BDNF prevents some of the pain-related behavior associated with either inflammation or NGF treatment (Mannion et al., 1999; Groth and Aanonsen, 2002).

Taken together, these data clearly implicate NGF and BDNF as peripherally- and centrally-acting modulators of persistent pain states. Given that the induction of gene expression is paramount to the lasting effects initiated by these neurotrophins within primary afferent and spinal neurons, we recently examined whether these effects were mediated by NFATc transcription factors (Groth et al., 2006b). In cultured adult dorsal root ganglion (DRG) cells, NGF increased NFAT-dependent transcription in a CaN-sensitive manner. Furthermore, in cultured spinal cells, BDNF increased the expression of NFAT-regulated gene expression. These results implicate NFAT signaling as a critical modulator of pain neurotransmission. In addition, NGF and BDNF contribute to the regulation of pro-nociceptive genes associated with inflammation-induced hyperalgesia (see below).

Much work still remains in determining the extent to which activation of NFATc transcription factors contributes to hyperalgesia. Although experiments have focused on primary afferent and spinal neurons, future studies should explore the interactions of neurotrophins and NFATc transcription factors within immune cells. This area is ripe for investigation given that NFAT-dependent transcription is fundamental to the expression of genes that mediate a coordinated immune response (Rao et al., 1997) and since many immune cells synthesize neurotrophins and express Trk receptors (Tabakman et al., 2004). In fact, a recent report demonstrates that NGF activates NFAT-dependent transcription in mast cells to regulate chemokine production (Ahamed et al., 2004). Undoubtedly many more exciting findings, and possibly therapeutic targets, will stem from studies of neurotrophin/NFAT signaling.

10.6 Other Routes of NFAT Activation in the Nervous System

G protein-coupled receptors (GPCRs) constitute the largest known group of signal transduction molecules. Those GPCRs that couple to the Gq-type heterotrimeric G proteins activate PLC signaling. As such, it is not surprising that activation of Gq-type GPCRs has been shown to trigger NFAT-dependent transcription, within both non-neuronal (Boss et al., 1996) and neuronal cells (Seybold et al., 2006). Since the list of Gq-type GPCRs includes muscarinic acetylcholine receptors, type-2 serotonin receptors and many peptide receptors, the ligands for each of these receptors have the potential to activate NFAT-dependent transcription. For example, recent evidence has indicated that SP, which activates the Gq-type G protein-coupled NK1R, induces NFAT-dependent transcription within spinal neurons (Seybold et al., 2006).

As previously mentioned, the first signaling pathway demonstrated to trigger NFAT-dependent gene expression in neurons was calcium entry through L-type calcium channels (Graef et al., 1999). The L-type calcium channel is a ubiquitous signaling molecule, previously shown to be the initiation site for activation of various other transcription factors such as CREB, MEF-2, SRF and NF-κB (Misra et al., 1994; Mao et al., 1999; Shen et al., 2002). Interactions with PDZ domain proteins are one mechanism that confers a privileged role for the L-type calcium channel (in comparison to other voltage-gated calcium channels) in regulating gene expression (Weick et al., 2003; Zhang et al., 2005). Interactions with these structural proteins appear to localize the L-type channels with the signaling intermediaries that bridge the gap between events that occur at the membrane surface with those that occur in the nucleus (Weick et al., 2005a). Notably, calcium entry through L-type calcium channels was once thought to preferentially activate protein kinases. However, recent work has suggested that the pattern of synaptic input determines whether the magnitude and duration of calcium entry through L-type calcium channels is optimal to elicit activation of calcium-dependent kinases and/or phosphatases (Groth et al., 2003). For NFAT-dependent gene expression, calcium entry through L-type calcium channels not only appears to activate calcineurin, but also stimulates several different kinase-dependent processes, resulting in both stimulation of NFATn, as well as inhibition of GSK-3β, prolonging NFATc localization in the nucleus (Graef et al., 1999).

While activation of L-type calcium channels and Gq GPCRs may stimulate NFAT signaling, there are several reasons to believe it is the Trk receptors that are optimal triggers in activating NFATc/NFATn transcriptional complexes. In addition to activation of PLC, Trk receptors also directly couple to the PI3K signaling pathway. As mentioned, NFAT-dependent transcription is terminated upon phosphorylation and subsequent nuclear export of NFATc proteins. Moreover, GSK-3β is inhibited by the effector of PI3K, Akt (Cross et al., 1995; Pap and Cooper, 1998). As such, activation of PI3K in addition to PLC is predicted to prolong the nuclear localization of NFATc proteins and potentiate NFAT-dependent transcription. Trk receptors also couple to activation of the SHC/Ras/ERK pathway. There is considerable cross-talk between SHC/Ras/ERK signaling and PKC signaling, as well as between SHC/Ras/ERK activation of NFATn (Ueda et al., 1996; Schreiber and Crabtree, 1995; Isakov and Altman, 2002). Thus, activation of the SHC/Ras/ERK pathway may also enhance NFAT-dependent transcription. The direct activation PI3K and SHC/Ras/ERK signaling are two rationales to suggest Trk receptors provide enhanced activation and stabilization of NFATc/NFATn in comparison to other signaling mechanisms (Fig. 1). A third potential rationale is that following Trk activation at the membrane surface, neurotrophin-Trk complexes can be internalized within endosomes that continue to signal as they are transported to the nucleus (Ginty and Segal, 2002; Miller and Kaplan, 2001; Reynolds et al., 2000; Sofroniew et al., 2001). Thus, neurotrophin-Trk signaling endosomes may also potentiate

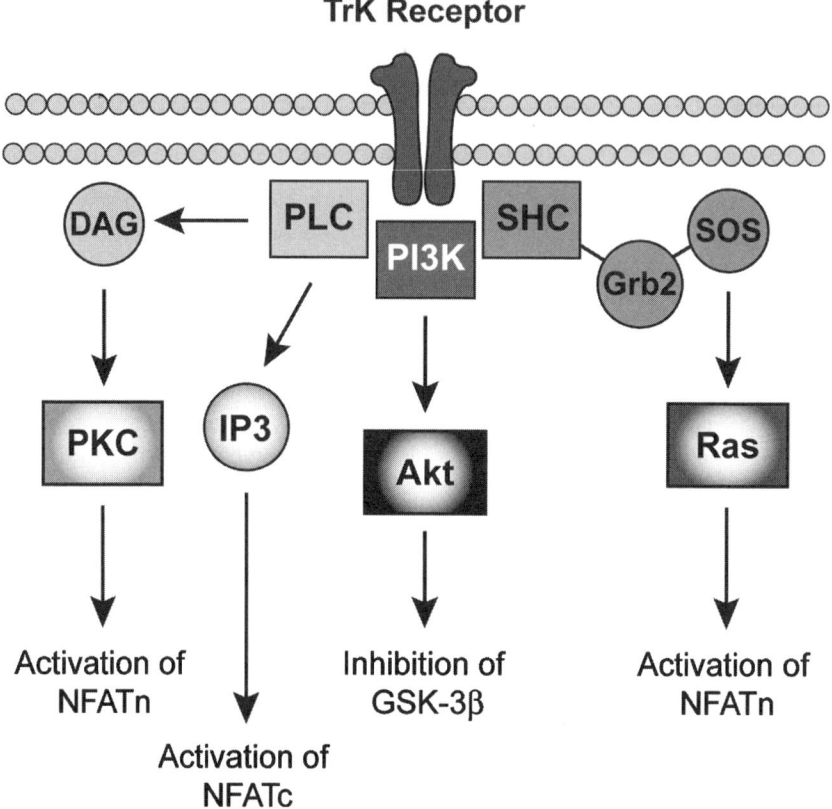

Fig. 1 Through multiple signaling cascades, activation of neurotrophin receptors leads to activation of NFAT-mediated gene expression

NFAT-dependent transcription both by prolonging NFATc and NFATn activation as well as overcoming the spatial limitations imposed by the compartmentalization of effector molecules.

10.7 Genes regulated by NFAT in the Nervous System

To better understand the role of NFAT-dependent transcription in nervous tissue, the genes that are regulated by NFATc in neurons need to be identified. The primary strategy employed thus far has been to test for an NFAT dependence on the expression of proteins whose genes have been previously identified to be regulated by (1) calcineurin or (2) stimuli recognized to trigger NFAT-dependent transcription. In addition, potential NFATc binding sites (GGAAA) must be identified within the promoter regions of these genes. Thus far, the neuronal genes known to be influenced by NFAT transcribe proteins that are critical in functions related to cell signaling and plasticity.

The first gene demonstrated to be NFAT-sensitive is that encoding IP$_3$R1. Initial suggestions that IP$_3$R1 may be regulated by NFAT were based on the fact that expression of this protein increased following calcium entry through L-type calcium channels, and furthermore, was dependent upon activation of calcineurin (Genazzani et al., 1999). Additional studies determined that NFATc/NFATn protein complexes bind to the promoter region of the IP$_3$R1 gene (Graef et al., 1999) and in transfected hippocampal neurons, NFAT regulates expression of a luciferase gene under the regulation of the IP$_3$R1 promoter (Groth and Mermelstein, 2003). Finally, inhibition of calcineurin diminishes both L-type calcium channel and BDNF induction of IP$_3$R1 protein (Graef et al., 1999; Groth and Mermelstein, 2003). Thus, IP$_3$R1 expression seems to be a common target of NFAT-dependent transcription, at least within the hippocampus.

Expression of BDNF is also influenced by NFAT-dependent transcription. The gene encoding BDNF was an attractive candidate for regulation by NFAT for several reasons. First, BDNF is known to regulate its own expression (Saarelainen et al., 2001), fulfilling the criteria that the gene be regulated by stimuli known to activate NFAT-dependent transcription (Groth and Mermelstein, 2003). Second, the rodent BDNF gene consists of several differentially regulated promoters that produce transcripts containing a unique 5' exon (I-VII) spliced to a common 3' exon (VIII) that encodes the mature protein (Liu et al., 2006). While the exon IV-containing transcript is primarily regulated by the transcription factor CREB (Shieh et al., 1998), the promoter regions for exon I- and exon II-containing transcripts contain several putative NFATc DNA binding sites. Consistent with the hypothesis that NFAT regulates BDNF expression, transfection of hippocampal neurons with constitutively nuclear NFATc4 results in an increase in BDNF mRNA. Further, BDNF-induced BDNF expression is strongly attenuated following inhibition of calcineurin (Groth and Mermelstein, 2003). In relation to NFAT regulation of nociception, NGF also promotes BDNF gene expression in DRG cells, an effect dependent on calcineurin (Groth et al., 2006b). Therefore, neurotrophins are able to drive NFAT-mediated expression of BDNF.

It is interesting to consider that neurotrophin activation of NFAT-dependent transcription leads to the increased expression of IP$_3$R1 and BDNF. Both proteins participate in the very signaling pathway that ultimately results in their increased expression. The increase in IP$_3$R1 expression may serve as a positive-feedback loop by increasing the excitability of the cell in which NFAT-dependent transcription was initiated. Alternatively, increased BDNF synthesis may lead to increased BDNF release presynaptically, which may strengthen the next synapse in the neural circuit to effect a feed-forward mechanism of plasticity. While an intriguing possibility, whether increased expression of IP$_3$R1 and BDNF actually leads to lasting changes in synaptic transmission has yet to be determined.

Endogenous negative feedback loops also appear to exist in which NFAT regulates the expression if proteins that inhibit further NFAT-dependent transcription. CaN activity increases expression of the CaN inhibitor, modulatory calcineurin interacting protein (MCIP1.4) (Yang et al., 2000a) presumably via activation of NFAT-dependent transcription, as several NFAT binding sites have been identified within the MCIP1.4 promoter (Rothermel et al., 2003). Altering the balance of positive and negative feedback loops may have dramatic effects on functioning. In Alzheimer's disease, expression of MCIP1 is increased several fold (Ermak et al., 2001) while BDNF expression levels are severely diminished (Phillips et al., 1991; Holsinger et al., 2000), suggesting that the balance was somehow tipped towards negative feedback circuits. The reverse might be true for persistent pain states, in which levels of BDNF are increased (Apfel et al., 1996; Cho et al., 1997; Michael et al., 1997), though it remains to be determined whether MCIP1.4 expression is decreased under these conditions.

Within the immune system, NFATc is known to regulate the expression of cyclooxygenase-2 (COX-2) (Iniguez et al., 2000; Yiu and Toker, 2006). Interestingly, COX-2 has gained attention for its importance within nervous tissue as well. For example, upregulation of COX-2 in spinal neurons has been suggested to play a role in enhanced nociception following peripheral inflammation (Hay and de Belleroche, 1997; Beiche et al., 1998; Samad et al., 2001; Seybold et al., 2003). Recent data from spinal neurons demonstrated that BDNF increased COX-2 mRNA, an effect blocked by inhibition of calcineurin (Groth et al., 2006b). Thus, NFAT signaling appears to underlie COX-2 expression within neurons as well as non-neuronal cells.

A fifth gene that appears to be regulated by NFAT is that encoding the AMPA receptor subunit, GluR2. Interestingly, this protein whose functional role is directly involved with synaptic neurotransmission appears to be negatively regulated by the transcription factor, at least in hippocampal neurons. The 5' untranslated region (5' UTR) of the gene encoding GluR2 (Myers et al., 1998) contains multiple putative NFATc binding sites downstream of the transcriptional initiation sites, making GluR2 a candidate for negative regulation by NFATc (Baksh et al., 2002). Consistent with this hypothesis, NFATc proteins bind to the GluR2 5' UTR to repress GluR2 gene expression in hippocampal neurons, leading to reduced GluR2 protein abundance, and an attenuation of glutamatergic neurotransmission (Weick et al., 2005b). These data provide the first evidence that NFATc proteins can directly suppress gene expression in brain, and further establish a role for NFAT in the regulation of neuronal function.

10.8 Conclusion

The mechanisms by which external signals are transduced to regulate gene expression are fundamental to the lasting changes in neuronal activity that underlie adaptation and plasticity. Herein we have reviewed data implicating members of the NFATc family of transcription factors as integral intermediaries in this process, particularly in response to neurotrophin signaling. Given the diverse functions subserved by members of the neurotrophin family, both during development and within the adult nervous system, the identification of a means by which neurotrophin signaling can extend to the nucleus to initiate long-term changes is of critical importance. Further, the widespread distribution of NFATc1-4 in the nervous system, coupled with the exquisite versatility through which NFAT-dependent transcription can be initiated suggests that the utility of these transcription factors to neuronal functioning has yet to be fully appreciated.

Acknowledgments Supported by DA017881 (P.G.M.) and training grant DA07234 (R.D.G.). The authors would like to thank Jessie Luoma for her comments on this chapter.

References

Ahamed, J., Venkatesha, R. T., Thangam, E. B. and Ali, H. (2004) C3a enhances nerve growth factor-induced NFAT activation and chemokine production in a human mast cell line, HMC-1. J. Immunol. 172, 6961–6968.

Aloe, L., Tuveri, M. A. and Levi-Montalcini, R. (1992) Studies on carrageenan-induced arthritis in adult rats: presence of nerve growth factor and role of sympathetic innervation. Rheumatol. Int. 12, 213–216.

Andreev, N. Y., Dimitrieva, N., Koltzenburg, M. and McMahon, S. B. (1995) Peripheral administration of nerve growth factor in the adult rat produces a thermal hyperalgesia that requires the presence of sympathetic post-ganglionic neurones. Pain. 63, 109–115.

Apfel, S. C., Wright, D. E., Wiideman, A. M., Dormia, C., Snider, W. D. and Kessler, J. A. (1996) Nerve growth factor regulates the expression of brain-derived neurotrophic factor mRNA in the peripheral nervous system. Mol. Cell Neurosci. 7, 134–142.

Asai, M., Iwasaki, Y., Yoshida, M., Mutsuga-Nakayama, N., Arima, H., Ito, M., Takano, K. and Oiso, Y. (2004) Nuclear factor of activated T cells (NFAT) is involved in the depolarization-induced activation of growth hormone-releasing hormone gene transcription in vitro. Mol. Endocrinol. 18, 3011–3019.

Ayer-LeLievre, C., Olson, L., Ebendal, T., Seiger, A. and Persson, H. (1988) Expression of the beta-nerve growth factor gene in hippocampal neurons. Science. 240, 1339–1341.

Baksh, S., Widlund, H. R., Frazer-Abel, A. A., Du, J., Fomsire, S., Fisher, D. E., DeCaprio, J. A., Modiano, J. F. and Burakoff, S. J. (2002) NFATc2-mediated repression of cyclin-dependent kinase 4 expression. Mol. Cell. 10, 1071–1081.

Barnes, C. A. (1979) Memory deficits associated with senescence: a neurophysiological and behavioral study in the rat. J. Comp. Physiol. Psychol.. 93, 74–104.

Beals, C. R., Sheridan, C. M., Turck, C. W., Gardner, P. and Crabtree, G. R. (1997) Nuclear export of NF-ATc enhanced by glycogen synthase kinase-3. Science. 275, 1930–1934.

Beiche, F., Brune, K., Geisslinger, G. and Goppelt-Struebe, M. (1998) Expression of cyclooxygenase isoforms in the rat spinal cord and their regulation during adjuvant-induced arthritis. Inflamm Res. 47, 482–487.

Benedito, A. B., Lehtinen, M., Massol, R., Lopes, U. G., Kirchhausen, T., Rao, A. and Bonni, A. (2005) The transcription factor NFAT3 mediates neuronal survival. J. Biol. Chem. 280, 2818–2825.

Bennett, G., al-Rashed, S., Hoult, J. R. and Brain, S. D. (1998) Nerve growth factor induced hyperalgesia in the rat hind paw is dependent on circulating neutrophils. Pain. 77, 315–322.

Bert, A. G., Burrows, J., Hawwari, A., Vadas, M. A. and Cockerill, P. N. (2000) Reconstitution of T cell-specific transcription directed by composite NFAT/Oct elements. J. Immunol. 165, 5646–5655.

Bodor, J., Bodorova, J. and Gress, R. E. (2000) Suppression of T cell function: a potential role for transcriptional repressor ICER. J. Leukoc. Biol. 67, 774–779.

Bonnington, J. K. and McNaughton, P. A. (2003) Signalling pathways involved in the sensitisation of mouse nociceptive neurones by nerve growth factor. J. Physiol. 551, 433–446.

Boss, V., Talpade, D. J. and Murphy, T. J. (1996) Induction of NFAT-mediated transcription by Gq-coupled receptors in lymphoid and non-lymphoid cells. J. Biol. Chem. 271, 10429–10432.

Bower, K. E., Zeller, R. W., Wachsman, W., Martinez, T. and McGuire, K. L. (2002) Correlation of transcriptional repression by p21(SNFT) with changes in DNA.NF-AT complex interactions. J. Biol. Chem. 277, 34967–34977.

Bradley, K. C., Groth, R. D. and Mermelstein, P. G. (2005) Immunolocalization of NFATc4 in the adult mouse brain. J. Neurosci. Res. 82, 762–770.

Castren, E., Pitkanen, M., Sirvio, J., Parsadanian, A., Lindholm, D., Thoenen, H. and Riekkinen, P. J. (1993) The induction of LTP increases BDNF and NGF mRNA but decreases NT-3 mRNA in the dentate gyrus. Neuroreport 4, 895–898.

Chen, L., Glover, J. N., Hogan, P. G., Rao, A. and Harrison, S. C. (1998) Structure of the DNA-binding domains from NFAT, Fos and Jun bound specifically to DNA. Nature 392, 42–48.

Chin, E. R., Olson, E. N., Richardson, J. A., Yang, Q., Humphries, C., Shelton, J. M., Wu, H., Zhu, W., Bassel-Duby, R. and Williams, R. S. (1998) A calcineurin-dependent transcriptional pathway controls skeletal muscle fiber type. Genes Dev. 12, 2499–2509.

Cho, H. J., Kim, S. Y., Park, M. J., Kim, D. S., Kim, J. K. and Chu, M. Y. (1997) Expression of mRNA for brain-derived neurotrophic factor in the dorsal root ganglion following peripheral inflammation. Brain Res. 749, 358–362.

Chow, C. W., Rincon, M., Cavanagh, J., Dickens, M. and Davis, R. J. (1997) Nuclear accumulation of NFAT4 opposed by the JNK signal transduction pathway. Science 278, 1638–1641.

Chytil, M. and Verdine, G. L. (1996) The Rel family of eukaryotic transcription factors. Curr. Opin. Struct. Biol. 6, 91–100.

Clipstone, N. A. and Crabtree, G. R. (1992) Identification of calcineurin as a key signalling enzyme in T-lymphocyte activation. Nature 357, 695–697.

Crabtree, G. R. and Olson, E. N. (2002) NFAT signaling: choreographing the social lives of cells. Cell 109 Suppl, S67–79.

Cross, D. A., Alessi, D. R., Cohen, P., Andjelkovich, M. and Hemmings, B. A. (1995) Inhibition of glycogen synthase kinase-3 by insulin mediated by protein kinase B. Nature 378, 785–789.

de la Pompa, J. L., Timmerman, L. A., Takimoto, H., Yoshida, H., Elia, A. J., Samper, E., Potter, J., Wakeham, A., Marengere, L., Langille, B. L., Crabtree, G. R. and Mak, T. W. (1998) Role of the NF-ATc transcription factor in morphogenesis of cardiac valves and septum. Nature 392, 182–186.

Decker, E. L., Nehmann, N., Kampen, E., Eibel, H., Zipfel, P. F. and Skerka, C. (2003) Early growth response proteins (EGR) and nuclear factors of activated T cells (NFAT) form heterodimers and regulate proinflammatory cytokine gene expression. Nucleic Acids Res. 31, 911–921.

Decker, E. L., Skerka, C. and Zipfel, P. F. (1998) The early growth response protein (EGR-1) regulates interleukin-2 transcription by synergistic interaction with the nuclear factor of activated T cells. J. Biol. Chem. 273, 26923–26930.

Deisseroth, K., Mermelstein, P. G., Xia, H. and Tsien, R. W. (2003) Signaling from synapse to nucleus: the logic behind the mechanisms. Curr. Opin. Neurobiol. 13, 354–365.

Dolmetsch, R. (2003) Excitation-transcription coupling: signaling by ion channels to the nucleus. Sci STKE. 2003, PE4.

Donnerer, J., Schuligoi, R. and Stein, C. (1992) Increased content and transport of substance P and calcitonin gene-related peptide in sensory nerves innervating inflamed tissue: evidence for a regulatory function of nerve growth factor in vivo. Neuroscience 49, 693–698.

Dragunow, M., Beilharz, E., Mason, B., Lawlor, P., Abraham, W. and Gluckman, P. (1993) Brain-derived neurotrophic factor expression after long-term potentiation. Neurosci. Lett. 160, 232–236.

Emmel, E. A., Verweij, C. L., Durand, D. B., Higgins, K. M., Lacy, E. and Crabtree, G. R. (1989) Cyclosporin A specifically inhibits function of nuclear proteins involved in T cell activation. Science 246, 1617–1620.

Ermak, G., Morgan, T. E. and Davies, K. J. (2001) Chronic overexpression of the calcineurin inhibitory gene DSCR1 (Adapt78) is associated with Alzheimer's disease. J. Biol. Chem. 276, 38787–38794.

Ernfors, P., Wetmore, C., Olson, L. and Persson, H. (1990) Identification of cells in rat brain and peripheral tissues expressing mRNA for members of the nerve growth factor family. Neuron 5, 511–526.

Figurov, A., Pozzo-Miller, L. D., Olafsson, P., Wang, T. and Lu, B. (1996) Regulation of synaptic responses to high-frequency stimulation and LTP by neurotrophins in the hippocampus. Nature 381, 706–709.

Finkbeiner, S. (2000) CREB couples neurotrophin signals to survival messages. Neuron 25, 11–14.

Flanagan, W. M., Corthesy, B., Bram, R. J. and Crabtree, G. R. (1991) Nuclear association of a T-cell transcription factor blocked by FK-506 and cyclosporin A. Nature 352, 803–807.

Foehr, E. D., Lin, X., O'Mahony, A., Geleziunas, R., Bradshaw, R. A. and Greene, W. C. (2000) NF-kappa B signaling promotes both cell survival and neurite process formation in nerve growth factor-stimulated PC12 cells. J. Neurosci. 20, 7556–7563.

Furstenau, U., Schwaninger, M., Blume, R., Jendrusch, E. M. and Knepel, W. (1999) Characterization of a novel calcium response element in the glucagon gene. J. Biol. Chem. 274, 5851–5860.

Genazzani, A. A., Carafoli, E. and Guerini, D. (1999) Calcineurin controls inositol 1,4,5-trisphosphate type 1 receptor expression in neurons. Proc. Natl. Acad. Sci. USA. 96, 5797–5801.

Giffin, M. J., Stroud, J. C., Bates, D. L., von Koenig, K. D., Hardin, J. and Chen, L. (2003) Structure of NFAT1 bound as a dimer to the HIV-1 LTR kappa B element. Nat. Struct. Biol. 10, 800–806.

Ginty, D. D. and Segal, R. A. (2002) Retrograde neurotrophin signaling: Trk-ing along the axon. Curr. Opin. Neurobiol. 12, 268–274.

Gomez del Arco, P., Martinez-Martinez, S., Maldonado, J. L., Ortega-Perez, I. and Redondo, J. M. (2000) A role for the p38 MAP kinase pathway in the nuclear shuttling of NFATp. J. Biol. Chem. 275, 13872–13878.

Graef, I. A., Chen, F., Chen, L., Kuo, A. and Crabtree, G. R. (2001) Signals transduced by Ca(2+)/calcineurin and NFATc3/c4 pattern the developing vasculature. Cell 105, 863–875.

Graef, I. A., Mermelstein, P. G., Stankunas, K., Neilson, J. R., Deisseroth, K., Tsien, R. W. and Crabtree, G. R. (1999) L-type calcium channels and GSK-3 regulate the activity of NF-ATc4 in hippocampal neurons. Nature 401, 703–708.

Graef, I. A., Wang, F., Charron, F., Chen, L., Neilson, J., Tessier-Lavigne, M. and Crabtree, G. R. (2003) Neurotrophins and netrins require calcineurin/NFAT signaling to stimulate outgrowth of embryonic axons. Cell 113, 657–670.

Groth, R. and Aanonsen, L. (2002) Spinal brain-derived neurotrophic factor (BDNF) produces hyperalgesia in normal mice while antisense directed against either BDNF or trkB, prevent inflammation-induced hyperalgesia. Pain 100, 171–181.

Groth, R.D., Bradley, K.C., Mermelstein, P.G., Nakagawa, Y. (2006a) Expression of NFATc1-4 mRNA Within the Adult Mouse Brain. Submitted.

Groth RD, Coicou LG, Seybold VS (2006b) Neurotrophin Activation of NFAT-Dependent Transcription Within Primary Afferent and Spinal Neurons Contributes to Expression of Pro-Nociceptive Genes. Submitted.

Groth, R. D., Dunbar, R. L. and Mermelstein, P. G. (2003) Calcineurin regulation of neuronal plasticity. Biochem. Biophys. Res. Commun. 311, 1159–1171.

Groth, R. D. and Mermelstein, P. G. (2003) Brain-derived neurotrophic factor activation of NFAT (nuclear factor of activated T-cells)-dependent transcription: a role for the transcription factor NFATc4 in neurotrophin-mediated gene expression. J. Neurosci. 23, 8125–8134.

Hay, C. and de Belleroche, J. (1997) Carrageenan-induced hyperalgesia is associated with increased cyclo-oxygenase-2 expression in spinal cord. Neuroreport. 8, 1249–1251.

Ho, I. C., Kim, J. H., Rooney, J. W., Spiegelman, B. M. and Glimcher, L. H. (1998) A potential role for the nuclear factor of activated T cells family of transcriptional regulatory proteins in adipogenesis. Proc. Natl. Acad. Sci. USA. 95, 15537–15541.

Ho, I. C., Hodge, M. R., Rooney, J. W. and Glimcher, L. H. (1996) The proto-oncogene c-maf is responsible for tissue-specific expression of interleukin-4. Cell 85, 973–983.

Ho, S., Timmerman, L., Northrop, J. and Crabtree, G. R. (1994) Cloning and characterization of NF-ATc and NF-ATp: the cytoplasmic components of NF-AT. Adv. Exp. Med. Biol. 365, 167–173.

Hoey, T., Sun, Y. L., Williamson, K. and Xu, X. (1995) Isolation of two new members of the NF-AT gene family and functional characterization of the NF-AT proteins. Immunity. 2, 461–472.

Hofer, M., Pagliusi, S. R., Hohn, A., Leibrock, J. and Barde, Y. A. (1990) Regional distribution of brain-derived neurotrophic factor mRNA in the adult mouse brain. EMBO J. 9, 2459–2464.

Hogan, P. G., Chen, L., Nardone, J. and Rao, A. (2003) Transcriptional regulation by calcium, calcineurin, and NFAT. Genes Dev. 17, 2205–2232.

Holsinger, R. M., Schnarr, J., Henry, P., Castelo, V. T. and Fahnestock, M. (2000) Quantitation of BDNF mRNA in human parietal cortex by competitive reverse transcription-polymerase chain reaction: decreased levels in Alzheimer's disease. Brain Res. Mol. Brain Res. 76, 347–354.

Horigome, K., Bullock, E. D. and Johnson, E. M. J. (1994) Effects of nerve growth factor on rat peritoneal mast cells. Survival promotion and immediate-early gene induction. J. Biol. Chem. 269, 2695–2702.

Hu, C. M., Jang, S. Y., Fanzo, J. C. and Pernis, A. B. (2002) Modulation of T cell cytokine production by interferon regulatory factor-4. J. Biol. Chem. 277, 49238–49246.

Huang, E. J. and Reichardt, L. F. (2001) Neurotrophins: roles in neuronal development and function. Annu. Rev. Neurosci. 24, 677–736.

Huxford, T., Huang, D. B., Malek, S. and Ghosh, G. (1998) The crystal structure of the IkappaBalpha/NF-kappaB complex reveals mechanisms of NF-kappaB inactivation. Cell 95, 759–770.

Iniguez, M. A., Martinez-Martinez, S., Punzon, C., Redondo, J. M. and Fresno, M. (2000) An essential role of the nuclear factor of activated T cells in the regulation of the expression of the cyclooxygenase-2 gene in human T lymphocytes. J. Biol. Chem. 275, 23627–23635.

Isakov, N. and Altman, A. (2002) Protein kinase C(theta) in T cell activation. Annu. Rev. Immunol. 20, 761–794.

Jacobs, M. D. and Harrison, S. C. (1998) Structure of an IkappaBalpha/NF-kappaB complex. Cell 95, 749–758.

Jayanthi, S., Deng, X., Ladenheim, B., McCoy, M. T., Cluster, A., Cai, N. S. and Cadet, J. L. (2005) Calcineurin/NFAT-induced up-regulation of the Fas ligand/Fas death pathway is involved in methamphetamine-induced neuronal apoptosis. Proc. Natl. Acad. Sci. USA. 102, 868–873.

Jin, L., Sliz, P., Chen, L., Macian, F., Rao, A., Hogan, P. G. and Harrison, S. C. (2003) An asymmetric NFAT1 dimer on a pseudo-palindromic kappa B-like DNA site. Nat. Struct. Biol. 10, 807–811.

Kang, H. and Schuman, E. M. (1995) Long-lasting neurotrophin-induced enhancement of synaptic transmission in the adult hippocampus. Science 267, 1658–1662.

Kang, H. and Schuman, E. M. (1996) A requirement for local protein synthesis in neurotrophin-induced hippocampal synaptic plasticity. Science 273, 1402–1406.

Kang, H., Welcher, A. A., Shelton, D. and Schuman, E. M. (1997) Neurotrophins and time: different roles for TrkB signaling in hippocampal long-term potentiation. Neuron 19, 653–664.

Kaplan, D. R., Martin-Zanca, D. and Parada, L. F. (1991) Tyrosine phosphorylation and tyrosine kinase activity of the trk proto-oncogene product induced by NGF. Nature 350, 158–160.

Kaplan, D. R. and Miller, F. D. (2000) Neurotrophin signal transduction in the nervous system. Curr. Opin. Neurobiol. 10, 381–391.

Kerr, B. J., Bradbury, E. J., Bennett, D. L., Trivedi, P. M., Dassan, P., French, J., Shelton, D. B., McMahon, S. B. and Thompson, S. W. (1999) Brain-derived neurotrophic factor modulates nociceptive sensory inputs and NMDA-evoked responses in the rat spinal cord. J. Neurosci. 19, 5138–5148.

Kincaid, R. L., Balaban, C. D. and Billingsley, M. L. (1987) Differential localization of calmodulin-dependent enzymes in rat brain: evidence for selective expression of cyclic nucleotide phosphodiesterase in specific neurons. Proc. Natl. Acad. Sci. USA. 84, 1118–1122.

Korte, M., Carroll, P., Wolf, E., Brem, G., Thoenen, H. and Bonhoeffer, T. (1995) Hippocampal long-term potentiation is impaired in mice lacking brain-derived neurotrophic factor. Proc. Natl. Acad. Sci. USA. 92, 8856–8860.

Korte, M., Griesbeck, O., Gravel, C., Carroll, P., Staiger, V., Thoenen, H. and Bonhoeffer, T. (1996) Virus-mediated gene transfer into hippocampal CA1 region restores long-term potentiation in brain-derived neurotrophic factor mutant mice. Proc. Natl. Acad. Sci. USA. 93, 12547–12552.

Lamballe, F., Klein, R. and Barbacid, M. (1991) The trk family of oncogenes and neurotro phin receptors. Princess Takamatsu Symp. 22, 153–170.

Lever, I. J., Bradbury, E. J., Cunningham, J. R., Adelson, D. W., Jones, M. G., McMahon, S. B., Marvizon, J. C. and Malcangio, M. (2001) Brain-derived neurotrophic factor is released in the dorsal horn by distinctive patterns of afferent fiber stimulation. J. Neurosci. 21, 4469–4477.

Levine, E. S., Dreyfus, C. F., Black, I. B. and Plummer, M. R. (1995) Brain-derived neurotrophic factor rapidly enhances synaptic transmission in hippocampal neurons via postsynaptic tyrosine kinase receptors. Proc. Natl. Acad. Sci. USA. 92, 8074–8077.

Liu, J., Farmer, J. D. J., Lane, W. S., Friedman, J., Weissman, I. and Schreiber, S. L. (1991) Calcineurin is a common target of cyclophilin-cyclosporin A and FKBP-FK506 complexes. Cell 66, 807–815.

Liu, Q. R., Lu, L., Zhu, X. G., Gong, J. P., Shaham, Y. and Uhl, G. R. (2006) Rodent BDNF genes, novel promoters, novel splice variants, and regulation by cocaine. Brain Res. 1067, 1–12.

Lonze, B. E., Riccio, A., Cohen, S. and Ginty, D. D. (2002) Apoptosis, axonal growth defects, and degeneration of peripheral neurons in mice lacking CREB. Neuron 34, 371–385.

Lopez-Rodriguez, C., Aramburu, J., Rakeman, A. S. and Rao, A. (1999) NFAT5, a constitutively nuclear NFAT protein that does not cooperate with Fos and Jun. Proc. Natl. Acad. Sci. USA. 96, 7214–7219.

Macian, F., Garcia-Rodriguez, C. and Rao, A. (2000) Gene expression elicited by NFAT in the presence or absence of cooperative recruitment of Fos and Jun. EMBO J. 19, 4783–4795.

Macian, F., Lopez-Rodriguez, C. and Rao, A. (2001) Partners in transcription: NFAT and AP-1. Oncogene. 20, 2476–2489.

Mannion, R. J., Costigan, M., Decosterd, I., Amaya, F., Ma, Q. P., Holstege, J. C., Ji, R. R., Acheson, A., Lindsay, R. M., Wilkinson, G. A. and Woolf, C. J. (1999) Neurotrophins: peripherally and centrally acting modulators of tactile stimulus-induced inflammatory pain hypersensitivity. Proc. Natl. Acad. Sci. USA. 96, 9385–9390.

Mao, Z., Bonni, A., Xia, F., Nadal-Vicens, M. and Greenberg, M. E. (1999) Neuronal activity-dependent cell survival mediated by transcription factor MEF2. Science. 286, 785–790.

Markus, A., Patel, T. D. and Snider, W. D. (2002) Neurotrophic factors and axonal growth. Curr. Opin. Neurobiol. 12, 523–531.

Mazurek, N., Weskamp, G., Erne, P. and Otten, U. (1986) Nerve growth factor induces mast cell degranulation without changing intracellular calcium levels. FEBS Lett. 198, 315–320.

McKinsey, T. A., Zhang, C. L. and Olson, E. N. (2002) MEF2: a calcium-dependent regulator of cell division, differentiation and death. Trends Biochem. Sci. 27, 40–47.

Mendell, L. M., Albers, K. M. and Davis, B. M. (1999) Neurotrophins, nociceptors, and pain. Microsc. Res. Tech. 45, 252–261.

Michael, G. J., Averill, S., Nitkunan, A., Rattray, M., Bennett, D. L., Yan, Q. and Priestley, J. V. (1997) Nerve growth factor treatment increases brain-derived neurotrophic factor selectively in TrkA-expressing dorsal root ganglion cells and in their central terminations within the spinal cord. J. Neurosci. 17, 8476–8490.

Millan, M. J. (1999) The induction of pain: an integrative review. Prog. Neurobiol. 57, 1–164.

Miller, F. D. and Kaplan, D. R. (2001) On Trk for retrograde signaling. Neuron 32, 767–770.

Minichiello, L., Calella, A. M., Medina, D. L., Bonhoeffer, T., Klein, R. and Korte, M. (2002) Mechanism of TrkB-mediated hippocampal long-term potentiation. Neuron 36, 121–137.

Misra, R. P., Bonni, A., Miranti, C. K., Rivera, V. M., Sheng, M. and Greenberg, M. E. (1994) L-type voltage-sensitive calcium channel activation stimulates gene expression by a serum response factor-dependent pathway. J. Biol. Chem. 269, 25483–25493.

Miyakawa, H., Woo, S. K., Dahl, S. C., Handler, J. S. and Kwon, H. M. (1999) Tonicity-responsive enhancer binding protein, a rel-like protein that stimulates transcription in response to hypertonicity. Proc. Natl. Acad. Sci. USA. 96, 2538–2542.

Molkentin, J. D., Lu, J. R., Antos, C. L., Markham, B., Richardson, J., Robbins, J., Grant, S. R. and Olson, E. N. (1998) A calcineurin-dependent transcriptional pathway for cardiac hypertrophy. Cell 93, 215–228.

Murer, M. G., Yan, Q. and Raisman-Vozari, R. (2001) Brain-derived neurotrophic factor in the control human brain, and in Alzheimer's disease and Parkinson's disease. Prog. Neurobiol. 63, 71–124.

Myers, S. J., Peters, J., Huang, Y., Comer, M. B., Barthel, F. and Dingledine, R. (1998) Transcriptional regulation of the GluR2 gene: neural-specific expression, multiple promoters, and regulatory elements. J. Neurosci. 18, 6723–6739.

Northrop, J. P., Ho, S. N., Chen, L., Thomas, D. J., Timmerman, L. A., Nolan, G. P., Admon, A. and Crabtree, G. R. (1994) NF-AT components define a family of transcription factors targeted in T-cell activation. Nature 369, 497–502.

Obermeier, A., Lammers, R., Wiesmuller, K. H., Jung, G., Schlessinger, J. and Ullrich, A. (1993) Identification of Trk binding sites for SHC and phosphatidylinositol 3'-kinase and formation of a multimeric signaling complex. J. Biol. Chem. 268, 22963–22966.

Ohmichi, M., Decker, S. J. and Saltiel, A. R. (1992b) Activation of phosphatidylinositol-3 kinase by nerve growth factor involves indirect coupling of the trk proto-oncogene with src homology 2 domains. Neuron 9, 769–777.

Ohmichi, M., Decker, S. J. and Saltiel, A. R. (1992a) Nerve growth factor stimulates the tyrosine phosphorylation of a 38-kDa protein that specifically associates with the src homology domain of phospholipase C-gamma 1. J. Biol. Chem. 267, 21601–21606.

Olson, E. N. and Williams, R. S. (2000) Remodeling muscles with calcineurin. Bioessays 22, 510–519.

Pap, M. and Cooper, G. M. (1998) Role of glycogen synthase kinase-3 in the phosphatidylinositol 3-Kinase/Akt cell survival pathway. J. Biol. Chem. 273, 19929–19932.

Patel, T. D., Jackman, A., Rice, F. L., Kucera, J. and Snider, W. D. (2000) Development of sensory neurons in the absence of NGF/TrkA signaling in vivo. Neuron 25, 345–357.

Patterson, S. L., Abel, T., Deuel, T. A., Martin, K. C., Rose, J. C. and Kandel, E. R. (1996) Recombinant BDNF rescues deficits in basal synaptic transmission and hippocampal LTP in BDNF knockout mice. Neuron 16, 1137–1145.

Patterson, S. L., Grover, L. M., Schwartzkroin, P. A. and Bothwell, M. (1992) Neurotrophin expression in rat hippocampal slices: a stimulus paradigm inducing LTP in CA1 evokes increases in BDNF and NT-3 mRNAs. Neuron 9, 1081–1088.

Phillips, H. S., Hains, J. M., Armanini, M., Laramee, G. R., Johnson, S. A. and Winslow, J. W. (1991) BDNF mRNA is decreased in the hippocampus of individuals with Alzheimer's disease. Neuron 7, 695–702.

Plyte, S., Boncristiano, M., Fattori, E., Galvagni, F., Paccani, S. R., Majolini, M. B., Oliviero, S., Ciliberto, G., Telford, J. L. and Baldari, C. T. (2001) Identification and characterization of a novel nuclear factor of activated T-cells-1 isoform expressed in mouse brain. J. Biol. Chem. 276, 14350–14358.

Polli, J. W., Billingsley, M. L. and Kincaid, R. L. (1991) Expression of the calmodulin-dependent protein phosphatase, calcineurin, in rat brain: developmental patterns and the role of nigrostriatal innervation. Brain Res. Dev. Brain Res. 63, 105–119.

Ranger, A. M., Gerstenfeld, L. C., Wang, J., Kon, T., Bae, H., Gravallese, E. M., Glimcher, M. J. and Glimcher, L. H. (2000) The nuclear factor of activated T cells (NFAT) transcription factor NFATp (NFATc2) is a repressor of chondrogenesis. J. Exp. Med. 191, 9–22.

Ranger, A. M., Grusby, M. J., Hodge, M. R., Gravallese, E. M., de la Brousse, F. C., Hoey, T., Mickanin, C., Baldwin, H. S. and Glimcher, L. H. (1998) The transcription factor NF-ATc is essential for cardiac valve formation. Nature 392, 186–190.

Rao, A., Luo, C. and Hogan, P. G. (1997) Transcription factors of the NFAT family: regulation and function. Annu. Rev. Immunol. 15, 707–747.

Rengarajan, J., Mowen, K. A., McBride, K. D., Smith, E. D., Singh, H. and Glimcher, L. H. (2002) Interferon regulatory factor 4 (IRF4) interacts with NFATc2 to modulate interleukin 4 gene expression. J. Exp. Med. 195, 1003–1012.

Reynolds, A. J., Bartlett, S. E. and Hendry, I. A. (2000) Molecular mechanisms regulating the retrograde axonal transport of neurotrophins. Brain Res. Brain Res. Rev. 33, 169–178.

Riccio, A., Ahn, S., Davenport, C. M., Blendy, J. A. and Ginty, D. D. (1999) Mediation by a CREB family transcription factor of NGF-dependent survival of sympathetic neurons. Science 286, 2358–2361.

Rothermel, B. A., Vega, R. B. and Williams, R. S. (2003) The role of modulatory calcineurin-interacting proteins in calcineurin signaling. Trends Cardiovasc. Med. 13, 15–21.

Saarelainen, T., Vaittinen, S. and Castren, E. (2001) trkB-receptor activation contributes to the kainate-induced increase in BDNF mRNA synthesis. Cell. Mol. Neurobiol. 21, 429–435.

Samad, T. A., Moore, K. A., Sapirstein, A., Billet, S., Allchorne, A., Poole, S., Bonventre, J. V. and Woolf, C. J. (2001) Interleukin-1beta-mediated induction of Cox-2 in the CNS contributes to inflammatory pain hypersensitivity. Nature. 410, 471–475.

Schreiber, S. L. and Crabtree, G. R. (1995) Immunophilins, ligands, and the control of signal transduction. Harvey Lect. 91, 99–114.

Seybold, V. S., Coicou, L. G., Groth, R. D. and Mermelstein, P. G. (2006) Substance P initiates NFAT-dependent gene expression in spinal neurons. J. Neurochem. 97, 397–407.

Seybold, V. S., Jia, Y. P. and Abrahams, L. G. (2003) Cyclo-oxygenase-2 contributes to central sensitization in rats with peripheral inflammation. Pain 105, 47–55.

Shaw, J. P., Utz, P. J., Durand, D. B., Toole, J. J., Emmel, E. A. and Crabtree, G. R. (1988) Identification of a putative regulator of early T cell activation genes. Science 241, 202–205.

Shen, W., Zhang, C. and Zhang, G. (2002) Nuclear factor kappaB activation is mediated by NMDA and non-NMDA receptor and L-type voltage-gated Ca(2+) channel following severe global ischemia in rat hippocampus. Brain Res. 933, 23–30.

Shieh, P. B., Hu, S. C., Bobb, K., Timmusk, T. and Ghosh, A. (1998) Identification of a signaling pathway involved in calcium regulation of BDNF expression. Neuron 20, 727–740.

Shioda, N., Moriguchi, S., Shirasaki, Y. and Fukunaga, K. (2006) Generation of constitutively active calcineurin by calpain contributes to delayed neuronal death following mouse brain ischemia. J. Neurochem. 98, 310–320.

Shu, X. and Mendell, L. M. (1999) Nerve growth factor acutely sensitizes the response of adult rat sensory neurons to capsaicin. Neurosci. Lett. 274, 159–162.

Sofroniew, M. V., Howe, C. L. and Mobley, W. C. (2001) Nerve growth factor signaling, neuroprotection, and neural repair. Annu. Rev. Neurosci. 24, 1217–1281.

Soppet, D., Escandon, E., Maragos, J., Middlemas, D. S., Reid, S. W., Blair, J., Burton, L. E., Stanton, B. R., Kaplan, D. R., Hunter, T. and et, a. (1991) The neurotrophic factors brain-derived neurotrophic factor and neurotrophin-3 are ligands for the trkB tyrosine kinase receptor. Cell 65, 895–903.

Stephens, R. M., Loeb, D. M., Copeland, T. D., Pawson, T., Greene, L. A. and Kaplan, D. R. (1994) Trk receptors use redundant signal transduction pathways involving SHC and PLC-gamma 1 to mediate NGF responses. Neuron 12, 691–705.

Tabakman, R., Lecht, S., Sephanova, S., Arien-Zakay, H. and Lazarovici, P. (2004) Interactions between the cells of the immune and nervous system: neurotrophins as neuroprotection mediators in CNS injury. Prog. Brain Res. 146, 387–401.

Takaishi, T., Saito, N., Kuno, T. and Tanaka, C. (1991) Differential distribution of the mRNA encoding two isoforms of the catalytic subunit of calcineurin in the rat brain. Biochem. Biophys. Res. Commun. 174, 393–398.

Tolwani, R. J., Buckmaster, P. S., Varma, S., Cosgaya, J. M., Wu, Y., Suri, C. and Shooter, E. M. (2002) BDNF overexpression increases dendrite complexity in hippocampal dentate gyrus. Neuroscience. 114, 795–805.

Tomita, M., Reinhold, M. I., Molkentin, J. D. and Naski, M. C. (2002) Calcineurin and NFAT4 induce chondrogenesis. J. Biol. Chem. 277, 42214–42218.

Tucker, K. L., Meyer, M. and Barde, Y. A. (2001) Neurotrophins are required for nerve growth during development. Nat. Neurosci. 4, 29–37.

Tyler, W. J. and Pozzo-Miller, L. D. (2001) BDNF enhances quantal neurotransmitter release and increases the number of docked vesicles at the active zones of hippocampal excitatory synapses. J. Neurosci. 21, 4249–4258.

Ueda, Y., Hirai, S., Osada, S., Suzuki, A., Mizuno, K. and Ohno, S. (1996) Protein kinase C activates the MEK-ERK pathway in a manner independent of Ras and dependent on Raf. J. Biol. Chem. 271, 23512–23519.

van Rooij, E., Doevendans, P. A., de Theije, C. C., Babiker, F. A., Molkentin, J. D. and de Windt, L. J. (2002) Requirement of nuclear factor of activated T-cells in calcineurin-mediated cardiomyocyte hypertrophy. J. Biol. Chem. 277, 48617–48626.

Vetter, M. L., Martin-Zanca, D., Parada, L. F., Bishop, J. M. and Kaplan, D. R. (1991) Nerve growth factor rapidly stimulates tyrosine phosphorylation of phospholipase C-gamma 1 by a kinase activity associated with the product of the trk protooncogene. Proc. Natl. Acad. Sci. USA. 88, 5650–5654.

Weick, J. P., Groth, R. D., Isaksen, A. L. and Mermelstein, P. G. (2003) Interactions with PDZ proteins are required for L-type calcium channels to activate cAMP response element-binding protein-dependent gene expression. J. Neurosci. 23, 3446–3456.

Weick, J. P., Kuo, S. P. and Mermelstein, P. G. (2005a) L-type calcium channel regulation of neuronal gene expression. Cellsci. Rev. 1, 44-49.

Weick, J. P., Kuo, S. P. and Mermelstein, P. G. (2005b) NFATc regulates GluR2 expression in the hippocampus. Soc. Neurosci. Abstr. 734.14

Weskamp, G. and Otten, U. (1987) An enzyme-linked immunoassay for nerve growth factor (NGF): a tool for studying regulatory mechanisms involved in NGF production in brain and in peripheral tissues. J. Neurochem. 48, 1779–1786.

West, A. E., Griffith, E. C. and Greenberg, M. E. (2002) Regulation of transcription factors by neuronal activity. Nat. Rev. Neurosci. 3, 921–931.

Woo, S. K., Lee, S. D. and Kwon, H. M. (2002) TonEBP transcriptional activator in the cellular response to increased osmolality. Pflugers Arch. 444, 579–585.

Woolf, C. J., Safieh-Garabedian, B., Ma, Q. P., Crilly, P. and Winter, J. (1994) Nerve growth factor contributes to the generation of inflammatory sensory hypersensitivity. Neuroscience. 62, 327–331.

Wooten, M. W., Seibenhener, M. L., Mamidipudi, V., Diaz-Meco, M. T., Barker, P. A. and Moscat, J. (2001) The atypical protein kinase C-interacting protein p62 is a scaffold for NF-kappaB activation by nerve growth factor. J. Biol. Chem. 276, 7709–7712.

Yang, J., Rothermel, B., Vega, R. B., Frey, N., McKinsey, T. A., Olson, E. N., Bassel-Duby, R. and Williams, R. S. (2000a) Independent signals control expression of the calcineurin inhibitory proteins MCIP1 and MCIP2 in striated muscles. Circ. Res. 87, E61–8.

Yang, T. T., Xiong, Q., Enslen, H., Davis, R. J. and Chow, C. W. (2002) Phosphorylation of NFATc4 by p38 mitogen-activated protein kinases. Mol. Cell Biol. 22, 3892–3904.

Yang, X. Y., Wang, L. H., Chen, T., Hodge, D. R., Resau, J. H., DaSilva, L. and Farrar, W. L. (2000b) Activation of human T lymphocytes is inhibited by peroxisome proliferator-activated receptor gamma (PPARgamma) agonists. PPARgamma co-association with transcription factor NFAT. J. Biol. Chem. 275, 4541–4544.

Yiu, G. K. and Toker, A. (2006) NFAT induces breast cancer cell invasion by promoting the induction of cyclooxygenase-2. J. Biol. Chem. 281, 12210–12217.

Yoshida, T. and Mishina, M. (2005) Distinct roles of calcineurin-nuclear factor of activated T-cells and protein kinase A-cAMP response element-binding protein signaling in presynaptic differentiation. J. Neurosci. 25, 3067–3079.

Zhang, H., Maximov, A., Fu, Y., Xu, F., Tang, T. S., Tkatch, T., Surmeier, D. J. and Bezprozvanny, I. (2005) Association of CaV1.3 L-type calcium channels with Shank. J. Neurosci. 25, 1037–1049.

Zhu, J., Shibasaki, F., Price, R., Guillemot, J. C., Yano, T., Dotsch, V., Wagner, G., Ferrara, P. and McKeon, F. (1998) Intramolecular masking of nuclear import signal on NF-AT4 by casein kinase I and MEKK1. Cell 93, 851–861.

Zhuang, Z. Y., Xu, H., Clapham, D. E. and Ji, R. R. (2004) Phosphatidylinositol 3-kinase activates ERK in primary sensory neurons and mediates inflammatory heat hyperalgesia through TRPV1 sensitization. J. Neurosci. 24, 8300–8309.

Chapter 11
Activity-Dependent Bigenomic Transcriptional Regulation of Cytochrome c Oxidase in Neurons

Margaret T. T. Wong-Riley, Huan Ling Liang, and Sakkapol Ongwijitwat

Abstract Cytochrome c oxidase, a sensitive indicator of neuronal metabolic capacity and activity, is one of only four proteins in mammalian cells that are bigenomically encoded. The mitochondrial and nuclear genomes have to coordinate to form a functional holoenzyme with 13 subunits. Nuclear respiratory factors 1 and 2 (NRF-1 and NRF-2) are viable candidates for such bigenomic coordination, most likely in association with coactivators, such as peroxisome proliferator-activated receptor gamma coactivator-1alpha (PGC-1α). These three factors are themselves regulated by neuronal activity.

11.1 Neuronal Activity and Energy Metabolism

Neuronal activity and energy metabolism are tightly coupled processes, and under normal conditions, neuronal activity controls energy expenditure (Lowry 1975). However, when energy metabolism becomes defective, it threatens the proper functioning and eventual survival of neurons. The major energy-generating enzymes in the mitochondria are four electron transport chain complexes (I, III, IV, and V). These four complexes are unique in that they are the only ones in mammalian cells that are bigenomically encoded, that is, each has nuclear-encoded subunits as well as mitochondrial-encoded subunits. Their bigenomic nature underscores the importance of a symbiotic relationship between the nucleus and the mitochondria through evolution. Coordination between the two genomes is necessary for proper functioning of these four enzymes, yet the mechanism of bigenomic regulation is poorly understood, especially in neurons.

This review will concentrate on complex IV or cytochrome c oxidase, the best studied of the four unique bigenomic enzymes in neurons.

11.2 Cytochrome c Oxidase

Cytochrome c oxidase or complex IV of the mitochondrial electron transport chain (COX or CO; EC 1.9.3.1) is the terminal enzyme that catalyzes the oxidation of its substrate, cytochrome c, and the reduction of molecular oxygen to water.

Without COX, oxidative metabolism simply cannot be carried to completion. The strict reliance of neurons on oxidative metabolism enables COX to be a sensitive and reliable indicator of neuron's metabolic capacity as well as its activity (Wong-Riley 1989; Wong-Riley, Nie, Hevner, and Liu 1998). COX is composed of 13 subunits; the largest 3 (I, II, III) forming the catalytic core of the enzyme are encoded in the mitochondrial genome, while the other 10 (IV, Va, Vb, VIa, b, c, VIIa, b, c, VIII) are encoded in the nuclear DNA (Kadenbach, Jarausch, Hartmann, and Merle 1983). Some of the subunits have tissue-specific isoforms (reviewed in Kadenbach, Hüttemann, Arnold, Lee, and Bender 2000). To form a functional holoenzyme, the two genomes have to coordinate to synthesize and assemble the 13 subunits in a 1:1 stoichiometry. Coordination of the two genomes poses a special challenge for neurons in that the mitochondrial genome in distal dendrites can be far away from the nuclear genome in the cell body. Dendrites consume the bulk of energy in the brain (Wong-Riley 1989), as they generally have the largest surface-to-volume ratio in a neuron, are the major receptive sites for excitatory (and inhibitory) synapses, and their membranes have to be constantly repolarized after depolarization, a highly energy-dependent process.

What are the sites of synthesis for mitochondrial and nuclear-encoded subunits of COX in neurons? By means of light and electron microscopic *in situ* hybridization, we found that the mRNAs for mitochondrial-encoded subunits reside in mitochondria throughout the cell, while the transcripts for nuclear-encoded subunits are restricted to the cell body (Hevner and Wong-Riley 1991; Wong-Riley, Mullen, Huang, and Guyer 1997). This implies that translation of nuclear-encoded COX subunits is confined to the perikarya. *How, then, do these subunits get to the dendrites, where they are most needed?* The evidence thus far suggests that neurons synthesize precursor proteins of nuclear-encoded COX subunits in the cell bodies and import them into the mitochondria, where the precursor proteins are not all fully processed, but rather, much of them are delivered to distal dendrites (as well as axon terminals), where they form a precursor pool, presumably to be processed locally when the energy demand is increased (Liu and Wong-Riley 1994). Neurons have, therefore, evolved a unique way to regulate their energy metabolism at a very local level.

Is there evidence for coordinated transcription of all 13 COX subunit genes in neurons, and is such transcription regulated by neuronal activity? In vitro experiments with cultured primary visual cortical neurons indicate that all 13 COX subunit transcripts were significantly up-regulated after 5 hours of depolarizing stimulation with potassium chloride (KCl) (Liang, Ongwijitwat, and Wong-Riley 2006) (Fig. 1). Thus, increased neuronal activity exerts a higher energy demand on neurons to repolarize their membrane for reactivation, necessitating the up-regulation of their energy-generating enzyme (Wong-Riley 1989).

Decreased neuronal activity with tetrodotoxin (TTX) blockade, however, leads to the down-regulation of COX gene expression. In this case, the three mitochondrial-encoded transcripts were down-regulated earlier than the ten nuclear ones (2 days versus 4 days) (Liang et al., 2006) (Fig. 2).

In vivo studies on monocular enucleation also revealed that all three mitochondrial-encoded subunit mRNAs were down-regulated earlier than the ten nuclear ones

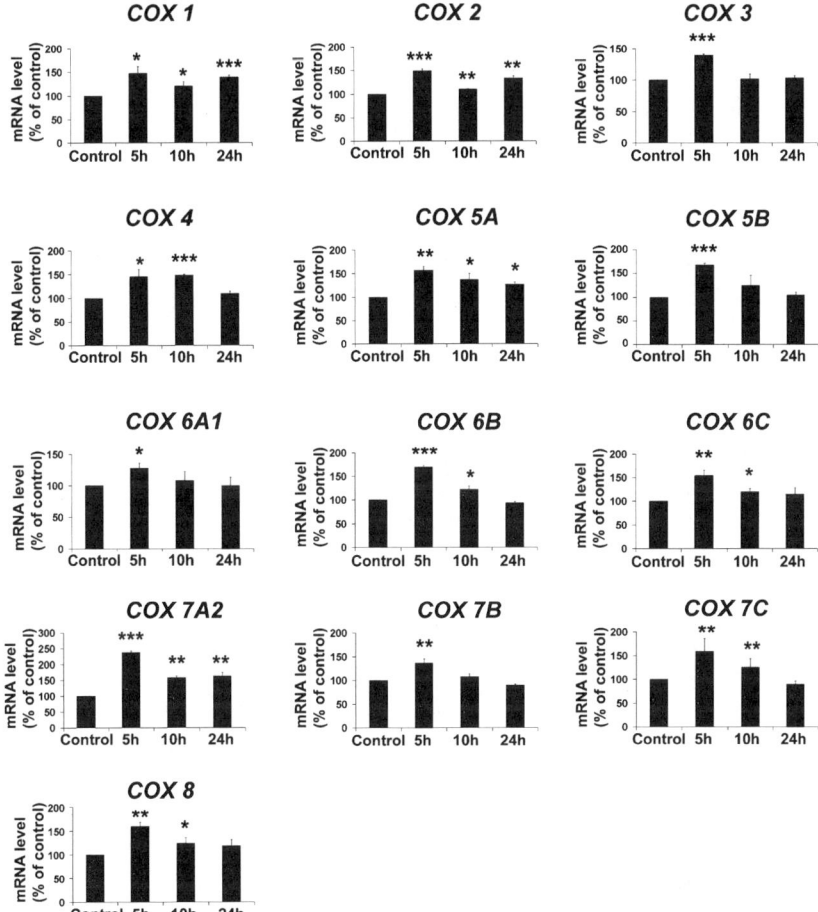

Fig. 1 Quantitative real time PCR analysis of relative changes in all 13 cytochrome c oxidase subunit mRNA levels *in vitro*. Primary neuronal cultures were depolarized for 5, 10, or 24 hours. All 13 COX subunits were significantly increased after 5 hours of depolarization ($P < 0.05 - 0.001$), the levels of *COX 1, 2, 4, 5A, 6B, 6C, 7A2, 7C*, and *8* remained significantly higher than controls after 10 hours of KCl exposure, and *COX 1, 2, 5A*, and *7A2* stayed elevated 24 hours after KCl treatment ($P < 0.05 - 0.001$). (N = 6 for each group). *$P < 0.05$, **$P < 0.01$, ***$P < 0.001$. (Modified from Liang et al., 2006)

in deprived visual cortex (4 days versus 1 week after monocular enucleation in rats). COX activity and protein levels were significantly decreased in parallel after 4 days of deprivation in vitro and 1 week in vivo (Liang et al., 2006). These results are consistent with a coordinated mechanism of up-regulation of all transcripts in response to functional stimulation, but an earlier and more severe down-regulation of the mitochondrial transcripts than the nuclear ones in response to functional deprivation. Thus, the mitochondrial subunits may play a more important role in regulating COX protein amount and activity in neurons (Hevner and Wong-Riley 1993).

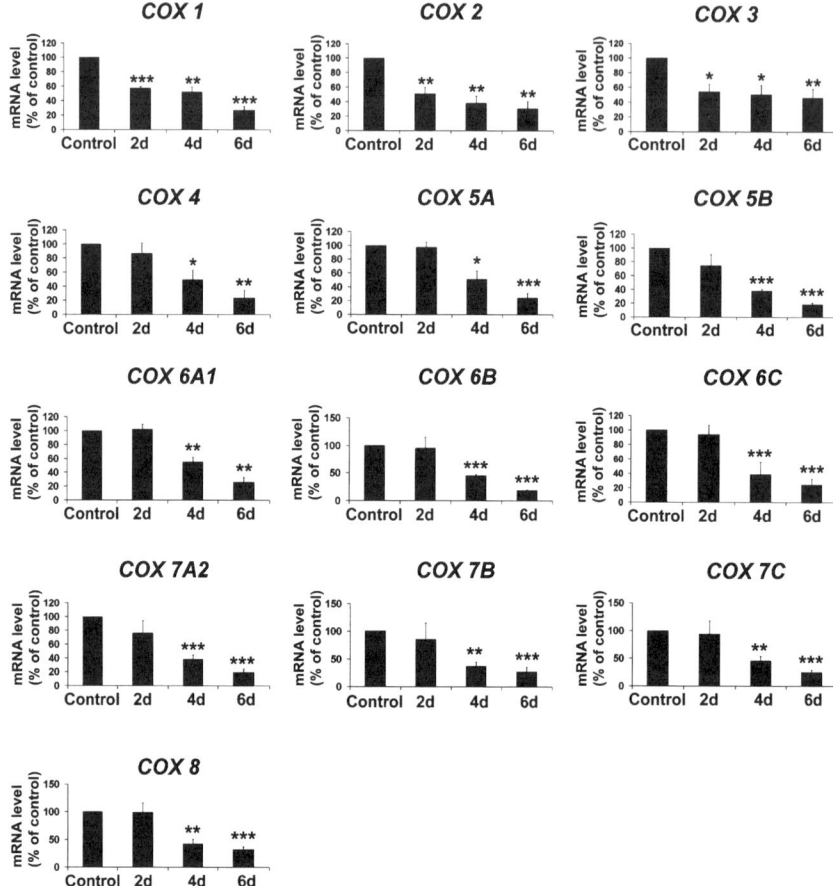

Fig. 2 Quantitative real time PCR analysis of relative changes in all 13 COX subunit mRNA levels in vitro. Primary neuronal cultures were functionally inactivated by exposure to TTX for 2, 4, and 6 days. Mitochondrial-encoded subunits (*COX 1, 2, 3*) were dramatically decreased after 2 days of TTX exposure ($P < 0.05 – 0.001$). All ten of the nuclear-encoded subunits were significantly decreased after 4 days of TTX ($P < 0.05 – 0.001$). (N = 6 for each group). $^*P < 0.05$, $^{**}P < 0.01$, $^{***}P < 0.001$. (Modified from Liang et al., 2006)

How are the mitochondrial and nuclear-encoded COX subunit genes regulated transcriptionally in neurons? The three largest subunits are transcribed as part of the polycistronic transcript in the mitochondria and are posttranscriptionally processed into *COX 1, 2,* and *3* (Clayton 1991). The ten nuclear-encoded subunits have genes in different loci and are transcribed independently. Before considering candidate factors for bigenomically coordinating the two genomes, we shall first briefly examine the mitochondrial transcription factors A and B that play important roles in mitochondrial DNA (mtDNA) transcription and replication.

11.3 Mitochondrial Transcription Factors A (mtTFA or TFAM) and B (mtTFB1 or TFB1M and mtTFB2 or TFB2M)

Mitochondrial DNA (mtDNA) encodes 2 rRNAs, 22 tRNAs, and only 13 polypeptides that comprise various subunits of complexes I, III, IV, and V, all of which also have nuclear-encoded subunits to help form each of the bigenomic holoenzymes (Anderson, Bankier, Barrell, de Bruijn, Coulson, Drouin, Eperon, Nierlich, Roe, Sanger, Schreier, Smith, Staden, and Young 1981). Three of the 13 polypeptides are COX subunits I, II, and III that form the redox center of cytochrome c oxidase. Transcription and replication of mtDNA are controlled by nuclear-encoded transcription factors, mitochondrial transcription factors A (mtTFA or TFAM) (Fisher and Clayton 1988; Larsson, Barsh, and Clayton 1997; Clayton 2000) and B (which has two forms, mtTFB1 or TFB1M and mtTFB2 or TFB2M), in association with mitochondrial RNA polymerase (Falkenberg, Gaspari, Rantanen, Trifunovic, Larsson, and Gustafsson 2002).

The human mtTFA was originally purified and characterized by Fisher and Clayton (1988). It is a monomeric protein of ~ 25 kDa, and contains 2 domains that are characteristic of high mobility group (HMG) proteins (Parisi and Clayton 1991). The C-terminal tail region is important for specific DNA recognition and is essential for transcriptional activation (Dairaghi, Shadel, and Clayton 1995). Inhibition of mtTFA transcription by antisense RNA in transfected COS-7 cells caused a down regulation of not only mtTFA proteins, but also COX subunits I and III that are encoded in the mtDNA (Inagaki, Kitano, Lin, Maeda, and Saito 1998). This indicates that the expression of mitochondrial genes is under the control of mtTFA. *Tfam* knockout mice have been generated and the homozygous embryos did not survive past embryonic day 10.5, whereas the heterozygous mice exhibit reduced mtDNA copy number and respiratory deficiency in the heart (Larsson, Wang, Wilhelmsson, Oldfors, Rustin, Lewandoski, Barsh, and Clayton 1998; Ekstrand, Falkenberg, Rantanen, Park, Gaspari, Hultenby, Rustin, Gustafsson, and Larsson 2004). Thus, mtTFA regulates mtDNA copy number *in vivo* and is essential for mitochondrial biogenesis and embryonic development. Less is known about TFB1M or TFB2M, other than the fact that one of them needs to interact with mtTFA and mitochondrial RNA polymerase for mtDNA transcription to take place (McCulloch and Shadel 2003).

What is the distribution of mtTFA in the central nervous system? A *Drosophila* form of mtTFA was detected in the fly brain throughout development (Takata, Inoue, Hirose, Murakami, Shimanouchi, Sakimoto, and Sakaguchi 2003), and mtTFA was isolated from a rat cerebellar cDNA library (Inagaki, Hayashi, Matsushima, Lin, Maeda, Ichihara, Kitagawa, and Saito 2000). However, the pattern of distribution of mtTFA in the brain remains unknown. Preliminary studies in our laboratory indicate that mtTFA is present in visual cortical neurons of rats and monkeys (data not shown), but a detailed study is still lacking. Even less is known about the distribution of TFB1M and TFB2M in the CNS. This dearth of information underscores our limited knowledge of the molecular mechanism of energy regulation in neurons.

Exercise training in humans caused an increase in the level of mtTFA and COX subunits I (mitochondrial) and IV (nuclear) in muscle biopsy samples, indicating a clear response to energy demand (Bengtsson, Gustafsson, Widegren, Jansson, and Sundberg 2001). In rats, increased muscular contractile activity with electrical stimulation induced an increase in mtTFA expression that preceded that of COX gene expression and enzyme activity (Gordon, Rungi, Inagaki, and Hood 2001). Virtually nothing is known about possible responses of mtTFA in neurons to increased or decreased neuronal activity.

11.4 Bigenomic Coordinators of Cytochrome c Oxidase Genes

What are the major transcriptional players in bigenomically regulating the expression of mitochondrial and nuclear-derived COX subunit genes? Currently, two candidate transcription factors can be considered to mediate bigenomic regulation of COX: Nuclear respiratory factors 1 and 2 (NRF-1 and NRF-2) (Evans and Scarpulla 1989; Virbasius, Virbasius, and Scarpulla 1993a,b). In non-neuronal and cell-free *in vitro* preparations, both factors activate the transcription of some of the nuclear-encoded COX subunit genes (reviewed in Grossman and Lomax 1997; Scarpulla 1997, 2002, 2006). They also indirectly regulate the expression of the three mitochondrial-encoded COX subunits by activating mitochondrial transcription factors A and B (both B1 and B2) (Virbasius and Scarpulla 1994; Gleyzer, Vercauteren, and Scarpulla (2005). In addition, a documented coactivator of NRF-1 and NRF-2, known as peroxisome proliferator-activated receptor gamma coactivator-1 or PGC-1, is thought to be critical for energy metabolism in non-neural tissues (Puigserver, Wu, Park, Graves, Wright, and Spiegelman 1998; Puigserver, Adelmant, Wu, Fan, Xu, O'Malley, and Spiegelman 1999). How these factors regulate the expression of cytochrome c oxidase in neurons is discussed below. The role of NRF-2 is much better understood, while those of NRF-1 and PGC-1 are only beginning to be probed.

11.4.1 Nuclear Respiratory Factor 2 (NRF-2)

NRF-2 was discovered as the human homologue of the murine GA-binding protein (GABP), a heteromeric transcription factor with α and β subunits (Thompson, Brown, and McKnight 1991; Virbasius and Scarpulla 1991; Virbasius et al., 1993a). The α subunit belongs to the E26 transformation-specific (Ets) family of proteins that bind the GGAA core DNA sequence of the consensus sequence motif (C/A)GGA(A/T)(A/G), whereas the β subunit contains the transactivating domain and the nuclear localizing signal (LaMarco, Thompson, Byers, Walton, and McKnight 1991; Virbasius and Scarpulla 1991; Virbasius et al., 1993a; Scarpulla 1997). The human NRF-2 is a five-subunit protein, comprising the α subunit and four other subunits (β_1, β_2, γ_1, and γ_2) (Virbasius et al., 1993b). The γ subunits are similar to the β ones, except that they lack the C-terminal homodimerization domain. Such domains enable the formation of a heterotetrameric complex ($\alpha_2\beta_2$) that binds to tandem NRF-2 binding sites in target genes (Scarpulla 2002).

Cytochrome c oxidase subunit *4* in rats was one of the first target genes found to be activated by NRF-2 (Virbasius and Scarpulla 1991; Virbasius et al., 1993a). This was confirmed in mice (Carter, Bhat, Basu, and Avadhani 1992; Carter and Avadhani 1994). The promoters of four other COX subunit genes were subsequently analyzed to functionally bind to NRF-2: *5B* in mice (Virbasius et al., 1993a; Sucharov, Basu, Carter, and Avadhani, 1995), *6A1* in humans (Wong-Riley, Guo, Bachman, and Lomax, 2000), and *7A2* and *7C* in cows (Seelan, Gopalakrishnan, Scarpulla, and Grossman 1996; Seelan and Grossman 1997). However, the role of NRF-2 in neurons remained obscure until 1999, when the distribution of NRF-2 in the monkey visual cortex was found to be virtually identical to that of COX (Nie and Wong-Riley 1999), a feature not shared by any other transcription factors studied thus far, such as c-Fos, Jun-B, and Zif268 (Chaudhuri, Matsubara, and Cynader 1995; Herdegen, Kovary, Buhl, Bravo, Zimmermann, and Gass 1995; Okuno, Kanou, Tokuyama, Li, and Miyashita 1997). NRF-2 is also more strongly expressed in cell types that have higher COX activity than those with lower activity (Wong-Riley, Yang, Liang, Ning, and Jacobs 2005). Moreover, NRF-2 responds to monocular impulse blockade by down-regulating its protein and message levels in deprived cortical columns in which COX activity is suppressed (Nie and Wong-Riley 1999; Guo, Nie, and Wong-Riley 2000; Wong-Riley et al., 2005). In response to KCl depolarization in cultured primary neurons, NRF-2 protein is up-regulated prior to the up-regulation of COX subunit message and activity (Zhang and Wong-Riley 2000), and both α and β subunits respond to increased neuronal activity by translocating from the cytoplasm to the nucleus, where they associate primarily with euchromatin and activate their target genes (Yang, Liang, Ning, and Wong-Riley 2004). These findings are consistent with NRF-2's proposed role as a transcriptional activator of COX and that its own expression is regulated by neuronal activity.

Prior to 2004, as discussed above, only one of the ten COX subunit genes (subunit *1*) in the rat nuclear genome was shown to be functionally activated by NRF-2 (Virbasius and Scarpulla 1991; Virbasius et al., 1993a). By means of promoter mutation analysis, chromatin immunoprecipitation assays (ChIP), electrophoretic mobility shift assays (EMSA), supershift assays, and dominant-negative experiments, we found that all ten COX nuclear subunit promoters in the rat have functional binding sites for NRF-2 (Ongwijitwat and Wong-Riley 2004, 2005). Thus, NRF-2 is the only transcription factor known thus far to be in a strong position to coordinate the transcriptional activity of all ten COX subunit genes located in different chromosomes (Ongwijitwat and Wong-Riley 2005). Moreover, NRF-2 indirectly affects the expression of the remaining three COX mitochondrial subunits by activating mtTFA, TFB1M, and TFB2M (Virbasius and Scarpulla 1994; Gleyzer et al., 2005). NRF-2, therefore, can be regarded as a theoretical master transcriptional regulator of all 13 subunits of COX, at least in the rat. When the promoter sequences of all ten nuclear-encoded COX subunit genes were compared between rat, mouse, and human, it was found that each subunit promoter has a subset of the NRF-2 binding sites that is highly conserved among the three species (Ongwijitwat and Wong-Riley 2005).

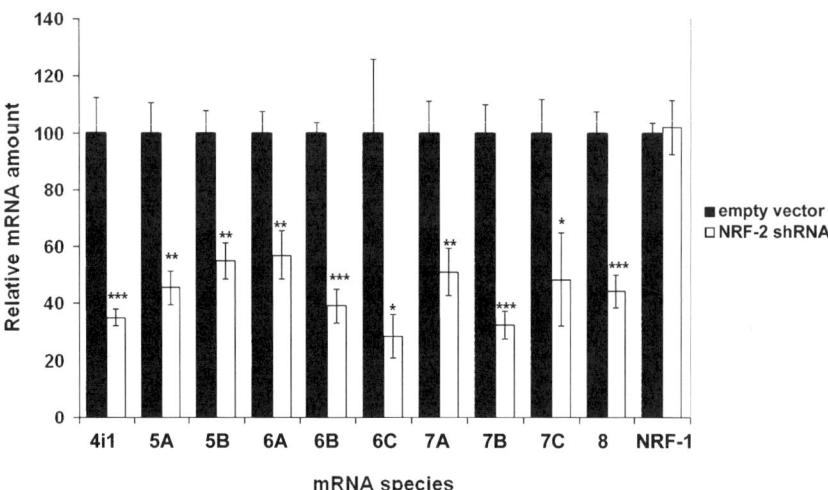

Fig. 3 Following shRNA-mediated silencing of NRF-2, all ten nuclear-encoded COX subunit mRNAs were reduced (clear bars). Empty vectors served as controls (dark bars). An additional control was NRF-1, whose mRNA remained unchanged. (N = 5-6 for each data point; $^*P < 0.05$; $^{**}P < 0.01$; $^{***}P < 0.001$). (Modified from Ongwijitwat et al., 2006)

Is NRF-2 absolutely necessary for the expression of COX subunit genes? To answer this question, the expression of NRF-2 was suppressed by vector-mediated short-hairpin RNA interference (shRNA), and all ten COX nuclear subunit mRNAs were found to be significantly reduced by 40–70% (Ongwijitwat, Liang, Graboyes, and Wong-Riley 2006) (see Fig. 3 above). Thus, NRF-2 is necessary for the transcription of all ten nuclear-encoded COX subunit genes. In addition, silencing of NRF-2 also significantly reduced the expressions of mtTFA and mtTFB1, thereby indirectly regulating the levels of the three mitochondrial COX subunit mRNAs (Ongwijitwat et al., 2006)

To determine if NRF-2 binding is necessary and sufficient for COX subunit promoters to respond to reduced energy demand in primary neurons with tetrodotoxin (TTX)-induced impulse blockade, promoter-reporter experiments were done with two exemplary COX promoters, one with NRF-2 alone and one with additional NRF-1 binding sites (*COX 4i1* and *6A1*, respectively) (Fig. 4 A–D; modified from Ongwijitwat et al., 2006). Results indicate that the transcriptional activity of *COX 4i1* as measured by luciferase activity in primary neurons was reduced with TTX treatment by ∼ 60% (Fig. 4A), and over-expression of NRF-2 prevented this reduction (Fig. 4B). Point mutation of a NRF-2-binding site on the *COX 4i1* promoter reduced the baseline activity by ∼ 50% and eliminated the TTX-induced down-regulation (Fig. 4A, B). On the other hand, *COX6A1* promoter did not respond to TTX (Fig. 4C), presumably because it also has binding sites for NRF-1 that contributes to and compensates for the activity of the basal promoter (Ongwijitwat and Wong-Riley 2004). To overcome this complication, artificial promoters were constructed with two sets of tandem NRF-2 binding sites previously

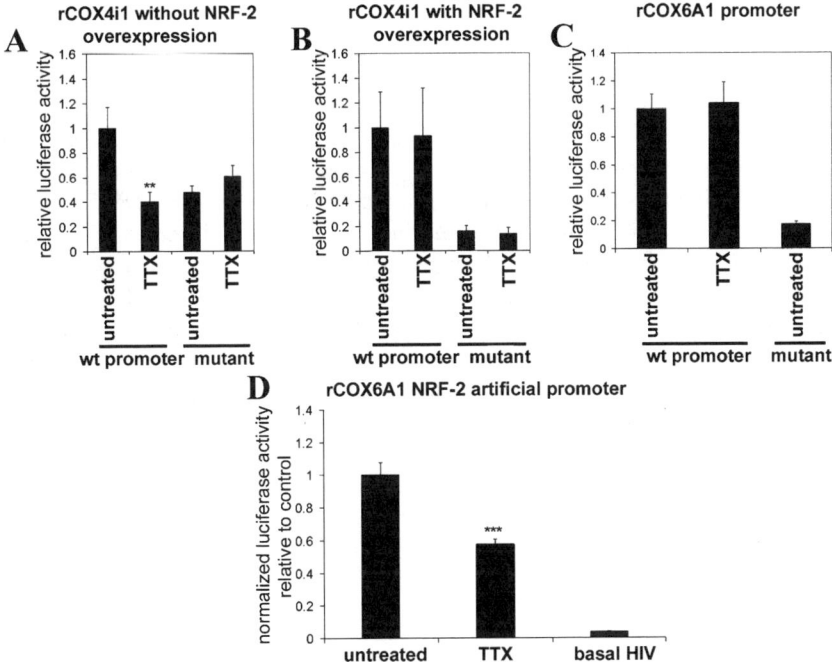

Fig. 4 NRF-2 –dependent response of *COX4i1* and *COX6A1* promoters to control and TTX treatment in primary neurons. (A) Activity of wild type (WT) versus mutant *COX4i1* promoter with and without TTX. (B) Over-expression of NRF-2 prevented the down-regulation of *COX4i1* promoter activity by TTX. (C) *COX6A1* promoter was not affected by TTX, but an artificial *COX6A1* promoter with only NRF-2 binding sites did respond (D). (Modified from Ongwijitwat et al., 2006)

identified on the *COX6A1* promoter (Ongwijitwat and Wong-Riley 2004) and the elimination of the NRF-1 site. This construct increased transcriptional activity of the basal HIV promoter by $\sim$ 100 fold and was able to respond to TTX with $\sim$ 40% reduction in transcriptional activity (Ongwijitwat et al., 2006) (Fig. 4D). Thus, NRF-2 binding sites alone are sufficient for the promoter's response to TTX

Besides cytochrome c oxidase and mitochondrial transcription factors A and B, NRF-2 also binds to promoters of other genes involved in mitochondrial biogenesis, such as subunits of complexes II and V (reviewed in Scarpulla 2002; Kelly and Scarpulla 2004). In addition, silencing of NRF-2 with small interference RNA results in a significant down-regulation of mRNAs for *SURF1* (surfeit 1, a COX assembly protein), *VDAC1* (voltage-dependent anion channel) and *TOM20* (transporter of outer mitochondrial membrane 20) (Ongwijitwat et al., 2006). Thus, NRF-2 exerts a strong influence on diverse mitochondrial functions.

11.4.2 Nuclear Respiratory Factor 1 (NRF-1)

NRF-1 was discovered during mutational and DNA binding analysis of the somatic cytochrome c promoter, indicating that it is involved in the transcriptional regulation of cytochrome c, the substrate for cytochrome c oxidase (Evans and Scarpulla 1989). NRF-1 binding sites have been found in a number of other genes, whose products operate within the mitochondria. Four COX subunit genes have been functionally characterized to be activated by NRF-1: *5B* in mice, *6A1* in rats and humans, *6C* in rats, and *7A2* in cows (Evans and Scarpulla, 1990; Virbasius et al., 1993a; Seelan et al., 1996; Wong-Riley et al., 2000; Ongwijitwat and Wong-Riley 2004). Putative NRF-1 binding sites have also been reported for the human *COX 5B*, the rat and bovine *6A2*, and the human and murine *7A2* (Smith and Lomax 1993; Mell, Seibel, and Kadenbach 1994, Bachman, Yang, Dasen, Ernst, and Lomax 1996, Hüttemann, Muhlenbein, Schmidt, Grossman, and Kadenbach, 2000; Chantrel-Groussard, Deply, Ratinaud, and Cogne 2001). Importantly, NRF-1 also activates mtTFA, TFB1M, and TFB2M (Virbasius and Scarpulla 1994; Gleyzer et al., 2005), which, as described above, are instrumental in generating the polycistronic transcript of mtDNA. Thus, NRF-1 is another viable, bigenomic regulator of COX. In addition, one or more subunit genes of all five respiratory complexes reportedly have NRF-1 recognition sites, as do genes encoding mitochondrial RNA processing (MRP) RNA, 5-aminolevulinate synthase for heme biosynthesis, and some of the mitochondrial translocases and ion channels (reviewed in Scarpulla 1997, 2002;

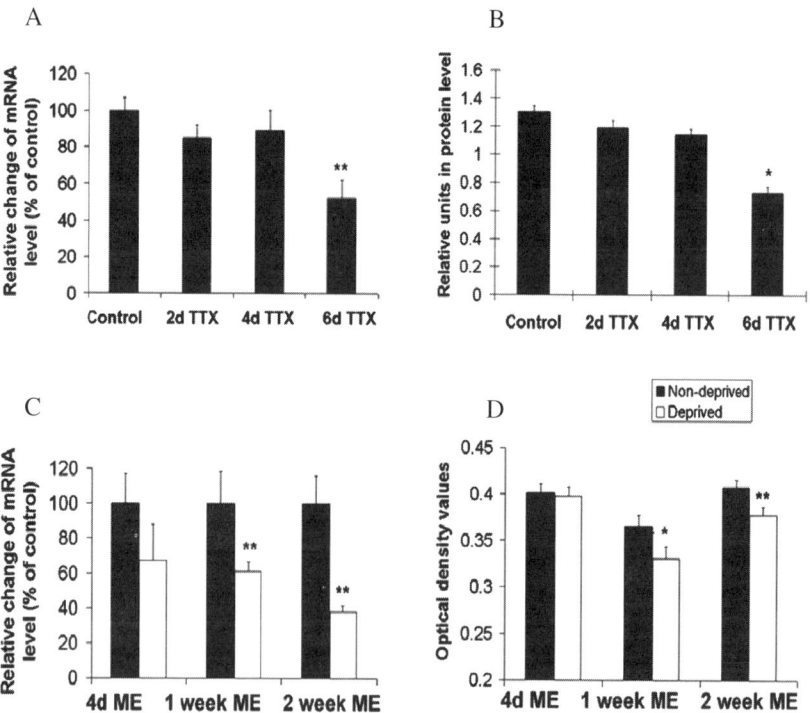

Fig. 5 The expression of NRF-1 mRNA measured with real-time quantitative PCR was significantly reduced after 6 days of TTX inactivation *in vitro* (A) and 1 or 2 weeks of monocular deprivation *in vivo* (C). NRF-1 protein level was likewise down-regulated *in vitro* (B) and *in vivo* (D). $^*P < 0.05$; $^{**}P < 0.01$. (Modified from Liang and Wong-Riley 2006)

Kelly and Scarpulla 2004). NRF-1, therefore, also plays a crucial role in mitochondrial biogenesis and functions.

Unlike NRF-2, NRF-1 is a single-gene product whose gene is approximately 104 kb in the human (Huo and Scarpulla 1999) and is mapped to human chromosome 7 (7q31) (Gopalakrishnan and Scarpulla 1995). The amino terminus contains the DNA binding domain as well as the nuclear localizing signal and is highly conserved, while the carboxy terminus is quite divergent among the species and is necessary for transcriptional activation (Virbasius et al., 1993b; Gugneja, Virbasius, and Scarpulla 1996). NRF-1 binds the palindromic consensus sequence (T/C)GCGCA(T/C)GCGC(A/G) (Evans and Scarpulla 1990; Virbasius et al., 1993b; Scarpulla 1997). Phosphorylated NRF-1 enhances DNA binding activity, suggesting that posttranslational modification may modulate transcriptional activity in response to the availability of ATP (Evans and Scarpulla 1990; Virbasius et al., 1993b; Scarpulla 1997). Homozygous NRF-1 knockout mice are embryonically lethal between E3.5 and E6.5, and the blastocysts have markedly reduced levels of mtDNA, consistent with the requirement of NRF-1 in the maintenance of mtDNA and respiratory chain function during early embryogenesis (Huo and Scarpulla 2001). Heterozygous mice develop normally, and no clear deficit has been detected (unpublished observations from our lab in collaboration with Scarpulla's lab). Relatively few studies focused on NRF-1's response to functional alterations, and all of them were performed on non-neural tissues, such as

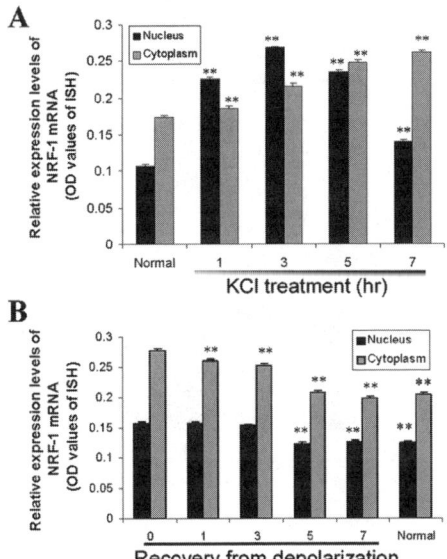

Fig. 6 Quantitative analysis of in situ hybridization data indicating the effects of (A) KCl depolarization for varying periods of time or (B) withdrawal of KCl for varying times after 7 hours of KCl treatment on the expression of NRF-1 mRNA. Optical densitometric values represent mean ± SEM for at least three different experiments, $^*P < 0.05$, $^{**}P < 0.01$ versus normal control in (A) and versus 0 hr recovery in (B) (Student's t test). (Modified from Yang et al., 2006)

cardiac myocytes and skeletal muscles (Xia, Buja, Scarpulla, and McMillin 1997; Murakami, Shimomura, Yoshimura, Sokabe, and Fujitsuka 1998). Recently, we examined this issue and found that neuronal activity regulates the protein and mRNA expressions of NRF-1 after functional inactivation *in vivo* and *in vitro* (Liang and Wong-Riley 2006) (Fig. 5).

In vitro impulse blockade was induced by tetrodotoxin in primary visual cortical neurons. This led to a reduction in the expression of NRF-1 mRNA (Fig. 5A) and protein (Fig. 5B) that reached statistical significance after 6 days of inactivation. *In vivo* functional deprivation was achieved in the visual cortex by monocular enucleation, and the expressions of both NRF-1 mRNA (Fig. 5C) and protein (Fig. 5D) were significantly decreased after 1 week of deprivation. The effect of depolarizing stimulation was also tested in primary neurons *in vitro* (Yang, Liang, and Wong-Riley 2006). We found that depolarizing KCl treatment progressively up-regulated both NRF-1 message (Fig. 6) and protein (Fig. 7) in a time-dependent manner, increasing above controls after 1 hour and remaining high at 3, 5, and 7 hours. Both nuclear and cytoplasmic mRNA levels increased with stimulation, with an apparent cytoplasmic-to-nuclear translocation of NRF-1 protein (as in the case of NRF-2 described above). Withdrawal of KCl led to a progressive decline in the

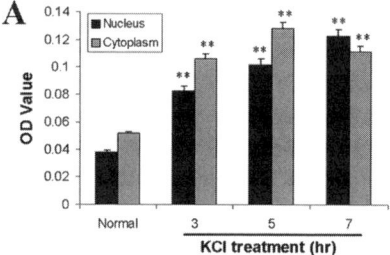

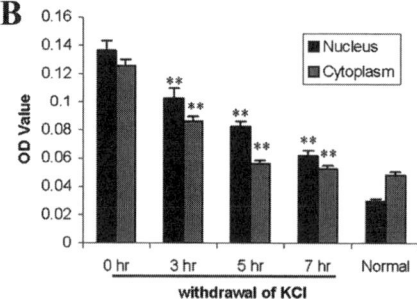

Fig. 7 A. Optical densitometric measurements of immunohistochemical labeling of NRF-1 protein in neurons grown in control medium or in medium containing 20 mM KCl for 3, 5, and 7 hours. NRF-1 expression progressively increased after depolarizing stimulation. By 7 hr of stimulation, labeling in the nucleus surpassed that in the cytoplasm. B. Starting at 7 hr of stimulation, NRF-1 protein values gradually decreased in both the nucleus and the cytoplasm with the withdrawal of KCl for 0, 3, 5, and 7 hr. Values in normal neurons are shown for comparison. Each value represents the mean ± SEM of combined data from three separate experiments. $^{**}P < 0.001$ versus starting values at 0 hr (Student's t test). (Modified from Yang et al., 2006)

levels of NRF-1 mRNA and protein. The message returned to basal levels by 5 hours and the protein by 7 hours. Thus, NRF-1 responds rapidly to functional stimulation but declines more slowly with functional impulse blockade in neurons. The close correlation between NRF-1 mRNA and protein levels suggests that the regulation of NRF-1 in neurons is primarily transcriptional. Our findings are also consistent with an activity-dependency of the synthesis, distribution, and possibly stability of NRF-1 mRNA and protein in neurons.

Unlike the case of NRF-2, where all ten nuclear-encoded COX subunit genes are found to be under its regulatory influence, more research is needed to determine if NRF-1 is another "master regulator" or only a partial coordinator of transcription for the 13 COX subunits. Moreover, possible interactions between NRF-1 and NRF-2, or NRF-1 with other transcription factors, such as Sp1 and YY-1 (Basu, Park, Atchison, Carter, and Avadhani, 1993; Hoshinaga, Amuro, Goto, and Okazaki 1994; Wong-Riley et al., 2000), need to be investigated.

11.5 Transcriptional Coactivators: PPARγ Coactivator-1α (PGC-1α)

PGC-1 or PGC-1α (peroxisome proliferator-activated receptor gamma (PPARγ) coactivator-1α) belongs to a recently defined class of coactivators that do not bind DNA directly, but rather, interact with DNA-bound transcription factors to regulate gene expression (reviewed in Scarpulla 2002, 2006). As a coactivator, PGC-1α is thought to play a critical role in regulating many aspects of energy metabolism, including mitochondrial biogenesis, adaptive thermogenesis, and fatty acid beta-oxidation (Puigserver et al., 1999; Knutti and Kralli 2001; Lin, Wu, Tarr, Zhang, Wu, Boss, Michael, Puigserver, Isotani, Olson, Lowell, Bassel-Duby, and Spiegelman 2002). PGC-1α gene expression is induced by cold exposure and endothelial nitric oxide synthase in adipocytes and endurance exercise in skeletal muscles (Puigserver et al., 1998; Goto, Terada, Kato, Katoh, Yokozeki, Tabata, and Shimokawa 2000; Baar, Wende, Jones, Marison, Nolte, Chen, Kelly, and Holloszy 2002; Nisoli, Clementi, Paolucci, Cozzi, Tonello, Sciorati, Bracale, Valerio, Francolini, Moncada, and Carruba 2003). It is also activated by key cellular signals that control energy and nutrient homeostasis, such as cAMP and cytokine pathways (reviewed in Puigserver and Spiegelman 2003). Once activated, PGC-1α reportedly induces and coordinates gene expression that stimulates mitochondrial oxidative metabolism in brown fat, fiber-type switching in skeletal muscle, and liver response to fasting (Baar et al., 2002; Irrcher, Adhihetty, Sheehan, Joseph, and Hood 2003; Puigserver et al., 1998; Puigserver and Spiegelman 2003; Short, Vittone, Bigelow, Proctor, Rizza, Coenen-Schimke, and Nair 2003). PGC-1α null mice have reduced mitochondrial function, but are paradoxically lean, hyperactive, and resistant to diet-induced obesity. Lesions were found in the striatum, but no explanation was given as to why the striatum should be selectively affected (Lin, Wu, Tarr, Lindenberg, St-Pierre, Zhang, Mootha, Jäger, Vianna, Reznick,

Cui, Manieri, Donovan, Wu, Cooper, Fan, Rohas, Zavacki, Cinti, Shulman, Lowell, Krainc, and Spiegelman 2004; Arany, Novikov, Chin, Ma, Rosenzweig, and
Speigelman 2005). Of the COX subunit genes, only *COX6A1* was analyzed in the
brain and found to be reduced in the knockout mice. NRF-1 and NRF-2 levels were
not tested in the brain. Although PGC 1α is expressed in the normal murine brain, it
is not regulated there by states of caloric deficiency, leptin, obesity or cold exposure
as it is in other tissues studied (Tritos, Mastaitis, Kokkotou, Puigserver, Spiegelman,
and Maratos-Flier 2003). Thus, the role of PGC-1α in neurons is unclear.

Recently, we found that neuronal activity directly regulates the protein and
mRNA expressions of PGC-1α after functional inactivation *in vivo* and *in vitro*
(Liang and Wong-Riley 2006) (Fig. 8). *In vitro* TTX blockade down-regulated the
expression of PGC-1α mRNA starting at 2 days (Fig. 8A) and protein commenc-
ing at 4 days (Fig. 8B) after inactivation. *In vivo* functional deprivation of visual

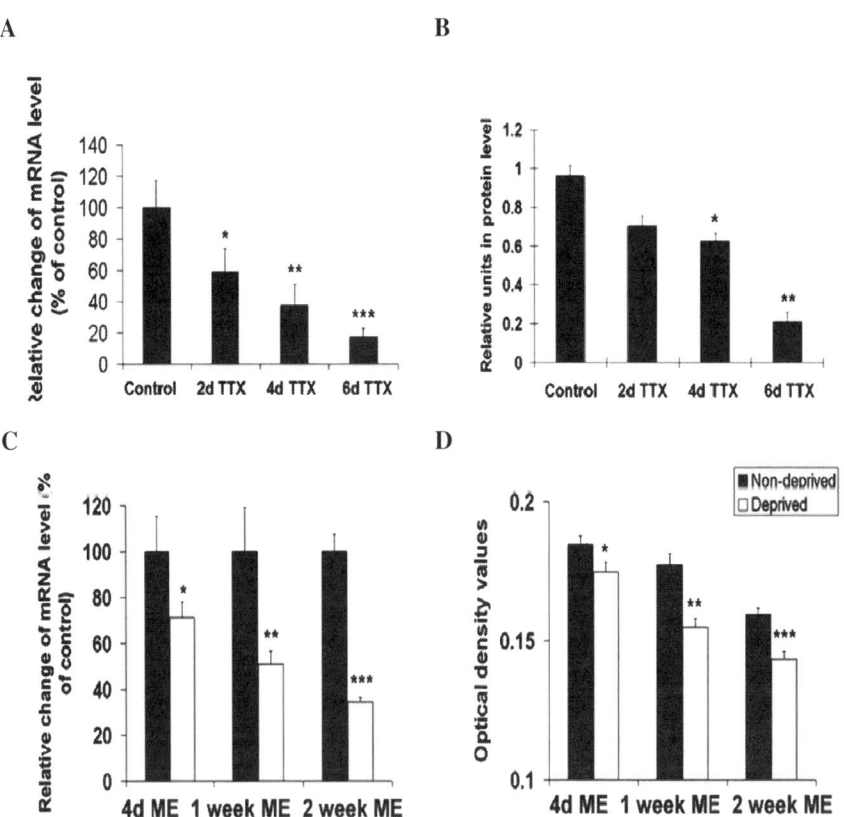

Fig. 8 The expression of PGC-1α mRNA measured with real-time quantitative PCR was significantly reduced after 2 days of TTX inactivation *in vitro* (A) and 4 days of monocular deprivation
in vivo (C). PGC-1α protein level was down-regulated after 4 days of TTX *in vitro* (B) and after 4
days of deprivation *in vivo* (D). These changes preceded those of NRF-1 (Fig. 5 above) and NRF-2
(Liang and Wong-Riley 2006). (Modified from Liang and Wong-Riley 2006)

cortex with monocular enucleation also suppressed PGC-1α message (Fig. 8C) and protein (Fig. 8D) at the earliest time-point examined (4 days). The most striking finding is that PGC-1α responds significantly earlier than NRF-1 (see Fig. 5) or NRF-2 (Liang and Wong-Riley 2006) to functional perturbations *in vitro* and *in vivo*. Thus, PGC-1α may act as a sensor in detecting environmental or functional changes in neurons and may coordinate the action of transcription factors to alter gene expressions. Indeed, in non-neural cells, PGC-1α is proposed to be an important link between external physiological stimuli and transcriptional activation necessary for mitochondrial functions (Puigserver and Spiegelman 2003). It reportedly stimulates an induction of both NRF-1 and NRF-2 gene expression in C2C12 cells (Wu, Puigserver, Andersson, Zhang, Adelmant, Mootha, Troy, Cinti, Lowell, Scarpulla, and Spiegelman 1999). It also binds to and coactivates with NRF-1 in regulating the expression of mtTFA (Wu et al., 1999). Its close relationship with both NRF-1 and NRF-2 in non-neuronal cells makes it a viable candidate as an early though indirect bigenomic coordinator of COX subunit gene expression in neurons. This point, however, has not been rigorously tested. Future studies need to determine if PGC-1α is critical in the transcriptional regulation of COX in neurons, a subject that is unknown at present. Other coactivators of the PGC-1 family, such as PGC-1β and PRC (Andersson and Scarpulla 2001; Scarpulla 2006), also deserve our attention.

11.6 Conclusions

The tight coupling between neuronal activity and energy metabolism necessitates a clear understanding of the mechanisms of energy regulation in neurons. The major energy-generating enzymes in mitochondria are four complexes of the electron transport chain, all of which are uniquely encoded in both the nuclear and the mitochondrial genomes. Precise coordination between the two genomes is needed to produce these enzymes for the support of neuronal activity and survival. Cytochrome c oxidase (COX) is one such bigenomic enzyme with 13 subunits. Previous *in vitro* and *in vivo* studies have documented that the coordinated regulation of its 13 subunits is activity-dependent. Yet, the signaling pathway linking changes in neuronal activity, hence, energy demands, and COX transcriptional adjustment remains poorly understood. In addition, the nature of the regulatory unit that enables bigenomic regulation of COX subunits is unclear. Our laboratory has recently uncovered the importance of nuclear respiratory factor 2 in activating all 10 nuclear COX subunit genes in neurons. This coupled with NRF-2's indirect regulation of the 3 mitochondrial COX subunits via its control of mitochondrial transcription factors A and B makes NRF-2 currently a dominant candidate as a bigenomic transcriptional coordinator of COX. NRF-2 itself is under the regulation of neuronal activity. The roles of NRF-1 and PGC-1α are beginning to be unraveled in neurons, but much remains to be investigated. Understanding the transcriptional regulation of cytochrome c oxidase in neurons is central to the understanding of how the brain regulates energy metabolism under normal and functionally altered states.

Acknowledgments Much of this work was supported by NIH R01 EY05439.

References

Anderson, S., Bankier, A.T., Barrell, B.G., de Bruijn, M.H., Coulson, A.R., Drouin, J., Eperon, I.C., Nierlich, D.P., Roe, B.A., Sanger, F., Schreier, P.H., Smith, A.J., Staden, R. and Young, I.G. (1981) Sequence and organization of the human mitochondrial genome. Nature 290, 457–465.

Andersson, U. and Scarpulla, R.C. (2001) PGC-1-related coactivator, a novel, serum-inducible coactivator of nuclear respiratory factor 1-dependent transcription in mammalian cells. Mol. Cell Biol. 21, 3738–3749.

Arany, Z., Novikov, M., Chin, S., Ma, Y., Rosenzweig, A. and Speigelman, B.M. (2006) Transverse aortic constriction leads to accelerated heart failure in mice lacking PPAR-γ coactivator 1α. Proc. Natl. Acad. Sci. 103, 10086–10091.

Baar, K., Wende, A.R., Jones, T.E., Marison, M., Nolte, L.A., Chen, M., Kelly, D.P. and Holloszy, J.O. (2002) Adaptations of skeletal muscle to exercise: rapid increase in the transcriptional coactivator PGC-1. Faseb. J. 16, 1879–1886.

Bachman, N.J., Yang, T.L., Dasen, J.S., Ernst, R.E. and Lomax, M.I. (1996) Phylogenetic footprinting of the human cytochrome c oxidase subunit VB promoter. Arch. Biochem. Biophys. 333, 152–162.

Basu, A., Park, K., Atchison, M.L., Carter, R.S. and Avadhani, N.G. (1993) Identification of a transcriptional initiator element in the cytochrome c oxidase subunit Vb promoter which binds to transcription factors NF-E1 (YY-1, delta) and Sp1. J. Biol. Chem. 268, 4188–4196.

Bengtsson, J., Gustafsson, T., Widegren, U., Jansson, E. and Sundberg, C. (2001) Mitochondrial transcription factor A and respiratory complex IV increase in response to exercise training in humans. Pflugers Arch. 443, 61–66.

Carter, R.S. and Avadhani, N.G. (1994) Cooperative binding of GA-binding protein transcription factors to duplicated transcription initiation region repeats of the cytochrome c oxidase subunit IV gene. J. Biol. Chem. 269, 4381–4387.

Carter, R.S., Bhat, N.K., Basu, A.. and Avadhani, N.G. (1992) The basal promoter elements of murine cytochrome c oxidase subunit IV gene consist of tandemly duplicated ets motifs that bind to GABP-related transcription factors. J. Biol. Chem. 267, 23418–23426.

Chantrel-Groussard, K., Deply, L., Ratinaud, M.H. and Cogne, M. (2001) Characterization of the murine gene for subunit VIIaL of cytochrome c oxidase. C.R. Adem. Sci. III 324, 1117–1123.

Chaudhuri, A., Matsubara, J.A. and Cynader, M.S. (1995) Neuronal activity in primate visual cortex assessed by immunostaining for the transcription factor Zif268. Vis. Neurosci. 12, 35–50.

Clayton, D.A. (1991) Replication and transcription of vertebrate mitochondrial DNA. Annu. Rev. Cell Biol. 7, 409–435.

Clayton, D.A. (2000) Transcription and replication of mitochondrial DNA. Hum. Reprod. 15 Suppl. 2 11–17

Dairaghi, D.J., Shadel, G.S. and Clayton, D.A. (1995) Addition of a 29 residue carboxyl-terminal tail converts a simple HMG box-containing protein into a transcriptional activator. J. Mol. Biol. 249, 11–28.

Ekstrand, M.I., Falkenberg, M., Rantanen, A., Park, C.B., Gaspari, M., Hultenby, K., Rustin, P., Gustafsson, C.M. and Larsson, N.G.. (2004) Mitochondrial transcription factor A regulates mtDNA copy number in mammals. Human Mol. Genetics 13, 935–944.

Evans, M.J. and Scarpulla, R.C. (1989) Interaction of nuclear factors with multiple sites in the somatic cytochrome c promoter. Characterization of upstream NRF-1, ATF, and intron Sp1 recognition sequences. J. Biol. Chem. 264, 14361–14368.

Evans, M.J. and Scarpulla, R.C. (1990) NRF-1: a trans-activator of nuclear-encoded respiratory genes in animal cells. Genes Dev. 4, 1023–1034

Falkenberg, M., Gaspari, M., Rantanen, A., Trifunovic, A., Larsson, N.G.. and Gustafsson, C.M. (2002) Mitochondrial transcription factors B1 and B2 activate transcription of human mtDNA. Nat. Genet. 31, 289–294.

Fisher, R.P. and Clayton, D.A. (1988) Purification and characterization of human mitochondrial transcription factor 1. Mol. Cell Biol. 8, 3496–3509.

Gleyzer, N., Vercauteren, K. and Scarpulla, R.C. (2005) Control of mitochondrial transcription specificity factors (TFB1M and TFB2M) by nuclear respiratory factors (NRF-1 and NRF-2) and PGC-1 family coactivators. Mol. Cell Biol. 25, 1354–1366.

Gopalakrishnan, L. and Scarpulla, R.C. (1995) Structure, expression, and chromosomal assignment of the human gene encoding nuclear respiratory factor 1. J. Biol. Chem. 270, 18019–18025.

Goto, M., Terada, S., Kato, M., Katoh, M., Yokozeki, T., Tabata, I. and Shimokawa, T. (2000) cDNA Cloning and mRNA analysis of PGC-1 in epitrochlearis muscle in swimming-exercised rats. Biochem. Biophys. Res. Commun. 274, 350–354.

Gordon, J.W., Rungi, A.A., Inagaki, H. and Hood, D.A. (2001) Effects of contractile activity on mitochondrial transcription factor A expression in skeletal muscle. J. Appl. Physiol. 90, 389–396.

Grossman, L.I., Lomax, M.I. (1997) Nuclear genes for cytochrome c oxidase. Biochim. Biophys. Acta 1352, 174–192.

Gugneja, S., Virbasius, C.M. and Scarpulla, R.C. (1996) Nuclear respiratory factors 1 and 2 utilize similar glutamine-containing clusters of hydrophobic residues to activate transcription. Mol. Cell Biol. 16, 5708–5716.

Guo, A.L., Nie, F. and Wong-Riley, M.T.T. (2000) Human brain nuclear respiratory factor (NRF) 2α cDNA: Isolation, subcloning, sequencing and in situ hybridization of transcripts in normal and visually deprived macaque visual system. J. Comp. Neurol. 417, 221–232.

Hevner, R.F. and Wong-Riley, M.T.T. (1991) Neuronal expression of nuclear and mitochondrial genes for cytochrome oxidase (CO) subunits analyzed by *in situ* hybridization: comparison with CO activity and protein. J. Neurosci. 11, 1942–1958.

Hevner, R.F. and Wong-Riley, M.T.T. (1993) Mitochondrial and nuclear gene expression for cytochrome oxidase subunits are disproportionately regulated by functional activity in neurons. J. Neurosci. 13, 1805–1819.

Herdegen, T., Kovary, K., Buhl, A., Bravo, R., Zimmermann, M. and Gass, P. (1995) Basal expression of the inducible transcription factors c-Jun, Jun B, Jun D, c-Fos, Fos B, and Krox-24 in the adult rat brain. J. Comp. Neurol. 354, 39–56.

Hoshinaga, H., Amuro, N., Goto, Y. and Okazaki, T. (1994) Molecular cloning and characterization of the rat cytochrome c oxidase subunit Vb gene. J. Biochem. (Tokyo) 115, 194–201.

Huo, L. and Scarpulla, R.C. (1999) Multiple 5'-untranslated exons in the nuclear respiratory factor 1 gene span 47 kb and contribute to transcript heterogeneity and translational efficiency. Gene 233, 213–224.

Huo, L., Scarpulla, R.C. (2001) Mitochondrial DNA instability and peri-implantation lethality associated with targeted disruption of nuclear respiratory factor 1 in mice. Mol. Cell Biol. 21, 644–654.

Hüttemann, M., Muhlenbein, N., Schmidt, T.R., Grossman, L.I. and Kadenbach, B. (2000) Isolation and sequence of the human cytochrome c oxidase subunit VIIaL gene. Biochim. Biophys. Acta 1492, 252–258.

Inagaki, H., Kitano, S., Lin, K.H., Maeda, S. and Saito, T. (1998) Inhibition of mitochondrial gene expression by antisense RNA of mitochondrial transcription factor A (mtTFA). Biochem. Mol. Biol. Int. 45, 567–573.

Inagaki, H., Hayashi, T., Matsushima, Y., Lin, K.H., Maeda, S., Ichihara, S., Kitagawa, Y. and Saito, T. (2000) Isolation of rat mitochondrial transcription factor A (r-Tfam) cDNA. DNA Seq. 11, 131–135.

Irrcher, I., Adhihetty, P.J., Sheehan, T., Joseph, A.M. and Hood, D.A. (2003) PPARgamma coactivator-1alpha expression during thyroid hormone- and contractile activity-induced mitochondrial adaptations. Am. J. Physiol. Cell Physiol. 284, C1669–1677.

Kadenbach, B., Jarausch, J., Hartmann, R. and Merle, P. (1983) Separation of mammalian cytochrome c oxidase into 13 polypeptides by a sodium dodecyl sulfate-gel eletrophoretic procedure. Analyt. Biochem. 129, 517–521.

Kadenbach, B., Hüttemann, M., Arnold, S., Lee, I. and Bender, E. (2000) Mitochondrial energy metabolism is regulated via nuclear-cpded subunits of cytochrome c oxidase. Free Rad. Biol. Med. 29, 211–221.

Kelly, D.P. and Scarpulla, R.C. (2004) Transcriptional regulatory circuits controlling mitochondrial biogenesis and function. Genes Dev. 18, 357–368.

Knutti, D. and Kralli, A. (2001) PGC-1, a versatile coactivator. Trends Endocrinol. Metab. 12, 360–365.

LaMarco, K., Thompson, C.C., Byers, B.P., Walton, E.M. and McKnight, S.L. (1991) Identification of Ets- and Notch-related subunits in GA-binding protein. Science 253, 789–792.

Larsson, N.G., Barsh, G.S. and Clayton, D.A. (1997) Structure and chromosomal localization of the mouse mitochondrial transcription factor A gene (Tfam). Mamm. Genome 8, 139–140.

Larsson, N.G., Wang, J., Wilhelmsson, H., Oldfors, A., Rustin, P., Lewandoski, M., Barsh, G.S. and Clayton, D.A. (1998) Mitochondrial transcription factor A is necessary for mtDNA maintenance and embryogenesis in mice. Nat. Genet. 18, 231–236.

Liang, H.L. and Wong-Riley, M.T.T. (2006) Activity-dependent regulation of nuclear respiratory factor-1, nuclear respiratory factor-2, and peroxisome proliferators-activated receptor gamma coactivator-1 in neurons. NeuroReport 17, 401–405.

Liang, H.L., Ongwijitwat, S. and Wong-Riley, M.T.T. (2006) Bigenomic functional regulation of all 13 cytochrome c oxidase subunit transcripts in rat neurons in vitro and in vivo. Neurosci. 140, 177–190.

Lin, J., Wu, H., Tarr, P.T., Zhang, C.Y., Wu, Z., Boss, O., Michael, L.F., Puigserver, P., Isotani, E., Olson, E.N., Lowell, B.B., Bassel-Duby. R. and Spiegelman, B.M. (2002) Transcriptional co-activator PGC-1 alpha drives the formation of slow-twitch muscle fibres. Nature 418, 797–801.

Lin, J., Wu, P.-H., Tarr, P.T., Lindenberg, K.S., St-Pierre, J., Zhang, C.Y., Mootha, V.K., Jäger, S., Vianna, C.R., Reznick, R.M., Cui, L., Manieri, M., Donovan, M.X., Wu, Z., Cooper, M.P., Fan, M.C., Rohas, L.M., Zavacki, A.M., Cinti, S., Shulman, G.I., Lowell, B.B., Krainc. D. and Spiegelman, B.M. (2004) Defects in adaptive energy metabolism with CNS-linked hyperactivity in PGC-1α null mice. Cell 119, 121–135.

Liu, S. and Wong-Riley, M.T.T. (1994) Nuclear-encoded mitochondrial precursor protein: Intramitochondrial delivery to dendrites and axon terminals of neurons and regulation by neuronal activity. J. Neurosci. 14, 5338–5351.

Lowry, O.H. (1975) Energy metabolism in brain and its control. In: D.H. Ingvar and N.A. Lassen (Eds.), Brain Work. Alfred Benzon Symposium VIII. Academic Press, New York, pp. 377–388.

McCulloch, V. and Shadel, G.S. (2003) Human mitochondrial transcription factor B1 interacts with the C-terminal activation region of h-mtTFA and stimulates transcription independent of its RNA methyltransferase activity. Mol. Cell. Biol. 23, 5816–5824.

Mell, O.C., Seibel, P. and Kadenbach, B. (1994) Structural organization of the rat genes encoding liver- and heart-type of cytochrome c oxidase subunit VIa and a pseudogene related to the COXVIa-L cDNA. Gene 140, 179–186.

Murakami, T., Shimomura, Y., Yoshimura, A., Sokabe, M. and Fujitsuka, N. (1998) Induction of nuclear respiratory factor-1 expression by an acute bout of exercise in rat muscle. Biochim. Biophys. Acta 1381, 113–122.

Nie, F. and Wong-Riley, M. (1999) Nuclear respiratory factor-2 subunit protein: Correlation with cytochrome oxidase and regulation by functional activity in the monkey primary visual cortex. J. Comp. Neurol. 404, 310–320.

Nisoli, E., Clementi, E., Paolucci, C., Cozzi, V., Tonello, C., Sciorati, C., Bracale, R., Valerio, A., Francolini, M., Moncada, S. and Carruba, M.O. (2003) Mitochondrial biogenesis in mammals: the role of endogenous nitric oxide. Science 299, 896–899.

Okuno, H., Kanou, S., Tokuyama, W., Li, Y.X. and Miyashita, Y. (1997) Layer-specific differential regulation of transcription factors Zif268 and Jun-D in visual cortex V1 and V2 of macaque monkeys. Neurosci. 81, 653–666.

Ongwijitwat, S. and Wong-Riley, M.T.T. (2004) Functional analysis of the rat cytochrome c oxidase subunit 6A1 promoter in primary neurons. Gene 337, 163–171.

Ongwijitwat, S. and Wong-Riley, M.T.T. (2005) Is nuclear respiratory factor 2 a master transcriptional coordinator for all ten nuclear-encoded cytochrome c oxidase subunits in neurons? Gene 360, 65–77.

Ongwijitwat, S., Liang, H.L., Graboyes, E.M. and Wong-Riley, M.T.T. (2006) Nuclear respiratory factor 2 senses changing cellular energy demands and its silencing down-regulates cytochrome oxidase and other target gene mRNAs. Gene 374, 39–49.

Parisi, M.A. and Clayton, D.A. (1991) Similarity of human mitochondrial transcription factor 1 to high mobility group proteins. Science 252, 965–969.

Puigserver, P., Wu, Z., Park, C.W., Graves, R., Wright, M. and Spiegelman, B.M. (1998) A cold-inducible coactivator of nuclear receptors linked to adaptive thermogenesis. Cell 92, 829–839.

Puigserver, P., Adelmant, G., Wu, Z., Fan, M., Xu, J., O'Malley, B. and Spiegelman, B.M. (1999) Activation of PPARgamma coactivator-1 through transcription factor docking. Science 286, 1368–1371.

Puigserver, P. and Spiegelman, B.M. (2003) Peroxisome proliferator-activated receptor-gamma coactivator 1 alpha (PGC-1 alpha): transcriptional coactivator and metabolic regulator. Endocr. Rev. 24, 78–90.

Scarpulla, R.C. (1997) Nuclear control of respiratory chain expression in mammalian cells. J. Bioenerg. Biomemb. 29, 109–119.

Scarpulla, R.C. (2002) Nuclear activators and coactivators in mammalian mitochondrial biogenesis. Biochim. Biophys. Acta 1576, 1–14.

Scarpulla, R.C. (2006) Nuclear control of respiratory gene expression in mammalian cells. J. Cell Biochem. 97, 673–683.

Seelan, R.S. and Grossman, L.I. (1997) Structural organization and promoter analysis of the bovine cytochrome c oxidase subunit VIIc gene. A functional role for YY1. J. Biol. Chem. 272, 19175–10181.

Seelan, R.S., Gopalakrishnan, L., Scarpulla, R.C. and Grossman, L.I. (1996) Cytochrome c oxidase subunit VIIa liver isoform. Characterization and identification of promoter elements in the bovine gene. J. Biol. Chem. 271, 2112–2120.

Short, K.R., Vittone, J.L., Bigelow, M.L., Proctor, D.N., Rizza, R.A., Coenen-Schimke, J.M. and Nair, K.S. (2003) Impact of aerobic exercise training on age-related changes in insulin sensitivity and muscle oxidative capacity. Diabetes 52, 1888–1896.

Smith, E.O. and Lomax, M.I. (1993) Structural organization of the bovine gene for the heart/muscle isoform of cytochrome c oxidase subunit VIa. Biochim. Biophys. Acta 1174, 63–71.

Sucharov, C., Basu, A., Carter, R.S. and Avadhani, N.G. (1995) A novel transcriptional initiator activity of the GABP factor binding ets sequence repeat from the murine cytochrome c oxidase Vb gene. Gene Expr. 5, 93–111.

Takata, K., Inoue, Y.H., Hirose, F., Murakami, S., Shimanouchi, K., Sakimoto, I. and Sakaguchi, K. (2003) Spatio-temporal expression of Drosophila mitochondrial transcription factor A during development. Cell Biol. Int. 27, 361–374.

Thompson, C.C., Brown, T.A. and McKnight, S.L. (1991) Convergence of Ets- and Notch-related structural motifs in a heteromeric DNA binding complex. Science 253, 762–768.

Tritos, N.A., Mastaitis, J.W., Kokkotou, E.G., Puigserver, P., Spiegelman, B.M. and Maratos-Flier, E. (2003) Characterization of the peroxisome proliferators activated receptor coactivator 1 alpha (PGC 1α) expression in the murine brain. Brain Res. 961, 255–260.

Virbasius, J.V. and Scarpulla, R.C. (1991) Transcriptional activation through ETS domain binding sites in the cytochrome c oxidase subunit IV gene. Mol. Cell Biol. 11, 5631–5638.

Virbasius, J.V. and Scarpulla, R.C. (1994) Activation of the human mitochondrial transcription factor A gene by nuclear respiratory factors: a potential regulatory link between nuclear and mitochondrial gene expression in organelle biogenesis. Proc. Natl. Acad. Sci. USA 91, 1309–1313.

Virbasius, J.V., Virbasius, C.A. and Scarpulla, R.C. (1993) Identity of GABP with NRF-2: a multisubunit activator of cytochrome oxidase expression, reveals a cellular role for an ETS domain activator of viral promoters. Gene Dev. 7, 380–392.

Virbasius, C.A., Virbasius, J.V. and Scarpulla, R.C. (1993b) NRF-1, an activator involved in nuclear-mitochondrial interaction, utilizes a new DNA-binding domain conserved in a family of developmental regulators. Gene Dev. 7, 2431–2445.

Wong-Riley, M.T.T. (1989) Cytochrome oxidase: an endogenous metabolic marker for neuronal activity. Trends Neurosci. 12, 94–101.

Wong-Riley, M.T.T., Mullen, M.A., Huang, Z. and Guyer, C. (1997) Brain cytochrome oxidase subunit genes: Isolation, subcloning, sequencing, light and EM *in situ* hybridization of transcripts, and regulation by neuronal activity. Neurosci. 79, 1035–1055.

Wong-Riley, M.T.T., Nie, F., Hevner, R.F. and Liu, S. (1998) Brain cytochrome oxidase: Functional significance and bigenomic regulation in the CNS. In: F. Gonzalez-Lima (Ed.) *Cytochrome oxidase in neuronal metabolism and Alzheimer's Disease*. Plenum Press, New York, PP 1–53.

Wong-Riley, M.T.T., Guo, A., Bachman, N.J. and Lomax, M.I. (2000) Human *COX6A1* gene: promoter analysis, cDNA isolation and expression in the monkey brain. Gene 247, 63–75.

Wong-Riley, M.T.T., Yang, S.J., Liang, H.L., Ning, G. and Jacobs, P. (2005) Quantitative immuno-electron microscopic analysis of nuclear respiratory factor 2 alpha and beta subunits: Normal distribution and activity-dependent regulation in mammalian visual cortex. Visual Neurosci. 22, 1–18.

Wu, Z., Puigserver, P., Andersson, U., Zhang, C., Adelmant, G., Mootha, V., Troy, A., Cinti, S., Lowell, B., Scarpulla, R.C. and Spiegelman, B.M. (1999) Mechanisms controlling mitochondrial biogenesis and respiration through the thermogenic coactivator PGC-1. Cell 98, 115–124.

Xia, Y., Buja, L.M., Scarpulla, R.C. and McMillin, J.B. (1997) Electrical stimulation of neonatal cardiomyocytes results in the sequential activation of nuclear genes governing mitochondrial proliferation and differentiation. Proc. Natl. Acad. Sci. USA, 94, 11399–11404.

Yang, S.J., Liang, H.L., Ning, G. and Wong-Riley, M.T.T. (2004) Ultrastructural study of depolarization-induced translocation of NRF-2 transcription factor in cultured rat visual cortical neurons. Eur. J. Neurosci. 19, 1153–1162.

Yang, S.J., Liang, H.L. and Wong-Riley, M.T.T. (2006) Activity-dependent transcriptional regulation of nuclear respiratory factor-1 in cultured rat visual cortical neurons. Neurosci. 141, 1181–1192.

Zhang, C., Wong-Riley, M. (2000) Depolarization stimulation upregulates GA-binding protein in neurons: a transcription factor involved in the bigenomic expression of cytochrome oxidase subunits. Eur. J. Neurosci. 12, 1013–1023.

Chapter 12
Not Just for Muscle Anymore: Activity and Calcium Regulation of MEF2-Dependent Transcription in Neuronal Survival and Differentiation

Aryaman Shalizi and Azad Bonni

Abstract Post-mitotic neurons of the central nervous system express one or more MEF2 proteins from the time of cell-cycle exit through adulthood. Furthermore, it is now evident that MEF2 regulates diverse aspects of neuronal development including cell survival and synaptogenesis. MEF2 proteins are bifunctional transcriptional regulators, a property that arises from the signal-dependent association of MEF2 proteins with distinct chromatin modifying activities. The goal of this chapter is to provide an account of the various control mechanisms that MEF2 proteins are subjected to within the central nervous system, placing a specific emphasis on the contribution of activity- and calcium-dependent signaling pathways.

12.1 A Brief Introduction to the MEF2 Family

The MEF2 proteins comprise a family of four transcriptional regulators, MEF2A-D, that are present throughout the metazoan lineage (Black and Olson 1998; Naya and Olson 1999; Shalizi and Bonni 2005). All MEF2 proteins share a common generic tripartite structure consisting of the MADS box, a conserved DNA binding domain that recognizes A/T-rich promoter sequences, the MEF2 domain which provides additional DNA binding sequence specificity and serves as a platform for interaction with proteins that directly regulate chromatin structure, and a C-terminal transcriptional regulatory domain that is subject to multiple post-translational modifications (Black and Olson 1998; Naya and Olson 1999; McKinsey, Zhang and Olson 2001). The MEF2 proteins act as transcriptional activators or transcriptional repressors, a bifunctionality that arises from the regulated association of MEF2 proteins with distinct chromatin-modifying activities, particularly those of histone acetyltransferases (HATs) and histone deacetylases (HDACs), as well as a spectrum of post-translational modifications (McKinsey et al., 2001).

12.2 MEF2 Proteins are Subject to Multiple Modes of Regulation

Essentially all post-mitotic neurons of the central nervous system express one or more MEF2 proteins (Leifer, Krainc, Yu, McDermott, Breitbart, Heng, Neve, Kosofsky, Nadal-Ginard and Lipton 1993; McDermott, Cardoso, Yu, Andres, Leifer, Krainc, Lipton and Nadal-Ginard 1993; Leifer, Golden and Kowall 1994; Ikeshima, Imai, Shimoda, Hata and Takano 1995; Lyons, Micales, Schwarz, Martin and Olson 1995; Lin, Shah and Bulleit 1996). Furthermore, it is now evident that MEF2 regulates diverse aspects of neuronal development including cell survival and synaptogenesis (Mao, Bonni, Xia, Nadal-Vicens and Greenberg 1999; Gaudilliere, Shi and Bonni 2002; Shalizi and Bonni 2005; Flavell, Cowan, Kim, Greer, Lin, Paradis, Griffith, Hu, Chen and Greenberg 2006; Shalizi, Gaudilliere, Yuan, Stegmuller, Shirogane, Ge, Tan, Schulman, Harper and Bonni 2006). The goal of this chapter is to provide an account of the layers of control MEF2 proteins are subjected to within the central nervous system, placing a special emphasis on the contribution of activity- and calcium-dependent regulatory mechanisms. We will discuss the modes of transcriptional and post-transcriptional control of MEF2 function, and highlight the major activity-dependent roles of MEF2 in the nervous system and the signaling cascades that regulate MEF2 function. We will also discuss the role of MEF2 in brain development.

12.2.1 Transcriptional Control of MEF2 Gene Expression

From the earliest stages of neurogenesis through adulthood, almost all neuronal populations of the mammalian central nervous system express at least one of the four MEF2 proteins (Leifer et al., 1993; McDermott et al., 1993; Leifer et al., 1994; Ikeshima et al., 1995; Lyons et al., 1995; Lin et al., 1996). However, the spatiotemporal distribution of each family member is highly variable (Leifer et al., 1994; Ikeshima et al., 1995; Lyons et al., 1995). MEF2 is also expressed in neurons in invertebrates including the model organism *C. elegans*, but the functional significance of this observation is only beginning to be explored (Schulz, Chromey, Lu, Zhao and Olson 1996; Dichoso, Brodigan, Chwoe, Lee, Llacer, Park, Corsi, Kostas, Fire, Ahnn and Krause 2000; van der Linden, Nolan and Sengupta 2007).

While recent studies have elaborated the role of the MEF2 proteins in neurons, the MEF2 proteins have established functions in muscle development and homeostasis. During muscle development, the myogenic bHLH proteins MyoD and Myogenin stimulate MEF2 gene expression, and MEF2 itself subsequently feeds back to potentiate its own expression cooperatively with the myogenic bHLH proteins (Kaushal, Schneider, Nadal-Ginard and Mahdavi 1994; Molkentin and Olson 1996; Black and Olson 1998; Cripps, Black, Zhao, Lien, Schulz and Olson 1998; Wang, Valdez, McAnally, Richardson and Olson 2001; Dodou, Xu and Black 2003). The importance of appropriate levels of MEF2 expression in development is readily observed in *Drosophila*, where both overexpression and inhibition of dMEF2 disrupts normal somatic, visceral and heart muscle development (Gunthorpe, Beatty and Taylor 1999).

It is unclear how MEF2 gene expression is controlled in neurons during nervous system development, how it is maintained through adulthood, or whether neurogenic bHLH proteins play a role in the induction and maintenance of MEF2 expression. An interesting feature of MEF2 gene expression is the tissue-specific expression of different MEF2 genes in distinct regions of the mammalian brain variable (Leifer et al., 1993; Ikeshima et al., 1995; Lyons et al., 1995). Whereas MEF2C is highly expressed in the cerebral cortex and hippocampus, it is found at lower levels in other regions including the cerebellum (Leifer et al., 1993; Lyons et al., 1995). Conversely, MEF2A is highly expressed in the cerebellar cortex and hippocampus and found at lower levels in the cerebral cortex (Lyons et al., 1995; Lin et al., 1996). It will be interesting to determine how this region-specific expression of MEF2 isoforms arises and to identify the region-specific functions of the different MEF2 isoforms in the brain.

12.2.2 Differential Splicing of MEF2

Differential splicing of MEF2 transcripts contributes an additional layer of complexity to the regulation of gene expression by MEF2 proteins, as summarized

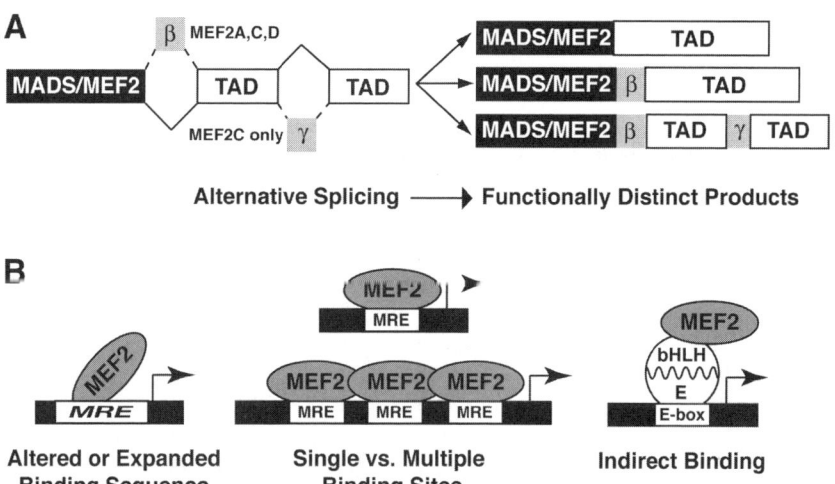

Fig. 1 Regulation of MEF2 Function by Alternative Splicing or Different Modes of Protein-DNA Interactions. A: Alternative Splicing: MEF2A, C and D genes encode an alternatively spliced exon, β, that is enriched in brain and muscle and enhances transcriptional activity of the mature MEF2 protein. The MEF2C gene encodes an additional alternatively spliced exon, γ, that confers altered signal responsiveness upon the mature protein. B: Different Modes of Protein-DNA Interaction. The response of a target gene to MEF2 can depend upon the sequence of the MRE in the promoter or enhancer, the number of MREs, or the indirect binding of MEF2 through interactions with bHLH transcription factors

in Fig. 1A. The MEF2A, C and D genes all contain a differentially spliced exon that encodes a 7-8 amino acid acidic "β" motif that enhances MEF2-dependent transcription (Martin, Miano, Hustad, Copeland, Jenkins and Olson 1994; Zhu, Ramachandran and Gulick 2005). Transcripts that include the β motif are enriched in muscle and brain, where high levels of MEF2-dependent transcription are detected (Zhu et al., 2005). This suggests that MEF2 transcripts detected in non-muscle and non-neuronal tissues may reflect the expression of a weak activation, or indeed, transcriptionally repressive form of MEF2, which may have functional significance in controlling the acquisition of a myogenic or neurogenic phenotype (Dodou, Sparrow, Mohun and Treisman 1995; Ornatsky and McDermott 1996).

In addition to the β motif, the MEF2C gene contains an additional alternatively spliced exon, the "γ" domain, which negatively regulates MEF2C-dependent gene expression (Leifer et al., 1993; McDermott et al., 1993; Zhu and Gulick 2004). The γ exon is preferentially included in mature neurons of the cerebral cortex, and contains the paired phosphorylation-dependent sumoylation-acetylation switch (SAS) motif that has recently been implicated in the control of postsynaptic dendritic morphogenesis as discussed below (Leifer et al., 1993; Zhu and Gulick 2004). The amino acids encoded by this alternative exon in MEF2C are not alternatively spliced in either MEF2A or MEF2D, suggesting that regulation of MEF2C splicing may have developmentally unique functions in cortical development (Leifer et al., 1993; Zhu and Gulick 2004). Nonetheless, the mechanisms controlling MEF2 alternative splicing, and their significance for neuronal development remain poorly understood.

12.2.3 MEF2 is a Target of Multiple Signaling Pathways

In addition to regulation at the transcriptional and post-transcriptional level, MEF2 is also subject to several forms post-translational regulation. MEF2 is a target of both neurotrophin- and activity-dependent signaling pathways in neurons (McKinsey, Zhang and Olson 2002; Cavanaugh 2004; Shalizi and Bonni 2005). This research has focused primarily on the control of serine/threonine-directed phosphorylation of MEF2 by calcium-dependent signaling pathways, but recent findings have highlighted the importance of lysine-directed modifications in the control of MEF2-dependent transcription (Gregoire and Yang 2005; Ma, Chan, Zhu and Wu 2005; Zhao, Sternsdorf, Bolger, Evans and Yao 2005; Shalizi et al., 2006). The following sections will discuss the major activity- and calcium-dependent signaling pathways that control MEF2 function. For a more detailed discussion of neurotrophin-mediated regulation of MEF2 function, please see recent reviews (Cavanaugh 2004; Shalizi and Bonni 2005).

As in muscle cells, MEF2 is a target of three major activity- and calcium-dependent signaling pathways (McKinsey et al., 2002). All three pathways stimulate MEF2-dependent transcription through distinct mechanisms: Calcium, calmodulin-dependent kinases (CaMKs) activate MEF2 by inhibiting the association of class IIa HDACs with MEF2 (Lu, McKinsey, Nicol and Olson 2000; McKinsey, Zhang, Lu and Olson 2000; McKinsey, Zhang and Olson 2000); the protein kinase p38MAPK activates MEF2 by direct phosphorylation of residues within the transactivation domain of MEF2 (Han, Jiang, Li, Kravchenko and Ulevitch 1997; Mao et al., 1999; Yang, Galanis and Sharrocks 1999; Zhao, New, Kravchenko, Kato, Gram, di Padova, Olson, Ulevitch and Han 1999; Okamoto, Krainc, Sherman and Lipton 2000); and the calcium-activated phosphatase calcineurin dephosphorylates MEF2 on at least one residue (Olson and Williams 2000; Wu, Naya, McKinsey, Mercer, Shelton, Chin, Simard, Michel, Bassel-Duby, Olson and Williams 2000; Flavell et al., 2006; Gregoire, Tremblay, Xiao, Yang, Ma, Nie, Mao, Wu, Giguere and Yang 2006; Shalizi et al., 2006). Glutamate-induced excitotoxic stimulation of neurons activates a fourth calcium-dependent signaling pathway, whereby the protein kinase Cdk5 inhibits MEF2-dependent transcription by promoting caspase-mediated degradation of MEF2 (Okamoto, Li, Ju, Scholzke, Mathews, Cui, Salvesen, Bossy-Wetzel and Lipton 2002; Gong, Tang, Wiedmann, Wang, Peng, Zheng, Blair, Marshall and Mao 2003; Tang, Wang, Gong, Tong, Park, Xia and Mao 2005). These pathways are summarized in Figs 2 and 3, and will be discussed in greater detail within the context of MEF2 function in activity-dependent survival and excitotoxic death of neurons, and the MEF2-dependent control of post-synaptic dendritic morphogenesis. Unless otherwise noted, amino acid numbering for modified sites refers to the position in human MEF2A.

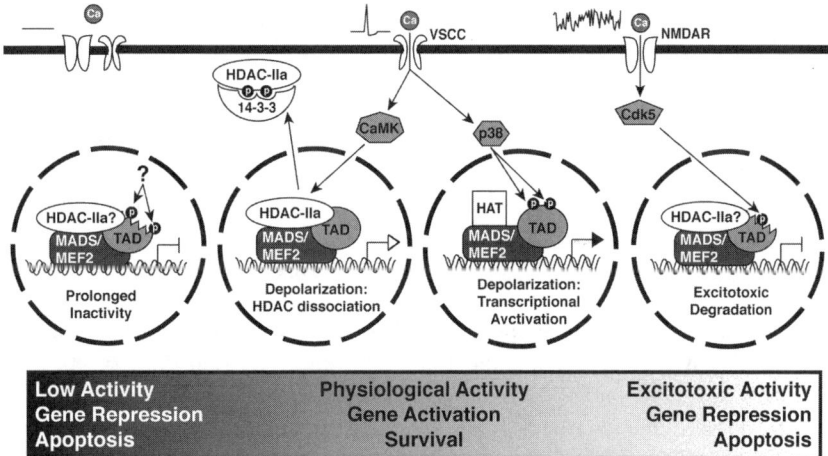

Fig. 2 Signaling Pathways Engaged by MEF2 in the Control of Activity-Dependent Survival. Prolonged hyperpolarization leads to MEF2 hyperphosphorylation, cleavage, transcriptional repression, and cell death. Physiological levels of depolarization promote MEF2-dependent transcription and cell survival through dissociation of class IIa HDACs and phosphorylation. Calcium overload leads to MEF2 hyperphosphorylation, cleavage, transcriptional repression, and cell death. VSCC = Voltage Sensitive Calcium Channel; NMDAR = N-Methyl-D-Aspartate receptor; HDAC-IIa = class IIa histone deacetylase; CaMK = Calcium/Calmodulin-dependent Kinase; p38 = p38 Mitogen Activated Protein Kinase; Cdk5 = Cyclin dependent kinase 5; TAD = MEF2 transactivation domain; HAT = Histone Acetyltransferase

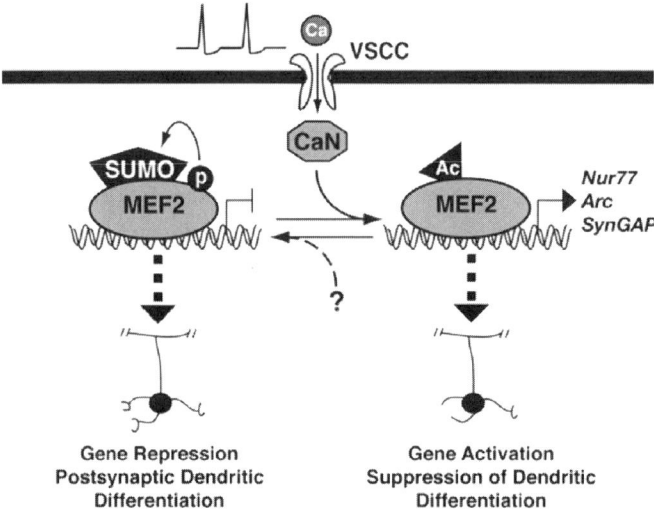

Fig. 3 MEF2 Regulates Activity-Dependent Postsynaptic Dendritic Morphogenesis. Sumoylated MEF2 promotes dendritic differentiation by repressing expression of genes that negatively regulate postsynaptic dendritic morphogenesis. Calcineurin-dependent dephosphorylation of MEF2 in response to calcium influx promotes MEF2-desumoylation and culminates in the expression of target genes that suppress postsynaptic dendritic differentiation. VSCC = Voltage Sensitive Calcium Channel; CaN = Calcineurin; SUMO = Small Ubiquitin-related Modifier

12.2.4 Regulation of MEF2 Function at the Protein-DNA Interface

A final critical layer of regulation of gene expression by MEF2 occurs at the interface of MEF2 with the genome itself. Differential affinity of MEF2 for particular binding sites, altered responsiveness of MEF2 on distinct binding sites and number of MEF2 binding sites, as well as indirect binding of MEF2 to gene promoters have all been reported to influence MEF2-dependent transcription. The distinct modes of controlling gene expression through differential MEF2-DNA interactions are summarized in Fig. 1B.

Sequence constraints may play a significant role in the control of MEF2-dependent transcription. While the four nucleotide A/T-rich core of the 10 bp consensus MEF2 response element (MRE) is essential for MEF2 binding and transcriptional activation, the 5' and 3' flanking nucleotides can alter both binding and signal responsiveness (Andres, Cervera and Mahdavi 1995; Wu et al., 2000). Comparison of the binding preferences of MEF2 complexes extracted from brain versus those extracted from cardiac and skeletal muscle revealed an expanded binding specificity for neuronal MEF2, with specific nucleotide preferences 5 nucleotide 5' and 4 nucleotides 3' (Andres et al., 1995). The expanded neural MEF2 binding specificity may reflect neuron-specific modifications of the MADS/MEF2 domains responsible for DNA binding, or the incorporation of neuron-specific factors in neuronal MEF2 complexes. Regardless of the underlying mechanism, the functional consequences of the expanded binding specificity of MEF2 in neurons are unknown.

In calcineurin-induced MEF2-dependent muscle fiber type switching, the expression of slow-twitch genes is regulated by a distinct MRE that is different by two nucleotides from MREs controlling genes encoding fast-twitch muscle proteins (Wu et al., 2000). Slow-twitch specific MREs mediate robust MEF2-dependent transcription in response to calcineurin, while fast-twitch specific MREs are only modestly activated (Wu et al., 2000). The difference in MRE sequence in these two classes of MEF2-induced genes in muscle cells illustrates how sequence-specific interactions between MEF2 and DNA can confer a selective response to distinct stimuli.

The number of MEF2 binding sites in a target promoter may also influence the spatiotemporal activation of MEF2-dependent gene expression. A comprehensive analysis of promoter occupancy and gene expression mediated by dMEF2 during *Drosophila* embryogenesis found that dMEF2 is required at all stages of myogenesis, rather than exclusively at later stages of differentiation (Sandmann, Jensen, Jakobsen, Karzynski, Eichenlaub, Bork and Furlong 2006). Correlation of gene expression patterns with dMEF2 chromatin occupancy revealed three major clusters of dMEF2-dependent genes: one cluster shows correlated transient expression and binding in early myogenesis, a second cluster shows correlated expression and binding throughout myogenesis, and the third cluster shows correlated expression and binding late in myogenesis (Sandmann et al., 2006). Interestingly, the first cluster of genes exhibit a significant enrichment of binding sites for *Twist*, a *Drosophila* myogenic bHLH protein, but no enrichment for MEF2 binding sites, suggesting the possibility that MEF2 regulates these genes by indirect binding (Molkentin, Black, Martin and Olson 1995; Sandmann et al., 2006). By contrast, the second and third clusters of genes show a significant enrichment for multiple MEF2 binding sites and no significant enrichment for *Twist* binding sites. It will be interesting to determine if similar clusters of MEF2 target genes exist in neurons based on the organization or type of their MREs.

Finally, a well-characterized feature of MEF2-dependent gene regulation is the ability of MEF2 proteins to activate gene expression without directly binding to DNA . In muscle development this occurs through the recruitment of MEF2 to DNA by interactions with myogenic bHLH proteins (Molkentin et al , 1995; Molkentin and Olson 1996; Arnold and Winter 1998). Neurogenic bHLH proteins have a similar capacity to activate MEF2-dependent transcription independently of MEF2-DNA binding in heterologous cells, but this regulatory mode has not yet been documented in primary neurons or *in vivo* (Black, Ligon, Zhang and Olson 1996; Mao ·
and Nadal-Ginard 1996).

12.2.5 Perspective: Layers of Complexity

As a key regulator of many aspects of neural and muscle development, it may not be surprising that MEF2 is controlled at many levels. What is of interest is how these distinct modes of regulation may confer upon MEF2 differing degrees of transcriptional activity or differential responsiveness to a variety of stimuli in distinct functional contexts. Our understanding of the contribution these different regulatory modes make to MEF2 function in neurons is still in its infancy and should be a major area of research in the future. For example, what mechanisms control the inclusion or exclusion of certain MEF2 exons within distinct neuronal populations, and what is the significance of alternative MEF2 exon usage for neuronal function? How does the abundance of particular MEF2 isoforms in a cell dictate promoter occupancy and target gene expression during neuronal development, and how is the frequency of genomic occupancy integrated with signaling networks? As we await these answers, a clear outline of how MEF2 controls certain aspects of neuronal development is coming into focus. The remainder of this review will discuss the role of MEF2 proteins in two major activity-dependent processes, the control of cell survival and the development of dendritic specialization's involved in synaptogenesis, with a particular emphasis on the signal-transduction pathways involved in regulating each function.

12.3 MEF2 and Activity Dependent Survival and Death

Membrane depolarization is a strong trophic stimulus for neurons, and exerts pro-survival effects via the influx of calcium through voltage-sensitive calcium channels (Ghosh and Greenberg 1995; Finkbeiner and Greenberg 1996; West, Chen, Dalva, Dolmetsch, Kornhauser, Shaywitz, Takasu, Tao and Greenberg 2001). The first, and to date most thoroughly, characterized function of MEF2 in neurons is the activity-dependent regulation of survival (Heidenreich and Linseman 2004; Shalizi and Bonni 2005). While membrane depolarization and calcium influx via voltage-gated calcium channels promotes neuronal survival in a MEF2-dependent manner, inactivation of voltage-gated calcium channels inhibit MEF2-dependent transcription, leading to apoptosis. Intriguingly, excitotoxic stimuli that induce calcium influx via glutamate receptors also inhibit MEF2 function and thereby trigger neuronal cell death. These pathways are summarized in Fig. 2.

Activation of voltage-sensitive calcium channels elevates cytoplasmic calcium concentrations and consequently activates calcium/calmodulin-dependent kinases (CaMKs) (Buonanno and Fields 1999; Soderling 2000; Soderling, Chang and Brickey 2001). In turn, CaMKs release MEF2 target genes from repression by class IIa HDACs (McKinsey et al., 2000). The class IIa HDACs, HDACs 4, 5, 7 and 9, are distantly related to the yeast protein *hda1*, and all are potent repressors of MEF2-dependent transcription (Fischle, Dequiedt, Hendzel, Guenther, Lazar, Voelter and Verdin 2002; Verdin, Dequiedt and Kasler 2003). Interestingly, the catalytic function of class IIa HDACs is dispensable for transcriptional repression, while the N-terminal MEF2-interacting domain is both necessary and sufficient for repressive function, since this region is able to recruit other repressors of transcription (Grozinger, Hassig and Schreiber 1999; Grozinger and Schreiber 2000; Lemercier, Verdel, Galloo, Curtet, Brocard and Khochbin 2000; Lu et al., 2000; Lu, McKinsey, Zhang and Olson 2000). Association between MEF2 and class IIa HDACs occurs through the MADS and MEF2 domains of MEF2 proteins (Lemercier et al., 2000; Lu et al., 2000; Lu et al., 2000).

Color Plates

Color Plate 1

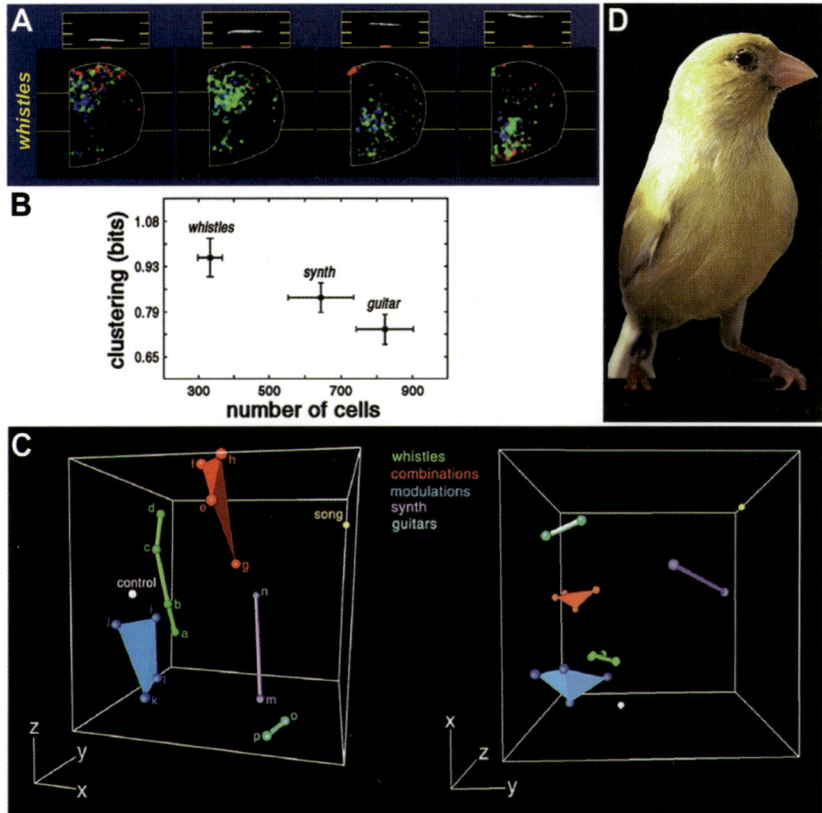

Chapter1, Fig. 4. Transcriptional activation analysis of syllabic representation in canary NCM (modified from Ribeiro et al 1998). A) ZENK expression maps in the NCM of adult female canaries resulting from the presentation of natural canary whistles. The sonograms (top) show the respective frequencies (left to right), namely 1.4, 2.2, 2.8 and 3.5 kHz. B) Patterns elicited by natural whistles and artificial stimuli of the same frequencies can be clearly separated by quantifying total cell number and spatial clustering. C) Principal Component Analysis of ZENK expression maps in NCM. The first three components (plotted on the x, y and z axes) provide for a clear separation of the ZENK patterns that accords to the various groups of stimuli in this study, indicating that a syllabic representation of canary song is present in NCM. D) Female canary (*See* page 10).

Color Plate 2

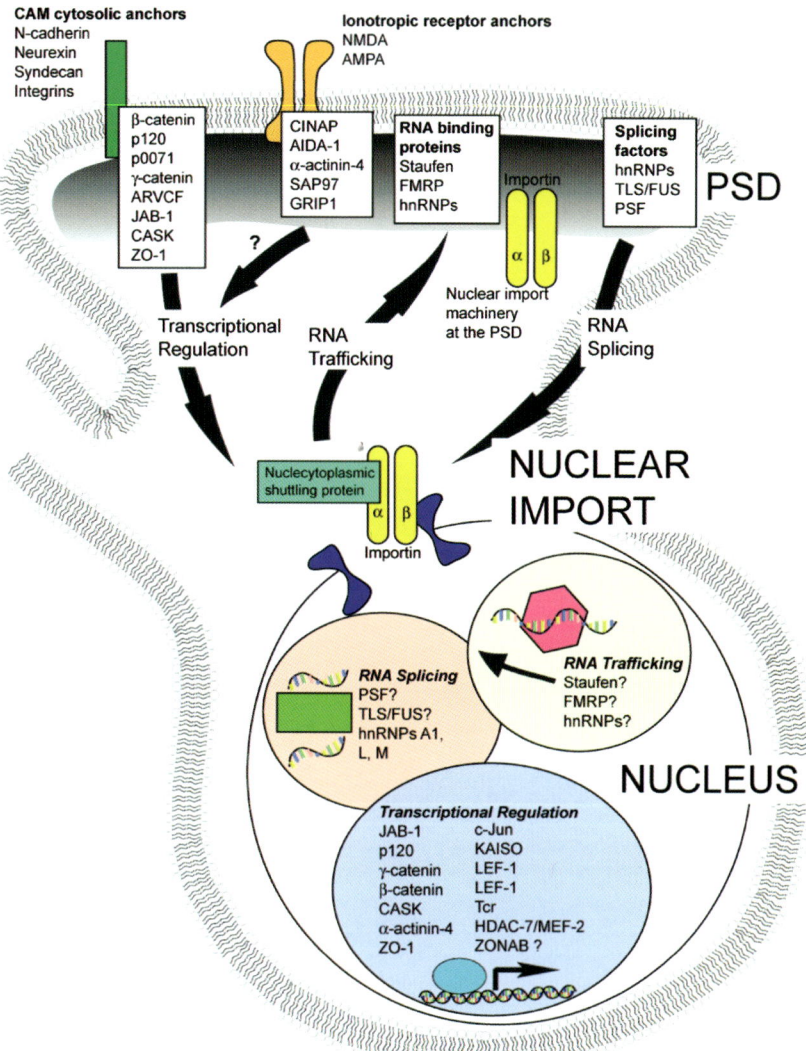

Chapter 2, Fig. 4. Schematic representation of the major nucleocytoplasmic shuttling proteins found at the PSD. PSDs contain CAM-associated transcriptional regulators such as catenins a, b, g and p120 as well as cytoskeletal/scaffolding proteins with newly identified roles in regulating transcription, such as ZO-1, GRIP1 and a-actinin-4. PSD-associated RNA binding proteins such as hnRNP A1, L and M, PSF and TLS/FUS may regulate activity-dependent alternative splicing. The RNA binding proteins Staufen, and other hnRNPs, which regulate RNA trafficking, may help localize newly synthesized mRNAs to synaptic junctions (*See* page 42).

Color Plate 3

Synaptically or mGluR Agonist Elicited Ca²⁺ Waves

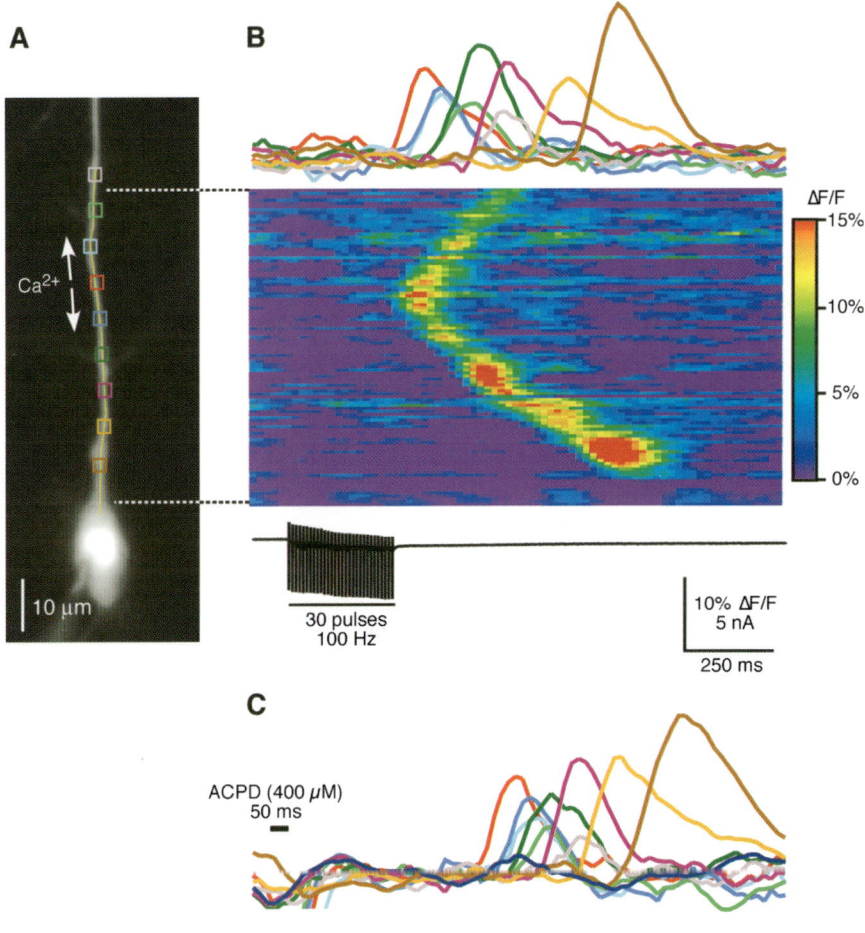

Chapter 4, Fig. 3. A Ca²⁺ Wave Triggered By Synaptic Stimulation. (A) A CA1 pyramidal cell filled with bis-fura-2. Rectangles define regions of interest, and the line along the primary apical dendrite is used for the kymograph. (B) A train of synaptic stimulation (30 impulses at 100 Hz in voltage clamp, bottom) triggered a Ca²⁺ wave that propagated along the primary apical dendrite. The wave is represented in the top and middle panels using different methods. In the top panel, the wave is depicted by using regions of interest. [Ca²⁺]ᵢ rises first in the black region and then sequentially in the other regions, indicating that the wave propagated toward the soma. In the middle panel the wave is depicted in a kymograph. In the kymograph the vertical dimension represents location along the line drawn in the image, and the horizontal dimension represents time. The wave appears as a diagonal band. The rise in [Ca²⁺]ᵢ occurred earliest at the most distal location along the line and then propagated as a wave along the dendrite toward the soma, stopping where the dendrite meets the soma. (*See* page 78)

Color Plate 4

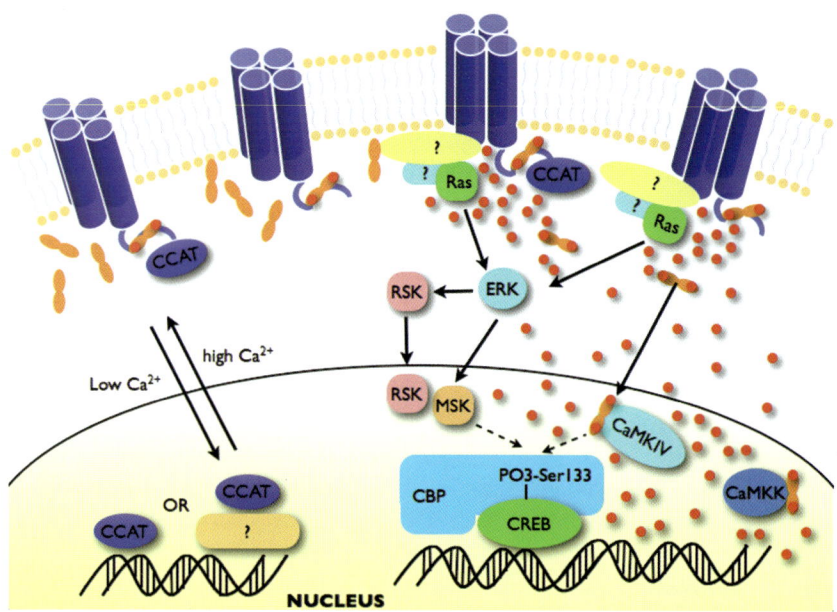

Chapter 6, Fig. 1. LTC regulation of gene expression. Calcium influx though LTCs activates signaling pathways that lead to the activation of CREB. Calcium (red circles) close to the channels triggers the activation of the Ras/MAPK cascade while nuclear calcium activates nuclear Calmodulin (orange) dependent kinases. Both of these pathways are necessary for the transcriptional activation of CREB in response to neuronal depolarization. Another pathway by which LTCs regulate transcription is through CCAT, a C-terminal protein derived from LTCs. CCAT moves in and out of the nucleus in response to changes in intracellular calcium and can directly activate transcription of target genes (*See* page 117).

Color Plate 5

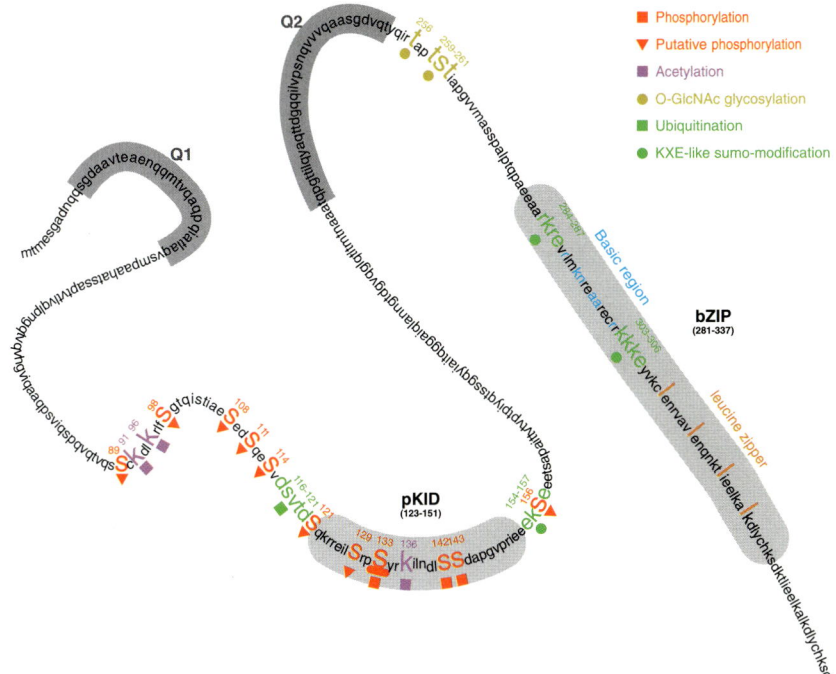

Chapter 7, Fig. 1. CREB structure and relevant residues. CREB has a highly conserved leucine zipper and adjacent basic region responsible for DNA-binding, a regulatory kinase inducible domain (KID), and two glutamine-rich regions (Q1 and Q2). CREB is substrate of various enzymatic activities that affect its capability to bind CRE sites and activate transcription. Different symbols indicate the location of those residues relevant for CREB function discussed in the text (*See* page 128).

Color Plate 6

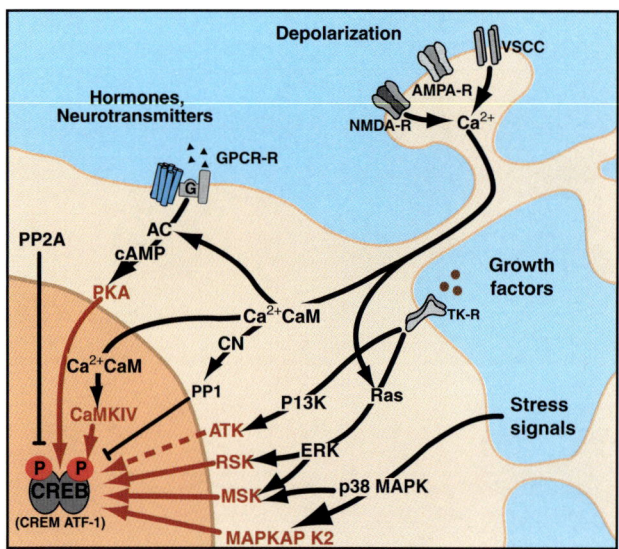

Chapter 7, Fig. 2. Regulation of CREB activation by synaptic activity. Diverse external stimuli, such as stress, the activation of G protein-coupled receptors (GPC-R) by neurotransmitters, the activation of receptor tyrosine kinases (RTK) by neurotrophins, or the opening of Ca^{2+} channels (NMDA-R and VSCC, voltage-sensitive calcium channels) activate protein-kinase pathways that converge on the phosphorylation of CREB at Ser133 (*See* page 130).

Color Plate 7

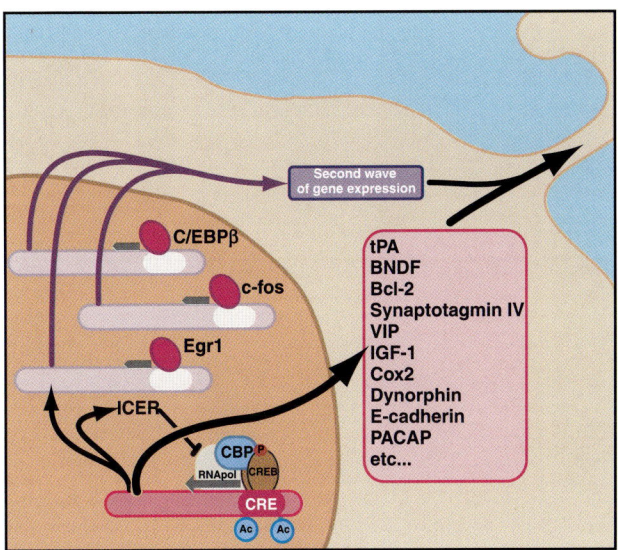

Chapter 7, Fig. 3. Regulation of synaptic function by CREB-dependent gene expression. The phosphorylation of CREB at Ser133 promotes the recruitment of the co-activator CBP and initiates the transcription of targets genes. These gene products contribute to the stabilization of the short-term process (tPA, BDNF, only a few examples are listed etc...), or are themselves transcription factors that trigger a second wave of gene expression (zif268, c-fos, C/EBPb). More extensive, but still non all-inclusive, lists of putative CREB target genes can be found in recent reviews (Barco and Kandel, 2005; Lonze and Ginty, 2002; Mayr and Montminy, 2001). In addition, the HAT activity of CBP can produce persistent changes in the acetylation state of the chromatin leading to long-term effects in the transcriptional activity of specific *loci* involved in synaptic plasticity (*See* page 136).

Color Plate 8

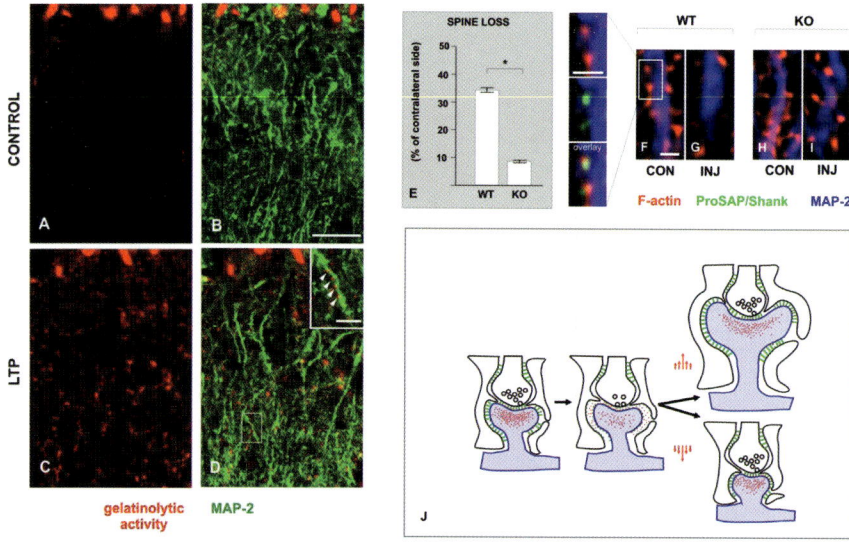

Chapter 14, Fig. 2. A–D) Confocal microscopic image of a hippocampal slice (area CA1) processed for fluorescence in situ zymography demonstrates that long-term potentiation (LTP) induces gelatinolytic activity (red, compare A and C (single color), B and D (overlay)) in the dendritic area (dendrites green, B, D). At high magnification (inset in D) foci of gelatinolytic activity appear to be positioned along individual dendrites. Please note also strong gelatinolytic activity over the blood vessels that does not change in response to LTP (top of each panel). Bar: 75 µM, bar in inset: 15 µm. E-I) Quantitative analysis and fluorescent confocal visualization of dendritic spines MMP-9 KO and WT mice, at 24 hours after intraamygdalar kainate (KA) injection. E) Estimates of dendritic spine loss at the injected side (relative to the contralateral side) in MMP-9 WT and KO mice at 24 hours after intraamygdalar (KA) injection. F-I) Representative dendritic segments from WT (F, G) and KO (H, I) animals stained for F-actin (red) and MAP-2 (blue). Bar: 5 µM. Middle: An area from E (marked with white rectangle) showing colocalization of F-actin (red) and a postsynaptic density protein ProSAP/Shank (green) that indentifies dendrit spines. Yellow in overlay indicates colocalization. J) Hypothetical involvement of MMP-9 in remodeling of dendritic spines. Neurotransmitter-evoked excitation of a postsynaptic membrane results in the release of MMP-9 (red dots) from the spine into the perisynaptic space. MMP-9 degrades cell adhesion- and extracellular matrix molecules (green bars) thereby removing mechanical constraints imposed on spines by the neighboring tissue elements. Now, the driving forces resulting from cytoskeletal rearrangements are free to cause either spine growth (upper picture) or retraction (lower picture) (*See* page 285).

Color Plate 9

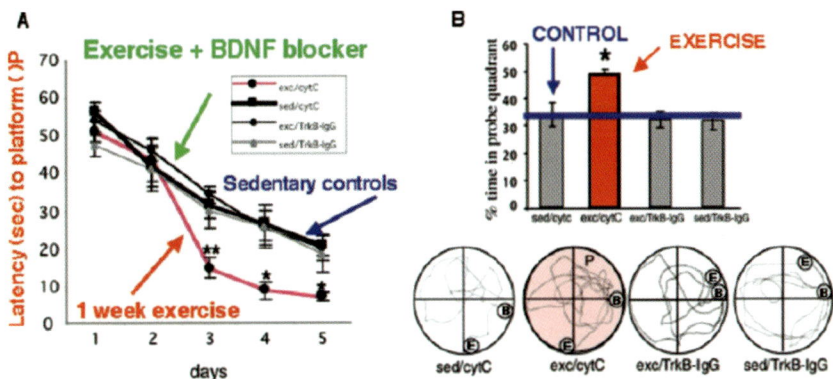

Chapter 17, Fig. 2. Exercise augments learning and memory on the Morris water maze task through the action of BDNF: (A) Animals who underwent a two week voluntary running wheel regimen had located the hidden platform in a significantly shorter time frame than sedentary control animals (exc/cutC demonstrated shorter escape latencies compared to sed/cytC controls). Blocking the action of BDNF with TrkBIgG during the exercise period successfully abolished the exercise-induced enhancement in learning ability (exc/TrkBIgG had escape latencies comparable to sed/cytC controls), to support the role of BDNF in advancing exercise-induced learning ability. Data are expressed as mean ± SEM (ANOVA, Fischer test, Scheffe F-test, HP < 0.05, HH P < 0.01). (B) Exercise enhanced memory retention on the MWM. Exercise animals performed significantly better on the probe trial of the MWM task as indicated by their increased time spent in the p quadrant where the platform used to be located (exc/cytC vs. sed/cytC). Blocking BDNF abolished the preference of exercise animals for the quadrant, illustrating the importance of BDNF mechanisms in mediating the effect of exercise on memory recall (time spent in the p quadrant of exc/TrkBIgG animals was not significantly different form sed/cytC controls). Blocking BDNF action in sedentary animals had no effect for p quadrant preference, thereby reinforcing the activity dependence of BDNF (sed/TrkBIgG vs. sed/cytC). Data are expressed as mean ± SEM (ANOVA, Fischer test, HP < 0.05). Representative samples of the times spent in the quadrants, illustrate the marked preference of the exercise animals for the P quadrant as compared to all other groups. (B, begin, E, end, P quadrant that previously housed the platform). Adopted from Vaynman et al., 2004 (*See* page 345).

Color Plate 10

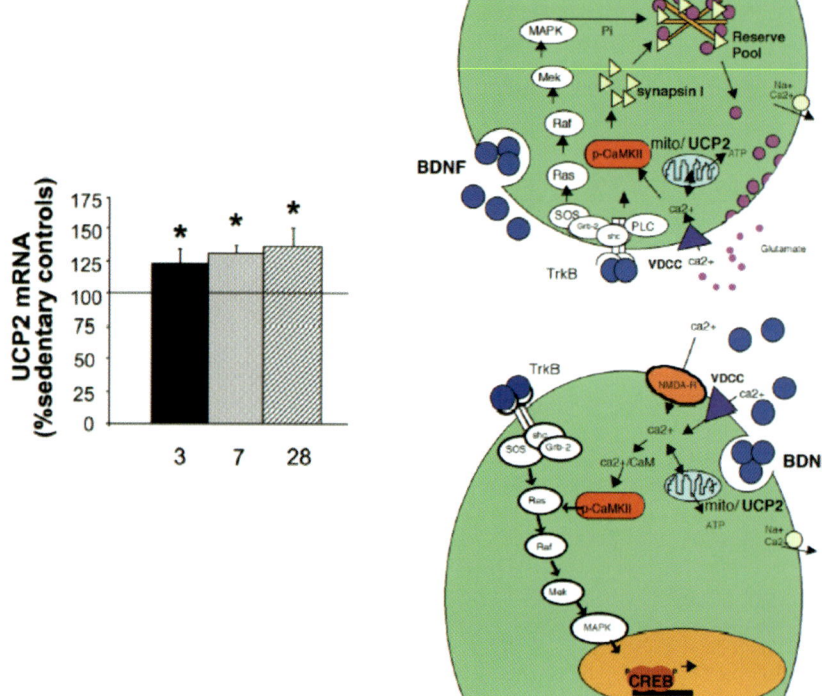

Chapter 17, Fig. 3. (A) Exercise increases UCP2 expression in the hippocampus. UCP2 expressions was significantly elevated after three days of exercise and remained significantly above sedentary after 7 and 28 days of exercise. Result as displayed as percentage of sedentary controls (represented by the 100% line). Each value represent the mean ± SEM (ANOVA, Fischer test, * P < 0.05). (B) Illustration of the potential mechanism through which elements central to energy metabolism, such as mitochondria and UCP2, interface with BDNF-mediated synaptic plasticity in the hippocampus during exercise. BDNF activates intracellular signal transduction cascades (MAPK and CAMKII) that recruit CREB and synapsin I. The uncoupling protein, UCP2 can interface with these signal transduction cascades to impact synaptic transmission and transcription, by modulating calcium homeostasis, OS production, and ATP production. Adopted from Ding et al., 2006 (*See* page 351).

Color Plate 11

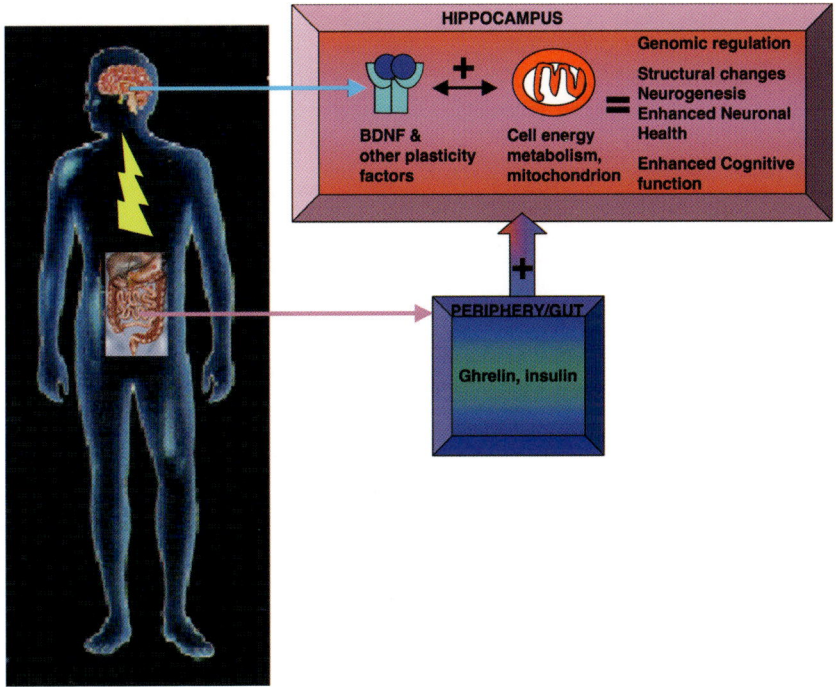

Chapter 17, Fig. 4. Metabolism: The interface between the gut and the brain: Factors produced in the periphery such as ghrelin and insulin are shown to impact BDNF-mediated plasticity and energy metabolism in the hippocampus. As a result programs orchestrating genomic regulation, structural changes, and neurogenesis are activated to promote neuronal and cognitive plasticity (*See* page 355).

Color Plate 12

A mechanism of epigenetic transcriptional repression

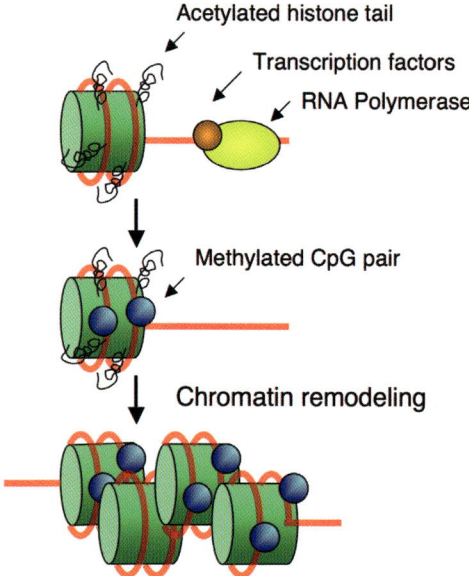

Chapter 17, Fig. 5. Model for transcriptional repression showing that the methylation of CpG pairs recruits protein complexes that cause histone deacytylastion and the repression of transcription (*See* page 358).

Color Plate 13

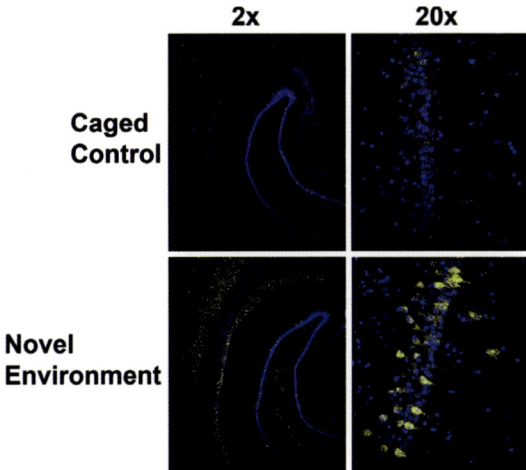

Chapter 19, Fig. 2. Dramatic induction of Arc mRNA expression by a brief exposure of a rat to a novel environment. The upper panel shows both DAPI and *Arc* RNA staining, detected using FISH, of a region of ventral hippocampus from a caged control rat taken with both a 2x and 20x objective. The lower panel shows both DAPI and *Arc* RNA staining from a rat that was exposed for 5 min to a novel environment, and then returned to its home cage for 25 min. For each the upper and lower panels, the 2x field encompasses a large portion of the hippocampus and areas of the neocortex, and the 20x field shows the CA1 region of hippocampus. Please note the dramatic increase of *Arc* RNA staining in the "novel environment" section as compared to the "caged control" in all fields of the hippocampus and in the neocortex (*See* page 406).

In both muscle cells and neurons, class IIa HDACs are predominantly nuclear until cytoplasmic calcium concentrations are increased by influx through calcium channels or release from intracellular stores (Lu et al., 2000; McKinsey et al., 2000; McKinsey et al., 2000; McKinsey, Zhang and Olson 2001; Zhang, McKinsey, Chang, Antos, Hill and Olson 2002; Chawla, Vanhoutte, Arnold, Huang and Bading 2003). In the basal state, the class IIa HDAC associates with promoter-bound MEF2 and represses MEF2 target gene activation (Lu et al., 2000; McKinsey et al., 2000; McKinsey et al., 2000). Increased cytoplasmic calcium results in the activation of CaMKs, which enter the nucleus and directly phosphorylate class IIa HDACs on a pair of conserved regulatory serines (Lu et al., 2000; McKinsey et al., 2000; McKinsey et al., 2000). This calcium-dependent phosphorylation generates a docking site for proteins of the 14-3-3 family that act as intracellular chaperone/shuttling proteins (Lu et al., 2000; McKinsey et al., 2000; McKinsey et al., 2000; Chawla et al., 2003). Binding to 14-3-3 results in the eviction of class IIa HDACs from the nucleus, and the activation of MEF2-dependent transcription by histone acetyltransferases (Lu et al., 2000; McKinsey et al., 2000; McKinsey et al., 2000; Chawla et al., 2003). Because HATs and HDACs compete for binding to the same region of the MADS/MEF2 domain, this allows for the signal-specific conversion of MEF2 from transcriptional repressor to transcriptional activator (McKinsey et al., 2001).

In conjunction with dissociation of class IIa HDACs, MEF2-dependent transcription is activated following membrane depolarization by p38MAPK-mediated phosphorylation within the transactivation domain (Mao et al., 1999). Activation of p38MAPK, and subsequent induction of MEF2-dependent transcription, has been observed in response to stimuli as diverse as growth factors and osmotic stress in a variety of cell types (Han et al., 1997; Marinissen, Chiariello, Pallante and Gutkind 1999; Yang et al., 1999; Zhao et al., 1999). However, the precise mechanism leading from neuronal activity and depolarization to the induction of p38 activity in neurons remains to be fully elucidated. A pathway linking CaMKII and the protein kinase ASK1 has been suggested to induce the activity of the p38MAPK activator MKK6 (Takeda, Matsuzawa, Nishitoh, Tobiume, Kishida, Ninomiya-Tsuji, Matsumoto and Ichijo 2004). However, the role of this pathway in depolarization-induced cell survival remains to be determined.

The ability of MEF2A, C and D to activate transcription is critically dependent upon the integrity of a pair of proline-directed threonine phosphorylation sites within the transactivation domain, Thr312 and Thr319 (Han et al., 1997; Ornatsky, Cox, Tangirala, Andreucci, Quinn, Wrana, Prywes, Yu and McDermott 1999; Yang et al., 1999; Zhao et al., 1999). This paired TP motif is a substrate for direct phosphorylation by members of the p38 family of MAP kinases (Han et al., 1997; Ornatsky et al., 1999; Yang et al., 1999; Zhao et al., 1999). Although the paired-TP motif is conserved in three of the four mammalian MEF2 genes, only MEF2A and C are targets of activation by p38 (Yang et al., 1999; Zhao et al., 1999). This specificity is conferred by a p38-docking domain present only in MEF2A and C, and absent from MEF2D (Yang et al., 1999). By contrast, ERK5, which mediates neurotrophin-dependent survival through MEF2 activation, can phosphorylate MEF2A, C and D, an event that may account for some degree of specificity in the activation of

MEF2 target genes by neurotrophins and calcium signaling, respectively (Kato, Kravchenko, Tapping, Han, Ulevitch and Lee 1997; Yang, Ornatsky, McDermott, Cruz and Prody 1998; Cavanaugh, Ham, Hetman, Poser, Yan and Xia 2001; Shalizi, Lehtinen, Gaudilliere, Donovan, Han, Konishi and Bonni 2003; Barsyte-Lovejoy, Galanis, Clancy and Sharrocks 2004).

In addition to the transactivating paired TP motif, several other sites of p38-dependent phosphorylation have been identified in MEF2 by both tandem mass spectrometry and phosphopeptide mapping (Yang et al., 1999; Zhao et al., 1999; Cox, Du, Marback, Yang, Chan, Siu and McDermott 2003). Phosphorylation of at least one conserved site, Ser255 in MEF2A, appears to regulate MEF2A stability and activity in gene reporter assays (Cox et al., 2003). At least three other sites of p38-mediated phosphorylation, Ser355, Ser453 and Ser479 have been mapped in MEF2A that appear to have no effect on p38-induced, MEF2A-dependent transcription, and the biological significance of these phosphorylation events remains poorly characterized (Yang et al., 1999; Zhao et al., 1999). However, mutation of the site corresponding to MEF2A-Ser453 in MEF2C, Ser387, completely abolishes MEF2C-dependent transcription in response to p38 activation (Han et al., 1997). These observations suggest that there may be specificity in signal-activated phosphorylation events at the level of MEF2 isoforms, with the potential to dictate distinct biological outcomes.

The concerted stimulation of CaMK and p38MAPK by neuronal activity derepresses MEF2-dependent transcription and leads to the expression of an ill-defined set of pro-survival MEF2 target genes (Mao et al., 1999; Gaudilliere et al., 2002; Linseman, Bartley, Le, Laessig, Bouchard, Meintzer, Li and Heidenreich 2003). Inhibition of either CaMK or p38MAPK activation by pharmacologic agents, antisense RNA or dominant-interfering mutants leads to a significant reduction in MEF2-dependent transcription and a concomitant increase in apoptosis (Mao et al., 1999; Gaudilliere et al., 2002; Linseman et al., 2003). Similarly, transcriptionally inert versions of MEF2, or MEF2-specific RNAi can inhibit activity-dependent survival, while constitutively active variants of MEF2 promote neuronal survival in the absence of activity, overall suggesting that transcriptional activation by MEF2 is necessary for neuronal survival (Mao et al., 1999; Gaudilliere et al., 2002; Linseman et al., 2003).

Neurons that remain hyperpolarized for a sustained interval will eventually succumb to apoptosis, and the destruction of MEF2 appears to be an important component of this process as well (Li, Linseman, Allen, Meintzer, Wang, Laessig, Wierman and Heidenreich 2001). In cerebellar granule neurons, MEF2A and MEF2D, but not MEF2C or MEF2B, undergo lithium-sensitive hyperphosphorylation following sustained activity withdrawal (Butts, Linseman, Le, Laessig and Heidenreich 2003; Linseman, Bartley, Le, Laessig, Bouchard, Meintzer, Li and Heidenreich 2003). This hyperphosphorylation precedes the caspase-mediated cleavage of MEF2, although a causative link between phosphorylation and caspase-mediated cleavage has not been established yet (Li et al., 2001; Linseman et al., 2003). Cleavage of MEF2 generates a population of truncated MEF2 that retains DNA binding competency (Li et al., 2001). This truncated MEF2 pool may then exert a dominant-negative effect the transactivating function of the remaining pool of MEF2 by competing for promoter occupancy (Li et al., 2001). Sustained hyperpolarization can thus initiate a self-reinforcing pro-apoptotic cascade of MEF2-dependent transcriptional inhibition, whereby MEF2 blocks the induction of its own target genes.

Interestingly, excitotoxic stimuli that cause oxidative stress, including the cytoplasmic elevation of calcium resulting from glutamate receptor overstimulation, inhibit MEF2 function prior to apoptosis through a mechanism remarkably similar to that observed following activity withdrawal (Okamoto et al., 2002; Gong et al., 2003; Tang et al., 2005; Smith, Mount, Shree, Callaghan, Slack, Anisman, Vincent, Wang, Mao and Park 2006). Calcium overload and oxidative stress culminate in the caspase-mediated truncation of MEF2 protein, which exerts dominant-interfering effects upon the transcriptional activity of uncleaved MEF2 (Okamoto et al., 2002; Tang et al., 2005). Excitotoxic stimuli may cause this in part by inducing the Cdk5-dependent phosphorylation of MEF2 (Gong et al., 2003).

Although Cdk5 itself is broadly expressed its kinase activity is dependent upon the neuronally restricted partner protein p35 (Dhavan and Tsai 2001; Cheung, Fu and Ip 2006). The Cdk5/p35 complex functions in both the cytoplasm and the nucleus, and has been implicated in a bevy of nervous system developmental and pathological processes, from neuronal migration to neurite outgrowth and neurodegeneration (Dhavan and Tsai 2001; Shelton and Johnson 2004; Cheung et al., 2006; Nebreda 2006). A diverse group of neurotoxic stimuli lead to Cdk5 dysregulation by initiating cleavage of p35 to p25 (Patrick, Zukerberg, Nikolic, de la Monte, Dikkes and Tsai 1999; Lee, Kwon, Li, Peng, Friedlander and Tsai 2000; Cruz, Tseng, Goldman, Shih and Tsai 2003). The p25/Cdk5 complex is constitutively active and phosphorylates multiple targets that promote apoptosis (Patrick et al., 1999). The cleavage of p35 to p25 is performed by the protease calpain, which is activated by sustained increases in cytoplasmic calcium levels (Lee et al., 2000; Cruz et al., 2003). Whereas p35/Cdk5 acts in the cytoplasm to promote cell survival, p25/Cdk5 is believed to act in the nucleus to promote cell death (O'Hare, Kushwaha, Zhang, Aleyasin, Callaghan, Slack, Albert, Vincent and Park 2005). Cdk5/p25 phosphorylates MEF2A, C and D on the residue corresponding to MEF2A Ser408, thereby inhibiting MEF2-dependent transcription (Gong et al., 2003).

Phosphorylation on Ser408 residue may render MEF2 a better substrate for caspase-mediated degradation, as MEF2D lacking this phosphorylation site is at least partially resistant to excitotoxic degradation, and MEF2C that, as a result of splicing, lacks the residue corresponding to Ser408 in MEF2A is not degraded at all following oxidative stress (Gong et al., 2003; Tang et al., 2005). Furthermore, maintaining the neuroprotective function of MEF2 proteins may be part of the mechanism whereby calpain inhibitors mitigate the long-term effects of ischemic damage to the brain (Verdaguer, Alvira, Jimenez, Rimbau, Camins and Pallas 2005; Smith et al., 2006).

That MEF2 serves as an integration point for regulating the survival response of neurons to different levels of activity is an attractive idea but raises many important questions. First and foremost, the studies discussed above have been undertaken almost exclusively in primary culture settings. The role of MEF2 in activity- and calcium-induced neuronal survival remains to be demonstrated *in vivo* in the intact organism. In addition, MEF2's role in activity-regulated responses may be subject to control by additional signaling pathways. Evidence that other signaling pathways can modulate MEF2 function in response to calcium flux has emerged recently from studies in primary hippocampal neurons, in which the cyclic AMP-dependent protein kinase PKA was found to reduce nuclear export of class II HDACs and MEF2-dependent transcription after depolarization, but enhances CREB-dependent transcription (Belfield, Whittaker, Cader and Chawla 2006). These findings raise additional questions of how MEF2 may collaborate with or antagonize other calcium-dependent signaling pathways that impinge on the transcription factors CREB, SRF and NFAT, all of which have reported functions in activity-dependent neuronal survival (Tao, Finkbeiner, Arnold, Shaywitz and Greenberg 1998; Chang, Poser and Xia 2004; Benedito, Lehtinen, Massol, Lopes, Kirchhausen, Rao and Bonni 2005). MEF2 may cooperatively regulate pro-survival gene expression with any or all of these factors in a manner specific to the spatial and temporal characteristics of a particular stimulus, as has been found with the cooperative regulation of BDNF-dependent survival by MEF2 and CREB (Shalizi et al., 2003). Finally, the set of pro-survival target genes regulated by MEF2 remains to be characterized. To date, one pro-survival MEF2 target gene has been identified for BDNF-dependent neuronal survival, the neurotrophin NT-3; but NT-3 is dispensable for activity-dependent survival, suggesting that neurotrophins and activity regulate the expression of distinct populations of MEF2-dependent targets (Shalizi et al., 2003). A comprehensive understanding of the contribution MEF2 makes to activity-dependent survival will require MEF2 knockdown *in vivo*, or the generation of conditional MEF2 knockout alleles targeted to specific developmental stages and neuronal population, in addition to further genetic and biochemical analyses of interactions between MEF2 and other signaling pathways.

12.4 MEF2 and Synaptic Dendritic Morphogenesis

A unique feature of neurons is their ability to communicate with each other through synapses. Information transfer via synapses together with the plasticity of synaptic connections between neurons may represent the physiological substrate for complex behaviors such as learning and memory (Kandel 2001; Chklovskii, Mel and Svoboda 2004). The postsynaptic dendritic membrane exhibits a tremendous degree of local specialization to allow for the proper receipt of neuronal impulses from the presynaptic cell. Assembly and disassembly of the postsynaptic apparatus is a dynamic process, and individual synapses can exhibit changes in shape and biochemical composition in response to different patterns of activity on a variety of time scales, from minutes and hours to days or months (Holtmaat, Trachtenberg, Wilbrecht, Shepherd, Zhang, Knott and Svoboda 2005; Zuo, Lin, Chang and Gan 2005; Zuo, Yang, Kwon and Gan 2005; Knott, Holtmaat, Wilbrecht, Welker and Svoboda 2006). While many of these changes occur in part through local regulation of the cytoskeleton and postsynaptic apparatus, there is mounting evidence that nuclear factors play a major role in orchestrating synapse development and remodeling.

Recently, MEF2 has catapulted to the forefront of nuclear factors that control the genesis and stability of postsynaptic dendritic specializations. The MEF2 proteins have been found to control key steps in synapse development, including postsynaptic dendritic differentiation and synapse elimination, through a mechanism summarized in Fig. 3. These advances have come in two systems, the cerebellar cortex and hippocampal neurons (Flavell et al., 2006; Shalizi et al., 2006).

The cerebellar cortex has long captivated neurobiologists in part because of its simple anatomical structure. Within the cerebellar cortex, granule neurons populate the internal granule layer (IGL) in very large numbers and represent the primary recipients of mossy fiber input from outside the cerebellum (Palay and Chan-Palay 1974; Altman and Bayer 1997). Developing granule neurons extend their dendrites within the IGL, and following a phase of remodeling and pruning, these neurons are left with three or four primary dendrites (Palay and Chan-Palay 1974; Ramón y Cajal 1995; Altman and Bayer 1997). At the end of each dendrite, a large structure, termed dendritic claw, forms with a characteristic sickle- or cuplike shape (Palay and Chan-Palay 1974; Ramón y Cajal 1995; Altman and Bayer 1997). Dendritic claws house synapses with mossy fiber terminals and Golgi neuron axons (Palay and Chan-Palay 1974; Ramón y Cajal 1995; Altman and Bayer 1997). Thus, dendritic claws represent sites of postsynaptic dendritic differentiation.

Loss of MEF2A function in granule neurons by RNAi-mediated knockdown impairs the formation of the postsynaptic dendritic claws both in cerebellar slice cultures and *in vivo* in the postnatal rat cerebellum (Shalizi et al., 2006). The MEF2A RNAi-induced dendritic claw phenotype is specific to the loss of MEF2A, as RNAi-resistant MEF2A can rescue claw differentiation in the presence of MEF2A-specific RNAi, as well as restore the density of PSD95 puncta, an indicator of synaptic differentiation, to control levels (Shalizi et al., 2006). Furthermore, MEF2 function is specifically required for dendritic claw formation, and not for more general aspects of dendritic morphogenesis, since the overall length of the granule neuron dendritic arbor is not reduced following MEF2A knockdown (Shalizi et al., 2006). The ability of MEF2 to promote claw formation is activity- dependent but surprisingly requires the transcriptional repression, rather than activation, function of MEF2A (Shalizi et al., 2006). Inhibition of voltage-sensitive calcium channels stimulates claw formation and a fusion of the MEF2 DNA-binding domain with the potent transcriptional repressor Engrailed enhances dendritic claw formation (Shalizi et al., 2006).

MEF2 plays a similar role in controlling the development of the major postsynaptic dendritic specialization of hippocampal pyramidal neurons. Dendritic spines are small protuberances from the main dendritic shaft (Bonhoeffer and Yuste 2002; Luo 2002; Nimchinsky, Sabatini and Svoboda 2002; Tada and Sheng 2006). A typical pyramidal cell may be studded with thousands of dendritic spines, each representing a unique synaptic contact. Individual spines exhibit varying degrees of temporal stability, from a few minutes to several weeks (Holtmaat et al., 2005; Zuo et al., 2005; Zuo et al., 2005; Knott et al., 2006). It is now apparent that control of spine morphogenesis and stability is the result of a combination of local regulatory mechanisms at work within the spine compartment and global mechanisms that include cell-intrinsic nuclear factors (Kandel 2001).

In hippocampal neurons, knockdown of MEF2A and D together increases the number of PSD95/synapsin coclusters, an indicator of synaptic abundance (Flavell et al., 2006). Activation of MEF2-dependent transcription leads to a corresponding reduction in the number of PSD95/synapsin coclusters suggesting MEF2 controls a transcriptional program that negatively regulates synaptic stability (Flavell et al., 2006). This possibility is supported by the observations that a tamoxifen-inducible, constitutively active MEF2 can restrict spine number within hours of induction, and that knockdown of MEF2A and D in hippocampal neurons abrogates the activity-dependent expression of Arc and Syngap, proteins that inhibit synaptic development (Flavell et al., 2006). The latter observation is consistent with findings that functional repression of a MEF2 target gene, the orphan nuclear receptor *nur77* increases the number of granule neuron dendritic claws in the cerebellar cortex (Shalizi et al., 2006). As with the granule neuron dendritic claw, the effect of MEF2 is specific to synapses in hippocampal neurons, as alterations in MEF2 function do not alter overall dendritic morphology or membrane excitability (Flavell et al., 2006).

Regulation of MEF2 function in postsynaptic dendritic morphogenesis is mediated by the activity- and calcium-sensitive protein phosphatase calcineurin (Flavell et al., 2006; Shalizi et al., 2006). Calcineurin consists of two subunits, a catalytic A chain and a regulatory B chain (Aramburu, Rao and Klee 2000). Elevated cytoplasmic calcium concentrations cause displacement of the B chain regulatory loop from the catalytic pocket of the A chain, and allow for substrate dephosphorylation (Aramburu et al., 2000). Calcineurin emerged as a key regulator of MEF2 transcriptional activity in pioneering studies of muscle fiber-type switching in response to the greater metabolic demand that occurs concomitantly with increased physical activity (Olson and Williams 2000). This work identified MEF2 as a direct substrate of calcineurin-dependent dephosphorylation, and illustrated the specificity of calcineurin-MEF2 signaling in controlling a subset of MEF2-dependent target genes required for the differentiation of slow-twitch myofibers (Chin, Olson, Richardson, Yang, Humphries, Shelton, Wu, Zhu, Bassel-Duby and Williams 1998; Wu et al., 2000; Wu, Rothermel, Kanatous, Rosenberg, Naya, Shelton, Hutcheson, DiMaio, Olson, Bassel-Duby and Williams 2001). Concomitantly, calcineurin activation in response to neuronal depolarization was shown to enhance the DNA binding affinity of MEF2 in cerebellar granule neurons (Mao and Wiedmann 1999). However, the specific site of calcineurin-dependent MEF2 dephosphorylation remained unknown.

In the control of synaptic dendritic morphogenesis, Ser408 of MEF2A has emerged as the major site of calcineurin-dependent dephosphorylation of MEF2 (Flavell et al., 2006; Shalizi et al., 2006). Dephosphorylation of Ser408 by calcineurin occurs within minutes of calcium influx, and MEF2 mutants that are constitutively dephosphorylated at that site exhibit increased transcriptional activity (Flavell et al., 2006; Shalizi et al., 2006). Furthermore, constitutively dephosphorylated MEF2 cannot restore dendritic claw formation in granule neurons following RNAi-mediated knockdown of endogenous MEF2A, and pharmacologic inhibition of calcineurin promotes dendritic claw formation (Shalizi et al., 2006).

The calcineurin-regulated phosphorylation state of Ser408 coordinates the differential modification of a sumoylation-acetylation switch (SAS) motif in the MEF2 transactivation domain in an activity-dependent manner (Shalizi et al., 2006). In the absence of calcium influx, MEF2 is phosphorylated on Ser408, which stimulates sumoylation at Lys403 (Shalizi et al., 2006). In this state, MEF2 functions as a transcriptional repressor that drives dendritic claw differentiation (Shalizi et al., 2006). These modifications are reversed upon calcium influx, with calcineurin-dependent Ser408 dephosphorylation resulting in the desumoylation and subsequent acetylation of Lys403 (Shalizi et al., 2006). The calcium-induced modifications at Ser408 and Lys403 thus convert MEF2A from a transcriptional repressor to a transcriptional activator (Shalizi et al., 2006). These changes inhibit postsynaptic dendritic claw differentiation in the cerebellar cortex (Shalizi et al., 2006). Consistent with this scheme, a constitutively transcriptionally active form of MEF2 inhibits synapse number in hippocampal neurons (Flavell et al., 2006). Further elucidation of the calcium-regulated SAS motif modification is likely to bring novel insights into the mechanisms that govern synapse differentiation.

Several groups have independently identified a role for Ser408 phosphorylation in enhancing MEF2 sumoylation, and thereby repressing MEF2-dependent transcription (Gregoire and Yang 2005; Zhao et al., 2005; Gregoire et al., 2006; Shalizi et al., 2006). A post-translation modification closely related to ubiquitylation, sumoylation involves the progressive transfer of a 12 kilodalton SUMO moiety from a trio of activating enzymes to the ϵ-amine group of a lysine residue within a consensus sumoylation motif of the target protein (Gill 2003; Seeler and Dejean 2003; Johnson 2004). Sumoylation of MEF2 at Lys403 is dependent upon phosphorylation at Ser408 (Gregoire and Yang 2005; Gregoire et al., 2006; Kang, Gocke and Yu 2006; Shalizi et al., 2006). However, no link has been established between the Cdk5-mediated phosphorylation at Ser408 observed in excitotoxic apoptosis and sumoylation at Lys403 or between MEF2 suymoylation and excitotoxicity. Although sumoylation converts MEF2A into a transcriptional repressor, there may be subtle but functionally significant differences in the degree and mechanism of transcriptional repression mediated by sumoylated MEF2, MEF2 bound by class IIa HDACs, and caspase-cleaved MEF2.

Sumoylation of MEF2A is critically important for its ability to promote dendritic claw formation (Shalizi et al., 2006). Although only a small fraction of MEF2A is sumoylated at any given time in neurons, this is a sufficient percentage to provide a substantial degree of transcriptional repression (Shalizi et al., 2006). That sumoylation of MEF2A is essential for the formation of dendritic claws was established by a combination of gain-of-function and loss-of function experiments. Non-sumoylatable, RNAi-resistant MEF2A mutants are unable to rescue dendritic claw formation, despite having normal subcellular distribution and stability (Shalizi et al., 2006). By contrast, a fusion protein of full-length MEF2A and SUMO potently represses MEF2-depdendent transcription and stimulates dendritic claw formation, even in the context of MEF2A-specific RNAi (Shalizi et al., 2006).

The SAS motif is conserved in MEF2C and D as well as in the single MEF2 genes of *Drosophila* and *C. elegans* (Gregoire and Yang 2005; Shalizi et al., 2006). Interestingly, the SAS motif is contained within the alternatively spliced exon of MEF2C discussed above, suggesting that only a subset of MEF2C protein is subject to regulation by calcineurin through this site (Leifer et al., 1993; McDermott

et al., 1993; Zhu and Gulick 2004). Whether acetylation of Lys403 is functionally important *per se* or simply renders MEF2 refractory to sumoylation is an open
question. Mutations of MEF2 that prevent sumoylation but not acetylation of Lys403
have effects on MEF2-dependent transcription comparable to mutations that prevent
both sumoylation and acetylation of Lys403 (Shalizi et al., 2006). By comparison,
mutation of other sites of acetylation in the MEF2 transactivation domain reduce
overall transcriptional activity (Ma et al., 2005). Precisely how calcineurin coordinates sumoylation-to-acetylation switching in MEF2, since several enzymatic activities are required for each modification, remains an open question with substantial
implications for both neurobiology and muscle physiology.

While this overall mechanism suggests that MEF2-dependent transcriptional
repression promotes synapse differentiation, and induction of MEF2-dependent
transcription stimulates synapse elimination in an activity-dependent, cell-
autonomous manner, many questions remain unresolved. First, it is unclear how
cells coordinate the same input— elevated cytoplasmic calcium concentrations—
affecting the same target—MEF2-dependent transcription—into discrete outcomes
of survival and synapse formation. One possibility is that sumoylation of MEF2
only affects the activity of a subset of MEF2 target genes, as seen in muscle
fiber type switching (Olson and Williams 2000). Indeed, the documented role of
sumoylation as a regulator of transcriptional synergy suggests that the number
of binding sites in the promoter of a given MEF2 target gene promoter may
differentially affect the response of that gene to alterations in MEF2 transcriptional activity (Iniguez-Lluhi and Pearce 2000; Holmstrom, Van Antwerp and
Iniguez-Lluhi 2003; Chupreta, Holmstrom, Subramanian and Iniguez-Lluhi 2005).
Alternatively, the relative intensity or temporal pattern of calcium influx could
coordinate the activity-dependent functions of MEF2 in neuronal survival and
synapse development. Another possibility is that the cooperation of MEF2 with
other classes of transcription factors, as with myogenic bHLH proteins and NFATs
in the control of muscle differentiation and fiber type switching, confers specificity
on activity-dependent regulation of MEF2 function (Molkentin and Olson 1996;
McKinsey et al., 2001). Of course, these possibilities are not mutually exclusive,
and all are worthy of exploration.

12.5 Concluding Remarks

The importance of the MEF2 proteins in nervous system development and function has become increasingly appreciated over the past decade. First, in the early
to mid-nineties, the expression pattern of the different isoforms of MEF2 proteins
in the distinct regions of the brain was described in several studies, providing the
foundation for analysis of MEF2 protein function in postmitotic neurons. In the late
nineties, the role of MEF2 in neuronal survival was identified, and just a year ago
the role of MEF2 proteins in synapse development was uncovered. In both neuronal
survival and synaptic dendritic morphogenesis, MEF2-regulated transcription is
robustly influenced by calcium signals, making MEF2 a key and attractive target for

activity-induced long-term changes in neuronal function. Research in both of these areas has raised many questions outlined in this chapter that remain to be addressed. Characterization of MEF2's roles in additional aspects of neuronal development and function is likely to come in the years ahead. Since abnormalities in neuronal survival and synapse dendritic morphogenesis underlie diverse set of neurologic and psychiatric disorders, perhaps some of the more exciting future studies will determine if and how deregulation of activity-regulated MEF2-dependent transcription contributes to the pathogenesis of disorders of the nervous system. Thus, as MEF2 research in neurobiology enters well into the second decade, research in this area will likely pick up pace and place MEF2 in the pantheon of major activity-regulated transcription factors in the nervous system.

Acknowledgments Supported by a NIH grant to A.B. (NS41021), and a NIH training grant to A.S. (AG000222) We thank members of the Bonni laboratory for critical reading of the manuscript.

References

Altman, J. and Bayer, S. A. (1997) *Development of the cerebellar system: in relation to its evolution, structure, and functions.* CRC Press, Boca Raton.

Andres, V., Cervera, M. and Mahdavi, V. (1995) Determination of the consensus binding site for MEF2 expressed in muscle and brain reveals tissue-specific sequence constraints. J. Biol. Chem. 270, 23246–23249.

Aramburu, J., Rao, A. and Klee, C. B. (2000) Calcineurin: from structure to function. Curr. Top. Cell Regul. 36, 237–295.

Arnold, H. H. and Winter, B. (1998) Muscle differentiation: more complexity to the network of myogenic regulators. Curr. Opin. Genet. Dev. 8, 539–544.

Barsyte-Lovejoy, D., Galanis, A., Clancy, A. and Sharrocks, A. D. (2004) ERK5 is targeted to myocyte enhancer factor 2A (MEF2A) through a MAPK docking motif. Biochem. J. 381, 693–699.

Belfield, J. L., Whittaker, C., Cader, M. Z. and Chawla, S. (2006) Differential effects of Ca2+ and cAMP on transcription mediated by MEF2D and cAMP-response element-binding protein in hippocampal neurons. J. Biol. Chem. 281, 27724–27732.

Benedito, A. B., Lehtinen, M., Massol, R., Lopes, U. G., Kirchhausen, T., Rao, A. and Bonni, A. (2005) The transcription factor NFAT3 mediates neuronal survival. J. Biol. Chem. 280, 2818–2825.

Black, B. L., Ligon, K. L., Zhang, Y. and Olson, E. N. (1996) Cooperative transcriptional activation by the neurogenic basic helix-loop-helix protein MASH1 and members of the myocyte enhancer factor-2 (MEF2) family. J. Biol. Chem. 271, 26659–26663.

Black, B. L. and Olson, E. N. (1998) Transcriptional control of muscle development by myocyte enhancer factor-2 (MEF2) proteins. Annu. Rev. Cell Dev. Biol. 14, 167–196.

Bonhoeffer, T. and Yuste, R. (2002) Spine motility. Phenomenology, mechanisms, and function. Neuron 35, 1019–1027.

Buonanno, A. and Fields, R. D. (1999) Gene regulation by patterned electrical activity during neural and skeletal muscle development. Curr. Opin. in Neurobio. 9, 110–120.

Butts, B. D., Linseman, D. A., Le, S. S., Laessig, T. A. and Heidenreich, K. A. (2003) Insulin-like growth factor-I suppresses degradation of the pro-survival transcription factor myocyte enhancer factor 2D (MEF2D) during neuronal apoptosis. Horm. Metab. Res. 35(11–12), 763–770.

Cavanaugh, J. E. (2004) Role of extracellular signal regulated kinase 5 in neuronal survival. Eur. J. Biochem. 271, 2056–2059.

Cavanaugh, J. E., Ham, J., Hetman, M., Poser, S., Yan, C. and Xia, Z. (2001) Differential regulation of mitogen-activated protein kinases ERK1/2 and ERK5 by neurotrophins, neuronal activity, and cAMP in neurons. J. Neurosci. 21, 434–443.

Chang, S. H., Poser, S. and Xia, Z. (2004) A novel role for serum response factor in neuronal survival. J. Neurosci. 24, 2277–2285.

Chawla, S., Vanhoutte, P., Arnold, F. J., Huang, C. L. and Bading, H. (2003) Neuronal activity-dependent nucleocytoplasmic shuttling of HDAC4 and HDAC5. J. Neurochem. 85, 151–159.

Cheung, Z. H., Fu, A. K. and Ip, N. Y. (2006) Synaptic roles of Cdk5: implications in higher cognitive functions and neurodegenerative diseases. Neuron 50, 13–18.

Chin, E. R., Olson, E. N., Richardson, J. A., Yang, Q., Humphries, C., Shelton, J. M., Wu, H., Zhu, W., Bassel-Duby, R. and Williams, R. S. (1998) A calcineurin-dependent transcriptional pathway controls skeletal muscle fiber type. Genes Dev. 12, 2499–2509.

Chklovskii, D. B., Mel, B. W. and Svoboda, K. (2004) Cortical rewiring and information storage. Nature 431, 782–788.

Chupreta, S., Holmstrom, S., Subramanian, L. and Iniguez-Lluhi, J. A. (2005) A small conserved surface in SUMO is the critical structural determinant of its transcriptional inhibitory properties. Mol. Cell. Biol. 25, 4272–4282.

Cox, D. M., Du, M., Marback, M., Yang, E. C., Chan, J., Siu, K. W. and McDermott, J. C. (2003) Phosphorylation motifs regulating the stability and function of myocyte enhancer factor 2A. J. Biol. Chem. 278, 15297–15303.

Cripps, R. M., Black, B. L., Zhao, B., Lien, C. L., Schulz, R. A. and Olson, E. N. (1998) The myogenic regulatory gene MEF2 is a direct target for transcriptional activation by Twist during Drosophila myogenesis. Genes Dev. 12, 422–434.

Cruz, J. C., Tseng, H. C., Goldman, J. A., Shih, H. and Tsai, L. H. (2003) Aberrant Cdk5 activation by p25 triggers pathological events leading to neurodegeneration and neurofibrillary tangles. Neuron 40, 471–483.

Dhavan, R. and Tsai, L. H. (2001) A decade of CDK5. Nat. Rev. Mol. Cell Biol. 2, 749–759.

Dichoso, D., Brodigan, T., Chwoe, K. Y., Lee, J. S., Llacer, R., Park, M., Corsi, A. K., Kostas, S. A., Fire, A., Ahnn, J. and Krause, M. (2000) The MADS-Box factor CeMEF2 is not essential for Caenorhabditis elegans myogenesis and development. Dev. Biol. 223, 431–440.

Dodou, E., Sparrow, D. B., Mohun, T. and Treisman, R. (1995) MEF2 proteins, including MEF2A, are expressed in both muscle and non-muscle cells. Nucleic Acids Res. 23, 4267–4274.

Dodou, E., Xu, S. M. and Black, B. L. (2003) MEF2c is activated directly by myogenic basic helix-loop-helix proteins during skeletal muscle development in vivo. Mech. Dev. 120, 1021–1032.

Finkbeiner, S. and Greenberg, M. E. (1996) Ca(2+)-dependent routes to Ras: mechanisms for neuronal survival, differentiation, and plasticity? Neuron 16, 233–236.

Fischle, W., Dequiedt, F., Hendzel, M. J., Guenther, M. G., Lazar, M. A., Voelter, W. and Verdin, E. (2002) Enzymatic activity associated with class II HDACs is dependent on a multiprotein complex containing HDAC3 and SMRT/N-CoR. Mol. Cell 9, 45–57.

Flavell, S. W., Cowan, C. W., Kim, T. K., Greer, P. L., Lin, Y., Paradis, S., Griffith, E. C., Hu, L. S., Chen, C. and Greenberg, M. E. (2006) Activity-dependent regulation of MEF2 transcription factors suppresses excitatory synapse number. Science 311, 1008–1012.

Gaudilliere, B., Shi, Y. and Bonni, A. (2002) RNA interference reveals a requirement for myocyte enhancer factor 2A in activity-dependent neuronal survival. J. Biol. Chem. 277, 46442–46446.

Ghosh, A. and Greenberg, M. E. (1995) Calcium signaling in neurons: molecular mechanisms and cellular consequences. Science 268, 239–247.

Gill, G. (2003) Post-translational modification by the small ubiquitin-related modifier SUMO has big effects on transcription factor activity. Curr. Opin. Genet. Dev. 13, 108–113.

Gong, X., Tang, X., Wiedmann, M., Wang, X., Peng, J., Zheng, D., Blair, L. A., Marshall, J. and Mao, Z. (2003) Cdk5-mediated inhibition of the protective effects of transcription factor MEF2 in neurotoxicity-induced apoptosis. Neuron 38, 33–46.

Gregoire, S., Tremblay, A. M., Xiao, L., Yang, Q., Ma, K., Nie, J., Mao, Z., Wu, Z., Giguere, V. and Yang, X. J. (2006) Control of MEF2 transcriptional activity by coordinated phosphorylation and sumoylation. J. Biol. Chem. 281, 4423–4433.

Gregoire, S. and Yang, X. J. (2005) Association with class IIa histone deacetylases upregulates the sumoylation of MEF2 transcription factors. Mol. Cell. Biol. 25, 2273–2287.

Grozinger, C. M., Hassig, C. A. and Schreiber, S. L. (1999) Three proteins define a class of human histone deacetylases related to yeast Hda1p. Proc. Natl. Acad. Sci. USA. 96 4868–4873.

Grozinger, C. M. and Schreiber, S. L. (2000) Regulation of histone deacetylase 4 and 5 and transcriptional activity by 14-3-3-dependent cellular localization. Proc. Natl. Acad. Sci. USA. 97, 7835–7840.

Gunthorpe, D., Beatty, K. E. and Taylor, M. V. (1999) Different levels, but not different isoforms, of the Drosophila transcription factor DMEF2 affect distinct aspects of muscle differentiation. Dev. Biol. 215, 130–145.

Han, J., Jiang, Y., Li, Z., Kravchenko, V. V. and Ulevitch, R. J. (1997) Activation of the transcription factor MEF2C by the MAP kinase p38 in inflammation. Nature 386, 296–299.

Heidenreich, K. A. and Linseman, D. A. (2004) Myocyte enhancer factor-2 transcription factors in neuronal differentiation and survival. Mol. Neurobiol. 29, 155–166.

Holmstrom, S., Van Antwerp, M. E. and Iniguez-Lluhi, J. A. (2003) Direct and distinguishable inhibitory roles for SUMO isoforms in the control of transcriptional synergy. Proc. Natl. Acad. Sci. USA. 100, 15758–15763.

Holtmaat, A. J., Trachtenberg, J. T., Wilbrecht, L., Shepherd, G. M., Zhang, X., Knott, G. W. and Svoboda, K. (2005) Transient and persistent dendritic spines in the neocortex in vivo. Neuron 45, 279–291.

Ikeshima, H., Imai, S., Shimoda, K., Hata, J. and Takano, T. (1995) Expression of a MADS box gene, MEF2D, in neurons of the mouse central nervous system: implication of its binary function in myogenic and neurogenic cell lineages. Neurosci. Lett. 200, 117–120.

Iniguez-Lluhi, J. A. and Pearce, D. (2000) A common motif within the negative regulatory regions of multiple factors inhibits their transcriptional synergy. Mol. Cell. Biol. 20, 6040–6050.

Johnson, E. S. (2004) Protein modification by SUMO. Annu Rev Biochem 73, 355–382.

Kandel, E. R. (2001) The molecular biology of memory storage: a dialogue between genes and synapses. Science 294, 1030–1038.

Kang, J., Gocke, C. B. and Yu, H. (2006) Phosphorylation-facilitated sumoylation of MEF2C negatively regulates its transcriptional activity. BMC biochemistry 7, 5.

Kato, Y., Kravchenko, V. V., Tapping, R. I., Han, J., Ulevitch, R. J. and Lee, J. D. (1997) BMK1/ERK5 regulates serum-induced early gene expression through transcription factor MEF2C. EMBO J. 16, 7054–7066.

Kaushal, S., Schneider, J. W., Nadal-Ginard, B. and Mahdavi, V. (1994) Activation of the myogenic lineage by MEF2A, a factor that induces and cooperates with MyoD. Science 266, 1236–1240.

Knott, G. W., Holtmaat, A., Wilbrecht, L., Welker, E. and Svoboda, K. (2006) Spine growth precedes synapse formation in the adult neocortex in vivo. Nat. Neurosci. 9, 1117–1124.

Lee, M. S., Kwon, Y. T., Li, M., Peng, J., Friedlander, R. M. and Tsai, L. H. (2000) Neurotoxicity induces cleavage of p35 to p25 by calpain. Nature 405, 360–364.

Leifer, D., Golden, J. and Kowall, N. W. (1994) Myocyte-specific enhancer binding factor 2C expression in human brain development. Neuroscience 63, 1067–1079.

Leifer, D., Krainc, D., Yu, Y. T., McDermott, J., Breitbart, R. E., Heng, J., Neve, R. L., Kosofsky, B., Nadal-Ginard, B. and Lipton, S. A. (1993) MEF2C, a MADS/MEF2-family transcription factor expressed in a laminar distribution in cerebral cortex. Proc. Natl. Acad. Sci. USA. 90, 1546–1550.

Lemercier, C., Verdel, A., Galloo, B., Curtet, S., Brocard, M. P. and Khochbin, S. (2000) mHDA1/HDAC5 histone deacetylase interacts with and represses MEF2A transcriptional activity. J. Biol. Chem. 275, 15594–15599.

Li, M., Linseman, D. A., Allen, M. P., Meintzer, M. K., Wang, X., Laessig, T., Wierman, M. E. and Heidenreich, K. A. (2001) Myocyte enhancer factor 2A and 2D undergo phosphorylation and caspase-mediated degradation during apoptosis of rat cerebellar granule neurons. J. Neurosci. 21, 6544–6552.

Lin, X., Shah, S. and Bulleit, R. F. (1996) The expression of MEF2 genes is implicated in CNS neuronal differentiation. Brain Res. Mol. Brain Res. 42, 307–316.

Linseman, D. A., Bartley, C. M., Le, S. S., Laessig, T. A., Bouchard, R. J., Meintzer, M. K., Li, M. and Heidenreich, K. A. (2003) Inactivation of the myocyte enhancer factor-2 repressor histone deacetylase-5 by endogenous Ca(2+) //calmodulin-dependent kinase II promotes depolarization-mediated cerebellar granule neuron survival. J. Biol. Chem. 278, 41472–41481.

Linseman, D. A., Cornejo, B. J., Le, S. S., Meintzer, M. K., Laessig, T. A., Bouchard, R. J. and Heidenreich, K. A. (2003) A myocyte enhancer factor 2D (MEF2D) kinase activated during neuronal apoptosis is a novel target inhibited by lithium. J. Neurochem. 85, 1488–1499.

Lu, J., McKinsey, T. A., Nicol, R. L. and Olson, E. N. (2000) Signal-dependent activation of the MEF2 transcription factor by dissociation from histone deacetylases. Proc. Natl. Acad. Sci. USA. 97, 4070–4075.

Lu, J., McKinsey, T. A., Zhang, C. L. and Olson, E. N. (2000) Regulation of skeletal myogenesis by association of the MEF2 transcription factor with class II histone deacetylases. Mol. Cell 6, 233–244.

Luo, L. (2002) Actin cytoskeleton regulation in neuronal morphogenesis and structural plasticity. Annu. Rev. Cell Dev. Biol. 18, 601–635.

Lyons, G. E., Micales, B. K., Schwarz, J., Martin, J. F. and Olson, E. N. (1995) Expression of MEF2 genes in the mouse central nervous system suggests a role in neuronal maturation. J. Neurosci. 15, 5727–5738.

Ma, K., Chan, J. K., Zhu, G. and Wu, Z. (2005) Myocyte enhancer factor 2 acetylation by p300 enhances its DNA binding activity, transcriptional activity, and myogenic differentiation. Mol. Cell. Biol. 25, 3575–3582.

Mao, Z., Bonni, A., Xia, F., Nadal-Vicens, M. and Greenberg, M. E. (1999) Neuronal activity-dependent cell survival mediated by transcription factor MEF2. Science 286, 785–790.

Mao, Z. and Nadal-Ginard, B. (1996) Functional and physical interactions between mammalian achaete-scute homolog 1 and myocyte enhancer factor 2A. J. Biol. Chem. 271, 14371–14375.

Mao, Z. and Wiedmann, M. (1999) Calcineurin enhances MEF2 DNA binding activity in calcium-dependent survival of cerebellar granule neurons. J. Biol. Chem. 274, 31102–31107.

Marinissen, M. J., Chiariello, M., Pallante, M. and Gutkind, J. S. (1999) A network of mitogen-activated protein kinases links G protein-coupled receptors to the c-jun promoter, a role for c-Jun NH2-terminal kinase, p38s, and extracellular signal-regulated kinase 5. Mol. Cell. Biol. 19, 4289–4301.

Martin, J. F., Miano, J. M., Hustad, C. M., Copeland, N. G., Jenkins, N. A. and Olson, E. N. (1994) A MEF2 gene that generates a muscle-specific isoform via alternative mRNA splicing. Mol. Cell. Biol. 14, 1647–1656.

McDermott, J. C., Cardoso, M. C., Yu, Y. T., Andres, V., Leifer, D., Krainc, D., Lipton, S. A. and Nadal-Ginard, B. (1993) hMEF2C gene encodes skeletal muscle- and brain-specific transcription factors. Mol. Cell. Biol. 13, 2564–2577.

McKinsey, T. A., Zhang, C. L., Lu, J. and Olson, E. N. (2000) Signal-dependent nuclear export of a histone deacetylase regulates muscle differentiation. Nature 408, 106–111.

McKinsey, T. A., Zhang, C. L. and Olson, E. N. (2000) Activation of the myocyte enhancer factor-2 transcription factor by calcium/calmodulin-dependent protein kinase-stimulated binding of 14-3-3 to histone deacetylase 5. Proc. Natl. Acad. Sci. USA. 97, 14400–14405.

McKinsey, T. A., Zhang, C. L. and Olson, E. N. (2001) Control of muscle development by dueling HATs and HDACs. Curr. Opin. Genet. Dev. 11, 497–504.

McKinsey, T. A., Zhang, C. L. and Olson, E. N. (2001) Identification of a signal-responsive nuclear export sequence in class II histone deacetylases. Mol. Cell. Biol. 21, 6312–6321.

McKinsey, T. A., Zhang, C. L. and Olson, E. N. (2002) MEF2: a calcium-dependent regulator of cell division, differentiation and death. Trends Biochem. Sci. 27, 40–47.

Molkentin, J. D., Black, B. L., Martin, J. F. and Olson, E. N. (1995) Cooperative activation of muscle gene expression by MEF2 and myogenic bHLH proteins. Cell 83, 1125–1136.

Molkentin, J. D. and Olson, E. N. (1996) Combinatorial control of muscle development by basic helix-loop-helix and MADS-box transcription factors. Proc. Natl. Acad. Sci. USA. 93, 9366–9373.

Naya, F. J. and Olson, E. (1999) MEF2: a transcriptional target for signaling pathways controlling skeletal muscle growth and differentiation. Curr. Opin. in Cell Biol. 11, 683–688.

Nebreda, A. R. (2006) CDK activation by non-cyclin proteins. Curr. Opin. in Cell Biol. 18, 192–198.

Nimchinsky, E. A., Sabatini, B. L. and Svoboda, K. (2002) Structure and function of dendritic spines. Ann. Rev. of Physiol. 64, 313–353.

O'Hare, M. J., Kushwaha, N., Zhang, Y., Aleyasin, H., Callaghan, S. M., Slack, R. S., Albert, P. R., Vincent, I. and Park, D. S. (2005) Differential roles of nuclear and cytoplasmic cyclin-dependent kinase 5 in apoptotic and excitotoxic neuronal death. J. Neurosci. 25, 8954–8966.

Okamoto, S., Krainc, D., Sherman, K. and Lipton, S. A. (2000) Antiapoptotic role of the p38 mitogen-activated protein kinase-myocyte enhancer factor 2 transcription factor pathway during neuronal differentiation. Proc. Natl. Acad. Sci. USA. 97, 7561–7566.

Okamoto, S., Li, Z., Ju, C., Scholzke, M. N., Mathews, E., Cui, J., Salvesen, G. S., Bossy-Wetzel, E. and Lipton, S. A. (2002) Dominant-interfering forms of MEF2 generated by caspase cleavage contribute to NMDA-induced neuronal apoptosis. Proc. Natl. Acad. Sci. USA. 99, 3974–3979.

Olson, E. N. and Williams, R. S. (2000) Remodeling muscles with calcineurin. Bioessays 22, 510–519.

Ornatsky, O. I., Cox, D. M., Tangirala, P., Andreucci, J. J., Quinn, Z. A., Wrana, J. L., Prywes, R., Yu, Y. T. and McDermott, J. C. (1999) Post-translational control of the MEF2A transcriptional regulatory protein. Nucleic Acids Res. 27, 2646–2654.

Ornatsky, O. I. and McDermott, J. C. (1996) MEF2 protein expression, DNA binding specificity and complex composition, and transcriptional activity in muscle and non-muscle cells. J. Biol. Chem. 271, 24927–24933.

Palay, S. L. and Chan-Palay, V. (1974) *Cerebellar cortex: cytology and organization.* Springer, New York.

Patrick, G. N., Zukerberg, L., Nikolic, M., de la Monte, S., Dikkes, P. and Tsai, L. H. (1999) Conversion of p35 to p25 deregulates Cdk5 activity and promotes neurodegeneration. Nature 402, 615–622.

Ramón y Cajal, S. (1995) *Histology of the nervous system of man and vertebrates.* Oxford University Press, New York.

Sandmann, T., Jensen, L. J., Jakobsen, J. S., Karzynski, M. M., Eichenlaub, M. P., Bork, P. and Furlong, E. E. (2006) A temporal map of transcription factor activity: MEF2 directly regulates target genes at all stages of muscle development. Dev. Cell 10, 797–807.

Schulz, R. A., Chromey, C., Lu, M. F., Zhao, B. and Olson, E. N. (1996) Expression of the D-MEF2 transcription in the Drosophila brain suggests a role in neuronal cell differentiation. Oncogene 12, 1827–1831.

Seeler, J. S. and Dejean, A. (2003) Nuclear and unclear functions of SUMO. Nat. Rev. Mol. Cell Biol. 4, 690–699.

Shalizi, A., Gaudilliere, B., Yuan, Z., Stegmuller, J., Shirogane, T., Ge, Q., Tan, Y., Schulman, B., Harper, J. W. and Bonni, A. (2006) A calcium-regulated MEF2 sumoylation switch controls postsynaptic differentiation. Science 311, 1012–1017.

Shalizi, A., Lehtinen, M., Gaudilliere, B., Donovan, N., Han, J., Konishi, Y. and Bonni, A. (2003) Characterization of a neurotrophin signaling mechanism that mediates neuron survival in a temporally specific pattern. J. Neurosci. 23, 7326–7336.

Shalizi, A. K. and Bonni, A. (2005) Brawn for Brains: The Role of MEF2 Proteins in the Developing Nervous System. Current topics in Dev. Biol. 69, 239–266.

Shelton, S. B. and Johnson, G. V. (2004) Cyclin-dependent kinase-5 in neurodegeneration. J. Neurochem. 88, 1313–1326.

Smith, P. D., Mount, M. P., Shree, R., Callaghan, S., Slack, R. S., Anisman, H., Vincent, I., Wang, X., Mao, Z. and Park, D. S. (2006) Calpain-regulated p35/cdk5 plays a central role in dopaminergic neuron death through modulation of the transcription factor myocyte enhancer factor 2. J. Neurosci. 26, 440–447.

Soderling, T. R. (2000) CaM-kinases: modulators of synaptic plasticity. Curr. Opin. in Neurobio. 10, 375–380.

Soderling, T. R., Chang, B. and Brickey, D. (2001) Cellular signaling through multifunctional Ca2+/calmodulin-dependent protein kinase II. J. Biol. Chem. 276, 3719–3722.

Tada, T. and Sheng, M. (2006) Molecular mechanisms of dendritic spine morphogenesis. Curr. Opin. in Neurobio. 16, 95–101.

Takeda, K., Matsuzawa, A., Nishitoh, H., Tobiume, K., Kishida, S., Ninomiya-Tsuji, J., Matsumoto, K. and Ichijo, H. (2004) Involvement of ASK1 in Ca2+-induced p38 MAP kinase activation. EMBO reports 5, 161–166.

Tang, X., Wang, X., Gong, X., Tong, M., Park, D., Xia, Z. and Mao, Z. (2005) Cyclin-dependent kinase 5 mediates neurotoxin-induced degradation of the transcription factor myocyte enhancer factor 2. J. Neurosci. 25, 4823–4834.

Tao, X., Finkbeiner, S., Arnold, D. B., Shaywitz, A. J. and Greenberg, M. E. (1998) Ca2+ influx regulates BDNF transcription by a CREB family transcription factor-dependent mechanism. Neuron 20, 709–726.

van der Linden, A. M., Nolan, K. M. and Sengupta, P. (2007) KIN-29 SIK regulates chemoreceptor gene expression via an MEF2 transcription factor and a class II HDAC. EMBO J. 26, 358–370.

Verdaguer, E., Alvira, D., Jimenez, A., Rimbau, V., Camins, A. and Pallas, M. (2005) Inhibition of the cdk5/MEF2 pathway is involved in the antiapoptotic properties of calpain inhibitors in cerebellar neurons. Brit. J. Pharmacol. 145, 1103–1111.

Verdin, E., Dequiedt, F. and Kasler, H. G. (2003) Class II histone deacetylases: versatile regulators. Trends Genet. 19, 286–293.

Wang, D. Z., Valdez, M. R., McAnally, J., Richardson, J. and Olson, E. N. (2001) The MEF2c gene is a direct transcriptional target of myogenic bHLH and MEF2 proteins during skeletal muscle development. Development 128, 4623–4633.

West, A. E., Chen, W. G., Dalva, M. B., Dolmetsch, R. E., Kornhauser, J. M., Shaywitz, A. J., Takasu, M. A., Tao, X. and Greenberg, M. E. (2001) Calcium regulation of neuronal gene expression. Proc. Natl. Acad. Sci. USA. 98, 11024–11031.

Wu, H., Naya, F. J., McKinsey, T. A., Mercer, B., Shelton, J. M., Chin, E. R., Simard, A. R., Michel, R. N., Bassel-Duby, R., Olson, E. N. and Williams, R. S. (2000) MEF2 responds to multiple calcium-regulated signals in the control of skeletal muscle fiber type. EMBO J. 19, 1963–1973.

Wu, H., Rothermel, B., Kanatous, S., Rosenberg, P., Naya, F. J., Shelton, J. M., Hutcheson, K. A., DiMaio, J. M., Olson, E. N., Bassel-Duby, R. and Williams, R. S. (2001) Activation of MEF2 by muscle activity is mediated through a calcineurin-dependent pathway. EMBO J. 20, 6414–6423.

Yang, C. C., Ornatsky, O. I., McDermott, J. C., Cruz, T. F. and Prody, C. A. (1998) Interaction of myocyte enhancer factor 2 (MEF2) with a mitogen-activated protein kinase, ERK5/BMK1. Nucleic Acids Res. 26, 4771–4777.

Yang, S. H., Galanis, A. and Sharrocks, A. D. (1999) Targeting of p38 mitogen-activated protein kinases to MEF2 transcription factors. Mol. Cell. Biol. 19, 4028–4038.

Zhang, C. L., McKinsey, T. A., Chang, S., Antos, C. L., Hill, J. A. and Olson, E. N. (2002) Class II histone deacetylases act as signal-responsive repressors of cardiac hypertrophy. Cell 110, 479–488.

Zhao, M., New, L., Kravchenko, V. V., Kato, Y., Gram, H., di Padova, F., Olson, E. N., Ulevitch, R. J. and Han, J. (1999) Regulation of the MEF2 family of transcription factors by p38. Mol. Cell. Biol. 19, 21–30.

Zhao, X., Sternsdorf, T., Bolger, T. A., Evans, R. M. and Yao, T. P. (2005) Regulation of MEF2 by histone deacetylase 4- and SIRT1 deacetylase-mediated lysine modifications. Mol. Cell. Biol. 25, 8456–8464.

Zhu, B. and Gulick, T. (2004) Phosphorylation and alternative pre-mRNA splicing converge to regulate myocyte enhancer factor 2C activity. Mol. Cell. Biol. 24, 8264–8275.

Zhu, B., Ramachandran, B. and Gulick, T. (2005) Alternative pre-mRNA splicing governs expression of a conserved acidic transactivation domain in myocyte enhancer factor 2 factors of striated muscle and brain. J. Biol. Chem. 280, 28749–28760.

Zuo, Y., Lin, A., Chang, P. and Gan, W. B. (2005) Development of long-term dendritic spine stability in diverse regions of cerebral cortex. Neuron 46, 181–189.

Zuo, Y., Yang, G., Kwon, E. and Gan, W. B. (2005) Long-term sensory deprivation prevents dendritic spine loss in primary somatosensory cortex. Nature 436, 261–265.

Part III
Activity-regulated Transcription: Role in Nervous System Function

Chapter 13
Synaptic Growth and Transcriptional Regulation in *Drosophila*

Cynthia Barber and J. Troy Littleton

Abstract The elucidation of the underlying molecular mechanisms that allow learning and memory storage in the nervous system is one of the most exciting challenges in current neuroscience research. While memory storage in vertebrates can persist for a short time in the presence of mRNA transcription or protein translation blockers, these memories last for only a few hours and do not undergo consolidation into longer-term information storage. Physiological studies in the hippocampus have demonstrated that the late phases of long-term potentiation (LTP) depend on transcription and translation (Kelleher, Govindarajan, Jung, Kang and Tonegawa, 2004; Kelleher, Govindarajan and Tonegawa, 2004). Transcription-dependent brain plasticity is also evident in invertebrates, as overexpression of a dominant-negative CREB (cAMP-responsive element binding protein) transgene is sufficient to block long-term memory storage in *Drosophila* (Yin, Wallach, Del Vecchio, Wilder, Zhou, Quinn and Tully 1994). Through a combination of genetic approaches using the *Drosophila* neuromuscular junction to study synaptic growth regulation, and genome-wide microarray studies in *Drosophila* mutants with altered neuronal activity, it has become clear that a critical target for transcriptional recoding during neuronal plasticity involves changes in synaptic wiring. Here we review both the mechanisms of activity-dependent transcriptional regulation in *Drosophila,* as well as how this regulation interfaces with the control of synaptic growth and function.

13.1 Exploring Neuronal Transcriptional Regulation in *Drosophila*

In the 1940's, Donald Hebb postulated that selectivity of memory coding is achieved through the stabilization and strengthening of productive synaptic connections, and the elimination of unproductive ones. Rapid short-term changes in synaptic strength, such as hippocampal paired-pulse facilitation, can occur through temporary changes to presynaptic vesicle release or postsynaptic receptor density. However, long-term plasticity, such as Long-term Potentiation (LTP) in the mammalian hippocampus, or Long-term Facilitation (LTF) in *Aplysia*, require lasting modification of synaptic excitability or structure (Glanzman, Kandel and Schacher 1990; Desmond and Levy 1986; Bailey and Chen 1988). To trigger such synaptic changes, transcriptional regulation by specific patterns of neuronal activity has been proposed to alter protein levels within the molecular environment of the synapse. The pathways regulating transcription, the identities of the targets, and how they facilitate structural change and strengthening of the synapse are areas of intense investigation. *Drosophila*, with its powerful genetic toolkit and testable behavioral repertoire, has proven to be an important model system for dissecting the role of transcriptional regulation in the control of functional connectivity between neurons.

Glutamatergic synapses, being the common excitatory connection in the mammalian CNS, are of obvious interest to neuroscientists. However, the difficulty of accessing individual synapses within the mammalian CNS, as well as the complex nature of the CNS, significantly complicates analysis. In contrast, *Drosophila* neuromuscular junctions (NMJs) formed along larval body wall muscles offer easily accessible glutamatergic synapses for both morphological and electrophysiological analysis. Like hippocampal synapses, the *Drosophila* NMJ is subject to experiential strengthening, with increased larval locomotion leading to increases in both the magnitude of evoked junctional potentials (EJPs), as well as the number of boutons formed on muscles (Sigrist, Reiff, Thiel, Steinert and Schuster 2003). During the larval stages, NMJs undergo significant synaptic elaboration, expanding >10 fold at the muscle 6/7 NMJ from the 1st to the 3rd larval instar to compensate for rapid muscle growth. Furthermore, forward and reverse genetic approaches available in *Drosophila* make the NMJ an accessible model synapse for following both structural and functional changes in synapse strength. Not only can NMJ morphology be used as a target for genetic screens, the UAS-Gal4 system allows selective expression of transgenes on one or both sides of the synapse to define their pre- or post-synaptic function.

Beyond the synapse, *Drosophila* display a number of complex behaviors ranging from courtship and grooming to Pavlovian olfactory learning. Screens for mutants defective in such natural behaviors have identified molecular components important to learning, improving our understanding of how synaptic function translates to behavior. Meanwhile, additional screening approaches have identified temperature-sensitive (TS) activity mutants that display seizures or paralysis due to changes in the excitability of neurons at high temperatures. These mutants provide tools to identify the mechanisms that link neuronal activity to synaptic modification by allowing conditional control of synaptic activity through simple manipulation of the animal's external temperature. The ability to move rapidly from mutational screens and molecular identification of mutated gene products, to the characterization of the underling synaptic and behavioral defects, has made *Drosophila* a popular model for exploring gene function in learning and memory. New studies are now combining the genetic approach with genome-wide DNA microarray screening technology to reveal previously unsuspected transcriptional targets of neuronal activity and learning.

13.2 Regulation of Neuronal Transcription

13.2.1 Memory

Efficient tests for *Drosophila* learning and memory behaviors were first developed in the 1980s, and styled after classical Pavlovian training paradigms with neutral stimuli (such as an odor) paired to negative reinforcement (electric shock) (Tully and Quinn 1985). Olfactory-based learning is assessed by the fly's avoidance (or lack there of) of an odor previously paired with electric shock. In wildtype flies, avoidance behavior normally persists for up to 24 hours, and is taken as a measure of the flies' memory (Tully et al. 1985). Such olfactory memories depend on mushroom body neurons in the CNS, which preferentially express a number of proteins important to learning. *Drosophila* can also learn to discriminate and remember visual landmarks in response to negative reinforcement. In contrast to olfactory learning, the anatomical sites important for memory formation and recall of visual-based behaviors have been mapped to a brain region in the central complex known as the fan-shaped body (Liu, Seiler, Wen, Zars, Ito, Wolf, Heisenberg and Liu 2006). Beside olfaction and visual-based forms of memory, the anatomical requirements for other forms of memory have not yet been fully explored (Isabel, Pascual and Preat 2004). The general conclusion from anatomical mapping of learning and memory centers is that memory traces are not universally stored in one central location, but rather distributed to substructures of the fly brain that receive multimodal stimuli for the specific behavior in question. Whether different types of sensory-based memory traces require the same molecular pathway is unknown, and most efforts have focused on dissection of the molecular requirements for olfaction-based memory formation.

Mutant screens testing for defects in olfactory learning and memory have produced a series of mutants that define the pathways of memory formation, including the role of CREB-regulated transcription in long-term memory (LTM). While spaced learning trials with intermittent rest periods leads to the formation of LTM, massed learning trials without breaks leads to a separate type of memory termed amnesia resistant memory (ARM). ARM is a form of memory likely found in all animals that can be disrupted immediately after training by treatments such as concussion, anesthesia, or electroconvulsive shock that disrupt the pattern of neural activity of the brain. Some disagreement still exists as to whether the molecular pathways of LTM and ARM interact with each other, although a recent study has suggested LTM formation leads to the extinction of ARM (Isabel et al. 2004).

Learning in both memory pathways is mediated by cAMP signaling and disrupted in the *dunce (dnc)* cAMP phosphodiesterase mutant (which increases neuronal cAMP levels) and the *amnesiac (amn)* pituitary adenylyl cyclase activating peptide (PACAP) mutant (which decreases cAMP levels). In these learning deficient mutants, a rapid decay in conditioned olfactory avoidance occurs after only 30 minutes (Quinn, Sziber and Booker 1979; Feany and Quinn, 1995; Tully et al. 1985). While the chronic alterations in cAMP levels in either direction are sufficient to disrupt learning and memory, the molecular and morphological effects on synapses are distinct (Zhong and Wu 2004). The high cAMP levels in *dnc* mutants cause more boutons to form at the NMJ, as well as greater quantal content during synaptic transmission (Davis, Schuster and Goodman 1996). In contrast, *rutabaga (rut)* mutants, which disrupt the activity of an adenylyl cyclase, display low cAMP levels and have fewer NMJ boutons, each of which is larger than normal (Budnik et al. 1990). Comparing the effects of the mutations at the NMJ to those on behavior support the underlying assumption that specific patterns of activity may contribute to CNS memory formation through structural remodeling of synaptic connections, similar to what is observed at the NMJ.

To trigger LTM for olfactory-based behavioral changes in *Drosophila*, neuronal circuits must progress through more intermediate, genetically separable phases of short-term memory (STM) and medium-term memory (MTM). The intergrins Fasciclin II (FasII) and Volado (Vol) are essential for STM formation, and mutants in these proteins display memory defects within three minutes of training (Cheng, Endo, Wu, Rodan, Heberlein and Davis 2001; Grotewiel, Beck, Wu, Zhu and Davis 1998). The transition from STM to MTM is abolished in the *rut* mutant, which eliminates a Ca^{++} responsive adenylyl cyclase, as well as in *protein kinase A (PKA)* mutants (Livingstone, Sziber and Quinn 1984; Li, Tully and Kalderon 1996). The final step to LTM from MTM requires transcriptional induction of genes through CREB activity (Yin et al. 1994; Yin, Wallach, Wilder, Klingensmith, Dang, Perrimon, Zhou, Tully and Quinn 1995). CREB-induced expression of immediate-early genes (IEGs) is predicted to trigger permanent changes to neuronal structure and/or function.

CREB, a member of the bZIP transcription factor superfamily, is predicted to be the cAMP target largely responsible for triggering long-lasting transcriptional changes that alter synaptic structure and function in *Drosophila*. Two *CREB* genes have been identified in flies. *dCREB2-a* is a PKA-regulated cAMP-responsive transcriptional activator homologous to mammalian CREB, CREM, and ATF-1, and is the only cAMP-responsive *CREB* known to be encoded in the fly genome (Yin et al. 1995). A second member of the CREB family, *dCREB2-b*, functions as a CREB repressor and blocks CREB-dependent transcriptional upregulation and LTM formation in flies (Yin et al. 1994). Transgenic expression of the CREB suppressor, *dCREB2-b*, can also reduce the increased quantal content observed in *dnc* mutants, but does not reverse the structural changes at the synapse (Davis et al. 1996). Such observations suggest two independent pathways for plasticity, one regulating synaptic release and another effecting morphological expansion of the synapse. As such, cAMP increases caused by neuronal activity are proposed to trigger two parallel but interlinked pathways necessary for synaptic growth and increased neurotransmitter release.

The role of regulated gene expression in the ARM pathway, which is specifically blocked in *Drosophila radish (rad)* mutants, is still unknown (Folkers, Drain and Quinn 1993). It has been reported that *rad* encodes a phopholipase-A2, which may trigger ARM through arachidonic acid induction of the atypical protein kinase C (aPKC) signaling pathway, previously implicated in regulation of synaptic bouton budding (Chiang, Blum, Barditch, Chen, Chiu, Regulski, Armstrong, Tully and

Dubnau 2004; Ruiz-Canada, Ashley, Moeckel-Cole, Drier, Yin and Budnik 2004). However, recent evidence indicates that the original gene identification of *rad* as a phospholipase-A2 is incorrect (McGuire, Deshazer and Davis 2005; Folkers, Waddel and Quinn 2006). Folkers, Quinn and colleagues have mapped the *rad* mutation to a novel gene product encoded by the CG15720 locus. CG15720 encodes a mushroom-body enriched protein with multiple PKA phosphorylation sites and weak homology to RNA splicing factors. While the molecular function of the Rad protein is unclear, it could potentially regulate gene splicing. For in depth reviews of olfactory learning, LTM and ARM in *Drosophila*, the reader is referred to McGuire, Deshazer and Davis (2005) and Margulies, Tully and Dubnau (2005).

13.2.2 Mutations Altering Neuronal Activity

While each *Drosophila* memory mutant may only disrupt specific molecular pathways, conditional *Drosophila* mutants that alter neuronal firing patterns throughout the nervous system allow a broader evaluation of the effects of chronic and acute activity changes on gene expression. The hyperexcitability mutants *Shaker* (*Sh*), *ether-a-go-go* (*eag*), and *Hyperkinetic* (*Hk*) were originally isolated in screens for ether-induced leg shaking (Kaplan and Trout 1969). TS activity mutants such as *paralytic* (*para^{TS1}*) and *seizure* (*sei^{TS1}*) were isolated in separate behavioral screens for adults displaying either paralysis or seizures at elevated temperatures (Ganetzky and Wu 1986; Titus, Warmke and Ganetzky 1997). The cloning of the neuronal activity mutations was originally important in defining the components of neuronal electrical responses (Na$^+$ channels in the case of *para*, and K$^+$ channels in the case of *Sh*). However, these mutants have also offered an opportunity to examine the molecular changes that occur at synapses under conditions of either reduced or enhanced activity. For example, the double mutant of *eag Sh*, which disrupts two classes of potassium channels, results in repetitive neuronal firing and enhanced neurotransmitter release, leading to increases in cAMP levels and synaptic growth at the NMJ (Schuster, Davis, Fetter and Goodman, 1996 II). Examination of NMJ structure in a number of conditional mutants that increase synaptic activity has revealed a strong correlation between activity levels and synaptic bouton number, a correlation that is also observed with experiential strengthening of the synapse following increased locomotion (Budnik, Zhong and Wu, 1990; Sigrist et al. 2003; Guan, Saraswati, Adolfsen and Littleton 2005). With strengthening following enhanced locomotor activity in larvae, reversible increases in evoked synaptic responses occur after relatively short periods of increased locomotion (peaking after 2 hours). However, morphological changes occur on the order of several hours and include increases in the synaptic translational machinery, bouton number, and clustering of the glutamate receptor DGluR-IIA (Sigrist et al. 2003). Thus, rapid short-term changes in neuronal activity are thought to drive stable long-term plasticity, with separate molecular pathways regulating neurotransmitter release and synaptic structure.

13.3 Signaling Pathways to the Nucleus

CREB plays a well-characterized role as an activity-dependent transcriptional factor in LTM formation. However, a number of signaling pathways in the cell are regulated by activity. Several well known pathways impinge upon CREB activation, while others activate CREB-independent transcription. Here we briefly introduce known cellular signaling pathways that are regulated by neuronal activity and that induce nuclear transcription.

13.3.1 PKA Signaling

A homolog of one of the well-known cAMP signaling targets, *PKA*, was identified in an enhancer trap screen for genes preferentially expressed in the mushroom bodies (Skoulakis, Kalderon and Davis 1993). The *Drosophila* PKA homolog, DCO, is not only preferentially expressed in learning centers, it has also been demonstrated to play critical roles in learning and MTM formation (Skoulakis et al. 1993; Li, Tully and Kalderon 1996). The requirement for PKA in MTM formation has recently been confirmed in honeybees (Friedrich, Thomas and Muller 2004), indicating that PKA is a critical downstream signaling module in cAMP-mediated induction of CREB transcriptional activity.

13.3.2 ERK/MAPK Pathway

It is well established that Ca^{++} influx through voltage-sensitive Ca^{++} channels can trigger vesicle fusion in response to neuronal activity. Additional cellular signaling cascades are also linked to intracellular Ca^{++} signaling, coupling transcriptional regulation to neuronal activity. In mammals, MAPK/ERK kinase (MEK) interacts directly with neuronal Ca^{++} channels. Following Ca^{++} influx, MEK activates ERK (Ras/extracellular signal-regulated kinase) through phosphorylation (Dolmetsch, Pajvani, Fife, Spotts and Greenberg 2001). Conservation of Ca^{++}-dependent MEK induction of ERK in *Drosophila* was demonstrated using the TS mutant *comt*[tp7]; $Ca-P60A^{Kum170}$, which display long-lasting seizure activity following heat pulses. After a 4-minute pulse at 40°C, these animals showed a robust MEK-dependent phosphorlylation of ERK to Dp-ERK (Hoeffer, Sanyal and Ramaswami, 2003). Additional studies at the larval NMJ showed that both pre- and post-synaptic activation of ERK requires MEK. Postsynaptic phosphorylation of ERK drives translocation of the protein to muscle nuclei, where it induces CREB-dependent transcription of IEGs such as Fos (*kayak*) and c/EBP (*slbo*) (Hoeffer et al. 2003). Nuclear ERK signaling also influences the function of Jun, one component of the transcription factor AP-1.

AP-1 is the conserved heterodimer of the basic leucine zipper transcriptional activators Fos and Jun. In *Drosophila*, AP-1 regulates synaptic growth at the larval NMJ and increases quantal content of evoked synaptic potentials (Sanyal, Sandstrom, Hoeffer and Ramaswami 2002). Expression of either the D-Fos repressor (Fbz-) or the D-Jun repressor (Jbz-) results in reduced synapse size and higher synaptic FasII expression (Sanyal, Sandstrom, Hoeffer and Ramaswami 2002). AP-1's effect on neurotransmitter release is blocked by *dCREB2-b*, while its effect of synaptic varicosity number is not, suggesting that AP-1 can act upstream of CREB, possibly through direct binding of the CREB regulatory sequence (Sanyal et al. 2002). AP-1-dependent synaptic growth can also be induced through positive feedback from CREB-dependent transcriptional changes (Sanyal et al. 2002). In addition to ERK, AP-1-dependent synaptic growth can be induced by the MAPK, JNK (*basket*) (Peverali, Isaksson, Papavassiliou, Staszewski, Mlodzik and Bohmann 1996; Sanyal et al. 2002). Thus, Ca^{++} signals accompanying neuronal activity are able to alter the function of at least two transcription factors through MAPK signaling cascades.

13.3.3 CaMKII Signaling

Ca^{++}/calmodulin-dependent protein kinase II (CaMKII) is another Ca^{++}-dependent signaling protein enriched in neuronal tissues, and localizes to pre- and postsynaptic compartments. At low Ca^{++} concentrations in resting neurons, CaMKII is autoinhibited and the kinase activity of the protein is Ca^{++}/calmodulin-dependent. However, when neurons are active and Ca^{++} levels increase, autophosphorylation occurs at the T287 residue (T286 in mammals) and the autoinhibitory domain is released. Release of the inhibitory domain allows the catalytic domain to become Ca^{++}-independent (for a review of CaMKII biochemistry, see Hanson and Schulman 1992). While the protein is known to translocate to the nucleus, the mechanism by which this occurs is currently unknown (for a review, see Griffith, Lu, Sun 2003).

CaMKII activity in *Drosophila* has been linked to both behavioral memory and NMJ plasticity (for a review of CaMKII function in memory, see Lisman, Schulman and Cline 2002). Flies expressing a peptide inhibitor of CaMKII activity (Ala) perform poorly in a number of learning tasks, whereas expression of constitutively active CaMKII can enhance learning (Griffith, Verselis, Aitken, Kyriacou, Danho and Greenspan 1993; Mehren and Griffith 2004). However, the effects of CaMKII expression differ depending on the neuronal population, suggesting some neurons may have a CaMKII saturation point beyond which additional expression has no effect (Mehren et al. 2004). At the NMJ, overexpression of constitutively active CaMKII results in enlarged individual boutons with abnormal spacing, while inhibiting CaMKII produces the opposite affect (Koh, Popova, Thomas, Griffith and Budnik 1999). The ability of CAMKII to modulate morphological synaptic changes appears to rely upon CaMKII-dependent regulation of intergrins, which are discussed in greater detail below.

During induction of LTM, the RNA interference (RISC) pathway has been implicated in regulation of synaptic components such as CaMKII (Ashraf, McLoon, Sclarsic and Kunes 2006). The 3' UTR of *Drosophila CaMKII* is both necessary and sufficient for mRNA localization to dendrites (as in mammals) and *CaMKII* mRNA is found in the antennal lobe of flies during olfactory training (Ashraf et al. 2006). The genetic interactions between CaMKII and components of the RISC pathway suggest that the expression and synaptic localization of CaMKII is under control of RISC components (Ashraf et al. 2006).

13.3.4 BMP Signaling

Given fluctuations in Ca^{++} and cAMP levels are relatively non-specific indicators of neuronal activity, the synaptic specificity required for memory formation may require additional signals, such as the TGF-β BMP signaling pathway, to reinforce specific connections. BMP signaling requires binding of a secreted ligand, triggering interactions between type I and type II BMP receptors. Activation results in downstream signaling that drives synaptic growth at *Drosophila* NMJs (Wrana, Attisano, Wieser, Ventura and Massague 1994).

Both receptor types are single pass transmembrane serine/ threonine kinases. Ligand binding by type II receptors results in phosphorylation of the GS box of type I receptors, triggering kinase activation and subsequent phosphorylation of receptor-mediated Smads (R-Smads) (Wrana et al. 1994). Phosphorylated R-Smads associate with co-Smads to form a complex that translocates into the nucleus and activates gene transcription (Wrana and Attisano 2000). The *Drosophila* genome includes three type I receptors (Tkv, Sax, Babo) and two type II receptors (Punt and Wit). Mutants for the R-Smad *mad*, the co-Smad *med*, and the type I receptors *tkv* and *sax,* all show reduced NMJ size and function, implicating BMP-dependent gene transcription in synaptic growth and function (McCabe, Hom, Aberle, Fetter, Marques, Haerry, Wan, O'Connor, Goodman and Haghighi 2004). At the NMJ, BMP receptors are required in the presynaptic terminal, while BMP ligands are secreted from the muscle, indicating a retrograde signaling pathway regulates synaptic growth via changes to motor neuron gene expression (McCabe et al. 2004). Release of BMPs from the muscle may provide a signaling mechanism that confirms productive synapse formation, triggering additional changes in neuronal gene expression that promote enhanced connectivity.

13.4 Potential Targets of Transcriptional Regulation

The activation of IEGs and CREB-mediated transcription has been shown to be important in LTM in both vertebrates and invertebrates (Dash, Hochner and Kandel 1990; Huang, Li and Kandel 1994; Bito, Deisseroth and Tsien 1996; Sanyal et al. 2002; Kelleher et al. 2004; Etter, Narayanan, Navratilova, Patel, Bohmann, Jasper and Ramaswami 2005). As described above, many studies have focused on the signaling pathways that are activated by neuronal activity and lead to changes in nuclear transcription. In contrast, relatively little is known about the target genes that are transcriptionally regulated. In section 13.5 we describe how new genomic technologies are being used to identify novel transcriptional targets regulated by neuronal activity. In this section, we discuss classes of proteins involved in synaptic morphology or signaling at the NMJ that represent potential target sites for altering synaptic strength.

13.4.1 Cellular Adhesion: Fasciclin II

One target of activity-dependent transcriptional regulation is enhanced synaptic growth. Neuronal cell adhesion molecules (NCAMs) and intergrins are two important regulators of cellular movement and growth. One of the molecular changes observed during experiential strengthening of the NMJ is downregulation of the NCAM Fasciclin II (FasII) (Sigrist et al. 2003). An inverse relationship between synaptic activity and FasII expression is also observed in a number of *Drosophila* epilepsy and learning mutants, including *eag Sh* and *dnc*. Both of these mutants exhibit enhanced neurotransmitter release and synaptic growth, increased cAMP concentrations, and a decrease in FasII levels (Schuster et al. 1996 II; Davis et al. 1996; Sanyal et al. 2002). Reduction in the levels of FasII expression by ~50% is sufficient to increase the number of boutons at larval NMJs (Schuster, Davis, Fetter and Goodman 1996 I). Overexpression of FasII in *eag Sh, dnc,* or AP-1 mutant backgrounds can prevent the increased bouton growth observed in these mutants, but does not rescue the enhanced neurotransmitter release phenotype (Schuster et al. 1996 II). These findings suggest that alteration in FasII levels is a key regulatory pathway that controls synaptic structure. CREB-dependent transcriptional changes are thought to occur in parallel to FasII modulation, allowing additional regulation of neurotransmitter release.

The *fasII* gene produces three isoforms differentially expressed in *Drosophila*; one is attached to the plasma membrane via GPI linkage, while two isoforms contain transmembrane domains with a short carboxy termini (one of which contains a PEST protein degradation sequence) (Grenningloh, Rehm and Goodman 1991; Lin and Goodman 1994). During axon outgrowth, transmembrane-containing FasII isoforms are expressed throughout motor neurons. By the 2nd instar larval stage, expression is limited to synaptic terminals (Schuster et al. 1996 I). Early synapse formation in *fasII* null mutants appears normal. However, a few hours after hatching, mutant larvae become sluggish and eventually die as synapses retract (Schuster et al. 1996 I). These findings suggest a model whereby a threshold level of FasII is necessary for synapse stabilization, with further asymmetric increases in FasII levels at the synapse serving to inhibit additional synaptic expansion. Elevated levels of FasII may curb NMJ growth by 'over-stabilizing' synapses, as overexpression of FasII leads to the persistence of ectopic motor neuron synapses at mismatched nerve-muscle connections in larval animals (Davis, Schuster and Goodman 1997).

FasII plays a similar stabilizing role at synapses in the larval CNS and on adult flight muscles (Baines, Seugnet, Thompson, Salvaterra and Bate 2002; Hebbar and Fernandes 2005). The effects of FasII levels on adult CNS function, and specifically in learning and memory, were assessed using *Drosophila fasII* hypomorphs lacking expression in the mushroom bodies. While mushroom body neurons appear to form normally in *fasII* hypomorphs, the mutants are deficient in olfactory memory formation (Cheng et al. 2001). Such data support the idea that FasII may play a similar role in the *Drosophila* CNS as observed at the larval NMJ, and that acute regulation of synaptic growth is important in olfactory-cued learning. Rescue of the *fasII* null phenotype at the NMJ requires expression of FasII on both sides of the synapse, consistent with data obtained in cell culture that indicates FasII forms transmembrane homophilic complexes that span the synaptic cleft (Schuster et al. 1996 I).

Recent results indicate that the amyloid precursor protein (Appl), whose mammalian homolog (App) is implicated in Alzheimer's Disease, acts downstream of FasII in the regulation of NMJ structure. Although APPL is not necessary for synaptic stabilization (as observed in *FasII* mutants), *appl* mutants have smaller boutons than wildtype and are not capable of the synaptic overgrowth that *fasII* mutants produce (Ashley, Packard, Ataman, and Budnik 2005). Appl is postulated to provide a link between FasII and cytoskeleton regulation (Brouillet, Trembleau, Galanaud, Volovitch, Bouillot, Valenza, Prochiantz and Allinquant 1999) that would be necessary for regulation of bouton budding that occurs during synaptic growth (Ruiz-Canada et al. 2004). One additional member of the pathway, dX11 (*mint*), has been identified as an interaction partner for both Appl and FasII via its PDZ domains, and may mediate transport of binding partners to FasII (Ashley et al. 2005). While most studies have explored the roles FasII plays as a homophilic cell adhesion molecule, FasII may also directly participate in cell signaling. This hypothesis is supported by data showing that the intercellular domain of FasII is essential for promoting synaptic growth. FasII has also been shown to control proneural gene expression in sensory organ development (Garcia-Alonso, VanBerkum, Grenningloh, Schuster and Goodman 1995). These findings argue that a FasII-APPL signaling pathway plays a critical role in the modulation of synaptic growth at the *Drosophila* NMJ.

As mentioned, FasII appears to act downstream of AP-1-dependent transcription, though it also subject to non-transcriptional regulation. Clustering of FasII on both sides of the synapse, facilitated by the *Drosophila* PSD-95 homolog Dlg (which is also responsible for the clustering of Sh K^+ channels), is important to its function, and represents one opportunity for post-transcriptional regulation (Thomas, Kim, Kuhlendahl, Koh, Gundelfinger, Sheng, Garner and Budnik 1997). Clustering of FasII is disrupted when Dlg is phosphorylated by CaMKII, resulting in increased synaptic growth (Koh et al. 1999). In addition to Ca^{++}, position specific intergrins, including αPS1 (*mew*), αPS2 (*if*), αPS3 (*vol*), and βPS (*mys*), can also interfere with CaMKII's ability to regulate FasII distribution within synapses (Beumer, Matthies, Bradshaw and Broadie 2002). In *mys* mutants, synaptic FasII levels are higher compared to control animals, and larva show synaptic growth defects. These synaptic growth defects can be rescued by changes in CaMKII expression (Beumer et al. 2002).

Analysis of AP-1 mutants also suggest that the MAPK encoded by the *jnk* locus contributes to regulation of FasII levels downstream of AP-1 induced transcription (Sanyal et al. 2002). ERK and the Ras1-MAPK pathway have also been implicated in regulation, as ERK activation correlates with reduced FasII activity and the induction of the IEGs Fos (*kayak*) and c/EBP (*slbo*) (Hoeffer et al. 2003). Both RasI and MAPK are present at synapses, and increases in RasI transgenes that activate MAPK promote synaptic growth (Koh, Ruiz-Canada, Gorczyca and Budnik 2002). In MAPK loss-of-function mutants, the amount of FasII co-immunoprecipitating with Dlg increases by 170% (Koh et al. 2002). Conversely, increasing the amount of MAPK leads to a reduction of FasII and corresponding increases in synaptic growth (Koh et al. 2002). Similar analysis of the Aplysia FasII homolog, apCAM, in long-term facilitation indicates that MAPK phosphorylates the apCAM PEST sequence, leading to ubiquitination and degradation of the protein (Michael, Martin, Seger, Ning, Baston and Kandel 1998; Bailey, Kaang, Chen, Martin, Lim, Casadio, and Kandel 1997; Martin, Michael, Rose, Barad, Casadio, Zhu and Kandel 1997). While the full details of FasII regulation are unknown, FasII function, along with that of intergrins and other NCAMs, critically regulate synaptic structure and growth. Several additional NCAMs have also been identified as transcriptional targets, as described in the genomic studies below.

13.4.2 The Cytoskeleton: Futsch

Cytoskeletal regulation is likely to be a critical target for modulation of synaptic growth. In particular, regulation of the microtubule-based cytoskeleton has been implicated in synapse proliferation in *Drosophila*. Mutations in *futsch*, which encodes a microtubule-binding protein of the MAP1B family, indicate that MAP1B positively regulates both axonal and dendritic growth (Hummel, Krukkert, Roos, Davis and Klambt 2000). The Futsch protein localizes to microtubule loops that occur in terminal boutons and at sites of NMJ branching (Hummel et al. 2000; Roos, Hummel, Ng, Klambt and Davis 2000). Futsch staining is strongest at the proximal end of microtubule loops and may function to stabilize microtubules (Roos et al. 2000). Although the exact role of MAP1B in bouton formation is not clear, regulation of the formation of microtubule loops at terminal boutons appears to be critical for bouton division and synaptic branching (Roos et al. 2000). Loss-of-function *futsch* mutations disrupt microtubule loops, resulting in fewer, larger boutons (Roos et al. 2000). A partial rescue can be achieved by expressing the N-terminal of the protein, which contains the putative microtubule-binding domain, demonstrating that it is the microtubule-related function that it is responsible for synaptic growth (Roos et al. 2000).

Futsch-dependent morphological changes at the synapse appear to be regulated by the *Drosophila* fragile X related (dFxr) protein. Fragile X mental retardation is the most common cause of mental retardation in humans, suggesting a critical role for the fragile X protein in higher brain function. The *Drosophila* homolog, dFxr, is expressed in the cytoplasm of most neurons and suppresses Futsch translation through direct binding of *futsch* mRNA (Zhang, Bailey, Matthies, Renden, Smith, Speese, Rubin and Broadie 2001). Overexpression of dFxr causes phenotypes that mimic those observed in *futsch* null mutants, with reduced NMJ complexity and larger synaptic boutons. *dfxr* null mutants have phenotypes affecting synaptic growth in the opposite direction, with larvae showing increased synaptic branching and bouton number (Zhang et al. 2001). The similarity of phenotypes indicates dFxr may regulate synaptic structure through direct regulation of *futsch* (Zhang et al. 2001). In addition to regulating *futsch*, dFxr may also control translation of factors modulating neurotransmitter release, as *dfxr* mutants have a significant reduction in evoked currents and a slight increase in mini amplitude (Zhang et al. 2001). In contrast, overexpression of dFxr can increase mEJC frequency and also slightly increases mEJC amplitude (Zhang et al. 2001). Similarly, loss-of-function and overexpression of dFxr in the central nervous system can disrupt synaptic transmission in the visual system (Zhang et al. 2001). These data indicate that synaptic signaling pathways can regulate both transcription and local mRNA abundance to modulate synaptic growth during larval development in *Drosophila*.

13.4.3 Post-translational Machinery: Ubiquitination, Highwire and Fat Facets

Regulation of the post-translational machinery at synapses provides additional control mechanisms that allow neurons to generate rapid changes in the molecular environment of the synapse. Ubiquitination is one of the prominent pathways through which protein half-life is regulated, and the activity of the conserved ubiquitin ligase Highwire (*hiw*) and the deubiquitin protease Fat Facets (*faf*) plays a critical role in controlling synaptic structure and function.

Hiw, a putative RING-H2 E3 ubiquitin-protein ligase, is localized to presynaptic terminals throughout development (Wu, Wairkar, Collins and DiAntonio 2005). Like *fasII* hypomorphs, *hiw* mutants have an increased number of type I boutons at NMJs, although each bouton is smaller and has reduced neurotransmitter release (Wan, DiAntonio, Fetter, Bergstrom, Strauss and Goodman 2000). Similar to *hiw* mutants, Faf overexpression leads to proliferation of boutons, an increase in the length of synaptic innervation along the muscle, and a decrease in neurotransmitter release (DiAntonio, Haghighi, Portman, Lee, Amaranto and Goodman 2001). *faf* loss-of-function alleles can rescue the *hiw* loss-of-function phenotypes, while overexpression of Faf is lethal in the *hiw* null background (DiAntonio et al. 2001). Hiw/Faf regulation may interface with several signaling systems in the presynaptic terminal that independently regulate synaptic growth and function, as low levels of the *hiw* transgene can rescue neurotransmitter release defects, while high transgene expression is required to rescue structural overgrowth (Wu et al. 2005). Together, the manipulations of these two components of the ubiquitination pathway suggest an important role for regulated protein degradation at synapses in the control of signaling pathways that mediate synaptic growth (DiAntonio et al. 2001).

One target pathway regulated by Hiw/Faf is the BMP signaling cascade. Mutants in the type II receptor *wit* have reduced quantal content, a smaller number of boutons, and misalignment of presynaptic active zones (Marques, Bao, Haerry, Shimell, Duchek, Zhang and O'Connor 2002; Aberle, Haghighi, Fetter, McCabe,

Magalhaes and Goodman 2002). Mutants in the gene for the Wit ligand *glass bottom boat* (*gbb*) exhibit many of the same defects, including the decrease in bouton number, altered synaptic ultrastructure, and the lack of activated Mad in motor neurons (McCabe, Marques, Haghighi, Fetter, Crotty, Haerry, Goodman and O'Connor 2003; Marques, Haerry, Crotty, Xue, Zhang and O'Connor 2003). In the background of either the BMP signaling mutant *wit* or *med,* the synaptic overgrowth observed in *hiw* loss-of-function mutants and Faf overexpression animals is reduced (McCabe et al. 2004), suggesting Hiw/Faf ubiquitination/deubiquitination regulates synaptic structure in part through the BMP signaling cascade. Although overexpression of TGF-β signaling components at the synapse is not sufficient to trigger synaptic overgrowth in wildtype terminals, overexpression in the *hiw* mutant background leads to increased synaptic proliferation (McCabe et al. 2004). These results suggest a model whereby Hiw ubiquitination provides a dampening system to modulate BMP-dependent signaling at the synapse.

Genetic interactions between *Hiw* and mutations in the gene for the MAPKKK *wallenda* (*wnd*), suggest that Hiw also regulates MAPK signaling cascades independent of *wit* (Collins, Wairkar, Johnson and DiAntonio 2006). Indeed, Wnd expression is directly regulated by Hiw ubiquitination, and is essential for the synaptic overgrowth observed in both *hiw* loss-of-function mutants and *faf* gain of function strains (Collins et al. 2006). Meanwhile, Wnd overexpression is sufficient to phenocopy the *hiw* loss-of-function phenotype, suggesting it is an essential target of ubiquitination regulation (Collins et al. 2006). In contrast to *wnd's* suppression of the *hiw* synaptic overgrowth phenotype, the physiology defects in *hiw* mutants are not rescued, suggesting additional molecular pathways are likely modulated by Hiw function (Collins et al. 2006). Wnd is a member of a family of MAPKKKs involved in JNK and p38 signaling. Mutations in the *Drosophila* JNK homolog, *basket (bsk),* can rescue either *hiw* loss-of-function or *wnd* overexpression phenotypes (Collins et al. 2006). In contrast, mutations in the p38 signaling pathway are not sufficient for rescue in these genetic backgrounds (Collins et al. 2006). *Drosophila* Fos (but not Jun) is also necessary for *hiw* synaptic overgrowth phenotypes, suggesting that transcriptional changes secondary to altered ubiquitination may act through a non-AP 1 pathway (Collins et al. 2006). Hiw has been suggested to regulate a number of other proteins including the hamartin and tuberin complex (in *Drosophila dTsc1-dTsc2*), as well as components of the rapamycin/S6K/4E-binding protein signaling pathway (Murthy, Han, Beauchamp, Smith, Haddad, Ito and Ramesh 2004). In summary, regulation of protein ubiquitination at NMJs can modulate multiple signaling pathways, providing an avenue for regulation of synaptic structure and function.

13.4.4 Intercellular Signaling: Synaptotagmin 4

An additional class of genes whose transcription is activity-regulated are those that encode proteins that modulate membrane trafficking and secretion within neurons. In particular, mRNA levels for several members of the Synaptotagmin family of vesicle Ca^{++} sensors have been shown to be activity regulated. Transcriptional upregulation of mRNA for Synaptotagmin 1 (Syt 1) has been identified in several *Drosophila* seizure mutants (Guan et al. 2005). Syt 1 localizes to presynaptic vesicles and acts as the Ca^{++} sensor for synchronous neurotransmitter release in *Drosophila* and mammals (Yoshihara and Littleton 2002; Nishiki and Augustine 2004). Given that synaptic vesicle release has been shown to be modulated by the levels of Syt 1 (Yoshihara et al. 2002), a direct regulation of its mRNA by activity provides an avenue for long-term control of synaptic vesicle fusion rates. Another member of the Synaptotagmin Family, Synaptotagmin 4 (Syt 4), has also been identified in rat brain and PC12 cells as an IEG induced by seizure activity, depolarization, or forskolin application (Vician, Lim, Ferguson, Tocco, Baudry and Herschman 1995; Ferguson, Thomas, Elferink and Herschman 1999). Syt 4 is a conserved member of the Synaptotagmin Ca^{++} sensor family and localizes to postsynaptic vesicles, suggesting a role in coupling postsynaptic Ca^{++} influx to release of retrograde signals (Adolfsen, Saraswati, Yoshihara and Littleton 2004; Yoshihara, Adolfsen, Galle and Littleton 2005). Mutations in Syt 4 disrupt the timing of normal NMJ synapse development, and abolish specific forms of synaptic plasticity at embryonic NMJs (Yoshihara et al. 2005). Overexpression of Syt 4 in the postsynaptic compartment causes an overproliferation of boutons at the NMJ, suggesting a role for Syt 4-regulated retrograde signaling in synaptic wiring (Yoshihara et al. 2005). *syt 4* null mutants in mice show deficiencies in hippocampal dependent learning and memory tasks, as well as problems with motor control (Ferguson, Anagnostaras, Silva and Herschman 2000; Ferguson, Wang, Herschman and Storm 2004; Ferguson, Wang, Herschman and Storm 2004). Together, these data suggest that activity-dependent regulation of transcription of members of the Synaptotagmin family may provide a long-term mechanism to modulate vesicle cycling in both pre- and post-synaptic compartments.

13.5 Genomic Approaches to Identifying Synaptic Targets of Transcriptional Regulation

Neurons are capable of remarkable adaptations to altered input via both homeostatic compensation that maintain neurotransmission within a dynamic range (Davis and Bezprozvanny 2001; Turrigiano and Nelson 2004), as well as the short- and long-term plastic changes associated with learning and memory. However, the differences in transcriptional mechanisms that underlie homeostatic adaptation to chronic changes in neuronal activity versus plasticity arising from acute changes in neuronal firing patterns are unknown. The development of new RNA screening technologies, including genome-wide microarrays and serial analysis of gene expression (SAGE), has provided a new means for the rapid identification of candidate genes that undergo transcriptional regulation in response to learning stimuli or altered neuronal activity. Several studies have employed these emerging technologies to screen for transcriptional targets that participate in neuronal plasticity in *Drosophila*. Although the function of most of these candidate effector molecules is still being analyzed, there is evidence that genes encoding proteins involved in local mRNA translation, cytoskeletal regulation, cell adhesion, and neurotransmission may be important targets for transcriptional modification of synaptic function. As the technologies and validation methods are still being developed, more effort is still needed to define the importance of the currently identified targets in synaptic plasticity. Here we briefly review the targets that have been identified in *Drosophila* models.

13.5.1 Activity Induction of Transcription

Synaptic changes that occur during acute or chronic alteration in brain activity have been intensively studied, although the transcriptional targets that participate in plasticity expression are poorly characterized. Two recent studies have tackled this question in *Drosophila* by performing DNA microarray analyses and SAGE mRNA profiling in *Drosophila* mutants that conditionally alter neuronal activity or AP-1 activation (Guan et al. 2005; Etter et al. 2005). Etter and colleagues used SAGE profiling of larval brains, as well as DNA microarray hybridization with adult brain mRNAs, to identify transcripts that were altered secondary to transient (6 hour) induction of an activated JNK kinase (*hemipterous*). As was discussed above, JNK directly regulates synaptic plasticity through AP-1 activation. The authors demonstrate that overexpression of an activated JNK in post-mitotic neurons results in a significant increase in synaptic growth at the NMJ. In contrast to the increase in structural plasticity, overexpression of activated JNK decreases synaptic transmission. Both AP-1 components *fos* and *jun* were induced by the expression of activated JNK, consistent with a MAPK-dependent activation of AP-1 transcription. Using SAGE analysis, the authors identified 345 tags as upregulated, and 271 as decreased following overexpression of activated JNK compared to controls. Using brain *in situs* and Q-PCR for target confirmation, only 3 of 61 tested targets were consistently changed using more stringent verification methods. Those loci included the previously described *appl* gene, the *white* locus, which encodes an ABC transporter, and *cheerio*, a locus encoding an actin binding filamin homolog important in cytoskeletal reorganization. Microarray screening of adult brains following overexpression of activated JNK yielded another 115 transcriptionally altered candidates. Among this group was the CG6044 locus, encoding a novel protein that has been previously implicated in olfactory memory in flies (Dubnau, Chiang, Grady, Barditch, Gossweiler, McNeil, Smith, Buldoc, Scott, Certa, Broger and Tully 2003). As with most mRNA profiling studies, it is unknown which of the identified targets are directly under the control of AP-1 transcription, and which occur secondary to unknown effects of excessive JNK activation. Further studies will be required to integrate these findings into a more complete picture of how AP-1 activation alters structural plasticity in flies, and which of the identified targets represent relevant candidate plasticity genes.

A second large-scale genome-wide microarray approach to identify activity-induced changes in gene transcription used temperature-sensitive (TS) seizure mutants in *Drosophila* to transiently induce excessive activity in adult brains

(Guan et al., 2005). Previous studies of epilepsy mutants, or seizure-inducing stimuli in wildtype animals, demonstrated that seizures can cause specific patterns of secondary neuronal plasticity changes similar to those observed during learning and memory, including the transcriptional activation of genes that alter neuronal connectivity and function (Nedivi, Hevroni, Naot, Israeli and Citri 1993; Qian, Gilbert, Colicos, Kandel and Kuhl 1993; Ben-Ari and Represa 1990). Guan and colleagues employed several newly identified TS seizure mutants, as well as previously characterized potassium channel or sodium pump mutants (Ganetzky et al. 1986). Periodic developmental heat shocks in the TS seizure mutants resulted in synaptic over-proliferation at NMJs, indicating that increased neuronal activity in the mutants promotes synaptic growth, similar to the effects found by overexpressing activated MAPK. Several of the TS mutants used in the study showed altered neuronal activity even at permissive temperatures, which allowed additional analysis of the effects of acute versus chronic neuronal activity changes on gene expression.

The microarray analysis of mRNAs isolated from adult heads in numerous seizure induction and recovery paradigms identified over 250 transcripts with differential transcription compared to control animals. Several of the upregulated genes were identified as *Drosophila* homologs of mammalian transcripts previously known to be activity-regulated, including *fos* and *jun*, the CCAAT element binding protein encoded by *c/EBP*, the cytoskeletal regulator *Arc*, and the early growth transcription factor *Stripe*. A second class of regulated transcripts included previously characterized genes with defined roles in the nervous system, but not previously known to be regulated by activity. These included effectors of cytoskeletal regulation and synaptic growth, such as Hiw, Fax, Cheerio, Bazooka, Trio, Kelch, RacGAP50C, C3G, VAP-33A, SCAR and Neurofibromin. Several components of membrane excitability turned up in the screen, including a sodium/calcium exchanger, the Ih hyperpolarization activated potassium channel, the L-type Ca^{++} channel and the sodium pump itself. Regulated genes implicated in neurotransmitter release included those encoding Syt 1 and the NSF adapter protein α-SNAP. Signaling proteins that were identified included PKC, adenylate cyclase, multiple GPCRs, and several components of the EGF, TGFβ and NF-κB signaling systems. However, the largest class of transcripts found in the study were those encoding novel proteins of unknown function. One such transcript was the *kek 2* gene, encoding an NCAM that contains a leucine-rich repeat (LRR) and an immunoglobulin (Ig) domain. *kek 2* was upregulated in several hyperexcitability mutants and downregulated under conditions of reduced excitability, suggesting its transcription could be bi-directionally controlled by changes in neuronal firing patterns. Similar to studies on the FasII NCAM described above, manipulations of the levels of Kek 2 by overexpression or RNAi-mediated disruption suggested it may function as an activity-regulated modulator of synaptic growth at the NMJ. Recently, a conserved mammalian member of the Kek 2-like family was demonstrated to undergo activity-dependent transcriptional regulation and to promote neurite outgrowth (Kuja-Panula, Kiiltomaki, Yamashiro, Rouhiainen and Rauvala 2003; Ono, Sekino-Suzuki, Kikkawa, Yonekawa and Kawashima 2003), indicating that this family of NCAMs may play a conserved role in activity-dependent modification of synaptic connectivity.

Although Guan and colleagues identified numerous candidate plasticity genes in the microarray study, the analysis was biased for detecting changes in genes that are highly expressed and found in many neurons, or expressed at low levels but abundantly overexpressed after seizure induction. As a result, some transcripts expressed at lower levels and previously implicated in plasticity, such as *FasII*, could not be identified with the statistical methods used in the study (Guan et al. 2005). In addition to this limitation, the study was also biased towards identifying genes that changed during global alterations in neuronal activity, as opposed to those that might be modulated by the specific patterned activity normally created during learning behaviors in the fly. A subset of the transcripts were also likely to be altered independent of activity changes, and occur secondary to the primary defects occurring in mutant animals. In spite of these caveats, the data indicate that activity-dependent changes in gene transcription in the fly nervous system alters the levels of numerous classes of genes that encode proteins that can alter membrane excitability and synaptic transmission, as well as cytoskeletal architecture and synaptic growth. Follow-up studies looking at the effects of overexpression of the transcripts identified, as well as mutant analysis of the individual loci, will be required to build a more complete picture of how transcriptional regulation functions during synaptic plasticity and information storage.

13.5.2 Transcriptional Control of Synaptic Translation During LTM

To specifically analyze the transcriptional effects of CREB-dependent learning and memory on gene expression, Dubnau and colleagues conducted DNA microarray experiments on fruit flies after olfaction-based LTM formation. The authors used adult flies following spaced olfactory learning paradigms (which induces CREB-dependent LTM), subtracting out any transcripts altered during massed learning trials (which induces CREB-independent ARM). Approximately 40 transcripts that were identified as significantly altered during the microarray experiments were confirmed by QPCR as being significantly regulated in LTM formation (Dubnau et al. 2003). To complement the microarray study, the authors also performed a large-scale screen for new learning and memory mutants in *Drosophila*. These two techniques converged to identify a novel role for the translational repressor *pumilio* (*pum*) in memory (Dubnau et al. 2003).

During embryonic development, *pum* acts together with the RNA binding protein, *nanos* (*nos*), as a translational repressor of the *hunchback* (*hb*) locus. Following embryonic development, *pum* is expressed in the cytosol of neurons throughout the CNS. At the larval NMJ, Pum localizes postsynaptically (Menon, Sanyal, Habara, Sanchez, Wharton, Ramaswami and Zinn 2004). In *pum* null mutants, there is a decrease in synaptic growth at the NMJ, with larger individual boutons (Menon et al. 2004). Overexpression of *pum* has the opposite effect, increasing the number of boutons, while decreasing individual bouton diameter. Translational regulation of dendritic development requires *pum* to act in concert with *nos* (Ye, Petritsch, Clark, Gavis, Jan and Jan 2004).

The microarray analysis of Dubnau and colleagues identified additional targets of transcriptional regulation from the *pum* pathway, including *staufen (stau),* which is involved in mRNA translocation, and *orb* (cytoplasmic polyadenylation element binding protein (CPEB) homolog), which functions in localized synaptic translation (Dubnau et al. 2003). Additional elements of a local mRNA localization and translation pathway were found in the behavioral screen for new learning mutants, including *eIF-2G* (a translation initiation factor) and the *oskar* mutants *norka* and *krasavietz* (Dubnau et al. 2003). Like CAMKII, *oskar*'s synaptic expression can be regulated by the RISC complex (Sontheimer 2005). All of the identified components of the *pum/stau* pathway from the study are expressed in the mushroom bodies, consistent with a role in olfaction-based LTM formation. One model consistent with the regulation of these proteins suggests that following CREB activation of transcription, specific mRNAs are produced and transported out to the synapse, where they undergo local translation.

How local translation directly alters synaptic plasticity is still unknown, but the levels of *eIF-4E* are much higher in *pum* mutants (Pum represses *eIF-4E* expression in an activity-dependent manner), resulting in higher postsynaptic glutamate receptor levels at the NMJ (Sigrist et al. 2003). Pum overexpression can also suppress eIF-4E induction of synaptic growth, suggesting postsynaptic Pum regulates synaptic plasticity by controlling expression of the eIF-2G translation initiation factor (Menon et al. 2004). As a result of activity-induced transcriptional regulation, the function of the translational machinery can be directly regulated at the synapse. Further interactions with the RISC machinery may fine-tune local translational regulation. Direct transcriptional regulation of the synaptic translational machinery may allow rapid upregulation of entire classes of synaptic proteins that could locally alter synaptic plasticity.

13.6 Conclusions

Genetic and molecular analysis of *Drosophila* learning mutants and epilepsy models has lead to new insights into how transcriptional changes in gene expression might contribute to synaptic plasticity. In response to learning and changes in neuronal activity, specific classes of gene products undergo transcriptional regulation, including synaptic cell adhesion proteins, cytoskeletal regulators and proteins mediating local protein translation at the synapse. It is clear that both transcriptional and post-translational mechanisms are activated by enhanced neuronal activity and contribute to long-term alterations in neuronal connectivity and function. An extensive analysis of the signaling pathways that mediate synaptic growth at the *Drosophila* NMJ has provided a foundation for understanding the mechanisms by which alterations in cell adhesion and transsynaptic signaling mediate structural changes in synaptic strength. The current challenge is to begin to characterize how activity-regulated transcripts interface with the established synaptic machinery to alter both functional and structural properties of the synapse during learning.

References

Aberle, H., Haghighi, A.P., Fetter, R.D., McCabe, B.D., Magalhaes, T.R. and Goodman C.S. (2002) *wishful thinking* encodes a BMP type II receptor that regulates synaptic growth in *Drosophila*. Neuron 33, 545–558.

Adolfsen, B., Saraswati, S., Yoshihara, M. and Littleton, J.T. (2004) Synaptotagmins are Trafficked to Distinct Subcellular Domains Including the Postsynaptic Compartment. J. Cell. Biol. 166, 249–260.

Andersen, R., Li, Y., Resseguie, M. and Brenman, J.E. (2005) Calcium/calmodulin-dependent protein kinase II alters structural plasticity and cytoskeletal dynamics in *Drosophila*. J. Neurosci. 25, 8878–8888.

Ashley, J., Packard, M., Ataman, B. and Budnik, V. (2005) Fasciclin II signals new synapse formation through amyloid precursor protein and the scaffolding protein dX11/Mint. J. Neurosci. 25, 5943–5955.

Ashraf, S.I., McLoon, A.L., Sclarsic, S.M., and Kunes S. (2006) Synaptic Protein Synthesis Associated with Memory Is Regulated by the RISC Pathway in *Drosophila*. Cell 124, 191–205.

Bailey, C.H., Kaang, B.K., Chen, M., Martin, K.C., Lim, C.S., Casadio, A. and Kandel, E.R. (1997) Mutation in the phosphorylation sites of MAP kinase blocks learning-related internalization of apCAM in Aplysia sensory neurons. Neuron 18, 913–924.

Bailey, C.H., and Chen, M. (1988) Long-term memory in *Aplysia* modulates the total number of varicosities of single identified sensory neurons. Proc. Natl. Acad. Sci. USA 85, 2373–2377.

Baines, R.A., Seugnet, L., Thompson, A., Salvaterra, P.M. and Bate, M. (2002) Regulation of synaptic connectivity: levels of Fasciclin II influence synaptic growth in the *Drosophila* CNS. J. Neurosci. 22, 6587–6595.

Ben-Ari, Y. and Represa, A. (1990) Brief Seizure Episodes Induce Long-Term Potentiation and Mossy Fibre Sprouting in the Hippocampus. Trends Neurosci 8, 312–318.

Beumer, K., Matthies, H.J., Bradshaw, A. and Broadie, K. (2002) Integrins regulate DLG/FAS2 via a CaM kinase II-dependent pathway to mediate synapse elaboration and stabilization during postembryonic development. Development 129, 3381–3391.

Bito, H., Deisseroth, K. and Tsien, R.W. (1996) CREB Phosphorylation and Dephosphorylation: a Ca(2+) and Stimulus Duration-Dependent Switch for Hippocampal Gene Expression. Cell 87, 1203–1214.

Brouillet, E., Trembleau, A., Galanaud, D., Volovitch, M., Bouillot, C., Valenza, C., Prochiantz, A. and Allinquant, B. (1999) The amyloid precursor protein interacts with Go heterotrimeric protein within a cell compartment specialized in signal transduction. J. Neurosci. 19, 1717–1727.

Budnik, V., Zhong, Y. and Wu, C.F. (1990) Morphological plasticity of motor axons in *Drosophila* mutants with altered excitability. J. Neurosci. 10, 3754–3768.

Chang, Q., and Balice-Gordon, R. (2000) *highwire, rpm-1*, and *futsch*: Balancing Synaptic Growth and Stability. Neuron 26, 287–290.

Cheng, Y., Endo, K., Wu, K., Rodan, A.R., Heberlein, U. and Davis, R.L. (2001) *Drosophila* fasciclinII is required for the formation of odor memories and for normal sensitivity to alcohol. Cell 105, 757–768.

Chiang, A.S., Blum, A., Barditch, J., Chen, Y.H., Chiu, S.L., Regulski, M., Armstrong, J.D., Tully, T. and Dubnau, J. (2004) *radish* encodes a phospholipase-A2 and defines a neural circuit involved in anesthesia-resistant memory. Curr. Biol. 14, 263–272.

Collins, C.A., Wairkar, Y.P., Johnson, S.L. and DiAntonio, A. (2006) *Highwire* restrains synaptic growth by attenuating a MAP kinase signal. Neuron 51, 57–69.

Curtis, J., and Finnkbeiner, S. (1999) Sending signals from the synapse to the nucleus: possible roles for CaMK, Ras/ERK, and SAPK pathways in the regulation of synaptic plasticity and neuronal growth. J. Neurosci. Res. 58, 88–95.

Dash, P.K., Hochner, B. and Kandel, E.R. (1990) Injection of the cAMP-Responsive Element into the Nucleus of *Aplysia* Sensory Neurons Blocks Long-Term Facilitation. Nature 345, 718–721.

Davis, G. and Bezprozvanny, I. (2001) Maintaining the Stability of Neuronal Function: a Homeostatic Hypothesis. Annu. Rev. Physiol. 63, 847–869.

Davis, G., Schuster, C. and Goodman, C. (1996) Genetic Dissection of Structural and Functional Components of Synaptic Plasticity. III. CREB Is Necessary for Presynaptic Functional Plasticity. Neuron 17, 669–679.

Davis, G., Schuster, C. and Goodman, C. (1997) Genetic Analysis of the Mechanisms Controlling Target Selection: Target-Derived Fasciclin II Regulates the Pattern of Synapse Formation. Neuron 19, 561–573.

Desmond, N.L. and Levy, W.B. (1986) Changes in the numerical density of synaptic contacts with long-term potentiation in the hippocampal dentate gyrus. J. Comp. Neurol. 253, 466–475.

DeZazzo, J., Sandstrom, D., de Belle, S., Velinzon, K., Smith, P., Grady, L., DelVecchio, M., Ramaswami, M. and Tully, T. (2000) nalyot, a mutation of the Drosophila myb-related Adf1 transcription factor, disrupts synapse formation and olfactory memory. Neuron 27, 145–158.

DiAntonio, A., Haghighi, A.P., Portman, S.L., Lee, J.D., Amaranto, A.M. and Goodman, C.S. (2001) Ubiquitination-dependent mechanisms regulate synaptic growth and function. Nature 412, 449–452.

Dolmetsch, R.E., Pajvani, U., Fife, K., Spotts, J.M. and Greenberg, M.E. (2001) Signaling to the Nucleus by an L-type Calcium Channel-Calmodulin Complex Through the MAP Kinase Pathway. Science 294, 333–339.

Dubnau, J., Chiang, A.S., Grady, L., Barditch, J., Gossweiler, S., McNeil, J., Smith, P., Buldoc, F., Scott, R., Certa, U., Broger, C. and Tully, T. (2003) The staufen/pumilio Pathway Is Involved in Drosophila Long-Term Memory. Curr. Biol. 13, 286–296.

Etter, P.D., Narayanan, R., Navratilova, Z., Patel, C., Bohmann, D., Jasper, H. and Ramaswami, M. (2005) Synaptic and genomic responses to JNK and AP-1 signaling in Drosophila neurons. BMC Neurosci. 6, 39.

Feany, M.B. and Quinn, W.G. (1995) A neuropeptide gene defined by the Drosophila memory mutant amnesiac. Science 268.

Ferguson, G.D., Anagnostaras, S.G., Silva, A.J. and Herschman, H.R. (2000) Deficits in Memory and Motor Performance in Synaptotagmin IV Mutant Mice. Proc. Natl. Acad. Sci. USA 97, 5598–5603.

Ferguson, G.D., Herschman, H.R. and Storm, D.R. (2004) Reduced Anxiety and Depression-Like Behavior in Synaptotagmin IV (-/-) Mice. Neuropharmacol. 47, 604–611.

Ferguson, G.D., Thomas, D.M., Elferink, L.A. and Herschman, H.R. (1999) Synthesis, Degradation and Subcellular Localization of Synaptotagmin IV, a Neuronal Immediate Early Gene Product. J. Neurochem. 72, 1821–1831.

Ferguson, G.D., Wang, H., Herschman, H.R. and Storm, D.R. (2004) Altered Hoppocampal Short-Term Plasticity and Associative Memory in Synaptotagmin IV (-/-) Mice. Hippocampus 14, 964–974.

Folkers E, Drain, P. and Quinn, W.G. (1993) Radish, a Drosophila mutant deficient in consolidated memory. Proc. Natl. Acad. Sci. USA 90, 8123–8127.

Friedrich, A., Thomas, U. and Muller, U. (2004) Learning at different satiation levels reveals parallel functions for the cAMP-protein kinase A cascade in formation of long-term memory. J. Neurosci. 24, 4460–4468.

Ganetzky, B. and Wu, C.F. (1986) Neurogenetics of Membrane Excitability in Drosophila. Annu. Rev. Genet. 20, 13–44.

Garcia-Alonso, L., VanBerkum, M.F., Grenningloh, G., Schuster, C. and Goodman, C.S. (1995) Fasciclin II controls proneural gene expression in Drosophila. Proc. Natl. Acad. Sci. USA 92, 10501–10505.

Glanzman, D.L., Kandel, E.R. and Schacher, S. (1990) Target-dependent structural changes accompanying long-term synaptic facilitation in Aplysia neurons. Science 249, 799–802.

Grenningloh, G., Rehm, E.J. and Goodman, C.S. (1991) Genetic analysis of growth cone guidance in Drosophila: Fasciclin II functions as a neuronal recognition molecule. Cell 67, 45–57.

Griffith, L.C., Lu, C.S. and Sun, X.X. (2003) CaMKII, an Enzyme on the Move: Regulation of Temporospatial Localization. Mol. Interven. 3, 386–403.

Griffith, L.C., Verselis, L.M., Aitken, K.M., Kyriacou, C.P., Danho, W. and Greenspan, R.J. (1993) Inhibition of calcium/calmodulin-dependent protein kinase in Drosophila disrupts behavioral plasticity. Neuron 10, 501–509.

Grotewiel, M.S., Beck, C.D., Wu, K.H., Zhu, X.R. and Davis, R.L. (1998) Integrin-mediated short-term memory in *Drosophila*. Nature 391, 455–460.

Guan, Z., Saraswati, S., Adolfsen, B. and Littleton, J.T. (2005) Genome-Wide Transcriptional Changes Associated with Enhanced Activity in the *Drosophila* Nervous System. Neuron 48, 91–107.

Hanson, P.I. and Schulman, H. (1992) Neuronal Ca2+/calmodulin-dependent protein kinases. Annu. Rev. Biochem. 61, 559–601.

Hebbar, S. and Fernandes, J.J. (2005) A role for Fas II in the stabilization of motor neuron branches during pruning in *Drosophila*. Dev. Biol. 285, 185–199.

Hoeffer, C.A., Sanyal, S. and Ramaswami, M. (2003) Acute induction of conserved synaptic signaling pathways in *Drosophila melanogaster*. J. Neurosci. 23, 6362–6372.

Huang, Y.Y., Li, X.C. and Kandel, E.R. (1994) cAMP Contributes to Mossy Fiber LTP by Initiating both a Covalently Mediated Early Phase and Macromolecular Synthesis-dependent Late Phase. Cell 79, 69–79.

Hummel, T., Krukkert, K., Roos, J., Davis, G. and Klambt, C. (2000) *Drosophila* Futsch/22C10 is a MAP1B-like protein required for dendritic and axonal development. Neuron 26, 357–370.

Isabel, G., Pascual, A. and Preat, T. (2004) Exclusive Consolidated Memory Phases in *Drosophila*. Science 304, 1024–1027.

Kaplan, W.D. and Trout, W.E. (1969) The Behavior of Four Neurological Mutants of *Drosophila*. Genetics 61, 399–409.

Kelleher, R.J., Govindarajan, A. and Tonegawa, S. (2004) Translational Regulatory Mechanisms in Persistent Forms of Synaptic Plasticity. Neuron 44, 59–73.

Kelleher, R.J., Govindarajan, A., Jung, H.Y., Kang, H. and Tonegawa, S. (2004) Translational Control by MAPK Signaling in Long-Term Synaptic Plasticity and Memory. Cell 116,467–479.

Koh, Y.H., Popova, E., Thomas, U., Griffith, L.C. and Budnik, V. (1999) Regulation of DLG localization at synapses by CaMKII-dependent phosphorylation. Cell 98, 353–363.

Koh, Y.H., Ruiz-Canada, C., Gorczyca, M. and Budnik, V. (2002) The Ras1-mitogen-activated protein kinase signal transduction pathway regulates synaptic plasticity through fasciclin II-mediated cell adhesion. J. Neurosci. 22, 2496–2504.

Kuja-Panula, J., Kiiltomaki, M., Yamashiro, T., Rouhiainen, A. and Rauvala, H. (2003) AMIGO, a Transmembrane Protein Implicated in Axon Tract Development, Defines a Novel Protein Family with Leucine-Rich Repeats. J. Cell. Biol 160, 963–973.

Li, W., Tully, T. and Kalderon, D. (1996) Effects of a conditional *Drosophila* PKA mutant on olfactory learning and memory. Learn. Mem. 2, 320–333.

Lin, D.M. and Goodman, C.S. (1994) Ectopic and increased expression of fasciclin II alters motoneuron growth cone guidance. Neuron 13, 507–523.

Lisman, J., Schulman, H. and Cline, H. (2002) The molecular basis of CaMKII function in synaptic and behavioural memory. Nat. Rev. Neurosci. 3, 175–190.

Liu, G., Seiler, H., Wen, A., Zars, T., Ito, K., Wolf, R., Heisenberg, M. and Liu, L. (2006) Nature 439, 551–556.

Livingstone, M.S., Sziber, P.P. and Quinn, W.G. (1984) Loss of calcium/calmodulin responsiveness in adenylate cyclase of *rutabaga*, a *Drosophila* learning mutant. Cell 37, 205–215.

MacLaren, C.M., Evans, T.A., Alvarado, D. and Duffy, J.B. (2004) Comparative Analysis of the Kekkon Molecules, Related Members of the LIG Superfamily. Dev. Genes Evol. 214, 360–366.

Margulies, C., Tully, T. and Dubnau, J. (2005) Deconstructing Memory in *Drosophila*. Curr. Biol. 15, R700-R713.

Marques, G., Bao, H., Haerry, T.E., Shimell, M.J., Duchek, P., Zhang, B. and O'Connor, M.B. (2002) The *Drosophila* BMP type II receptor Wishful Thinking regulates neuromuscular synapse morphology and function. Neuron 33, 529–543.

Marques, G., Haerry, T.E., Crotty, M.L., Xue, M., Zhang, B. and O'Connor, M.B. (2003) Retrograde Gbb signaling through the Bmp type 2 receptor wishful thinking regulates systemic FMRFa expression in *Drosophila*. Devolopment 130, 5457–5470.

Martin, K.C., Michael, D., Rose, J.C., Barad, M., Casadio, A., Zhu, H. and Kandel, E.R. (1997) MAP kinase translocates into the nucleus of the presynaptic cell and is required for long-term facilitation in *Aplysia*. Neuron 18, 899–912.

McCabe, B.D., Hom, S., Aberle, H., Fetter, R.D., Marques, G., Haerry, T.E., Wan, H., O'Connor, M.B., Goodman, C.S. and Haghighi, A.P. (2004) Highwire Regulates Presynaptic BMP Signaling Essential for Synaptic Growth. Neuron 41, 891–905.

McCabe, B.D., Marques, G., Haghighi, A.P., Fetter, R.D., Crotty, M.L., Haerry, T.E., Goodman, C.S. and O'Connor, M.B. (2003) The BMP Homolog Gbb Provides a Retrograde Signal that Regulates Synaptic Growth at the *Drosophila* Neuromuscular Junction. Neuron 39, 241–254.

McGuire, S.E., Deshazer, M. and Davis, R.L. (2005) Thirty years of olfactory learning and memory research in *Drosophila melanogaster*. Progr. Neurobiol. 76, 328–347.

Mehren, J.E. and Griffith, L.C. (2004) Calcium-independent calcium/calmodulin-dependent protein kinase II in the adult *Drosophila* CNS enhances the training of pheromonal cues. J. Neurosci. 24, 10584–10593.

Menon, K.P., Sanyal, S., Habara, Y., Sanchez, R., Wharton, R.P., Ramaswami, M. and Zinn, K. (2004) The Translational Repressor Pumilio Regulates Presynaptic Morphology and Controls Postsynaptic Accumulation of Translation Factor eIF-4E. Neuron 44, 663–676.

Michael, D., Martin, K.C., Seger, R., Ning, M.M., Baston, R. and Kandel, E.R. (1998) Repeated pulses of serotonin required for long-term facilitation activate mitogen-activated protein kinase in sensory neurons of *Aplysia*. Proc. Natl. Acad. Sci. USA 95, 1864–1869.

Murthy, V., Han, S., Beauchamp, R.L., Smith, N., Haddad, L.A., Ito, N. and Ramesh, V. (2004) *Pam* and its ortholog *highwire* interact with and may negatively regulate the TSC1.TSC2 complex. J. Biol. Chem. 279, 1351–1358.

Nedivi, E., Hevroni, D., Naot, D., Israeli, D. and Citri, Y. (1993) Numerous Candidate Plasticity-Related Genes Revealed by Differential cDNA Cloning. Nature 363, 718–722.

Nguyen, M., Park, S., Marques, G. and Arora, K. (1998) Interpretation of a BMP activity gradient in *Drosophila* embryos depends on synergistic signaling by two type I receptors, SAX and TKV. Cell 95, 495–506.

Nishiki, T. and Augustine, G.J. (2004) Dual Roles of the C2B Domain of Synaptotagmin I in Synchronizing Ca2+-Dependent Neurotransmitter Release. J. Neurosci. 24, 8542–8550.

Ono, T., Sekino-Suzuki, N., Kikkawa, Y., Yonekawa, H. and Kawashima, S. (2003) Alivin 1, a Novel Neuronal Activity-Dependent Gene, Inhibits Apoptosis and Promotes Survival of Cerebellar Granule Neurons. J. Neurosci. 23, 5887–5896.

Peverali, F.A., Isaksson, A., Papavassiliou, A.G., Staszewski, L.M., Mlodzik, M. and Bohmann, D. (1996) Phosphorylation of *Drosophila* Jun by the MAP kinase rolled regulates photoreceptor differentiation. EMBO J 15, 3943–3950.

Qian, Z., Gilbert, M.E., Colicos, M.A., Kandel, E.R. and Kuhl, D. (1993) Tissue-plasminogen Activator is Induced as an Immediate-Early Gene During Seizure, Kindling and Long-Term Potentiation. Nature 361, 453–457.

Quinn, W.G., Sziber, P.P. and Booker, R. (1979) The *Drosophila* memory mutant *amnesiac*. Nature 277, 212–214.

Roos, J., Hummel, T., Ng, N., Klambt, C. and Davis, G.W. (2000) *Drosophila* Futsch regulates synaptic microtubule organization and is necessary for synaptic growth. Neuron 26, 371–382.

Ruiz-Canada, C., Ashley, J., Moeckel-Cole, S., Drier, E., Yin, J. and Budnik, V. (2004) New synaptic bouton formation is disrupted by misregulation of microtubule stability in aPKC mutants. Neuron 42, 567–580.

Sanyal, S., Kim, S.M. and Ramaswami, M. (2004) Retrograde Regulation in the CNS: Neuron-Specific Interpretations of TGF-beta Signaling. Neuron 41, 845–848.

Sanyal, S., Sandstrom, D.J., Hoeffer, C.A. and Ramaswami, M. (2002) AP-1 functions upstream of CREB to control synaptic plasticity in *Drosophila*. Nature 416, 870–874.

Schuster, C., Davis, G., Fetter, R. and Goodman C.S. (1996) Genetic Dissection of Structural and Functional Components of Synaptic Plasticity. II. Fasciclin II Controls Presynaptic Structural Plasticity. Neuron 17, 655–667.

Schuster, C., Davis, G., Fetter, R. and Goodman, C.S. (1996) Genetic Dissection of Structural and Functional Components of Synaptic Plasticity. I. Fasciclin II Controls Synaptic Stabilization and Growth. Neuron 17, 641–654.

Sigrist, S.J., Reiff, D.F., Thiel, P.R., Steinert, J.R. and Schuster, C.M. (2003) Experience-dependent strengthening of *Drosophila* neuromuscular junctions. J. Neurosci. 23, 6546–6556.

Sigrist, S.J., Thiel, P.R., Reiff, D.F., Lachance, P.E., Lasko, P., Schuster, C.M. (2000) Postsynaptic translation affects the efficacy and morphology of neuromuscular junctions. Nature 405, 1062–1065.

Skoulakis, E.M., Kalderon, D. and Davis, R.L. (1993) Preferential expression in mushroom bodies of the catalytic subunit of protein kinase A and its role in learning and memory. Neuron 11, 197–208.

Sontheimer, E.J. (2005) Assembly and function of RNA silencing complexes. Rev. Mol. Cell. Biol. 6, 127–138.

Thomas, U., Kim, E., Kuhlendahl, S., Koh, Y.H., Gundelfinger, E.D., Sheng, M., Garner, C.C. and Budnik, V. (1997) Synaptic clustering of the cell adhesion molecule fasciclin II by discs-large and its role in the regulation of presynaptic structure. Neuron 19, 787–799.

Titus, S.A., Warmke, J.W. and Ganetzky, B. (1997) The *Drosophila* erg K+ channel polypeptide is encoded by the *seizure* locus. J. Neurosci. 17, 875–881.

Tully, T. and Quinn, W.G. (1985) Classical conditioning and retention in normal and mutant *Drosophila melanogaster*. J. Comp. Physiol. 157, 263–277.

Turrigiano, G.G. and Nelson, S.B. (2004) Homeostatic Plasticity in the Developing Nervous System. Nat. Rev. Neurosci. 5, 97–107.

Vician, L., Lim, I.K., Ferguson, G., Tocco, G., Baudry, M. and Herschman, H.R. (1995) Synaptotagmin IV is an Immediate Early Gene Induced by Depolarization in PC12 Cells and in Brain. Proc. Natl. Acad. Sci. 92, 2164–2168.

Waddell, S. and Quinn, W.G. (2001) *Fas*-acting memory. Dev. Cell 1, 8–9.

Wan, H.I., DiAntonio, A., Fetter, R.D., Bergstrom, K., Strauss, R. and Goodman, C.S. (2000) Highwire regulates synaptic growth in *Drosophila*. Neuron 26, 313–329.

Wrana, J.L., Attisano, L., Wieser, R., Ventura, F. and Massague, J. (1994) Mechanism of activation of the TGF-beta receptor. Nature 370, 341–347.

Wrana, J.L. and Attisano, L. (2000) The Smad Pathway. Cytokine Growth Fact. Rev. 11, 5–13.

Wu, C., Wairkar, Y.P., Collins, C.A. and DiAntonio, A. (2005) Highwire function at the *Drosophila* neuromuscular junction, spatial, structural and temporal requirements. J. Neurosci. 25, 9557–9566.

Ye, B., Petritsch, C., Clark, I.E., Gavis, E.R., Jan, L.Y. and Jan, Y.N. (2004) *nanos* and *pumilio* Are Essential for Dendrite Morphogenesis in *Drosophila* Peripheral Neurons. Curr. Biol. 14, 314–321.

Yin, J.C., Wallach, J.S., Del Vecchio, M., Wilder, E.L., Zhou, H., Quinn, W.G. and Tully, T. (1994) Induction of a dominant negative CREB transgene specifically blocks long-term memory in *Drosophila*. Cell 79, 49–58.

Yin, J.C., Wallach, J.S., Wilder, E.L., Klingensmith, J., Dang, D., Perrimon, N., Zhou, H., Tully, T. and Quinn, W.G. (1995) A *Drosophila* CREB/CREM homolog encodes multiple isoforms, including a cyclic AMP-dependent protein kinase-responsive transcriptional activator and antagonist. Mol. Biol. Cell 15, 5123–5130.

Yoshihara, M., Adolfsen, B., Galle, K.T. and Littleton, J.T. (2005) Retrograde Signaling by Syt 4 Induces Presynaptic Release and Synapse-Specific Growth. Science 310, 858–863.

Yoshihara, M. and Littleton, J.T. (2002) Synaptotagmin I functions as a Calcium Sensor to Synchronize Neurotransmitter Release. Neuron 36, 897–908.

Zhang, Y.Q., Bailey, A.M., Matthies, H.J., Renden, R.B., Smith, M.A., Speese, S.D., Rubin, G.M. and Broadie, K. (2001) *Drosophila* fragile X-related gene regulates the MAP1B homolog Futsch to control synaptic structure and function. Cell 107, 591–603.

Zhong, Y., Budnik, V. and Wu C.F. (1992) Synaptic plasticity in *Drosophila* memory and hyperexcitable mutants, role of cAMP cascade. J. Neurosci. 12, 644–651.

Zhong, Y. and Wu, C.F. (2004) Neuronal activity and adenylyl cyclase in environment-dependent plasticity of axonal outgrowth in Drosophila. J. Neurosci. 24, 1439–1445.

Chapter 14
MMP-9/TIMP-1 Extracellular Proteolytic System as AP-1 Target in Response to Neuronal Activity

Grzegorz M. Wilczynski and Leszek Kaczmarek

Abstract For the last several years, a considerable body of evidence has been collected indicating that MMP-9-mediated TIMP-1-regulated extracellular proteolysis is a novel mechanism contributing to synaptic plasticity. An outstanding feature of this system is that it is kept under the genetic control of a transcription factor AP-1. Both MMP-9 and TIMP-1 are AP-1 target genes in brain and their expression correlates with a spatiotemporal expression pattern of c-Fos, the major neuronal activity-driven component of AP-1. Interestingly, following the gene transcription, MMP-9 mRNA undergoes activity-driven dendritic transport, a feature that is widely accepted as one of the fundamental mechanisms of plasticity. The MMP-9/TIMP-1 system appears to be involved in a broad range of physiological and pathological phenomena in various brain regions. These include developmental reorganization of the cerebellum and visual cortex, hippocampus-dependent learning and long-term-potentiation, as well as pathogenesis of temporal lobe epilepsy. In the hippocampus, MMP-9 appears to function through a direct effect on dendritic spines. We propose an existence of a functional axis, consisting of AP-1/MMP-9/TIMP-1, together with a yet-to-be-discovered synaptic substrate(s) of MMP-9 that has a special liaison with neuronal activity. First, it is activity-driven, and second, it carries on the activity signal to produce its effects on synaptic remodeling and subsequent plasticity.

14.1 AP-1 as Prototypical Transcription Factor Driven by Neuronal Activity

Reported in 1987 and 1988, discovery of a massive, rapid and transient accumulation of c-fos mRNA and protein in response to a variety of stimuli provided the first examples of rapid induction of gene expression in the CNS (Curran and Morgan, 1987; Dragunow and Robertson, 1987; Hunt et al., 1987; Kaczmarek et al., 1988). These findings opened a completely new avenue of research and have soon been followed by an extended list of other instances of c-fos activation as well as induction of a number of other genes (see for review: Hughes and Dragunow, 1995; Herrera and Robertson, 1996; Kaczmarek and Chaudhuri, 1997; Herdegen and Leah, 1998; Kovacs, 1998; Knapska and Kaczmarek, 2004)

S. M. Dudek (ed.), *Transcriptional Regulation by Neuronal Activity.*
© Springer 2008

The great interest that has surrounded *c-fos* in the brain has originally been spurred by the hypothesis that its protein product ("master switch") may control a variety of phenomena of long term change in cellular (and thus also neuronal) functioning, including in particular learning and memory (Berridge, 1986; Goelet et al., 1986; Kaczmarek, 1986; Curran and Morgan, 1987; Kaczmarek and Kaminska, 1989). Very soon after these hypotheses were formulated, first experimental studies showed that indeed *c-fos* activation follows behavioral training and correlates with an acquisition of the behaviorally measured responses, i.e., with learning (Maleeva et al., 1989; Kaczmarek, 1990; Kaczmarek and Nikolajew, 1990; Maleeva et al., 1990; Tischmeyer et al., 1990). These results were next reproduced and extended in dozens of research reports (see for reviews: Kaczmarek, 1993; Dragunow, 1996; Anokhin, 1997; Tischmeyer and Grimm, 1999; Kaczmarek et al., 2002).

c-fos mRNA and c-Fos protein levels are negligible at basal conditions, i.e., in a variety of non-stimulated cells, including neurons. Increased neuronal expression results from stimulation of membrane receptors and subsequent rise in second messengers and kinase activation. The temporal pattern of this activation is quite uniform. An increase in mRNA is observed within a few minutes after the signal arrives at the cell membrane, and then the protein accumulates, occurring roughly between 30 and 90 minutes. Both mRNA and protein increases are transient. *c-fos* transcription is soon shut off and, furthermore, the *c-fos* products' half-lives are very short: less than 20 minutes for *c-fos* mRNA and less than 2 hours for the protein (Vosatka et al., 1989; Greenberg et al., 1990; Werlen et al., 1993). More protracted time-courses of *c-fos* expression were sometimes reported for the brain (see e.g. Nikolaev et al., 1992), however, they result, most probably, from sequential activation of various cells within the heterogeneous nervous tissue.

As far as the neuronal activity is concerned, glutamate receptors, especially calcium-permeable NMDA receptors and AMPA receptors—producing depolarization and subsequent opening of the voltage gated calcium channel—have been the most often implicated in controlling *c-fos* expression (Xia et al., 1996; Fleischmann et al., 2003; Platenik et al., 2000). In both cases Ca^{2+} is believed to be the major second messenger (Dolmetsch, 2003). Next to its rise, however, the situation is so complicated that it is difficult at present to decipher the variety of kinases involved. Then, however, the *c-fos* regulation is apparently simplified at the level of the gene regulatory regions. Two major promoter sequences were identified in control of *c-fos*: CRE (Ca^{2+}/cAMP responsive element) occupied by CREB (CRE binding proteins) regulated by phosphorylation, and SRE (serum responsive element) to which SRF (SRE-binding factors) together with TCF (ternary complex factors) bind, with Elk-1 transcription factor protein identified as the major TCF in the brain, activated most probably by ERK kinase pathways (see e.g. Kaminska et al., 1999). Very recently, another (non-TCF) ERK-dependent activity-regulated neuronal cofactor of SRF, presumably playing important role in *c-fos* gene expression, has been identified as MKL1 (megakaryocytic leukemia) transcription factor (Kalita et al., 2007).

As soon as c-Fos protein is produced, it may interact with Jun proteins to form AP-1 transcription factor that acts as a protein dimer. There are three Juns identified: c-Jun, JunB and JunD. Their ability to form dimers with c-Fos in the brain has been investigated only marginally. However, the studies on cultured (mostly non-neuronal) cells suggest that c-Fos may interact with each of the Jun proteins. It has also been observed that AP-1 of various compositions may bring about differential effects on DNA binding and gene expression (see: Kaminska et al., 2000; Hess et al., 2004). The functional significance of this phenomenon in the brain is virtually unknown.

Despite multiple efforts, very little is known about AP-1 target genes in the brain (see: Rylski and Kaczmarek, 2004). These genes are so difficult to identify, because there are no good methods available to influence, in a controllable manner, the AP-1 function in the living animal. Below, we present evidence that genes coding for such components of extracellular matrix remodeling system as TIMP-1 (tissue inhibitor of matrix metalloproteinases) and MMP-9 (matrix metalloproteinase-9) are both neuronal activity-driven and AP-1-dependent.

14.2 TIMP-1 and MMP-9 as AP-1 Target Genes in the Brain

TIMP-1 was originally identified by Nedivi, et al., (1993) in a methodical search of kainate-responsive genes as being driven by neuronal activity. Kainate is a glutamate analog that can be injected intra-peritoneally and after penetration into the brain potently activates neurons, especially within the limbic system (Ben-Ari and Cossart, 2000). In result, the activated cells release their own neurotransmitter glutamate that activates other neurons. Consequently, a massive neuronal activity manifested by seizures is observed. The seizures are recurrent, lasting up to at least a few hours, supported by overlapping waves of increased neuronal function. Because of this recurrent nature, kainate-evoked enhancement of neuronal activity is very strong and produced effects, including the phenomena of the gene expression, are multiple, overlapping and not synchronous. Furthermore, because of the severity of neuronal over-excitation, the pyramidal neurons of the CA1 and CA3 subfields of the hippocampus undergo cell death that is partially apoptotic (Filipkowski et al., 1994; Kaminska et al., 1994; Pollard et al., 1994; Sperk, 1994). On the other hand, the granule neurons of the dentate gyrus not only do survive the insult, but after losing their input (from the entorhinal cortex that also deteriorates) and target (CA3) produce aberrant synaptic connections on themselves, that can be taken as a (pathological) kind of neuronal plasticity (Sutula et al., 1998; Ben-Ari, 2001). Thus, studying the effects of peripheral administration of the kainate to rodents offers a unique opportunity to dissect hippocampal neuronal cell loss and plasticity (Zagulska-Szymczak et al., 2001).

The finding of neuronal activity-driven TIMP-1 expression has been confirmed and extended by Rivera et al., (1997) and Jaworski et al., (1999). Notably, the latter authors showed, in addition to kainate effect, that elevated TIMP-1 expression follows also the seizures elicited by the treatment with pentylenetetrazole (PTZ). PTZ acts by blocking an inhibitory neurotransmitter GABAA receptor, thus potentiating an endogenous excitation (Huang et al., 2001). Most importantly, the PTZ evokes just a single burst of seizures, presenting a model of a synchronous neuronal responsiveness to rapid enhancement of neuronal activity. Furthermore, the PTZ seizures do not produce neuronal cell death, but they are capable of eliciting long term neuronal plasticity as indicated by a phenomenon of chemical kindling (repeated injections of subconvulsive doses of PTZ that eventually result in fully blown seizures).

Jaworski et al. (1999) combined a variety of approaches to show that gene encoding TIMP-1 may be an AP-1 target in the rodent hippocampus in response to seizures. The evidence has been based on the fact that *timp-1* mRNA expression is: (i) promoter activity-driven (alternatively it could be via mRNA stabilization); (ii) it is subsequent to *c-fos* mRNA accumulation and (iii) prior protein synthesis-dependent as well as (iv) it is spatially and temporally correlated with c-Fos protein expression; (v) furthermore, the AP-1 transcription factor, containing c-Fos, as well as c-Jun, JunB, and JunD, is accumulated in stimulated hippocampi and is capable of binding the *timp-1* AP-1 responsive DNA regulatory element in a sequence-dependent manner; (vi) finally, glutamate is capable to activate expression of a reporter gene driven by *timp-1* promoter containing intact AP-1 site in cultured neurons of the dentate gyrus, and mutation of this promoter element abolishes the expression. These data are in a good agreement with findings reported for non-neuronal cells (Logan et al., 1996; Clark et al., 1997; Botelho et al., 1998).

Our interest in MMP-9 has been spurred by the abovementioned findings on TIMP-1 and the well recognized notion that this protein is an endogenous inhibitor of enzymatic activity of various matrix metalloproteinases (MMPs), with a high affinity for MMP-9 in particular. Hence, we have investigated whether some of the MMPs may be also neuronal activity-driven. In the study by (Szklarczyk et al., 2002) we have used the kainate model to show that indeed MMP-9 mRNA is massively accumulated (selectively) in the granule neurons of the dentate gyrus under those conditions.

Interestingly, MMP-9 is a well-defined AP-1 target in non-neuronal cells (Sato and Seiki, 1993; Fini et al., 1994; Chakraborti et al., 2003). Thus, we have recently looked whether MMP-9 gene is also regulated by AP-1 in neurons. We have employed the PTZ seizure model and found that (i) neuronal activity drives MMP-9 accumulation in the rat hippocampus; (ii) the MMP-9 gene promoter is foot-printed in the AP-1 binding region located within the gene promoter; (iii) the DNA sequence from this region is bound, as indicated by electrophoretic mobility shift/supershift assay, by AP-1 transcription factor; (iv) chromatin immunoprecipitated (ChIP) with anti-JunB antibody contains MMP-9 promoter DNA (Rylski et al., 2006). Thus, both TIMP-1 and MMP-9, which are functionally and intimately linked, are *in vivo* AP-1 targets in the brain. This raises an intriguing question about their biological roles in the brain phenomena driven by neuronal activity. We have addressed these recently, focusing on neuronal physiological and pathological (epileptogenesis) plasticity, learning and memory.

14.3 Functions of MMP-9 and TIMP-1 in Neural Plasticity – Lessons from Experimental Models

Although from a biochemical point of view MMP-9 and TIMP-1 perform apparently opposite actions, they both participate in the same set of mechanisms associated with remodeling of the brain extracellular matrix (Dzwonek et al., 2004; Shiosaka, 2004; Yong, 2005). Other molecules implicated in these mechanisms are serine proteases (e.g. tPA, neuropsin, neurotrypsin) (see: Molinari et al., 2002; Salles and Strickland, 2002; Matsumoto-Miyai et al., 2003) together with their inhibitors (neuroserpin) (see: Galliciotti and Sonderegger, 2006). For the last decade, a large body of evidence has been accumulated indicating that this whole system plays important roles in both physiological and pathological synaptic plasticity (Dzwonek et al., 2004; Shiosaka, 2004). Yet, MMP-9 and TIMP-1 remain unique as they are regulated by AP-1.

MMP-9 is a zinc-dependent endopeptidase that shares a modular design and a high degree of sequence homology with other matrix metalloproteinases (Sternlicht and Werb, 2001; Visse and Nagase, 2003). Together with most closely related MMP-2, it forms a subfamily of gelatinases named after their ability to digest gelatin, i.e. denatured form of collagen type I. There are multiple substrates of MMP-9 identified *in vitro*, however only a few of them have been confirmed as *bona fide in vivo* targets (Sternlicht and Werb, 2001; Visse and Nagase, 2003). For example, in the brain, only laminin and β-dystroglycan have been proven to be cleaved by MMP-9 (Gu et al., 2005; Agrawal et al., 2006). In a way similar to other MMPs, MMP-9 is a secreted enzyme that either dissipates into the extracellular spaces or remains pericellular, associated with poorly defined cell surface molecules (e.g., integrins, CD44) (see: Sternlicht and Werb, 2001; Visse and Nagase, 2003). The regulation of MMP-9 expression and activity involves multiple mechanisms, including transcriptional regulators, mRNA stability, posttranslational processing (e.g., glycosylation), activation by other enzymes (for example MMP-3 and plasmin *in vitro*) and by autocatalysis (that can be influenced by reactive oxygen species), and finally allosteric inhibition by TIMP-1 (Sternlicht and Werb, 2001; Visse and Nagase, 2003).

MMP-9 is expressed ubiquitously, albeit at low levels, in various organs and tissues, including the brain (Pagenstecher et al., 1997; Vecil et al., 2000), where it is produced mainly by neurons, and, to some extent, in astrocytes, and oligodendrocytes (Oh et al., 1999; Vaillant et al., 1999; Szklarczyk et al., 2002). There are conflicting pieces of evidence regarding its presence in resident microglia (Rosenberg et al., 2001; Jourquin et al., 2003, Wilczynski et al., unpublished results). Recently, we studied a subcellular localization of MMP-9 in the rat hippocampus, using high resolution morphological as well as biochemical approaches. At the ultrastructural level, MMP-9 immunoreactivity was present in a subset of dendritic spines bearing asymmetric (e.g. excitatory, glutamatergic) synapses (Fig. 1). At very high electron-microscopic magnification, colloidal gold particles indicating the presence of MMP-9 antibody molecules were observed as associated with the postsynaptic spine membrane, extending into the synaptic cleft (Fig. 1) (Wilczynski et al., 2004). The presence of MMP-9 in spines (based on light-microscopic study) was reported

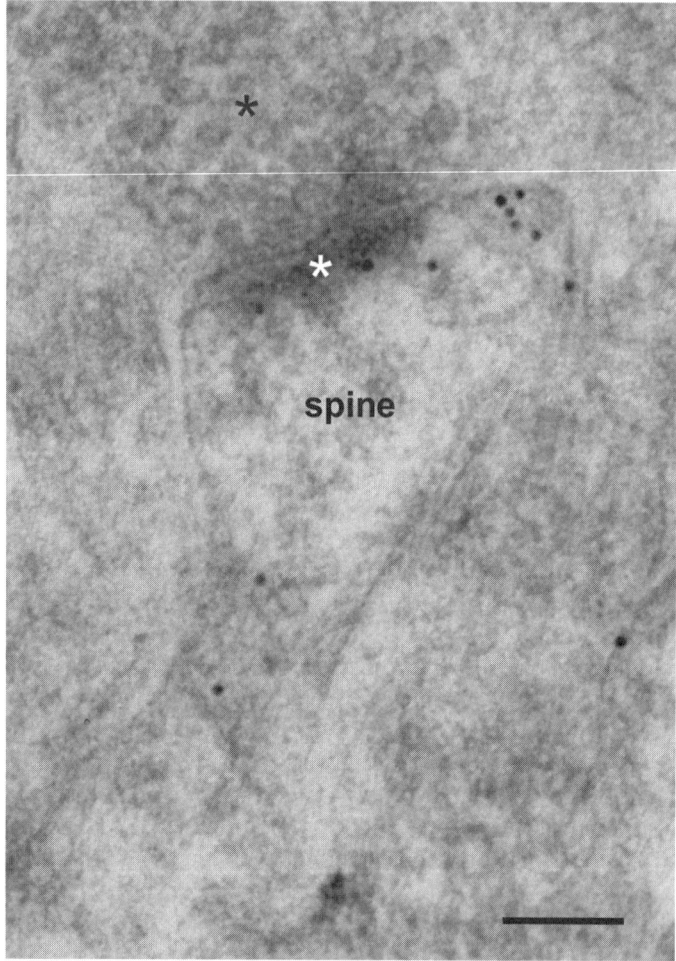

Fig. 1 Immunogold electron-microscopic visualization of MMP-9 in the hippocampal dendritic spine. Bar – 100 nm. Gold particle size – 10 nm. White asterisk – postsynaptic density. Black asterisk – synaptic vesicles

also in the cerebellum (Vaillant et al., 1999), whereas in neuronal culture MMP-9 was detected in growth cones (Shubayev and Myers, 2004).

TIMP-1 belongs to the family of tissue inhibitors of matrix metalloproteinases that has four members (Baker et al., 2002; Visse and Nagase, 2003). It preferentially binds to—and inhibits—MMP-9, although it has some influence on other MMPs too. In the brain, the major sources of TIMP-1 are astrocytes and neurons (Rivera et al., 1997). Unfortunately, the ultrastructural localization of TIMP-1 has not been determined, mainly because of the lack of suitable antibodies.

The prototypical situation in which MMP-9/TIMP-1 duo plays important roles in neural plasticity is brain development. MMP-9 and TIMP-1 mRNA levels in the whole brain are high immediately after birth, but they decrease to very low levels at the end of 1st and 3rd week respectively (Ayoub et al., 2005; Ulrich et al., 2005). It is of interest that this course of MMP-9/TIMP-1 expression is very similar to the course of the overall AP-1 binding activity in the brain (Pennypacker et al., 1993).

In the cerebellum, a decrease in MMP-9 mRNA was shown to be followed by a gradual decrease of its immunoreactivity and gelatinolytic activity (Ayoub et al., 2005). Notably, first three postnatal weeks are the period of intense tissue morphogenesis, including migration of neuronal precursor, their maturation and apoptotic death, as well as synaptogenesis (Luo, 2005). The specific role of MMP-9 in the development of cerebellum was revealed by studies of Vaillant et al., (2003) and Ayoub et al., (2005) who showed that blocking of MMP-9 activity either by the gene knockout (*in vivo*) or through the use of synthetic inhibitor or antibody (*in vitro*), hampers significantly the disappearance of cerebellar external granular layer. This is accomplished through the inhibition of both migration and apoptosis of granule cell precursors (Vaillant et al., 2003).

In the developing primary visual cortex of the rodents we have observed enhanced activity of MMP-9 (presumably synaptic as particularly pronounced within the neuropil) at the time of ocular dominance formation (3-5 weeks post-natally), characterized by massive refinement of neuronal connections (Szklarczyk and Kaczmarek, 2005). Incidentally, this was also in a good temporal correlation with AP-1 expression in the same brain area (Kaminska et al., 1995)

Seizures are very robust, albeit a pathophysiological, experimental paradigm to study cellular and molecular mechanisms of plasticity. *Status epilepticus* evoked through the injection of kainate (either systemic or intracerebral) results in very strong activation of limbic neuronal networks that causes either an extensive neuronal death (hippocampal Ammon's horn, entorhinal and piriform cortices, and amygdala) or intense and prolonged plastic response that occurs specifically in the hippocampal dentate gyrus (Sperk, 1994; Ben-Ari and Cossart, 2000). The seizures affect dentate granule cells in several ways including (i) a protracted synaptic potentiation (closely resembling LTP) that lasts for at least 1 day (Abegg et al., 2004), (ii) the elimination of dendritic spines, concomitant with the dissolution of presynaptic terminals originating from entorhinal neurons, persisting for several days (Nadler et al., 1980; Isokawa, 1998), and (iii) a sprouting of granule cells axons (mossy fibers) that seems to result from the death of their target (Ammon horn) cells; this phase is the longest; it starts after one week and continues for several months (Cantallops and Routtenberg, 1996; Suzuki et al., 1997; Sutula et al., 1998). Importantly, the sprouting is followed by autaptic synaptogenesis on the granule cells dendrites, thus forming recurrent excitatory connections that are believed to be an anatomical substrate of remote post-kainate epilepsy (Cronin and Dudek, 1988; Ben-Ari, 2001).

We and others have found that the plasticity occurring in the dentate gyrus is spatially and temporally correlated with a prolonged increases of MMP-9 as well as TIMP-1 expression/ activity (Rivera et al., 1997; Jaworski et al., 1999; Szklarczyk et al., 2002). Synaptic MMP-9 immunoreactivity and activity starts to be elevated 6 hours after kainate injection, peaks at 72 hours, and still remains very high after one week (Zhang et al., 1998; Szklarczyk et al., 2002;

Wilczynski et al., 2004). The MMP-9 mRNA increase does not precede, but follows (with a peak level about 24 hours after kainate administration) the rise in protein/activity and occurs not only in the granule cells bodies but also in their dendrites (Szklarczyk et al., 2002, Konopacki et al., submitted). Using electron microscopic in situ hybridization as well as quantitative RT-PCR assay on synaptosomal mRNA, we found the increase in MMP-9 mRNA levels even at the most distal branches of the dendritic tree, including spines, very close to the postsynaptic membranes (Konopacki et al., submitted). Thus, it seems that MMP-9 mRNA undergoes an activity-dependent dendritic transport that serves local protein expression at active (undergoing remodeling) synapses. It is of note, that the phenomenon of the dendritic mRNA translocation is widely recognized as being one of the most fundamental mechanisms underlying synaptic plasticity (Job and Eberwine, 2001; Steward and Schuman, 2003).

To explain an apparent reverse kinetics of MMP-9 protein vs. mRNA we propose the following scenario: (i) first, upon neuronal excitation, MMP-9 is released outside the cells from vesicular stores in spines to perform its task of synaptic modifications; (ii) this is followed by a homeostatic response of local translation of synaptic MMP-9 mRNA that replenishes synaptic protein content; (iii) the signal of the utilization of synaptic/dendritic mRNA is transported into the cell body and nucleus, thereby driving enhanced MMP-9 gene expression; (iv) the newly transcribed mRNA is transported upward the dendrites; (v) the mRNA cargo is delivered to the active synapses, thus not only replenishing MMP-9 content in those synapses in which it was utilized in the first place, but also equipping other active synapses with a molecule that serves to fulfill increased (upon seizures) demand of structural remodeling.

Regarding the functional aspects, our studies indicate that at the early period after kainate, MMP-9 is involved in the elimination of dendritic spines (see color Wilczynski Fig. 2E-I) (Wilczynski et al., submitted). This is evidenced by a much lower percentage of eliminated spines at 24 hours after kainate injection in MMP-9 knockout mice compared to wild-type ones (see color Wilczynski Fig. 2E). Most interestingly, the role of MMP-9 is apparently not restricted to early spine elimination. Our studies on the organotypic hippocampal cultures treated with kainate indicate that MMP-9 is also required for the formation of aberrant synapses between granule cells' axons and their own dendrites. (Wilczynski et al., submitted). Interestingly, this synaptogenesis have been shown to involve dendritic spine hypertrophy and/or proliferation (Suzuki et al., 1997).

The involvement of MMP-9 in two opposite processes of (i) spine elimination (early after kainate administration) and, (ii) spine growth (after several days) is puzzling. Nevertheless, it might be explained if the actual job of MMP-9 at spines was to release structural constrains imposed by the extracellular matrix and adhesion molecules on the spine flexibility. Therefore, in effect, MMP-9 action would facilitate either spine proliferation or spine pruning, depending on the net direction of the forces driving spine changes (see color Wilczynski Fig. 2J).

The deleterious consequence of events that are initiated by kainate-induced seizures is the development of secondary epilepsy, considered to be an adequate model of human temporal lobe epilepsy (TLE) (Ben-Ari, 2001). Another condition

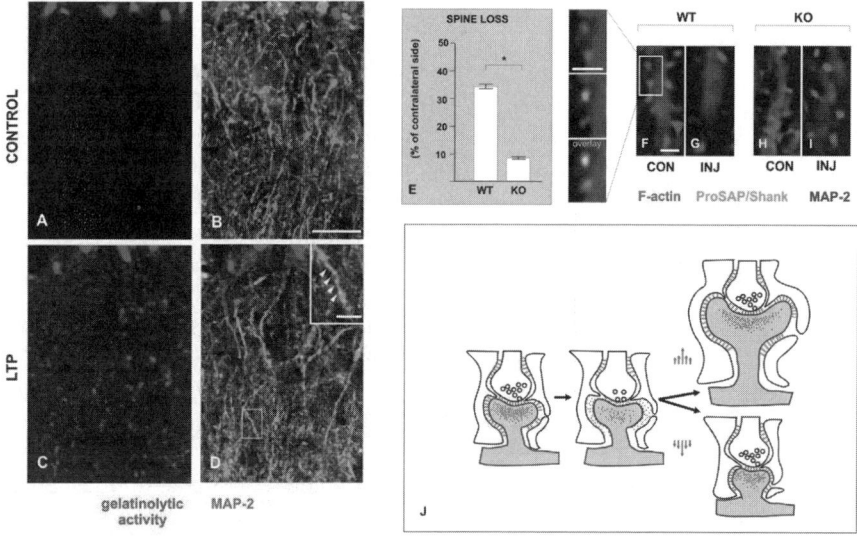

Fig. 2 A-D) Confocal microscopic image of a hippocampal slice (area CA1) processed for fluorescence in situ zymography demonstrates that long-term potentiation (LTP) induces gelatinolytic activity (red, compare A and C (single color), B and D (overlay)) in the dendritic area (dendrites green, B, D). At high magnification (inset in D) foci of gelatinolytic activity appear to be positioned along individual dendrites. Please note also strong gelatinolytic activity over the blood vessels that does not change in response to LTP (top of each panel). Bar: 75 μM, bar in inset: 15 μm. E-I) Quantitative analysis and fluorescent confocal visualization of dendritic spines MMP-9 KO and WT mice, at 24 hours after intraamygdalar kainate (KA) injection. E) Estimates of dendritic spine loss at the injected side (relative to the contralateral side) in MMP-9 WT and KO mice at 24 hours after intraamygdalar (KA) injection. F-I) Representative dendritic segments from WT (F, G) and KO (H, I) animals stained for F-actin (red) and MAP-2 (blue). Bar: 5 μM. Middle: An area from E (marked with white rectangle) showing colocalization of F-actin (red) and a postsynaptic density protein ProSAP/Shank (green) that indentifies dendrit spines. Yellow in overlay indicates colocalization. J) Hypothetical involvement of MMP-9 in remodeling of dendritic spines. Neurotransmitter-evoked excitation of a postsynaptic membrane results in the release of MMP-9 (red dots) from the spine into the perisynaptic space. MMP-9 degrades cell adhesion- and extracellular matrix molecules (green bars) thereby removing mechanical constraints imposed on spines by the neighboring tissue elements. Now, the driving forces resulting from cytoskeletal rearrangements are free to cause either spine growth (upper picture) or retraction (lower picture). (*See* Color Plate 8).

that may result in epilepsy is kindling that is repeated subthreshold excitation mediated either electrically or pharmacologically, leading to architectural changes in the hippocampus (Stafstrom and Sutula, 2005). It appears that MMP-9 plays an important role in the aforementioned processes. In particular, we found that in MMP-9 deficient animals, the rate of developing epilepsy upon pharmacological kindling (using PTZ, see above) is considerably slowed down (Wilczynski et al., submitted). This result also argues against the possibility that anti-epileptogenic effect of MMP-9 ablation is due to inhibition of MMP-9 mediated neurodegeneration (see below), as kindling is not associated with any major hippocampal degeneration.

Intriguingly, the spatiotemporal pattern of TIMP-1 expression upon kainate seizures is roughly similar to that of MMP-9 (Rivera et al., 1997; Jaworski et al., 1999). This brings about an important, yet unresolved, question why these two molecules, apparently working against each other, are induced simultaneously. While the subcellular localization of TIMP-1 in the hippocampus has not been revealed yet, it might be that the two proteins are spatially and/or temporally separated and TIMP-1 serves to fine-tune the action of MMP-9. Especially plausible is a hypothesis that TIMP-1 is released/activated just after the MMP-9 to prevent spreading the MMP-9 function too far and for too long. Additional, not exclusive, possibility is that the abundant expression of TIMP-1 in the dentate gyrus protects granule cells from the toxic influence of MMP-9. In fact, the death-promoting activity of MMP-9 has good experimental support from both *in vitro* and *in vivo* models as well as human diseases (Yong, 2005). Regarding the epileptogenesis, the recent study by Jourquin et al., (2005) has demonstrated that TIMP-1 knockout mice, that would rather be expected to be epilepsy-prone, appear in fact to be protected against kainate-evoked epilepsy. Certainly, more work is needed to elucidate complex relationship between MMP-9 and its inhibitor in the context of excitotoxic neuronal degeneration and epileptogenesis. Importantly, all the aforementioned conditions of increased MMP-9 and TIMP-1 expression/activity/action are accompanied by a persistently elevated expression of AP-1 constituent, and their increased binding activity (Kaminska et al., 1994; Jaworski et al., 1999). For example, it was shown that even one year after kainate-induced *status epilepticus* there is still substantially increased c-Fos-like immunoreactivity in the rat hippocampus (Bing et al., 1997).

The most direct experimental approach to study synaptic plasticity is an electrophysiological stimulation and recording of synaptic currents in a population of synapses in a brain slice or in the living animal. This way a long-term-potentiation (LTP) paradigm has been established as a model of activity dependent synaptic plasticity (Malenka and Bear, 2004). It is commonly believed that studies of LTP can give an insight into the biophysical and molecular mechanisms of learning and memory (Eichenbaum, 1996). Importantly, also LTP, especially in its late phase, appears to involve *c-fos* gene-expression (Kaczmarek, 1992). Thus, a hypothesis about the possible role of MMP-mediated extracellular proteolysis in synaptic plasticity, learning and memory has to be verified using electrophysiological methods. Accordingly, Nagy et al. (2006) have shown that MMP-9 is required for the development of the late-phase LTP (L-LTP) in the hippocampal slices The findings are documented by the following lines of evidence: (i) high-frequency stimulation that results in LTP, induces MMP-9 expression and activity (see color Wilczynski Fig. 2 A-D); (ii) inhibition of MMP-9 enzymatic activity prevents induction of L-LTP; (iii) in slices from MMP-9 knockout mice, LTP is impaired as far as both the magnitude and duration are concerned; (iv) addition of recombinant active MMP-9 to null-mutant slices restores the magnitude and duration of LTP to wild-type levels. Notably, MMP-9 appears to be required for L-LTP specifically, but not for short-term synaptic potentiation or long-term synaptic depression (LTD). Furthermore, inhibition of MMP-9 activity similarly prevents development of L-LTP in slices of the prefrontal cortex and TIMP-1 overexpression resulting from adenoviral delivery blocks L-LTP in the medial prefrontal cortex (evoked by subiculum stimulation) in the freely moving rats (Okulski et al., submitted). Importantly, stimuli that produce LTP also increase TIMP-1 expression (Nedivi, et al., 1993; Okulski et al., 2007).

Finally, the MMP-9/TIMP-1 system has recently been shown to play a role in learning and memory as defined in the behaving animals. First, Wright et al. (2003) have observed enhanced MMP-9 levels in the hippocampi of rats trained the water maze task. Next, Nagy et al. (2006) and Meighan et al., (2006) have shown that blocking of MMP-9 by various means (antisense oligonucleotides, broad-specificity MMP inhibitors, MMP-9 KO mice) impaired hippocampal learning in Morris water maze in rats and contextual fear conditioning in mice. Interestingly, the cued fear conditioning that is believed to be hippocampus-independent (Kim and Fanselow, 1992) was not affected in the MMP-9 KO mice (Nagy et al., 2006). As the cue-conditioning relies on the lateral amygdala, we have recently investigated whether L-LTP in this structure requires MMP-9 activity, indeed showing that it does not (Balcerzyk et al., 2006). Finally, TIMP-1 knockout mice also show learning deficits in the olfactory training paradigm that is also hippocampus dependent (Jourquin et al., 2005).

The question now arises, what are the molecular mechanisms of MMP-9/TIMP-1 dependent plasticity. The answer would have not only a great significance for advancing our knowledge on the brain function, but would also be helpful for the development of new pharmacological therapies to treat epilepsies. The pharmacological inhibition of MMP-9 should be especially feasible, since this is a molecule that acts in the extracellular space and it does not require the drug to be taken up by the neuronal cells. At the molecular level, MMP-9 action probably involves cleavage of peri- or intrasynaptic extracellular matrix, and/or synaptic adhesion molecules. In fact, degradation of laminin, a principal component of brain ECM has been shown to occur in various plasticity models, including kainate seizures (Chen and Strickland, 1997) and the development of the visual cortex (Oray et al., 2004). Notably, laminin is one of MMP-9 substrates (Gu et al., 2005). Other putative synaptic substrates of this enzyme are adhesion molecules. Dendritic spines are known to be anchored to other tissue elements by cadherins and integrins (Benson et al., 2000). Interestingly, outside the brain, these molecules appear to function as MMP substrates and/or receptors (Sternlicht and Werb, 2001). Furthermore, Nagy et al., (2006) have recently showed that an intact integrin function is required for MMP-9-mediated LTP. According to another scenario, MMP-9 could cleave, and thereby activate (or inhibit) a molecule involved in some aspects of functional plasticity. An obvious candidate for such a substrate is pro-BDNF, which is known to be cleaved by MMP-9 *in vitro* (Hwang et al., 2005). Finally, our most recent studies suggest that synaptic MMP-9 substrate in the brain is β-dystroglycan, a major component of dystrophin-associated transmembrane complex that is involved in both mechanical and signaling processes (Blake et al., 2002). A function of this complex, which is involved in the pathogenesis of Duchenne muscular dystrophy, in brain is largely unknown. We found that (i) MMP-9 colocalizes with β-dystroglycan at the postsynaptic domain of the hippocampal synapses; (ii) synaptic stimulation of cortical neurons in culture induces β-dystroglycan cleavage, which is prevented by MMP-9-specific inhibitor; (iii) enhanced β-dystroglycan cleavage occurs upon seizures in wild-type but not MMP-9 knockout animals (Kaczmarek et al., 2002; Michaluk et al., 2007). We have proposed previously that upon

MMP-9 action on β-dystroglycan, a conformation-encoded signal is transmitted in both post- and presynaptic directions *via* dystroglycan-neurexin interaction (Kaczmarek et al., 2002).

14.4 Summary and Conclusion

We have presented herein several lines of evidence indicating that MMP-9-mediated TIMP-1-regulated extracellular proteolysis is a novel important mechanism contributing to synaptic plasticity. An outstanding feature of this system is its dependence on the transcription factor AP-1. Both MMP-9 and TIMP-1 are AP-1 target genes in brain and the prominent feature of their expression is tight correlation with c-Fos protein expression and function. The MMP-9/TIMP-1 system appears to be involved in a broad range of physiological and pathological phenomena in various brain regions. These include developmental reorganization of the cerebellum and visual cortex, hippocampus-dependent learning, long-term potentiation, as well as pathogenesis of temporal lobe epilepsy. MMP-9 (modulated by TIMP-1) appears to exert its synaptic effects through a very direct influence on the shape of dendritic spines. However, at the molecular level, this action is still poorly understood; for example, although several candidates for synaptic MMP-9 substrate have been proposed, including β-dystroglycan, none of them has been demonstrated to exert a direct cleavage-related influence on synaptic efficacy. Finally, an unusual timecourse of the expression of MMP-9 mRNA vs. protein (reverse kinetics) suggests that AP-1-influenced MMP-9 gene transcription serves to replenish the synaptic pool of MMP-9, which is being rapidly depleted under the conditions of enhanced synaptic activity. This notion is further supported by the evidence of activity-driven dendritic transport of MMP-9 mRNA. The concept of "replenishment" as the major explanation for c-Fos function in the brain has already been discussed elsewhere (Kaczmarek and Chaudhuri, 1997; Kaczmarek, 2002).

In conclusion, we would like to propose an existence of a functional axis, consisting of AP-1/MMP-9/TIMP-1, together with a yet-to-be-discovered synaptic substrate(s) of MMP-9 that has a special liaison with neuronal activity. Firstly, it is activity-driven, and secondly, it carries on the activity signal to produce its effects on synaptic remodeling and subsequent plasticity.

Acknowledgments We thank Dr. Ole Petter Ottersen from the University of Oslo, Norway, for his help with electron-microscopic immunocytochemistry. This work has been supported by KBN grant P05A 22 024.

References

Abegg MH, Savic N, Ehrengruber MU, McKinney RA, Gahwiler BH (2004) Epileptiform activity in rat hippocampus strengthens excitatory synapses. J. Physiol. 554, 439–448.

Agrawal S, Anderson P, Durbeej M, van Rooijen N, Ivars F, Opdenakker G, Sorokin LM (2006) Dystroglycan is selectively cleaved at the parenchymal basement membrane at sites of leukocyte extravasation in experimental autoimmune encephalomyelitis. J. Exp. Med. 203, 1007–1019.

Anokhin KV (1997) Towards synthesis of systems and molecular genetics approaches to memory consolidation. J. Higher Nerv. Activ. 47, 157–169.

Ayoub AE, Cai TQ, Kaplan RA, Luo J (2005) Developmental expression of matrix metalloproteinases 2 and 9 and their potential role in the histogenesis of the cerebellar cortex. J. Comp. Neurol. 481, 403–415.

Baker AH, Edwards DR, Murphy G (2002) Metalloproteinase inhibitors: biological actions and therapeutic opportunities. J. Cell Sci. 115, 3719–3727.

Balcerzyk M, Konopacki F, Wilczynski GM, Kaczmarek L (2006) The role of matrix metalloproteinase – 9 in amygdalar long-term potentiation. In: 5th Forum of European Neuroscience. Vienna: FENS Abstr., vol.3, A085.1.

Ben-Ari Y (2001) Cell death and synaptic reorganizations produced by seizures. Epilepsia 42 Suppl 3, 5–7.

Ben-Ari Y, Cossart R (2000) Kainate, a double agent that generates seizures: two decades of progress. Trends Neurosci. 23, 580–587.

Benson DL, Schnapp LM, Shapiro L, Huntley GW (2000) Making memories stick: cell-adhesion molecules in synaptic plasticity. Trends Cell Biol. 10, 473–482.

Berridge M (1986) Second messenger dualism in neuromodulation and memory. Nature 323, 294–295.

Bing G, Wilson B, Hudson P, Jin L, Feng Z, Zhang W, Bing R, Hong JS (1997) A single dose of kainic acid elevates the levels of enkephalins and activator protein-1 transcription factors in the hippocampus for up to 1 year. Proc. Natl. Acad. Sci. USA 94, 9422–9427.

Blake DJ, Weir A, Newey SE, Davies KE (2002) Function and genetics of dystrophin and dystrophin-related proteins in muscle. Physiol. Rev. 82, 291–329.

Botelho FM, Edwards DR, Richards CD (1998) Oncostatin M stimulates c-Fos to bind a transcriptionally responsive AP-1 element within the tissue inhibitor of metalloproteinase-1 promoter. J. Biol. Chem. 273, 5211–5218.

Cantallops I, Routtenberg A (1996) Rapid induction by kainic acid of both axonal growth and F1/GAP-43 protein in the adult rat hippocampal granule cells. J. Comp. Neurol. 366, 303–319.

Chakraborti S, Mandal M, Das S, Mandal A, Chakraborti T (2003) Regulation of matrix metalloproteinases: an overview. Mol. Cell Biochem. 253, 269–285.

Chen ZL, Strickland S (1997) Neuronal death in the hippocampus is promoted by plasmin-catalyzed degradation of laminin. Cell 91, 917–925.

Clark IM, Rowan AD, Edwards DR, Bech-Hansen T, Mann DA, Bahr MJ, Cawston TE (1997) Transcriptional activity of the human tissue inhibitor of metalloproteinases 1 (TIMP 1) gene in fibroblasts involves elements in the promoter, exon 1 and intron 1. Biochem J. 324 (Pt 2), 611–617.

Cronin J, Dudek FE (1988) Chronic seizures and collateral sprouting of dentate mossy fibers after kainic acid treatment in rats. Brain Res. 474, 181–184.

Curran T, Morgan JI (1987) Memories of fos. Bioessays 7, 255–258.

Dolmetsch R (2003) Excitation-transcription coupling: signaling by ion channels to the nucleus. Sci. STKE 2003, PE4.

Dragunow M (1996) A role for immediate-early transcription factors in learning and memory. Behav. Genet. 26, 293–299.

Dragunow M, Robertson HA (1987) Kindling stimulation induces c-fos protein(s) in granule cells of the rat dentate gyrus. Nature 329, 441–442.

Dzwonek J, Rylski M, Kaczmarek L (2004) Matrix metalloproteinases and their endogenous inhibitors in neuronal physiology of the adult brain. FEBS Lett. 567, 129–135.

Eichenbaum H (1996) Learning from LTP: a comment on recent attempts to identify cellular and molecular mechanisms of memory. Learn. Mem. 3, 61–73.

Filipkowski RK, Hetman M, Kaminska B, Kaczmarek L (1994) DNA fragmentation in rat brain after intraperitoneal administration of kainate. Neuroreport 5, 1538–1540.

Fini ME, Bartlett JD, Matsubara M, Rinehart WB, Mody MK, Girard MT, Rainville M (1994) The rabbit gene for 92-kDa matrix metalloproteinase. Role of AP1 and AP2 in cell type-specific transcription. J. Biol. Chem. 269, 28620–28628.

Fleischmann A, Hvalby O, Jensen V, Strekalova T, Zacher C, Layer LE, Kvello A, Reschke M, Spanagel R, Sprengel R, Wagner EF, Gass P (2003) Impaired long-term memory and NR2A-type NMDA receptor-dependent synaptic plasticity in mice lacking c-Fos in the CNS. J. Neurosci. 23, 9116–9122.

Galliciotti G, Sonderegger P (2006) Neuroserpin. Front Biosci 11, 33–45.

Goelet P, Castellucci VF, Schacher S, Kandel ER (1986) The long and the short of long-term memory–a molecular framework. Nature 322, 419–422.

Greenberg ME, Shyu AB, Belasco JG (1990) Deadenylylation: a mechanism controlling c-fos mRNA decay. Enzyme 44, 181–192.

Gu Z, Cui J, Brown S, Fridman R, Mobashery S, Strongin AY, Lipton SA (2005) A highly specific inhibitor of matrix metalloproteinase-9 rescues laminin from proteolysis and neurons from apoptosis in transient focal cerebral ischemia. J. Neurosci. 25, 6401–6408.

Herdegen T, Leah JD (1998) Inducible and constitutive transcription factors in the mammalian nervous system: control of gene expression by Jun, Fos and Krox, and CREB/ATF proteins. Brain Res. Brain Res. Rev. 28, 370–490.

Herrera DG, Robertson HA (1996) Activation of c-fos in the brain. Prog. Neurobiol. 50, 83–107.

Hess J, Angel P, Schorpp-Kistner M (2004) AP-1 subunits: quarrel and harmony among siblings. J. Cell Sci. 117, 5965–5973.

Huang RQ, Bell-Horner CL, Dibas MI, Covey DF, Drewe JA, Dillon GH (2001) Pentylenetetrazole-induced inhibition of recombinant gamma-aminobutyric acid type A (GABA(A)) receptors: mechanism and site of action. J. Pharmacol. Exp. Ther 298, 986–995.

Hughes P, Dragunow M (1995) Induction of immediate-early genes and the control of neurotransmitter-regulated gene expression within the nervous system. Pharmacol. Rev. 47, 133–178.

Hunt SP, Pini A, Evan G (1987) Induction of c-fos-like protein in spinal cord neurons following sensory stimulation. Nature 328, 632–634.

Hwang JJ, Park MH, Choi SY, Koh JY (2005) Activation of the Trk signaling pathway by extracellular zinc. Role of metalloproteinases. J. Biol. Chem. 280, 11995–12001.

Isokawa M (1998) Remodeling dendritic spines in the rat pilocarpine model of temporal lobe epilepsy. Neurosci. Lett. 258, 73–76.

Jaworski J, Biedermann IW, Lapinska J, Szklarczyk A, Figiel I, Konopka D, Nowicka D, Filipkowski RK, Hetman M, Kowalczyk A, Kaczmarek L (1999) Neuronal excitation-driven and AP-1-dependent activation of tissue inhibitor of metalloproteinases-1 gene expression in rodent hippocampus. J. Biol. Chem. 274, 28106–28112.

Job C, Eberwine J (2001) Localization and translation of mRNA in dendrites and axons. Nat. Rev. Neurosci. 2, 889–898.

Jourquin J, Tremblay E, Decanis N, Charton G, Hanessian S, Chollet AM, Le Diguardher T, Khrestchatisky M, Rivera S (2003) Neuronal activity-dependent increase of net matrix metalloproteinase activity is associated with MMP-9 neurotoxicity after kainate. Eur. J. Neurosci. 18, 1507–1517.

Jourquin J, Tremblay E, Bernard A, Charton G, Chaillan FA, Marchetti E, Roman FS, Soloway PD, Dive V, Yiotakis A, Khrestchatisky M, Rivera S (2005) Tissue inhibitor of metalloproteinases-1 (TIMP-1) modulates neuronal death, axonal plasticity, and learning and memory. Eur. J. Neurosci. 22, 2569–2578.

Kaczmarek L (1986) Protooncogene expression during the cell cycle. Lab. Invest. 54, 365–376.

Kaczmarek L (1990) Molecular biology of long lasting memory formation. ESF Scientific Networks: Zeist 1–3 December, 1989.

Kaczmarek L (1992) Expression of c-fos and other genes encoding transcription factors in long-term potentiation. Behav. Neural Biol. 57, 263–266.

Kaczmarek L (1993) Molecular biology of vertebrate learning: Is c-fos a new beginning? J. Neurosci. Res., 34, 377–381.

Kaczmarek L, Kaminska B (1989) Molecular biology of cell activation. Exp. Cell Res. 183, 24–35.

Kaczmarek L, Nikolajew E (1990) c-fos protooncogene expression and neuronal plasticity. Acta Neurobiol Exp (Wars) 50, 173–179.

Kaczmarek L, Chaudhuri A (1997) Sensory regulation of immediate-early gene expression in mammalian visual cortex: implications for functional mapping and neural plasticity. Brain Res. Brain Res. Rev. 23, 237–256.

Kaczmarek L, Siedlecki JA, Danysz W (1988) Proto-oncogene c-fos induction in rat hippocampus. Brain Res. 427, 183–186.

Kaczmarek L, Lapinska-Dzwonek J, Szymczak S (2002) Matrix metalloproteinases in the adult brain physiology: a link between c-Fos, AP-1 and remodeling of neuronal connections? EMBO J. 21, 6643–6648.

Kalita K, Kharebava G, Zheng J-J, Hetman M (2006) Role of MKL1 in ERK1/2-dependent stimulation of SRF-driven transcription in response to BDNF or increased synaptic activity. J. Neurosci. 26, 10020–10032.

Kaminska B, Kaczmarek L, Zangenehpour S, Chaudhuri A (1999) Rapid phosphorylation of Elk-1 transcription factor and activation of MAP kinase signal transduction pathways in response to visual stimulation. Mol. Cell Neurosci. 13, 405–414.

Kaminska B, Pyrzynska B, Ciechomska I, Wisniewska M (2000) Modulation of the composition of AP-1 complex and its impact on transcriptional activity. Acta Neurobiol Exp (Wars) 60, 395–402.

Kaminska B, Mosieniak G, Gierdalski M, Kossut M, Kaczmarck L (1995) Elevated AP-1 transcription factor DNA binding activity at the onset of functional plasticity during development of rat sensory cortical areas. Brain Res. Mol. Brain Res. 33, 295–304.

Kaminska B, Filipkowski RK, Zurkowska G, Lason W, Przewlocki R, Kaczmarek L (1994) Dynamic changes in the composition of the AP-1 transcription factor DNA-binding activity in rat brain following kainate-induced seizures and cell death. Eur J. Neurosci. 6, 1558–1566.

Kim JJ, Fanselow MS (1992) Modality-specific retrograde amnesia of fear. Science 256, 675–677.

Knapska E, Kaczmarek L (2004) A gene for neuronal plasticity in the mammalian brain: Zif268/Egr-1/NGFI-A/Krox-24/TIS8/ZENK? Prog. Neurobiol. 74, 183–211.

Kovacs KJ (1998) c-Fos as a transcription factor: a stressful (re)view from a functional map. Neurochem. Int. 33, 287–297.

Logan SK, Garabedian MJ, Campbell CE, Werb Z (1996) Synergistic transcriptional activation of the tissue inhibitor of metalloproteinases-1 promoter via functional interaction of AP-1 and Ets-1 transcription factors. J. Biol. Chem. 271, 774–782.

Luo J (2005) The role of matrix metalloproteinases in the morphogenesis of the cerebellar cortex. Cerebellum 4, 239–245.

Maleeva NE, Ivolgina GL, Anokhin KV, Limborskaia SA (1989) [Analysis of the expression of the c-fos proto-oncogene in the rat cerebral cortex during learning]. Genetika 25, 1119–1121.

Maleeva NE, Bikbulatova LS, Ivolgina GL, Anokhin KV, Limborskaia SA, Kruglikov RI (1990) [Activation of the c-fos proto-oncogene in different structures of the rat brain during training and pseudoconditioning]. Dokl Akad Nauk SSSR 314, 762–764.

Malenka RC, Bear MF (2004) LTP and LTD: an embarrassment of riches. Neuron 44, 5–21.

Matsumoto-Miyai K, Ninomiya A, Yamasaki H, Tamura H, Nakamura Y, Shiosaka S (2003) NMDA-dependent proteolysis of presynaptic adhesion molecule L1 in the hippocampus by neuropsin. J. Neurosci. 23, 7727–7736.

Michaluk P, Kolodziej L, Mioduszewska B, Wilczynski GM, Dzwonek J, Jaworski J, Gorecki DC, Ottersen OP, Kaczmarek L (2007) Beta-dystroglycan as a target for MMP-9, in response to enhanced neuronal activity. J. Biol. Chem. 282, 16036–16041.

Molinari F, Rio M, Meskenaite V, Encha-Razavi F, Auge J, Bacq D, Briault S, Vekemans M, Munnich A, Attie-Bitach T, Sonderegger P, Colleaux L (2002) Truncating neurotrypsin mutation in autosomal recessive nonsyndromic mental retardation. Science 298, 1779–1781.

Nadler JV, Perry BW, Gentry C, Cotman CW (1980) Loss and reacquisition of hippocampal synapses after selective destruction of CA3-CA4 afferents with kainic acid. Brain Res. 191, 387–403.

Nedivi E, Hevroni D, Naot D, Israeli D, Citri Y (1993) Numerous candidate plasticity-related genes revealed by differential cDNA cloning. Nature 363, 718–722.

Nikolaev E, Werka T, Kaczmarek L (1992) C-fos protooncogene expression in rat brain after long-term training of two-way active avoidance reaction. Behav. Brain Res. 48, 91–94.

Oh LY, Larsen PH, Krekoski CA, Edwards DR, Donovan F, Werb Z, Yong VW (1999) Matrix metalloproteinase-9/gelatinase B is required for process outgrowth by oligodendrocytes. J. Neurosci. 19, 8464–8475.

Okulski P, Jay TM, Jaworski J, Duniec K, Dzwonek J, Konopacki FA, Wilczynski GM, Sánchez-Capelo A, Mallet J, Kaczmarek L (2007) TIMP-1 abolishes MMP-9-dependent long-lasting longterm potentiation in the prefrontal cortex. Biol. Psych. [Epub ahead of print].

Oray S, Majewska A, Sur M (2004) Dendritic spine dynamics are regulated by monocular deprivation and extracellular matrix degradation. Neuron 44, 1021–1030.

Pagenstecher A, Stalder AK, Campbell IL (1997) RNAse protection assays for the simultaneous and semiquantitative analysis of multiple murine matrix metalloproteinase (MMP) and MMP inhibitor mRNAs. J. Immunol. Meth. 206, 1–9.

Pennypacker KR, Dreyer D, Hong JS, McMillian MK (1993) Elevated basal AP-1 DNA binding activity in developing rat brain. Brain Res. Mol. Brain Res. 19, 349–352.

Platenik J, Kuramoto N, Yoneda Y (2000) Molecular mechanisms associated with long-term consolidation of the NMDA signals. Life Sci. 67, 335–364.

Pollard H, Charriaut-Marlangue C, Cantagrel S, Represa A, Robain O, Moreau J, Ben-Ari Y (1994) Kainate-induced apoptotic cell death in hippocampal neurons. Neuroscience 63, 7–18.

Rivera S, Tremblay E, Timsit S, Canals O, Ben-Ari Y, Khrestchatisky M (1997) Tissue inhibitor of metalloproteinases-1 (TIMP-1) is differentially induced in neurons and astrocytes after seizures: evidence for developmental, immediate early gene, and lesion response. J. Neurosci. 17, 4223–4235.

Rosenberg GA, Cunningham LA, Wallace J, Alexander S, Estrada EY, Grossetete M, Razhagi A, Miller K, Gearing A (2001) Immunohistochemistry of matrix metalloproteinases in reperfusion injury to rat brain: activation of MMP-9 linked to stromelysin-1 and microglia in cell cultures. Brain Res. 893, 104–112.

Rylski M, Kaczmarek L (2004) Ap-1 targets in the brain. Front. Biosci. 9, 8–23.

Rylski M, Saganek R, Bielinska B, Konopacki F, Wilczynski GM, Kaczmarek L (2006) Yin yang 1 is expressed in the brain and regulates in vivo matrix metalloproteinase 9 expression in the rat hippocampus. In: 5th Forum of European Neuroscience. Vienna: FENS Abstr., vol.3, A085.19

Salles FJ, Strickland S (2002) Localization and regulation of the tissue plasminogen activator-plasmin system in the hippocampus. J. Neurosci. 22, 2125–2134.

Sato H, Seiki M (1993) Regulatory mechanism of 92 kDa type IV collagenase gene expression which is associated with invasiveness of tumor cells. Oncogene 8, 395–405.

Shiosaka S (2004) Serine proteases regulating synaptic plasticity. Anat. Sci. Int. 79, 137–144.

Shubayev VI, Myers RR (2004) Matrix metalloproteinase-9 promotes nerve growth factor-induced neurite elongation but not new sprout formation in vitro. J. Neurosci. Res 77, 229–239.

Sperk G (1994) Kainic acid seizures in the rat. Prog. Neurobiol. 42, 1–32.

Stafstrom CE, Sutula TP (2005) Models of epilepsy in the developing and adult brain: implications for neuroprotection. Epilepsy Behav. 7 Suppl 3, S18–24.

Sternlicht MD, Werb Z (2001) How matrix metalloproteinases regulate cell behavior. Annu. Rev. Cell Dev. Biol. 17, 463–516.

Steward O, Schuman EM (2003) Compartmentalized synthesis and degradation of proteins in neurons. Neuron 40, 347–359.

Sutula T, Zhang P, Lynch M, Sayin U, Golarai G, Rod R (1998) Synaptic and axonal remodeling of mossy fibers in the hilus and supragranular region of the dentate gyrus in kainate-treated rats. J. Comp. Neurol. 390, 578–594.

Suzuki F, Makiura Y, Guilhem D, Sorensen JC, Ontenient B (1997) Correlated axonal sprouting and dendritic spine formation during kainate-induced neuronal morphogenesis in the dentate gyrus of adult mice. Exp. Neurol. 145, 203–213.

Szklarczyk A, Kaczmarek L (2005) Physiology of matrix MMPs and their tissue inhibitors in the brain. BIOTECH Intl. 17, 15–18.

Szklarczyk A, Lapinska J, Rylski M, McKay RD, Kaczmarek L (2002) Matrix metalloproteinase-9 undergoes expression and activation during dendritic remodeling in adult hippocampus. J. Neurosci. 22, 920–930.

Tischmeyer W, Grimm R (1999) Activation of immediate early genes and memory formation. Cell. Mol. Life Sci. 55, 564–574.

Tischmeyer W, Kaczmarek L, Strauss M, Jork R, Matthies H (1990) Accumulation of c-fos mRNA in rat hippocampus during acquisition of a brightness discrimination. Behav. Neural Biol. 54, 165–171.

Ulrich R, Gerhauser I, Seeliger F, Baumgartner W, Alldinger S (2005) Matrix metalloproteinases and their inhibitors in the developing mouse brain and spinal cord: a reverse transcription quantitative polymerase chain reaction study. Dev. Neurosci. 27, 408–418.

Vaillant C, Didier-Bazes M, Hutter A, Belin MF, Thomasset N (1999) Spatiotemporal expression patterns of metalloproteinases and their inhibitors in the postnatal developing rat cerebellum. J. Neurosci. 19, 4994–5004.

Vaillant C, Meissirel C, Mutin M, Belin MF, Lund LR, Thomasset N (2003) MMP-9 deficiency affects axonal outgrowth, migration, and apoptosis in the developing cerebellum. Mol. Cell Neurosci. 24, 395–408.

Vecil GG, Larsen PH, Corley SM, Herx LM, Besson A, Goodyer CG, Yong VW (2000) Interleukin-1 is a key regulator of matrix metalloproteinase-9 expression in human neurons in culture and following mouse brain trauma in vivo. J. Neurosci. Res. 61, 212–224.

Visse R, Nagase H (2003) Matrix metalloproteinases and tissue inhibitors of metalloproteinases: structure, function, and biochemistry. Circ. Res. 92, 827–839.

Vosatka RJ, Hermanowski-Vosatka A, Metz R, Ziff EB (1989) Dynamic interactions of c-fos protein in serum-stimulated 3T3 cells. J. Cell Physiol. 138, 493–502.

Werlen G, Belin D, Conne B, Roche E, Lew DP, Prentki M (1993) Intracellular Ca2+ and the regulation of early response gene expression in HL-60 myeloid leukemia cells. J. Biol. Chem. 268, 16596–16601.

Wilczynski GM, Konopacki FA, Wilczek E, Rylski M, Z. Lasiecka L (2004) Expression and subcellular localization of matrix metalloproteinase 9 (MMP-9) and its mRNA in rat brain after experimental status epilepticus. In: Society for Neuroscience Annual Meeting. San Diego: Abstract Viewer/Itinerary Planner, http://sfn.scholarone.com/itin2004/, Online.

Xia Z, Dudek H, Miranti CK, Greenberg ME (1996) Calcium influx via the NMDA receptor induces immediate early gene transcription by a MAP kinase/ERK-dependent mechanism. J. Neurosci. 16, 5425–5436.

Yong VW (2005) Metalloproteinases: mediators of pathology and regeneration in the CNS. Nat. Rev. Neurosci. 6, 931–944.

Zagulska-Szymczak S, Filipkowski RK, Kaczmarek L (2001) Kainate-induced genes in the hippocampus: lessons from expression patterns. Neurochem. Int. 38, 485–501.

Zhang JW, Deb S, Gottschall PE (1998) Regional and differential expression of gelatinases in rat brain after systemic kainic acid or bicuculline administration. Eur. J. Neurosci. 10, 3358–3368.

Chapter 15
Activity-dependent Gene Transcription in Neurons: Defining the Plasticity Transcriptome

Alison L. Barth and Lina Yassin

Abstract Transcription and translation are required for consolidation of long-lasting changes in synaptic function and are required for learning and memory. The targets of activity-dependent transcription in neurons have been of great interest. Despite this, the ultimate consequences of an activity-dependent change in programs of gene expression with respect to neural function have been surprisingly elusive. For example, experimental data have not clearly established how gene expression is required for memory-associated events, such as synapse-specific strengthening or changes in the input-output function of the neuron. Many activity-regulated genes are transcriptional factors which themselves modify programs of downstream gene expression, and the short- and long-term consequences of these changes in gene expression remain largely unknown. Although there have been many complementary approaches to experimentally and computationally define the plasticity transcriptome, the specific gene targets that have been identified using different experimental approaches show surprisingly little overlap. The purpose of this review is three-fold: 1) to discuss what is known about activity-dependent transcription during learning, 2) to review efforts to identify genes in the plasticity transcriptome, and 3) to develop a hypothesis about how transcription might be required for information coding at the synaptic and cellular level.

15.1 History

More than thirty years of experimental work has firmly established that neural activity is an effective stimulus at driving nuclear gene transcription, and that gene transcription and translation are required for learning and memory. In early *in vivo* experiments using both invertebrates, injection of transcription inhibitors was sufficient to block facilitation in Aplysia (Barzilai et al., 1989). In some mammalian experiments, systemic injection of inhibitors was used, and although the treatment was somewhat drastic, repetition by a large number of independent investigators using different compounds and treatment paradigms made it possible to draw some strong conclusions from these studies (see for example, Barondes and Jarvik, 1964; Flexner et al., 1963). From early on, it was clear that the identity of genes whose

S. M. Dudek (ed.), *Transcriptional Regulation by Neuronal Activity.*
© Springer 2008

expression was required for memory storage would be of substantial interest in understanding the cellular mechanisms that underlie memory (Goelet et al., 1986).

15.1.2 The Role of Transcription and Translation in Plasticity

Although both transcription and translation are required for long-lasting synaptic change, the identity and translational regulation of dendritically localized RNAs has been the subject of particular focus over the past decade. There is an expanding number of protein-encoding mRNAs that have been shown to actively transport to the dendrites, as well as a perhaps larger group of non-coding RNAs, such as miRNAs, that can repress translation of specific targets at the synapse. However, most of these RNAs appear to be constitutively expressed, with only a few exceptions showing both activity-dependent induction and activity-dependent increases in protein levels at the synapse. Thus, it remains unclear how activity-dependent transcription and local translation interact to facilitate synapse-specific strengthening. This issue is central to our understanding of how events at the nucleus are coupled to input specific changes, and will be discussed in more detail at the end of this review.

15.1.3 Activity-dependent Induction of Transcription Factor mRNAs

Extracellular stimulation results in gene expression, and these "immediate-early genes" (IEGs) are defined as being those transcripts that were upregulated in the presence of protein translation inhibitors, implying that they were the products of an initial wave of transcriptional activation (Cochran et al., 1983). Early studies of stimulus-transcription coupling identified a number of transcription factors that could be upregulated by growth factor application in different types of cultured cells. Later, Morgan and Curran (Morgan et al., 1987) showed that neural activity and voltage-dependent Ca^{++}-influx could be potent inducers of the transcription factor Fos (c-fos) in cultured PC12 cells. Further studies established that neural activity could induce expression of specific genes ranging from transcription factors such as *zif268*, *junB*, and *c-jun* (Saffen et al., 1988) to structural genes such as *clathrin heavy chain*, *secretogranin* (Nedivi et al., 1993), Arc (Link et al., 1995; Lyford et al., 1995), and *neuritin/cpg-15* (Nedivi et al., 1998) within relatively short time scales, ranging from minutes to hours after stimulation onset (Table 1).

   Although early studies were carried out in neural-like cell lines, chemoconvulsant-induced seizures were soon employed as an inducing stimulus *in vivo* (Morgan et al., 1987), and later, sensory experience was used after it was noted that it could also be a potent inducer of some gene products, specifically *arc* (Lyford et al., 1995) and *zif268* (Kaminska et al., 1996). The substantial diversity between induction protocols (system-wide, coordinated activity induced by chemoconvulsants) versus

Table 1 Immediate-Early Genes

Protein	Reference
Transcription factors	
c-Jun	(Sheng and Greenberg, 1990)
c-Fos	(Sheng and Greenberg, 1990)
FosB	(Sheng and Greenberg, 1990)
Zif268	(Sheng and Greenberg, 1990)
CREM	(Foulkes et al., 1991)
Synaptic	
Arc	(Lyford et al., 1995)
HOMER	(Xiao et al., 1998)
Synaptotagmin-related protein	(Babity et al., 1997)
Intracellular signaling	
MKP1 phosphatase	(Qian et al., 1994)
RGS2 (GTPase Activating Protein)	(Ingi et al., 1998)
Neurosecretion	
Proenkephalin	(Sonnenberg et al., 1989)
BDNF	(Lauterborn et al., 1996)
Prostaglandin synthase (COX2)	(Yamagata et al., 1993)
Extracellar adhesion	
Tissue plasminogen activator	(Qian et al., 1993)

more naturalistic stimuli such as modified visual experience raises the question of whether programs of gene expression induced by such varying stimuli may be directly comparable (see below).

15.2 Activity-induced mRNAs

The transcriptional requirement for long-lasting forms of memory begs the question of what are the specific genes that are regulated by activity? Approaches to identify the genes that are regulated as part of the plasticity transcriptome have varied, and there are likely to be stimulus-specific and time-dependent effects. In this section, we will review a number of different approaches to identify genes that are associated with plasticity, ranging from cDNA subtraction assays, microarray analysis, chromatin-immunoprecipitation assays, and computational genomics.

From a functional standpoint, analysis of mRNA transcripts that are enriched by activity has been a very direct way to identify potential genes of interest, and this method was employed in many early studies that identified some of the most well-known activity-regulated genes. A number of independent investigations have carried out screens to identify activity-regulated genes, using cDNA subtraction or more recently, DNA microarray analysis. In many cases, chemoconvulsant-induced seizures have been employed (rather than some training protocol that is more directly tied to behaviorally-relevant learning). However, more that a decade of work suggests that the genes that are induced following epileptic seizures are overlapping with those genes induced by synaptic activity that occurs during development, behavioral experience, or plasticity (i.e. long-term potentiation; Corriveau et al., 1999; Elliott et al., 2003; Nedivi et al., 1998). For example, the candidate plasticity gene *cpg-15*, also known as neuritin, is a secreted protein that facilitates synaptic formation, and was first identified by a cDNA subtraction screen on hippocampal tissue from kainic acid-injected rats (Nedivi et al., 1993). Another gene identified in this search, *cpg-2*, is involved in the endocytosis of glutamate receptors at the synapse, a process thought to be important for maintenance of synaptic strength onto a given cell (Cottrell et al., 2004).

Although many of these screens have identified a common subset of regulated genes, small differences in the induction protocol or the timing of sample collection have led to some diversity in the results. Programs of induced gene expression can be highly dynamic; for example, following chemoconvulsant-induced seizures in rodents, microarray analysis shows that gene expression remain abnormal for hours to weeks following the stimulus (Lukasiuk et al., 2003). In addition, genes that show less than a threshold value of regulation (typically, several-fold) may have been overlooked using this type of analysis. Thus, divergence in experimental procedures and analysis complicate efforts for a large-scale synthesis of these data.

15.3 Target Prediction using Transcription Factor-DNA Interactions

The identification of specific transcription factors associated with activity-dependent gene expression has enabled use of the specific binding sites for these transcription factors to isolate potential downstream gene targets. The most prominent transcription factor whose activation is linked to neural activity and learning has been CREB. Although CREB is constitutively expressed, its phosphorylation is required for transcriptional activation (reviewed by Shaywitz and Greenberg 1999), and this has been linked to activity-dependent gene transcription required for learning in *Aplysia* (Dash et al., 1990), *Drosophila* (Yin et al., 1994), and mice (Barco et al., 2002; Barco et al., 2005; Glazewski et al., 1999; Josselyn et al., 2004; Kida et al., 2002). Thus, it has been of particular interest to identify the gene targets regulated by CREB. Target identification has been carried out using case-by-case approaches, which have identified dozens of candidate genes (reviewed by West et al., 2002), and microarray analyses of genes expressed following induction of a constitutively activated form of CREB, VP-16 CREB (Barco et al., 2005).

Other approaches have included ChIP-on chip assays using anti-CREB antibodies against a subset of the genome (Euskirchen et al., 2004), unbiased approaches such as serial analyses of chromatin occupancy across the entire genome (SACO; (Impey et al., 2004), or purely computational analysis of gene promoter regions (Zhang et al., 2005; and Pfenning et al., 2007). Interestingly, both Impey et al., and *in silico* analyses identified a large number of novel CREB targets, but these novel targets have not been identified in all studies. This suggests that CREB site occupancy in some cell types is broader than might be expected by a site-prediction analysis. In some cases this may be due to CREB's promiscuous binding to AP-1 sites – a similar but non-identical binding consensus motif. CREB binding site occupancy may depend on the identity of the starting tissue (Cha-Molstad et al., 2004) where genes that have well-documented CREs nonetheless do not show CREB binding (i.e. somatostatin; see Impey *et al.,*) . Indeed, established CREB target genes such as tyrosine hydroxylase are not uniformly upregulated in all tissue types with stimuli that result in sustained CREB activation, suggesting that the transcriptional code for CREB-mediated gene expression is more complex than commonly conceived. These results imply that expression is not initiated at all CREB gene targets simply because the transcription factor is activated by phosphorylation. In short, just because a promoter has a CREB site doesn't mean that expression will be induced when nuclear CREB is phosphorylated.

Despite the enormous body of data that has accumulated using these different methods, synthesis of results from these searches has been difficult because of differences in gene nomenclature between analyses and between species (see also Lukasiuk and Pitkanen, 2004). Furthermore, differences between experimental approaches make it difficult to directly compare results. For example, in computational analyses of CREB sites in promoter regions, the particular consensus motif used and the stringency criterion employed to designate a gene as a putative target can differ widely, resulting in variability between studies (Zhang et al., 2005). Even in case-by-case investigations showing cAMP regulation of particular target genes, experiments have frequently been done under highly derived experimental conditions that can lead to the overestimation of potential CREB regulation of a particular target gene (for example, stimulation using the adenylate cyclase activator forskolin or prolonged overexpression of a superactivated form of CREB; Barco et al., 2002). Importantly, almost all approaches have not been sensitive to CREB binding sites that lie within intronic regions or that employ an internal promoter, a fact that can lead to a systematic underestimation of CREB target genes (although the function of CREB sites outside a promoter region may be distinct).

Despite all these caveats, a few genes have consistently come up as likely CREB targets and are worth summarizing. In almost all studies, *c-fos* and *BDNF* have been identified as strong CREB targets. Although somatostatin and tyrosine hydroxylase unequivocally show CREB-dependent regulation, they have not appeared in all analyses.

Binding site identification for transcription factors, such as *c-fos* and *zif268*, is an additional way to identify genes that can be regulated by activity, although care must be taken to use binding site consensus motifs that are not biased toward a few specific examples or are too degenerate to yield meaningful results (as has been the case with AP-1, the Fos-containing transcription factor whose binding site looks similar to a CREB site but has considerably more consensus flexibility, likely to be due to the fact that AP-1 is a dimer that can be composed of many different subunits).

Other constitutively-expressed transcription factors whose activity has been linked to synaptic function and neural activity are the serum response factor

(SRF/Elk-1). SRF is a transcription factor that binds to the serum response element, is sensitive to increases in intracellular Ca^{++} (Miranti et al., 1995) and is activated during long-lasting forms of LTP (Davis et al., 2000). The activity of MEF-2, another transcription factor that is regulated by increases in intracellular Ca^{++}, has been found to suppress synapse number, especially during development (Flavell et al., 2006). Studies to computationally determine binding sites for activity-related transcription factors can thus yield useful insights into directions for further analysis.

15.4 Dynamics of the Plasticity Transcriptome

How dynamic is the plasticity transcriptome? Is it a moving target that can vary from minutes to hours following an inducing stimulus? Does it depend on tissue type and genetic background? The answer to these questions appears to be that the subset of genes that are induced by neural activity can be incredibly variable, responding not only to the immediate stimulus but also taking into account the prior activity history of the cell, and likely depends on developmental age and cell type.

In an impressive study of five different genetically seizure-susceptible or chemo-convulsant induced *Drosophila* strains, microarray analysis was performed on CNS tissue from each group using multiple induction protocols, at different times following the stimulus. Because the four mutant lines carried temperature-sensitive alleles, the animals were presumably normal before the heat-shock stimulus that induced seizures. In a comparison between genes induced after seizure versus baseline levels of expression, only one gene was significantly elevated in all four mutant strains as well as the chemoconvulsant-induced wild-type animals: kayak, the *Drosophila* homologue of *c-fos*. In almost every other case, particular mRNAs showing significant alteration in abundance levels were observed in only 2 out of the 5 different seizure models employed (Guan et al., 2005).

The cellular context in which activity occurs can also change the content of the response, a kind of genomic metaplasticity. For example, expression of the IEG *arc* can be induced by exploration of a novel environment, but its expression is attenuated by prior neural activity, again suggesting that the cellular context in which gene expression might be initiated can profoundly influence the downstream candidates that are induced (Guzowski et al., 2006). Although neural firing is a potent inducer of IEG expression, layer 5 neurons in the cerebral cortex show very high firing rates compared to cells in other layers (in some cases, 10-100 times higher spontaneous activity; Shruti and Barth, unpublished observations), they conspicuously do not express c-fos *in vivo* under basal conditions (Barth et al., 2004).

A potential mechanism of how activity history can change transcriptional response profiles can be found examining the targets of zif268-dependent transcription. A predicted transcriptional target of zif268 is the repressor protein NGFI-A binding protein (Nab-1), which can heterodimerize with zif268 to suppress transcription of gene targets with a zif268 binding site (Swirnoff et al., 1998). The zif268-dependent transcription of Nab-1 suggests a model where genes within the plasticity transcriptome may be temporarily upregulated and then suppressed by continuing elevated neural activity. Indeed, in a study where zif268 was overexpressed in cultured PC12 cells, 99% of genes that showed statistically significant regulation by microarray analysis were actually downregulated compared to control untransfected cells (James et al., 2005). Subsequent studies on a subset of these gene targets showed that indeed, zif268 overexpression results in the reduction of gene products associated with protein degradation (James et al., 2006), and other studies suggest that activity-dependent repression of other zif268 gene targets may occur (Shruti et al., in revision). This may also be true for CREB-regulated genes; a predicted target of CREB is the CREB repressor (Foulkes et al., 1991), a protein which can itself suppress CREB-dependent transcription.

Additionally, different IEGs appear to have different thresholds for induction. Low frequency electrical stimulation is sufficient to induce zif268 but not c-fos and junB, for example, but high frequency stimulation induces a broader array of IEGs (Worley et al., 1993). Thus, depending on the duration and intensity of a stimulus, different subsets of genes may be activated, with the ensuing cascades of gene expression resulting in different perturbations of cell function. These results are troubling for experimentalists because they imply that there is not a common program of gene expression that is induced by activity, and that the particular type of activity as well as subtleties in genetic background can significantly influence the search for genes that define the plasticity transcriptome.

15.4.1 Sensory Experience and Transcription Factor Overexpression Studies

In a particularly prominent example of systems-level plasticity, involving a change in cellular responses to visual stimulation from the left or right eye in rodents after monocular deprivation, microarray analyses were performed after prolonged periods of monocular deprivation (Majdan and Shatz, 2006; Tropea et al., 2006). These studies showed upregulation of genes involved in synaptic transmission, electrical activity and growth while others genes, such as that encoding paravalbumin, were downregulated. However, these results are significantly complicated by the fact that the same cells improving their responses to open eye stimulation are simultaneously reducing their responses to deprived eye stimulation, and that changes in gene expression might be balanced in this case. Furthermore, the analyses were performed at time intervals sufficiently long (7+ days) that initial subsets of genes that were expressed have been replaced by the products of cascades of altered gene expression. This has been the case for other studies that attempted to look at CREB-dependent gene expression in the hippocampus using a superactivated form of this transcription factor under the control of the tet-off system (Barco et al., 2005). Nonetheless, in both types of studies, a few activity-regulated gene candidates have consistently shown up: most prominently, *BDNF* and *c-fos*.

15.4.2 Impact of Activity-dependent Gene Transcription: Synapses

The final question that we will consider in this review is how transcriptional events at the nucleus can be converted into changes in the cellular input-output function, especially with regard to synapse specificity. Here we discuss some constraints of immediate-early gene expression and the delivery of mRNAs or protein products to the synapse with respect to stimulus timing and the potential for synaptic change. Nuclear events, such as phosphorylation of the activity-dependent transcription factor CREB, can be detected within seconds to minutes of synaptic stimulation or depolarization, and transcription can occur within minutes following stimulation. Importantly, the duration of depolarization can influence how long CREB is activated, which in turn affects the target genes that will be regulated (Bito et al., 1996). Processed mRNAs can travel at rates of 0.1 mm/s (Kohrmann et al., 1999), and fast dendritic transport can range from 2–5 mm/s, meaning that the products of activity-dependent gene expression could conceivably reach synaptic sites within 15 minutes following stimulation. For gene products that are active in the soma, timing from stimulus to change in cell function is accelerated.

Many of these dendritically-localized mRNAs (summarized in Table 2) are already present at the synapse, implying that their transport may be constitutive, but in some cases, there is evidence for new mRNA transcripts that congregate at synaptic sites after stimulation (i.e. arc mRNA; Steward et al., 1998). These mRNAs may be redistributed from distal synaptic sites or may be delivered from the soma. Interestingly, mRNAs that have been localized to dendrites show a higher than expected probability of containing CREB or zif268 sites in their promoter regions (Table 3 and 4; based on data from Pfenning et al.,; summarized in Fig. 1). This suggests that activity-regulated transcription may indeed have the capacity to affect synaptic transmission site-specifically. Such a process could occur by synapse-specific capture of newly synthesized mRNAs or by a sort of priming strategy, whereby newly synthesized mRNAs are broadly available to all synapses throughout a cell which can individually undergo subsequent strengthening. Along these lines, in experiments where activated CREB was acutely overexpressed in hippocampal neurons *in vivo*, an increase in NMDAR-mediated synaptic transmission and the number of silent synapses were observed (Marie et al., 2005). Although a role for synapse-specificity in these studies was not determined, it is clear from these and other studies that CREB- and activity-dependent gene transcription can influence synaptic transmission.

15.4.3 Functional Outcome of Activity-dependent Gene Transcription: Firing Output

Many of the genes that show activity-dependent regulation do not have a clear synaptic function, but rather appear to alter cellular metabolism, protein stability, or neuronal excitability. For example, zif268 overexpression suppresses activity of the neuronal proteasome, a process that may enhance protein stability across the

Table 2 Summary of dendritically-localized mRNAs

Accession number	Gene Symbol	Gene name	Reference
NM_009599	Ache	acetylcholinesterase	(Meshorer et al., 2002)
NM_007393	Actb	actin, beta, cytoplasmic	(Kleiman et al., 1994)
NM_018790	Arc	activity regulated cytoskeletal-associated protein	(Link et al., 1995; Lyford et al., 1995)
NM_153534	Adcy2	adenylate cyclase 2	(Laurent-Demir et al., 2000)
NM_177034	Apba1	amyloid beta (A4) precursor protein binding, family A, member 1	(Strong et al., 1990)
BC051997	Avp	arginine vasopressin	(Bloch et al., 1990)
NM_016847	Avpr1a	arginine vasopressin receptor 1A	(Mohr et al., 1995)
NM_007540	Bdnf	brain derived neurotrophic factor	(Crino and Eberwine, 1996; Tongiorgi et al., 1997)
NM_177407	Camk2a	calcium/calmodulin-dependent protein kinase II alpha	(Burgin et al., 1990; Crino and Eberwine, 1996)
NM_009790	Calm1	calmodulin 1	(Berry and Brown, 1996)
NM_013497	Creb3	cAMP responsive element binding protein 3	(Crino et al., 1998)
NM_007913	Egr1	early growth response 1/zif268	(Eberwine et al., 2001)
NM_031261	Fthl17	ferritin, heavy polypeptide-like 17	(Ishimoto et al., 2000)
NM_010250	Gabra1	gamma-aminobutyric acid (GABA-A) receptor, subunit alpha 1	(Job and Eberwine, 2001; Sperk et al., 1998)
NM_008066	Gabra2	gamma-aminobutyric acid (GABA-A) receptor, subunit alpha 2	(Job and Eberwine, 2001; Sperk et al., 1998)
NM_008067	Gabra3	gamma-aminobutyric acid (GABA-A) receptor, subunit alpha 3	(Sperk et al., 1998)
NM_010251	Gabra4	gamma-aminobutyric acid (GABA-A) receptor, subunit alpha 4	(Job and Eberwine, 2001; Sperk et al., 1998)
NM_008069	Gabrb1	gamma-aminobutyric acid (GABA-A) receptor, subunit beta 1	(Job and Eberwine, 2001; Sperk et al., 1998)
NM_008070	Gabrb2	gamma-aminobutyric acid (GABA-A) receptor, subunit beta 2	(Sperk et al., 1998)
NM_008071	Gabrb3	gamma-aminobutyric acid (GABA-A) receptor, subunit beta 3	(Sperk et al., 1998)
NM_010252	Gabrg1	gamma-aminobutyric acid (GABA-A) receptor, subunit gamma 1	(Job and Eberwine, 2001)

(continued)

Table 2 (continued)

Accession number	Gene Symbol	Gene name	Reference
NM_008073	Gabrg2	gamma-aminobutyric acid (GABA-A) receptor, subunit gamma 2	(Sperk et al., 1998)
NM_008124	Gjb1	gap junction membrane channel protein beta 1	(Job and Eberwine, 2001)
NM_008125	Gjb2	gap junction membrane channel protein beta 2	(Job and Eberwine, 2001)
NM_013540	Gria2	glutamate receptor, ionotropic, AMPA2 (alpha 2)	(Miyashiro et al., 1994)
NM_016886	Gria3	glutamate receptor, ionotropic, AMPA3 (alpha 3)	(Miyashiro et al., 1994)
NM_019691	Gria4	glutamate receptor, ionotropic, AMPA4 (alpha 4)	(Miyashiro et al., 1994)
NM_008166	Grid1	glutamate receptor, ionotropic, delta 1	(Miyashiro et al., 1994)
NM_146072	Grik1	glutamate receptor, ionotropic, kainate 1	(Miyashiro et al., 1994)
AK034147	Grik2	glutamate receptor, ionotropic, kainate 2 (beta 2)	(Miyashiro et al., 1994)
NM_010349	Grik2	glutamate receptor, ionotropic, kainate 2 (beta 2)	(Miyashiro et al., 1994)
NM_175481	Grik4	glutamate receptor, ionotropic, kainate 4	(Miyashiro et al., 1994)
NM_008169	Grin1	glutamate receptor, ionotropic, NMDA1 (zeta 1)	(Miyashiro et al., 1994)
NM_008170	Grin2a	glutamate receptor, ionotropic, NMDA2A (epsilon 1)	(Miyashiro et al., 1994)
D10651	Grin2b	glutamate receptor, ionotropic, NMDA2B (epsilon 2)	(Miyashiro et al., 1994)
NM_010350	Grin2c	glutamate receptor, ionotropic, NMDA2C (epsilon 3)	(Miyashiro et al., 1994)
AY129229	Glra1	glycine receptor, alpha 1 subunit	(Racca et al., 1997)
NM_183427	Glra2	glycine receptor, alpha 2 subunit	(Racca et al., 1997)
NM_008083	Gap43	growth associated protein 43	(Chicurel et al., 1993) & (Landry et al., 1994)
AK042240	Itpr1	inositol 1,4,5-triphosphate receptor 1	(Furuichi et al., 1993)
NM_010838	Mapt	microtubule-associated protein tau	(Kosik et al., 1989)

(continued)

Table 2 (continued)

Accession number	Gene Symbol	Gene name	Reference
NM_153058	Mapre2	microtubule-associated protein, RP/EB family, member 2	(Garner et al., 1988)
V00836	Ngfb	nerve growth factor, beta	(Tongiorgi et al., 1997)
NM_008744	Ntn1	netrin 1	(Crino et al., 1998)
NM_008691	Nef3	neurofilament 3, medium	(Crino and Eberwine, 1996)
X17647	Ntrk2	neurotrophic tyrosine kinase, receptor, type 2	(Crino and Eberwine, 1996; Tongiorgi et al., 1997)
NM_011025	Oxt	oxytocin	(Bloch et al., 1990)
NM_019697	Kcnd2	potassium voltage-gated channel, Shal-related family, member 2	(Crino and Eberwine, 1996)
NM_011103	Prkcd	protein kinase C, delta	(Moriya et al., 1994)
L42339	Scn1a	sodium channel, voltage-gated, type I, alpha	(Job and Eberwine, 2001)
NM_011654	Tuba2	tubulin, alpha 2	(Kleiman et al., 1994)
NM_009450	Tubb2	tubulin, beta 2	(Kleiman et al., 1994)
NM_009377	Th	tyrosine hydroxylase	(Dumas et al., 1990)

Table 3 Dendritically-localized predicted zif268 targets

Accession number	Gene Symbol	Gene name	Reference
NM_153534	Adcy2	adenylate cyclase 2	(Laurent-Demir et al., 2000)
NM_007913	Egr1	early growth response 1/zif268	(Eberwine et al., 2001)
NM_031261	Fthl17	ferritin, heavy polypeptide-like 17	(Ishimoto et al., 2000)
NM_010250	Gabra1	gamma aminobutyric acid (GABA-A) receptor, subunit alpha 1	(Sperk et al., 1998 & Job and Eberwine 2001)
NM_013540	Gria2	glutamate receptor, ionotropic, AMPA2 (alpha 2)	(Miyashiro et al., 1994)
NM_008169	Grin1	glutamate receptor, ionotropic, NMDA1 (zeta 1)	(Miyashiro et al., 1994)
NM_010350	Grin2c	glutamate receptor, ionotropic, NMDA2C (epsilon 3)	(Miyashiro et al., 1994)
NM_153058	Mapre2	microtubule-associated protein, RP/EB family, member 2	(Garner et al., 1988)
NM_009377	Th	tyrosine hydroxylase	(Dumas et al., 1990)

Table 4 Dendritically-localized predicted CREB targets

Accession number	Gene Symbol	Gene name	Reference
NM_008073	*Gabrg2*	*gamma-aminobutyric acid (GABA-A) receptor, subunit gamma 2*	(Sperk et al., 1998)
NM_008124	*Gjb1*	*gap junction membrane channel protein beta 1*	(Job and Eberwine, 2001)
NM_175481	*Grik4*	*glutamate receptor, ionotropic, kainate 4*	(Miyashiro et al., 1994)
V00836	*Ngfb*	*nerve growth factor, beta*	(Tongiorgi et al., 1997)
NM_009450	*Tubb2*	*tubulin, beta 2*	(Kleiman et al., 1994)
NM_009377	*Th*	*tyrosine hydroxylase*	(Dumas et al., 1990)

cell (James et al., 2006). Other candidate genes that have been identified appear to have a more global effect on cell metabolism, such as *Hsp27* and *Hsp70* (Nedivi et al., 1993; Tang et al., 2002). Although few ion channel genes have been directly identified as regulated targets following acute induction of neural activity, there are some notable exceptions. Activity can alter expression of hyperpolarization-activated, non-specific cation channels (HCN) 1 and 2 (Bender et al., 2003; Shah et al., 2004; Surges et al., 2006), and can reduce expression of the BK channel modulatory subunit $\beta4$ (Shruti et al., in revision), important changes that can alter

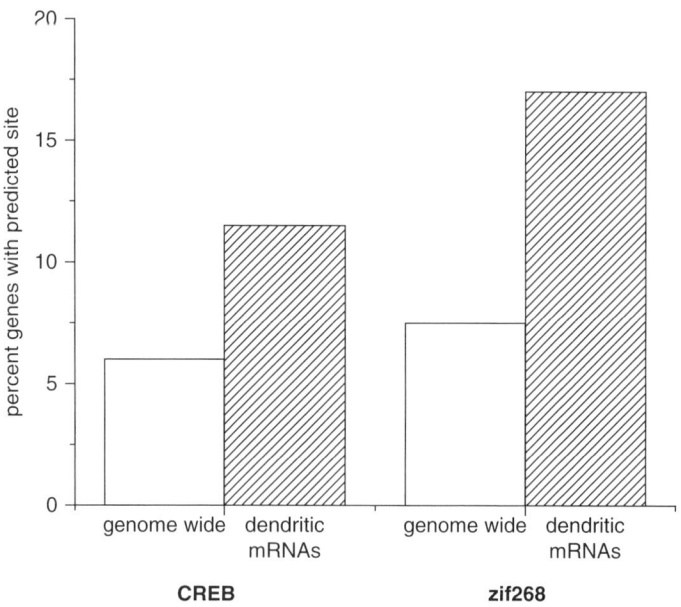

Fig. 1 Promoter regions for genes encoding dendritically–localized mRNAs are enriched for zif268 and CREB binding sites. The percent of genes with zif268 and CREB binding sites in their promoter region (1000 nt upstream of transcription start, Pfenning et al., 2007) across all annotated mouse genes (white bars) versus the frequency of CREB and zif268 sites in dendritically-localized mRNAs (grey bars; derived from Table 2 and Pfenning et al.,)

neuronal output and firing rate. Synaptic glutamate receptor mRNA can be altered by activity as well (Grooms et al., 2006), and GABA receptor expression can be reduced by seizure activity (Banerjee et al., 1998; Kamphuis et al., 1995).

15.5 Activity-dependent Gene Transcription and the Input-Output Function of the Neuron

Changes in intrinsic neural excitability and synaptic function can couple to significantly alter the input-output function of a neuron following periods of high activity. We propose that nuclear events increase the concentration of mRNAs for a variety of genes that are then distributed broadly across the cell, and that subsequently regulate the translation of specific mRNAs as required by later events. This model suggests that nuclear transcription primes the cell to respond in some programmed way to future stimuli, perhaps in a pathway-specific manner. The model is particularly attractive because it combines local control of translation without requiring a highly specific mechanism for sorting activated mRNAs to particular, individual synapses. Because the late phase of LTP is typically designated as >3 hrs after the potentiating stimulus, this allows more than adequate time to allow the transport of gene products to distant synaptic sites. Thus, it is not unreasonable to conclude that activity-dependent gene transcription has the capacity to alter the properties of synaptic neurotransmission at individual synapses.

Neural activity can further alter the way that synaptic neurotransmission leads to cell firing by globally changing ion channel properties that affect the generation or maintenance of action potential firing (Barth et al., 2004; Bernard et al., 2004; Desai et al., 1999; Shah et al., 2004). Together, the regulation of genes involved in synaptic properties or neural excitability can transform the input-output function of a neuron in both specific and very general ways. Determining how changes in activity-dependent programs of gene expression and their affect on neural activity persist over time are critical issues in the study of learning and memory as well as pathological states such as ischemia and epilepsy disorders.

References

Babity, J. M., Armstrong, J. N., Plumier, J. C., Curric, R. W., and Robertson, H. A. (1997) A novel seizure-induced synaptotagmin gene identified by differential display. Proc. Natl. Acad. Sci. USA 94, 2638–2641.

Banerjee, P. K., Tillakaratne, N. J., Brailowsky, S., Olsen, R. W., Tobin, A. J., and Snead, O. C., 3rd (1998) Alterations in GABAA receptor alpha 1 and alpha 4 subunit mRNA levels in thalamic relay nuclei following absence-like seizures in rats. Exp. Neurol. 154, 213–223.

Barco, A., Alarcon, J. M., and Kandel, E. R. (2002) Expression of constitutively active CREB protein facilitates the late phase of long-term potentiation by enhancing synaptic capture. Cell 108, 689–703.

Barco, A., Patterson, S., Alarcon, J. M., Gromova, P., Mata-Roig, M., Morozov, A., and Kandel, E. R. (2005) Gene expression profiling of facilitated L-LTP in VP16-CREB mice reveals that BDNF is critical for the maintenance of LTP and its synaptic capture. Neuron 48, 123–137.

Barondes, S. H., and Jarvik, M. E. (1964) The Influence Of Actinomycin-D On Brain Rna Synthesis And On Memory. J. Neurochem. 11, 187–195.

Barth, A. L., Gerkin, R. C., and Dean, K. L. (2004) Alteration of neuronal firing properties after in vivo experience in a FosGFP transgenic mouse. J. Neurosci. 24, 6466–6475.

Barzilai, A., Kennedy, T. E., Sweatt, J. D., and Kandel, E. R. (1989) 5-HT modulates protein synthesis and the expression of specific proteins during long-term facilitation in *Aplysia* sensory neurons. Neuron 2, 1577–1586.

Bender, R. A., Soleymani, S. V., Brewster, A. L., Nguyen, S. T., Beck, H., Mathern, G. W., and Baram, T. Z. (2003) Enhanced expression of a specific hyperpolarization-activated cyclic nucleotide-gated cation channel (HCN) in surviving dentate gyrus granule cells of human and experimental epileptic hippocampus. J. Neurosci. 23, 6826–6836.

Bernard, C., Anderson, A., Becker, A., Poolos, N. P., Beck, H., and Johnston, D. (2004) Acquired dendritic channelopathy in temporal lobe epilepsy. Science 305, 532–535.

Berry, F. B., and Brown, I. R. (1996) CaM I mRNA is localized to apical dendrites during postnatal development of neurons in the rat brain. J. Neurosci. Res. 43, 565–575.

Bito, H., Deisseroth, K., and Tsien, R. W. (1996) CREB phosphorylation and dephosphorylation: a Ca(2+)- and stimulus duration-dependent switch for hippocampal gene expression. Cell 87, 1203–1214.

Bloch, B., Guitteny, A. F., Normand, E., and Chouham, S. (1990) Presence of neuropeptide messenger RNAs in neuronal processes. Neurosci. Lett. 109, 259–264.

Burgin, K. E., Waxham, M. N., Rickling, S., Westgate, S. A., Mobley, W. C., and Kelly, P. T. (1990) In situ hybridization histochemistry of Ca2+/calmodulin-dependent protein kinase in developing rat brain. J. Neurosci. 10, 1788–1798.

Cha-Molstad, H., Keller, D. M., Yochum, G. S., Impey, S., and Goodman, R. H. (2004) Cell-type-specific binding of the transcription factor CREB to the cAMP-response element. Proc. Natl. Acad. Sci. USA 101, 13572–13577.

Chicurel, M. E., Terrian, D. M., and Potter, H. (1993) mRNA at the synapse: analysis of a synaptosomal preparation enriched in hippocampal dendritic spines. J. Neurosci. 13, 4054–4063.

Cochran, B. H., Reffel, A. C., and Stiles, C. D. (1983) Molecular cloning of gene sequences regulated by platelet-derived growth factor. Cell 33, 939–947.

Corriveau, R. A., Shatz, C. J., and Nedivi, E. (1999) Dynamic regulation of cpg15 during activity-dependent synaptic development in the mammalian visual system. J. Neurosci. 19, 7999–8008.

Cottrell, J. R., Borok, E., Horvath, T. L., and Nedivi, E. (2004) CPG2: a brain- and synapse-specific protein that regulates the endocytosis of glutamate receptors. Neuron 44, 677–690.

Crino, P., Khodakhah, K., Becker, K., Ginsberg, S., Hemby, S., and Eberwine, J. (1998) Presence and phosphorylation of transcription factors in developing dendrites. Proc. Natl. Acad. Sci. USA 95, 2313–2318.

Crino, P. B., and Eberwine, J. (1996) Molecular characterization of the dendritic growth cone: regulated mRNA transport and local protein synthesis. Neuron 17, 1173–1187.

Dash, P. K., Hochner, B., and Kandel, E. R. (1990) Injection of the cAMP-responsive element into the nucleus of Aplysia sensory neurons blocks long-term facilitation. Nature 345, 718–721.

Davis, S., Vanhoutte, P., Pages, C., Caboche, J., and Laroche, S. (2000) The MAPK/ERK cascade targets both Elk-1 and cAMP response element-binding protein to control long-term potentiation-dependent gene expression in the dentate gyrus in vivo. J. Neurosci. 20, 4563–4572.

Desai, N. S., Rutherford, L. C., and Turrigiano, G. G. (1999) Plasticity in the intrinsic excitability of cortical pyramidal neurons. Nat. Neurosci. 2, 515–520.

Dumas, S., Javoy-Agid, F., Hirsch, E., Agid, Y., and Mallet, J. (1990) Tyrosine hydroxylase gene expression in human ventral mesencephalon: detection of tyrosine hydroxylase messenger RNA in neurites. J. Neurosci. Res. 25, 569–575.

Eberwine, J., Miyashiro, K., Kacharmina, J. E., and Job, C. (2001) Local translation of classes of mRNAs that are targeted to neuronal dendrites. Proc. Natl. Acad. Sci. USA 98, 7080–7085.

Elliott, R. C., Miles, M. F., and Lowenstein, D. H. (2003) Overlapping microarray profiles of dentate gyrus gene expression during development- and epilepsy-associated neurogenesis and axon outgrowth. J. Neurosci. 23, 2218–2227.

Euskirchen, G., Royce, T. E., Bertone, P., Martone, R., Rinn, J. L., Nelson, F. K., Sayward, F., Luscombe, N. M., Miller, P., Gerstein, M., et al., (2004) CREB binds to multiple loci on human chromosome 22. Mol. Cell Biol. 24, 3804–3814.

Flavell, S. W., Cowan, C. W., Kim, T. K., Greer, P. L., Lin, Y., Paradis, S., Griffith, E. C., Hu, L. S., Chen, C., and Greenberg, M. E. (2006) Activity-dependent regulation of MEF2 transcription factors suppresses excitatory synapse number. Science 311, 1008–1012.

Flexner, J. B., Flexner, L. B., and Stellar, E. (1963) Memory in mice as affected by intracerebral puromycin. Science 141, 57–59.

Foulkes, N. S., Borrelli, E., and Sassone-Corsi, P. (1991) CREM gene: use of alternative DNA-binding domains generates multiple antagonists of cAMP-induced transcription. Cell 64, 739–749.

Furuichi, T., Simon-Chazottes, D., Fujino, I., Yamada, N., Hasegawa, M., Miyawaki, A., Yoshikawa, S., Guenet, J. L., and Mikoshiba, K. (1993) Widespread expression of inositol 1,4,5-trisphosphate receptor type 1 gene (Insp3r1) in the mouse central nervous system. Receptors Channels 1, 11–24.

Garner, C. C., Tucker, R. P., and Matus, A. (1988) Selective localization of messenger RNA for cytoskeletal protein MAP2 in dendrites. Nature 336, 674–677.

Glazewski, S., Barth, A. L., Wallace, H., McKenna, M., Silva, A., and Fox, K. (1999) Impaired experience-dependent plasticity in barrel cortex of mice lacking the alpha and delta isoforms of CREB. Cereb. Cortex 9, 249–256.

Goelet, P., Castellucci, V. F., Schacher, S., and Kandel, E. R. (1986) The long and the short of long-term memory–a molecular framework. Nature 322, 419–422.

Grooms, S. Y., Noh, K. M., Regis, R., Bassell, G. J., Bryan, M. K., Carroll, R. C., and Zukin, R. S. (2006) Activity bidirectionally regulates AMPA receptor mRNA abundance in dendrites of hippocampal neurons. J. Neurosci. 26, 8339–8351.

Guan, Z., Saraswati, S., Adolfsen, B., and Littleton, J. T. (2005) Genome-wide transcriptional changes associated with enhanced activity in the Drosophila nervous system. Neuron 48, 91–107.

Guzowski, J. F., Miyashita, T., Chawla, M. K., Sanderson, J., Maes, L. I., Houston, F. P., Lipa, P., McNaughton, B. L., Worley, P. F., and Barnes, C. A. (2006) Recent behavioral history modifies coupling between cell activity and Arc gene transcription in hippocampal CA1 neurons. Proc. Natl. Acad. Sci. USA 103, 1077–1082.

Impey, S., McCorkle, S. R., Cha-Molstad, H., Dwyer, J. M., Yochum, G. S., Boss, J. M., McWeeney, S., Dunn, J. J., Mandel, G., and Goodman, R. H. (2004) Defining the CREB regulon: a genome wide analysis of transcription factor regulatory regions. Cell 119, 1041–1054.

Ingi, T., Krumins, A. M., Chidiac, P., Brothers, G. M., Chung, S., Snow, B. E., Barnes, C. A., Lanahan, A. A., Siderovski, D. P., Ross, E. M., et al., (1998) Dynamic regulation of RGS2 suggests a novel mechanism in G-protein signaling and neuronal plasticity. J. Neurosci. 18, 7178–7188.

Ishimoto, T., Fujimori, K., Kasai, M., and Taguchi, T. (2000) Dendritic translocation of the rat ferritin H chain mRNA. Biochem. Biophys. Res. Commun. 272, 789–793.

James, A. B., Conway, A. M., and Morris, B. J. (2005) Genomic profiling of the neuronal target genes of the plasticity-related transcription factor – Zif268. J. Neurochem. 95, 796–810.

James, A. B., Conway, A. M., and Morris, B. J. (2006) Regulation of the neuronal proteasome by Zif268 (Egr1) J. Neurosci. 26, 1624–1634.

Job, C., and Eberwine, J. (2001) Localization and translation of mRNA in dendrites and axons. Nat. Rev. Neurosci. 2, 889–898.

Josselyn, S. A., Kida, S., and Silva, A. J. (2004) Inducible repression of CREB function disrupts amygdala-dependent memory. Neurobiol. Learn. Mem. 82, 159–163.

Kaminska, B., Kaczmarek, L., and Chaudhuri, A. (1996) Visual stimulation regulates the expression of transcription factors and modulates the composition of AP-1 in visual cortex. J. Neurosci. 16, 3968–3978.

Kamphuis, W., De Rijk, T. C., and Lopes da Silva, F. H. (1995) Expression of GABAA receptor subunit mRNAs in hippocampal pyramidal and granular neurons in the kindling model of epileptogenesis: an in situ hybridization study. Brain Res. Mol. Brain Res. 31, 33–47.

Kida, S., Josselyn, S. A., de Ortiz, S. P., Kogan, J. H., Chevere, I., Masushige, S., and Silva, A. J. (2002) CREB required for the stability of new and reactivated fear memories. Nat. Neurosci. 5, 348–355.

Kleiman, R., Banker, G., and Steward, O. (1994) Development of subcellular mRNA compartmentation in hippocampal neurons in culture. J. Neurosci. 14, 1130–1140.

Kohrmann, M., Luo, M., Kaether, C., DesGroseillers, L., Dotti, C. G., and Kiebler, M. A. (1999) Microtubule-dependent recruitment of Staufen-green fluorescent protein into large RNA-containing granules and subsequent dendritic transport in living hippocampal neurons. Mol. Biol. Cell 10, 2945–2953.

Kosik, K. S., Crandall, J. E., Mufson, E. J., and Neve, R. L. (1989) Tau in situ hybridization in normal and Alzheimer brain: localization in the somatodendritic compartment. Ann. Neurol. 26, 352–361.

Landry, C. F., Watson, J. B., Kashima, T., and Campagnoni, A. T. (1994) Cellular influences on RNA sorting in neurons and glia: an in situ hybridization histochemical study. Brain Res. Mol. Brain Res. 27, 1–11.

Laurent-Demir, C., Decorte, L., Jaffard, R., and Mons, N. (2000) Differential regulation of Ca(2+)-calmodulin stimulated and Ca(2+)-insensitive adenylyl cyclase messenger RNA in intact and denervated mouse hippocampus. Neuroscience 96, 267–274.

Lauterborn, J. C., Rivera, S., Stinis, C. T., Hayes, V. Y., Isackson, P. J., and Gall, C. M. (1996) Differential effects of protein synthesis inhibition on the activity-dependent expression of BDNF transcripts: evidence for immediate-early gene responses from specific promoters. J. Neurosci. 16, 7428–7436.

Link, W., Konietzko, U., Kauselmann, G., Krug, M., Schwanke, B., Frey, U., and Kuhl, D. (1995) Somatodendritic expression of an immediate early gene is regulated by synaptic activity. Proc. Natl. Acad. Sci. USA 92, 5734–5738.

Lukasiuk, K., Kontula, L., and Pitkanen, A. (2003) cDNA profiling of epileptogenesis in the rat brain. Eur. J. Neurosci. 17, 271–279.

Lukasiuk, K., and Pitkanen, A. (2004) Large-scale analysis of gene expression in epilepsy research: is synthesis already possible? Neurochem. Res. 29, 1169–1178.

Lyford, G. L., Yamagata, K., Kaufmann, W. E., Barnes, C. A., Sanders, L. K., Copeland, N. G., Gilbert, D. J., Jenkins, N. A., Lanahan, A. A., and Worley, P. F. (1995) Arc, a growth factor and activity-regulated gene, encodes a novel cytoskeleton-associated protein that is enriched in neuronal dendrites. Neuron 14, 433–445.

Majdan, M., and Shatz, C. J. (2006) Effects of visual experience on activity-dependent gene regulation in cortex. Nat. Neurosci. 9, 650–659.

Marie, H., Morishita, W., Yu, X., Calakos, N., and Malenka, R. C. (2005) Generation of silent synapses by acute in vivo expression of CaMKIV and CREB. Neuron 45, 741–752.

Meshorer, E., Erb, C., Gazit, R., Pavlovsky, L., Kaufer, D., Friedman, A., Glick, D., Ben-Arie, N., and Soreq, H. (2002) Alternative splicing and neuritic mRNA translocation under long-term neuronal hypersensitivity. Science 295, 508–512.

Miranti, C. K., Ginty, D. D., Huang, G., Chatila, T., and Greenberg, M. E. (1995) Calcium activates serum response factor-dependent transcription by a Ras- and Elk-1-independent mechanism that involves a Ca2+/calmodulin-dependent kinase. Mol. Cell Biol. 15, 3672–3684.

Miyashiro, K., Dichter, M., and Eberwine, J. (1994) On the nature and differential distribution of mRNAs in hippocampal neurites: implications for neuronal functioning. Proc. Natl. Acad. Sci. USA 91, 10800–10804.

Mohr, E., Meyerhof, W., and Richter, D. (1995) Vasopressin and oxytocin: molecular biology and evolution of the peptide hormones and their receptors. Vitam. Horm. 51, 235–266.

Morgan, J. I., Cohen, D. R., Hempstead, J. L., and Curran, T. (1987) Mapping patterns of c-fos expression in the central nervous system after seizure. Science 237, 192–197.

Moriya, M., and Tanaka, S. (1994) Prominent expression of protein kinase C (gamma) mRNA in the dendrite-rich neuropil of mice cerebellum at the critical period for synaptogenesis. Neuroreport 5, 929–932.

Nedivi, E., Hevroni, D., Naot, D., Israeli, D., and Citri, Y. (1993) Numerous candidate plasticity-related genes revealed by differential cDNA cloning. Nature 363, 718–722.

Nedivi, E., Wu, G. Y., and Cline, H. T. (1998) Promotion of dendritic growth by CPG15, an activity-induced signaling molecule. Science 281, 1863–1866.

Pfenning, A. R., Schwartz, R., and Barth, A. L. (2007) A comparative genomics approach to identifying the plasticity transcriptome. BMC Neurosci. 13(8), 20.

Qian, Z., Gilbert, M., and Kandel, E. R. (1994) Temporal and spatial regulation of the expression of BAD2, a MAP kinase phosphatase, during seizure, kindling, and long-term potentiation. Learn. Mem. 1, 180–188.

Qian, Z., Gilbert, M. E., Colicos, M. A., Kandel, E. R., and Kuhl, D. (1993) Tissue-plasminogen activator is induced as an immediate-early gene during seizure, kindling and long-term potentiation. Nature 361, 453–457.

Racca, C., Gardiol, A., and Triller, A. (1997) Dendritic and postsynaptic localizations of glycine receptor alpha subunit mRNAs. J. Neurosci. 17, 1691–1700.

Saffen, D. W., Cole, A. J., Worley, P. F., Christy, B. A., Ryder, K., and Baraban, J. M. (1988) Convulsant-induced increase in transcription factor messenger RNAs in rat brain. Proc. Natl. Acad. Sci. USA 85, 7795–7799.

Shah, M. M., Anderson, A. E., Leung, V., Lin, X., and Johnston, D. (2004) Seizure-induced plasticity of h channels in entorhinal cortical layer III pyramidal neurons. Neuron 44, 495–508.

Shaywitz, A. J., and Greenberg, M. E. (1999) CREB: a stimulus-induced transcription factor activated by a diverse array of extracellular signals. Annu. Rev. Biochem. 68, 821–861.

Sheng, M., and Greenberg, M. E. (1990) The regulation and function of c-fos and other immediate early genes in the nervous system. Neuron 4, 477–485.

Sonnenberg, J. L., Rauscher, F. J., 3rd, Morgan, J. I., and Curran, T. (1989) Regulation of proenkephalin by Fos and Jun. Science 246, 1622–1625.

Sperk, G., Schwarzer, C., Tsunashima, K., and Kandlhofer, S. (1998) Expression of GABA(A) receptor subunits in the hippocampus of the rat after kainic acid-induced seizures. Epilepsy Res. 32, 129–139.

Steward, O., Wallace, C. S., Lyford, G. L., and Worley, P. F. (1998) Synaptic activation causes the mRNA for the IEG Arc to localize selectively near activated postsynaptic sites on dendrites. Neuron 21, 741–751.

Strong, M. J., Svedmyr, A., Gajdusek, D. C., and Garruto, R. M. (1990) The temporal expression of amyloid precursor protein mRNA in vitro in dissociated hippocampal neuron cultures. Exp. Neurol. 109, 171–179.

Surges, R., Brewster, A. L., Bender, R. A., Beck, H., Feuerstein, T. J., and Baram, T. Z. (2006) Regulated expression of HCN channels and cAMP levels shape the properties of the h current in developing rat hippocampus. Eur. J. Neurosci. 24, 94–104.

Swirnoff, A. H., Apel, E. D., Svaren, J., Sevetson, B. R., Zimonjic, D. B., Popescu, N. C., and Milbrandt, J. (1998) Nab1, a corepressor of NGFI-A (Egr-1), contains an active transcriptional repression domain. Mol. Cell Biol. 18, 512–524.

Tang, Y., Lu, A., Aronow, B. J., Wagner, K. R., and Sharp, F. R. (2002) Genomic responses of the brain to ischemic stroke, intracerebral haemorrhage, kainate seizures, hypoglycemia, and hypoxia. Eur. J. Neurosci. 15, 1937–1952.

Tongiorgi, E., Righi, M., and Cattaneo, A. (1997) Activity-dependent dendritic targeting of BDNF and TrkB mRNAs in hippocampal neurons. J. Neurosci. 17, 9492–9505.

Tropea, D., Kreiman, G., Lyckman, A., Mukherjee, S., Yu, H., Horng, S., and Sur, M. (2006) Gene expression changes and molecular pathways mediating activity-dependent plasticity in visual cortex. Nat. Neurosci. 9, 660–668.

West, A. E., Griffith, E. C., and Greenberg, M. E. (2002) Regulation of transcription factors by neuronal activity. Nat. Rev. Neurosci. 3, 921–931.

Worley, P. F., Bhat, R. V., Baraban, J. M., Erickson, C. A., McNaughton, B. L., and Barnes, C. A. (1993) Thresholds for synaptic activation of transcription factors in hippocampus: correlation with long-term enhancement. J. Neurosci. 13, 4776–4786.

Xiao, B., Tu, J. C., Petralia, R. S., Yuan, J. P., Doan, A., Breder, C. D., Ruggiero, A., Lanahan, A. A., Wenthold, R. J., and Worley, P. F. (1998) Homer regulates the association of group 1 metabotropic glutamate receptors with multivalent complexes of homer-related, synaptic proteins. Neuron 21, 707–716.

Yamagata, K., Andreasson, K. I., Kaufmann, W. E., Barnes, C. A., and Worley, P. F. (1993) Expression of a mitogen-inducible cyclooxygenase in brain neurons: regulation by synaptic activity and glucocorticoids. Neuron 11, 371–386.

Yin, J. C., Wallach, J. S., Del Vecchio, M., Wilder, E. L., Zhou, H., Quinn, W. G., and Tully, T. (1994) Induction of a dominant negative CREB transgene specifically blocks long-term memory in Drosophila. Cell 79, 49–58.

Zhang, X., Odom, D. T., Koo, S. H., Conkright, M. D., Canettieri, G., Best, J., Chen, H., Jenner, R., Herbolsheimer, E., Jacobsen, E., et al., (2005) Genome-wide analysis of cAMP-response element binding protein occupancy, phosphorylation, and target gene activation in human tissues. Proc. Natl. Acad. Sci. USA 102, 4459–4464.

Chapter 16
Transcriptional Mechanisms Underlying the Mammalian Circadian Clock

Hai-Ying Mary Cheng and Karl Obrietan

Abstract The mammalian 'master' pacemaker is seated in the suprachiasmatic nuclei (SCN), a bilateral hypothalamic structure above the optic chiasm that is a heterogeneous conglomerate of ~ 20,000 neurons. Within the nuclei of SCN cells lies the molecular basis of circadian timekeeping: clock proteins that form interlocking transcriptional feedback loops, and that drive and sustain the rhythmic expression of their cognate genes as well as other clock-controlled genes (*ccg*). In addition to the SCN, this intracellular molecular clock is a common element in the many oscillating tissues of the central nervous system (CNS) and periphery that form the clock hierarchy. In this review we discuss the nature of the molecular clock, the neurotransmitter systems that actuate clock entrainment, and the intracellular signaling events leading to activation of transcriptional programs.

16.1 Introduction

So much of what defines the 'Living Planet' is rooted in the daily rotation of the Earth on its axis. Central to this is the 24-hr day/night cycle, which permeates nearly all aspects of physiology and behavior of most organisms, such as (in the case of mammals) the daily fluctuations in body temperature, cardiovascular function, liver metabolism and sleep-wake cycles. In fact, these and many other biological processes are driven by an internal timekeeping device, which allows their rhythms to persist with near 24-hr precision in the absence of external time cues (e.g., light/dark). Like a metronome, 'circadian' clocks keep tempo *autonomously*, but are still able to react and respond to the environment by resetting their phase, a process known as entrainment. The principle 'zeitgeber' ('zeit'=time, 'geber'=giver in *German*) is light, but non-photic cues such as social interaction, food, and novelty are also able to entrain circadian clocks in mammals.

The mammalian 'master' pacemaker is seated in the suprachiasmatic nuclei (SCN), a bilateral hypothalamic structure above the optic chiasm that is a heterogeneous conglomerate of ~ 20,000 neurons. Within the nuclei of SCN cells lies the molecular basis of circadian timekeeping: clock proteins that form interlocking transcriptional feedback loops, and that drive and sustain the rhythmic expression of their cognate genes as well as other clock-controlled genes (*ccg*). In addition

S. M. Dudek (ed.), *Transcriptional Regulation by Neuronal Activity.*
© Springer 2008

to the SCN, this intracellular molecular clock is a common element in the many oscillating tissues of the central nervous system (CNS) and periphery that form the clock hierarchy.

With respect to light entrainment of the SCN clock, specialized retinal ganglion cells (RGCs), which express the pigment melanopsin, project monosynaptically to a subset of SCN neurons via the retinohypothalamic tract (RHT) and convey photic information through the release of neurotransmitters. In addition, the SCN receives afferents from other brain regions, such as the thalamic intergeniculate leaflet (IGL) and midbrain raphe, that communicate photic and nonphotic cues. The evoked release of neurotransmitters triggers a sequence of signal transduction events within SCN neurons that ultimately impinge upon and alter the phasing of the core clock transcriptional loop. Both the transcriptional basis of inherent pacemaker activity, and the synaptic physiology that regulate the genetic clock are of significant interest to the field of timing and will be covered here.

This review will discuss the following topics: 1) the molecular clock; 2) neurotransmitter systems that actuate clock entrainment; 3) intracellular signaling events that transduce entraining inputs to the cell nucleus; and 4) activation of transcriptional programs.

16.2 The Molecular Clock

An interlocking positive and negative transcriptional feedback loop represents the minimal unit needed for daily 24-hr oscillations. As part of the positive limb, the basic helix-loop-helix (bHLH)-PAS transcription factors, CLOCK and BMAL1, form functional heterodimers that recognize upstream *cis*-acting elements, E-box motifs (CACGTG), and induce transcription of E-box-containing genes, including three *period* (*mper1*, *mper2* and *mper3*) and two *cryptochrome* (*mcry1* and *mcry2*) genes in mice (King, Zhao, Sangoram, Wilsbacher, Tanaka, Antoch, Steeves, Vitaterna, Kornhauser, Lowrey, Turek, and Takahashi 1997; Gekakis, Staknis, Nguyen, Davis, Wilsbacher, King, Takahashi, and Weitz 1998; Hogenesch, Gu, Jain, and Bradfield 1998; Bunger, Wilsbacher, Moran, Clendenin, Radcliffe, Hogenesch, Simon, Takahashi, and Bradfield 2000). The histone acetyltransferase p300, which physically binds CLOCK, is recruited to the promoters of E-box-bearing target genes and induces chromatin remodeling (as well as transcription) via acetylation of H3 histones (Etchegaray, Lee, Wade, and Reppert 2003). Following protein translation in the cytoplasm, PER-CRY complexes translocate to the nucleus and negatively regulate their own expression via inhibition of CLOCK/BMAL1-dependent transcription. This inhibitory feedback loop takes approximately 24 hrs to complete. In addition to their rhythmic regulation, both *mper1* and *mper2* are responsive to photic entrainment cues via an E-Box/CLOCK-BMAL1-independent mechanism that is described below (Travnickova-Bendova, Cermakian, Reppert, and Sassone-Corsi 2002).

With the exception of mCRY1 and mCRY2, core clock proteins belong to the PAS (Per-Arnt-Sim) family of transcription factors (reviewed in Taylor and

Zhulin 1999). Structurally, PAS domains are defined as a region of 100-120 amino acids consisting of two 50-residue conserved sequences (termed PAS-A and PAS-B). PAS domains are found in a plethora of proteins, including signal transduction molecules, ion channels and transcription factors, and work in concert with other regulatory modules within multidomain proteins to exert distinct biological responses. PAS domains may serve to bind cofactors that are essential for the regulation of transcriptional activity: for example, the PAS domains of NPAS2, a homolog of CLOCK, have been shown to bind heme as a prosthetic group (Dioum, Rutter, Tuckerman, Gonzalez, Gilles-Gonzalez, and McKnight 2002). The DNA-binding activity of heme-bound NPAS2 is sensitive to inhibition by carbon monoxide (CO) at micromolar concentrations *in vitro* (Dioum et al., 2002). In addition, CO inhibited the heterodimerization of functional BMAL1-NPAS2 (Diou et al., 2002). The significance of the PAS motif in the core clock timing processes was made evident in a mutant mouse strain in which the PAS domain was deleted from mPER2: these mPer2 mutant mice exhibited a shorter circadian period with a gradual loss of circadian rhythmicity under constant dark (DD) conditions (Zheng, Larkin, Albrecht, Sun, Sage, Eichele, Lee, and Bradley 1999).

Mammalian mCRY1 and mCRY2 are members of the family of plant blue-light receptors (cryptochrome) and DNA repair enzymes (photolyases), although they lack photolyase activity and their effects on CLOCK/BMAL1 are independent of light (Hsu, Zhao, Zhao, Kazantsev, Wang, Todo, Wei, and Sancar 1996; Griffin, Staknis, and Weitz 1999). mCry1/mCry2 double-null mice exhibit immediate and complete loss of circadian rhythmicity under DD, indicating an essential role of CRY proteins in the maintenance of circadian rhythms (van der Horst, Muijtjens, Kobayashi, Takano, Kanno, Takao, de Wit, Verkerk, Eker, van Leenen, Buijs, Bootsma, Hoeijmakers, and Yasui 1999).

A pivotal study in the field of mammalian timing was the characterization of the circadian timing gene *clock*. Using N-ethyl-N-nitrosourea (ENU) mutagenesis, Vitaterna et al., (Vitaterna, King, Chang, Kornhauser, Lowrey, McDonald, Dove, Pinto, Turek, and Takahashi 1994) generated the *"clock"* mutant mouse strain in which the CLOCK protein lacked residues encoded by exon 19 (51 aa within the C-terminal activation domain) and competed with wild-type CLOCK for binding to BMAL1. Mice carrying one or two copies of the dominant-negative *clock* allele exhibited a period lengthening of 1 hr or 4 hr, respectively, and eventual loss of circadian rhythmicity after 2 weeks in DD (Vitaterna et al., 1994). *Clock* heterozygous mice displayed high-amplitude (> 6 hr) phase resetting in response to a 6-hr light pulse in the mid subjective night, in contrast with low-amplitude resetting (< 6 hr) observed in wild-type controls. While light-induced *Per1* and *Per2* expression in the SCN was not affected by the mutation, *Clock* homozygous and heterozygous mice showed significantly dampened *Per1* and *Per2* rhythms in the SCN (Vitaterna et al., 1994). In a subsequent study, Low-Zeddies and Takahashi (2001) generated chimeric mice from the pairing of wild-type mouse embryos to *Clock* mutant mouse embryos. The circadian behaviour of the 100+ *Clock/Clock* ↔ wild-type chimeric mice analyzed covered a broad spectrum delimited by the parental strains, and correlated with the percentage contribution of wild-type vs. *clock* mutant cells in SCN tissue (Low-Zeddies et al., 2001).

Recently, a Clock knockout (null) mouse strain was generated using the CRE-loxP system, and, in contrast with the Clock mutant mice developed by Takahashi's group, exhibited robust circadian patterns of locomotor behaviour under DD with a mean period length that was only 20 min shorter than that of wild-type controls (De Bryune, Noton, Lambert, Maywood, Weaver, and Reppert 2006). However, unlike wild-type controls which exhibited strong early night phase delays and modest late-night phase advances in response to photic stimulation, Clock-deficient mice showed no phase delays to a 4-hr light exposure in the early night (CT 12-16) and exaggerated phase advances to late-night light administration (CT 20-24) (De Bryune et al., 2006). Circadian oscillations of the mRNA and protein levels of most core clock genes within the SCN were retained in the knockouts, albeit with a reduced amplitude. These data suggest that CLOCK is dispensable for the robust circadian rhythms at the behavioural and molecular level, although it appears to play a role in regulating light responsiveness of the SCN. Although compensatory mechanisms may allow for rhythms to persist in clock null mice, upregulation of NPAS2, a CLOCK homolog (Reick, Garcia, Dudley, and McKnight 2001) that could function in a compensatory manner, was not observed in the SCN (De Bryune et al., 2006).

The precise role of PER1 in clock timing processes remains unclear from the studies of three independent *mPer1*-null mouse strains. Cermakian et al., (Cermakian, Monaco, Pando, Dierich, and Sassone-Corsi 2001) deleted exons 4-10 of *mPer1*, which encode the PAS domain, producing mutant mice that nonetheless exhibited circadian rhythms under DD, with a period length that was shorter by 1 hr than the free-running period of wild-type controls. The amplitude and phasing of (truncated) *mPer1* and *mPer2* mRNA expression in the knockout SCN were comparable to that of wild-type mice, as were light-induced phase delays and advances and photic induction of *c-fos* in the SCN (Cermakian et al., 2001). Similarly, the null mutant generated by Zheng et al., (Zheng, Albrecht, Kaasik, Sage, Lu, Vaishnav, Li, Sun, Eichele, Bradley, and Lee 2001), involving a deletion of exons 4-18, exhibited a 1-hr shortening of the mean circadian period: however, there was a broader range of period length between *mPer1*-deficient animals, as well as reduced stability of the period in any given *mPer1*-null mouse. Interestingly, while mPer2 mRNA rhythms were comparable in the SCN of wild-type and *mPer1*-deficient mice (Zheng et al., 2001). Elevated levels of PER2 protein were observed in the SCN of the mutant mice at all circadian times tested, suggesting a role for PER1 in regulating PER2 protein stability. Finally, deletion of exons 2-12, encompassing the coding region for the PER1 bHLH and PAS domains, resulted in knockout mice that gradually lost circadian rhythms of locomotor activity after 2 weeks in DD (Bae, Jin, Maywood, Hastings, Reppert, and Weaver 2001). Rhythms of mPer2, Cry1 and Bmal1 transcripts were unaltered in these *mPer1*-deficient mice. However, in contrast with the observation of Zheng et al., (2001), the expression of mPER2 and mCRY1 was significantly reduced and their rhythms blunted (Bae et al., 2001). These findings were suggested to indicate that PER1 affects protein stability and/or nuclear entry of mPER2 and mCRY1 (Bae et al., 2001).

Two strains of *Per2* mutant mice were independently generated by Bae et al., (2001) and Zheng et al., (1999, 2001). The targeting strategy of Bae et al., (2001), which eliminated all of exon 5 and a portion of exon 6, resulted in mutant mice which presumably expressed mPer2 at the mRNA but not at the protein level: these animals became arrhythmic either immediately upon transfer to DD or within 3 weeks of onset of DD. The Per2 mutant mice generated by Zheng et al., (1999) displayed a markedly shorter circadian period (t $\sim$ 22 hr) under DD compared with wild-type animals, progressing to complete loss of circadian rhythmicity after several weeks. This residual clock function observed in Per2 single mutants was abolished in Per1/Per2 double mutants (Zheng et al., 2001), which became immediately arrhythmic upon transfer to DD; moreover, the apparent entrainment of double mutants under an LD schedule was, in fact, due to the masking effect of light (ie., the ability of light to suppress locomotor activity acutely) rather than photic regulation of a functional clock. The authors concluded that mPer1 and mPer2 together are necessary for proper functioning of the mammalian circadian clock (Zheng et al., 2001).

In the same study by Bae et al., (2001), Per3 was shown to have a non-essential role in clock timing processes: double knockouts of Per1/Per3 or Per2/Per3 were essentially the same as the single Per1- or Per2-null mutants with respect to circadian locomotor behaviour. Notably, an earlier analysis of Per3-deficient mice (Shearman, Jin, Lee, Reppert, and Weaver 2000) also revealed that loss of Per3 had no effect on activity rhythms or expression of other core clock components.

bmal1-deficient mice exhibited immediate loss of behavioural rhythms when transferred to DD conditions (Bunger et al., 2000). Consistent with this, expression of *mPer1* and *mPer2* within the SCN was not rhythmic and was comparable to baseline levels (Bunger et al., 2000). These data identified BMAL1 as a critical component of the core clock and an essential regulator of *Per* transcription.

It is unlikely that the autoregulatory transcription-translation feedback loop described above can account for the stability and precision of circadian rhythms exhibited at the cellular and organismal level. Rather, multiple secondary transcription feedback processes are thought to refine both the periodicity and amplitude of the CLOCK/BMAL1-driven *per1/per2* rhythm and thus constitute key components of the clock machinery. For example, CLOCK/BMAL1-induced transcription is regulated by DEC1 and DEC2, which are structurally related to Hairy/Enhancer of Split (IIES) bHLH transcriptional repressors (Honma, Kawamoto, Takagi, Fujimoto, Sato, Noshiro, Kato, and Honma 2002). DEC1 and DEC2 inhibit CLOCK/BMAL1-dependent transcription through protein-protein association with BMAL1, as well as competitive binding to E-box elements (Honma et al., 2002). Interestingly, both DEC1 and DEC2 exhibit a circadian rhythm of expression and are themselves transactivated by CLOCK-BMAL1 by virtue of E-box motifs in their promoters (Honma et al., 2002; Hamaguchi, Fujimoto, Kawamoto, Noshiro, Maemura, Takeda, Nagai, Furukawa, Honma, Honma, Kurihara, and Kato 2004; Kawamoto, Noshiro, Sato, Maemura, Takeda, Nagai, Iwata, Fujimoto, Furukawa, Miyazaki, Honma, Honma, and Kato 2004).

In addition to rhythmic regulation of *Cry* and *Per*, transcriptional regulation of *Bmal1* via REV-ERBα represents another regulatory checkpoint of the circadian timing machinery. REV-ERBα belongs to a family of retinoic acid-related nuclear orphan receptors (ROR), which includes other members of the REV-ERB (alpha, beta) and ROR (alpha, beta, gamma) transcription factors.

REV-ERBα represses *Bmal1* transcription by direct binding to two ROR elements (A[A/T]NT[A/G]GGTCA, where N is any nucleotide) residing in the Bmal1 gene promoter (Preitner, Damiola, Lopez-Molina, Zakany, Duboule, Albrecht, and Schibler 2002). To add another level of complexity, the REV-ERBα gene is also regulated by the clock; the promoter region of Rev-Erbα contains three consensus E-box elements, and *Rev-Erbα* transcription is positively regulated by BMAL1/CLOCK and negatively regulated by PER/CRY. As expected, rhythmic expression of *Rev-Erbα* is antiphasic to that of *Bmal1* (Preitner et al., 2002). Recently, RORα was shown to activate *Bmal1* transcription, and together with REV-ERBα, antagonistically regulate *Bmal1* expression via physical competition for the Bmal1 ROR element (Sato, Panda, Miraglia, Reyes, Rudic, McNamara, Naik, Fitz-Gerald, Kay, and Hogenesch 2004). In mouse mutants which lack Rev-Erbα (knockouts) or RORα (staggerer mice), *Bmal1* transactivation was elevated or suppressed, respectively (Preitner et al., 2002; Akashi and Takumi 2005). Interestingly, both genetic mutants exhibit a shorter mean period length under DD, greater variability in period length, and a stronger response to photic manipulation, suggesting that reduced amplitude in *Bmal1* rhythm may underlie a common behavioural phenotype (Preitner et al., 2002; Akashi and Takumi 2005).

A third class of transcription factors implicated in circadian timing processes is the proline and acidic amino acid-rich basic leucine zipper (PAR bZIP) family, to which DBP (albumin D-site-binding protein), TEF (thyrotroph embryonic factor) and HLF (hepatic leukemia factor) belong (Mueller, Maire, and Schibler 1990; Drolet, Scully, Simmons, Wegner, Chu, Swanson, and Rosenfeld 1991; Inaba, Roberts, Shapiro, Jolly, Raimondi, Smith, and Look 1992). All three genes exhibit robust rhythms in the SCN and several peripheral tissues (Lopez-Molina, Conquet, Dubois-Dauphin, and Schibler 1997, Mitsui, Yamaguchi, Matsuo, Ishida, and Okamura 2001). In the case of DBP, rhythmic expression appears to be mediated by CLOCK/BMAL1 via putative E-box motifs in enhancer regions on the first and second introns (Ripperger, Shearman, Reppert, and Schibler 2000). DBP activates transcription of mPer1 at a DBP binding site ([G/A]T[G/T]A[T/C]GTAA[T/C]) within the 5' flanking region of the gene (Yamaguchi, Mitsui, Yan, Yagita, Miyake, and Okamura 2000). The structurally related basic leucine zipper transcription factor, E4BP4, is also positively and negatively regulated by CLOCK/BMAL1 and PER/CRY, respectively, but suppresses *mPer1* transcription via competitive association with the DBP binding motif (Mitsui et al., 2001). Rhythms of mRNA and protein expression of the PAR bZIP transcription factors and E4BP4 are antiphasic with respect to each other, and may contribute to high amplitude expression of target genes such as mPer1 (Mitsui et al., 2001). However, *Ddp* single knockouts or *tef/hlf* double knockouts show subtle differences in period length compared with wild-type mice, and the triple knockout display normal rhythmic expression of core clock genes, prompting the suggestion that PAR bZIP transcription factors act as regulators of outputs of the mammalian circadian timing system rather than core components of the clock.

Although not a central part of this review, it should be noted that the core clock timing is regulated post-translationally by factors that affect protein stability and subcellular localization. For example, the *tau* mutation in hamsters, which results in dramatic shortening of period length, is a missense mutation in the casein kinase I epsilon (CKIε) gene leading to attenuated kinase activity (Lowrey, Shimomura, Antoch, Yamazaki, Zemenides, Ralph, Menaker, and Takahashi 2000). CKIε phosphorylates PER proteins and targets them for degradation by the ubiquitin-proteasome machinery, resulting in a slower rate of PER accumulation and delayed transport into the nucleus (Akashi, Tsuchiya, Yoshino, and Nishida 2002). Recent studies indicate that another serine/threonine kinase, glycogen synthase kinase-3β (GSK-3β), phosphorylates and stabilizes REV-ERBα (Yin, Wang, Klein, and Lazar 2006). Treatment with lithium, a potent inhibitor of GSK-3β activity, markedly lengthens circadian period, possibly by promoting *Bmal1* expression and activity of the positive limb (Yin et al., 2006). The role of GSK-3β in circadian timing mechanisms in Droshila has been defined using genetic approaches: overexpression or reduction in the activity of its ortholog, shaggy (sgg), shortens and lengthens circadian period, respectively, via sgg-dependent phosphorylation of timeless (the dimerization partner for Period in Drosophila) (Martinek, Inonog, Manoukian, and Young 2001). The potential involvement of GSK-3β in mammalian clock timing remains to be established in future work, as lithium may have an effect on circadian period as a result of its interaction with downstream targets other than GSK-3β.

16.3 Communicating with the SCN: Neurotransmitter Systems

The SCN is subdivided into two anatomically and functionally distinct compartments (reviewed in Abrahamson and Moore 2001). The 'core' or ventrolateral SCN receives afferent projections from the RHT, IGL and the median raphe nucleus, and is considered the entry point for photic and nonphotic entrainment cues. In contrast, innervations of the 'shell' or dorsomedial SCN are distinct and originate from the thalamus, brainstem, limbic cortex, basal forebrain and other hypothalamic nuclei. A preponderance of core SCN neurons express of neuropeptides vasoactive intestinal polypeptide (VIP) and gastrin releasing peptide (GRP), while arginine vasopressin (AVP) expression is localized to shell SCN neurons. The neurotransmitter gamma-aminobutyic acid (GABA) is present in nearly all SCN neurons.

Upon photic stimulation, RHT terminals release glutamate and pituitary adenylate cyclase activating peptide (PACAP) (Hannibal, Moller, Ottersen, and Fahrenkrug 2000; Hannibal, Hindersson, Knudsen, Georg, and Fahrenkrug 2002). A large body of evidence points to glutamate as the central mediator of photic entrainment. Blockade of the ionotropic glutamate receptors (iGluRs), N-methyl-D-aspartate (NMDA) or 2-amino-3-(3-hydroxy-(-methylisoaxol-4-yl)propanoic acid (AMPA)/kainate (KA), abolishes light-induced behavioural phase shifts (Colwell and Menaker 1992). Conversely, intra-SCN injections of NMDA *in vivo* (Mintz, Marvel, Gillespie, Price, and Albers 1999), or application of NMDA or glutamate on SCN slices *in vitro* (Shibata, Watanabe, Hamada, Ono, and Watanabe 1994; Ding, Chen, Weber, Faiman, Rea, and Gillette 1994), elicit photic-like phase shifts in locomotor behaviour or SCN neuronal firing rhythms, respectively, with the resulting phase response curve (PRC) of NMDA closely resembling that of light. Similar to the effects of light, NMDA-evoked currents and Ca^{2+} transients in the SCN are both rhythmic and gated, with peak effects observed in the subjective night (Colwell 2001, Pennartz, Hamstra, and Geurtsen 2001).

AMPA/KA GluRs have been implicated in excitatory synaptic transmission in the SCN, since the AMPA/KA GluR antagonist CNQX can reduce the spontaneous firing rate and resting Ca^{2+} levels of some SCN neurons (Michel, Itri, and Colwell 2002). However, whereas AMPA-induced Ca^{2+} transients in the SCN peak in the subjective night, AMPA-evoked currents were not observed strictly in the retinorecipient region of the SCN, nor did they exhibit a circadian rhythm (Michel et al., 2002). While the cumulative evidence supports a role of NMDA GluRs as the principle mediator of photic entrainment, the action of AMPA GluRs may be required in concert to set the neuronal membrane potential to a voltage range that permits subsequent activation of NMDA GluRs by light.

Metabotropic glutamate receptors (mGluRs) expressed in the SCN may act as negative modulators of glutamatergic neurotransmission at iGluRs. Type I/II mGluR agonists have been shown to suppress NMDA- and KA-evoked Ca^{2+} transients in neurons of SCN explant cultures (Haak 1999). Based on their subcellular distribution, mGluRs may alter cellular excitability by presynaptically regulating neurotransmitter release (van den Pol, Kogelman, Ghosh, Liljelund, and Blackstone 1994; Romano, Sesma, McDonald, O'Malley, van den Pol, and Olney 1995).

Pituitary adenylate cyclase activating polypeptide (PACAP) is also a potent regulator of the circadian clock (reviewed by Hannibal ct al., 2002). PACAP was initially isolated from ovine hypothalamic tissues and found to stimulate adenylate cyclase in pituitary cells (Miyata, Arimura, Dahl, Minamino, Uehara, Jiang, Culler, and Coy 1989). It is a member of the glucagon/VIP/secretin/GHRH family of peptides and is found in two forms, a 38- and a 27-amino acid residue peptide (Miyata et al., 1989; Miyata, Jiang, Dahl, Kitada, Kubo, Fujino, Minamino, and Arimura 1990). PACAP-38 is the principal isoform expressed in the brain (Arimura, Somogyvari-Vigh, Miyata, Mizuno, Coy, and Kitada 1991; Ghatei, Takahashi, Suzuki, Gardiner, Jones, and Bloom 1993; Hannibal, Mikkelsen, Clausen, Holst, Wulff, and Fahrenkrug 1995) and it binds to two classes of receptors: the PACAP type 1 (PAC1) receptor (Spengler, Waeber, Pantaloni, Holsboer, Bockaert, Seeburg, and Journot 1993) and VIP/PACAP receptors (VPAC1 and VPAC2) (Harmar, Arimura, Gozes, Journot, Laburthe, Pisegna, Rawlings, Robbrecht, Said, Sreedharan, Wank, and Waschek 1998). In the SCN, the PAC1 receptor appears to be the primary route by which PACAP conveys photic information (Barrie, Clohessy, Buensucso, Rogers, and Allen 1996; Tanaka, Shibuya, Nagamoto, Yamashita, and Kanno 1996; Tanaka, Shibuya, Harayama, Nomura, Kabashima, Ueta, and Yamashita 1997; Hannibal, Ding, Chen, Fahrenkrug, Larsen, Gillette, and Mikkelsen 1997; von Gall, Duffield, Hastings, Kopp, Dhghani, Korf, and Stehle 1998; Dziema and Obrietan 2002). PACAP stimulates signaling through a large number of second messenger pathways, including cAMP/PKA, the release of Ca^{2+} from intracellular stores, PKC and the MAPK cascade (Barrie et al., 1996, Dziema et al., 2002).

In an initial series of studies that examined PACAP physiology in the SCN, the administration of PACAP to SCN slice was found to entrain the neuronal firing rhythm (Hannibal et al., 1997; Hannibal, Ding, Chen, Fahrenkrug, Larsen, Gillette, and Mikkelsen 1998; Harrington, Hoque, Hall, Golombek, and Biello 1999). Interestingly, unlike glutamate, the entraining effects of PACAP are not phase restricted; the administration of PACAP during both the subjective day and night is an effective entrainment cue (Hannibal et al., 1997; Hannibal et al., 1998; Harrington et al., 1999). On a related note, circadian clock light responsiveness is dramatically altered in the PAC1 receptor-deficient mice: both early-night and late-night light trigger a phase delay (Hannibal, Jamen, Nielsen, Journot, Brabet, and Fahrenkrug 2001). This contrasts with wild-type mice, where late-night photic stimulation serves as a phase-advancing entrainment cue. Interestingly, in a further contrast with wild-type mice, subjective day photic stimulation elicits a strong, albeit non-significant, phase advance of behavioural rhythms in PAC1 knockout mice (Hannibal et al., 2001). The complexities of PACAP/PAC1 signaling are evident in the discordant results obtained from the various knockout models: PACAP-deficient mice generated independently by Kawaguchi et al., (Kawaguchi, Tanaka, Isojima, Shintani, Hashimoto, Baba, and Nagai 2003) and Colwell et al., (Colwell, Michel, Itri, Rodriguez, Tam, Lelievre, Hu, and Waschek 2004) exhibited attenuated phase advances during the late night, but only Colwell et al., (2004) reported a significant reduction in early-night phase delays.

The precise set of cellular signaling events that couple PACAP to the clock have not been elucidated. At the level of clock gene expression, PACAP has been shown to trigger transcription of *per1* and *per2* (Akiyama, Minami, Nakajima, Moriya, and Shibata 2001; Nielsen, Hannibal, Knudsen, and Fahrenkrug 2001). Interestingly, the capacity to effect clock gene expression is inversely related to PACAP concentration. Along these lines, Nielsen et al., (2001) showed that micromolar concentrations of PACAP administered to SCN slice did not induce clock gene expression, but did facilitate the transcriptional efficacy of glutamate. Conversely, nanomolar concentrations of PACAP effectively induced expression of both *per1* and *per2*. In some respects these results parallel the phase-shifting effects of PACAP detected using spontaneous SCN firing. Thus, Harrington et al., (1999) showed that nanomolar concentrations of PACAP entrained the clock, whereas micromolar levels were not as effective. In the same study, picomolar levels of PACAP were found to potentiate NMDA channel conductance, whereas at higher doses (> 10 nM) PACAP attenuated NMDA receptor conductance in SCN neurons, indicating a modulatory role for PACAP. In a related study, our lab reported that PACAP dramatically potentiates depolarization-evoked calcium transients in SCN neurons (Dziema et al., 2002). These modulatory effects have been placed in a circadian context in studies showing that PACAP blocks glutamate-induced phase advances during the late subjective night in SCN slice and facilitates glutamate-evoked phase delays during the early subjective night (Chen, Buchanan, Ding, Hannibal, and Gillette 1999). It should also be noted that the phase shifting effects of PACAP during the subjective day occur via a mechanism that is independent of glutamate (Hannibal et al., 1997). These complex time-of-day- and concentration-specific effects are the likely result of the capacity of PACAP to affect multiple signaling pathways which could directly affect the clock via transcriptional activation or indirectly through the modulation of glutamate-evoked cell excitability. The finding that glutamate and PACAP are co-stored in RHT terminals (Hannibal et al., 2000; Hannibal et al., 2002) lends credence to the idea that PACAP and glutamate work in combination to effect SCN physiology. It should also be noted that PACAP has been shown to exert modulatory effects via a presynaptic mechanism (Roberto and Brunelli 2000). Thus, an examination of potential modulatory effects of PACAP on RHT transmitter release may provide additional insight into how PACAP modulates clock entrainment.

An indirect route by which photic information is relayed to the SCN is via efferent connections from the IGL (reviewed in Moore 1996). The IGL receives projections from melanopsin-producing RGCs (Hannibal et al., 2002) and is the principle source of neuropeptide Y (NPY)-containing terminals, which co-store the inhibitory neurotransmitter GABA, to the SCN (Moore 1996). In addition, neuropeptides such as neurotensin and enkephalin, which are also secreted by IGL neurons, may influence clock entrainment (Morin and Blanchard 2001). The IGL has been suggested to be the site of integration of photic and nonphotic inputs to the circadian system: it is activated by both classes of entraining cues (Janik, Mikkelsen, and Mrosovsky 1995; Thankachan and Rusak 2005), and has widespread efferent and afferent projections to and from multiple loci in the brain that are potentially involved in the processing of nonphotic entrainment (Vrang, Mrosovsky, and Mikkelsen 2003). Moreover, IGL ablation has been shown to block activity-induced phase advances (Wickland and Turek 1994) and modify photic responses (Pickard, Ralph, and Menaker 1987; Edelstein and Amir 1999).

16.4 CREB and Entrainment of the SCN Clock

Clock entrainment ultimately requires activation of transcriptional networks that can change the phase of the molecular clock. The CREB/CRE transcriptional pathway plays a pivotal role in coupling the myriad of signaling events actuated by light to changes in gene expression in the SCN that underlie entrainment. Using a CRE-β-galactosidase transgenic reporter mouse strain, Obrietan et al., (Obrietan, Impey, Smith, Athos, and Storm 1999) demonstrated that CRE-mediated gene expression in the SCN is induced by light exposure during the subjective night but not during the subjective day. In keeping with this, nocturnal light causes the rapid phosphorylation of CREB at Ser-133, which is essential for promoting the recruitment of its co-activator CBP and p300 (Ginty, Kornhauser, Thompson, Bading, Mayo, Takahashi, and Greenberg 1993). A number of protein kinases which have been implicated in clock entrainment can phosphorylate CREB at Ser-133, including PKA, PKC, PKG, CaMKII, CaMKIV, Msk1, and Msk2 (reviewed in Johannessen, Delghandi, and Moens 2004). CaMKII can additionally phosphorylate CREB at Ser-142 (Sun, Enslen, Myung, and Maurer 1994; Sun and Maurer 1995). One study showed that a light stimulus in the early subjective night triggers phosphorylation of CREB at Ser-142 in SCN neurons *in vivo*, as does glutamate application to hypothalamic brain slices at the same circadian time *in vitro* (Gau, Lemberger, von Gall, Kretz, Le Minh, Gass, Schmid, Schibler, Korf, and Schutz 2002). In addition to the phase-restricted photic activation of the CREB/CRE pathway, CRE-mediated transcription and

CREB phosphorylation exhibited a pronounced circadian rhythm: levels of Ser-133 phosphorylated CREB and Ser-142 phosphorylated CREB peaked in the late subjective night (CT 18-22) and early subjective day (CT 1), respectively, and preceded the peak levels of CRE-mediated gene expression (CT 6) (Obrietan et al., 1999; Gau et al., 2002). While there is no genetic model to date that addresses the significance of Ser-133 phosphorylation on clock entrainment, both light-induced phase delays and advances were attenuated in a knock-in mouse strain lacking the Ser-142 phosphorylation site (CREB^{S142A}) (Gau et al., 2002). In an attempt to address the role of the CREB/CRE pathway in entrainment, Tischkau et al., (Tischkau, Mitchell, Tyan, Buchanan, and Gillette 2003) used synthetic oligodeoxynucleotide (ODN) decoys corresponding to a double-stranded cAMP response element (CRE) to inhibit CRE-mediated transcription via sequestration of CREB. The ability of the CRE decoy to block binding of endogenous CREB to a CRE probe and to inhibit CRE-mediated transcription was verified by electromobility shift assays (EMSA) and luciferase reporter assays, respectively, using extracts prepared from CRE decoy-transfected SCN2.2 cells in culture (Tischkau et al., 2003). Infusion of CRE decoy into the SCN 15 min prior to a CT 22 light pulse eliminated the behavioural phase advances normally observed in subjects which received an infusion of vehicle or of mismatched ODN decoy (Tischkau et al., 2003). When applied to a brain slice preparation *in vitro*, CRE decoys inhibited glutamate-induced phase advances (at CT 20) of SCN neuronal firing rhythms as well as glutamate-induced elevation of mPer1 mRNA abundance (Tischkau et al., 2003).

A number of core clock genes possess cAMP response elements in their promoter regions and are potential targets of light-induced activation by CREB. *mPer1* and *mPer2* each have a single canonical CRE (TGACGTCA) centered at -1728 and -1606, respectively, upstream of the +1 transcription start site (Travnickova-Bendova et al., 2002). While both *mPer1* and *mPer2* CREs can bind to CREB in SCN nuclear extracts, only *mPer1*, promoter activity was elevated in response to synergistic stimulation of cAMP and MAPK pathways (forskolin in combination with EGF or TPA) (Travnickova-Bendova et al., 2002). Stimuli-dependent activation of *mPer1* promoter activity was absent in *mPer1* promoter constructs carrying a mutated, non-functional CRE (Travnickova-Bendova et al., 2002), indicating an essential role of the CREB/CRE pathway in inducible *mPer1* expression. The biological significance of the mPer2 CRE remains an outstanding issue. This study also suggested that CLOCK/BMAL1-mediated *mPer1* transcription is distinct from (ie., does not require) CREB/CRE transactivation. (NOTE: More recent evidence points to signal-dependent regulation of CLOCK/BMAL1 transcription [Yujnovsky et al., 2006]. In cultured neuroglioma cell lines [NG108-15], dopamine D2 receptor-dependent activation of the mPer1 promoter *in vitro* involved recruitment of CBP to the CLOCK/BMAL1 heterocomplex. In this experimental paradigm, D2R-mediated *mPer1* activation required intact E-box motifs but not the CRE in the *mPer1* promoter, and was abolished by treatment with the MEK inhibitor U0126 [Yujnovsky et al., 2006]). Expression of *Dec1*, but not *Dec2*, within the SCN is induced following light exposure in the subjective night (Honma et al., 2002). Although mechanistic data are lacking, the 5' flanking region of the *Dec1* gene contains a CRE site and activation of cAMP-dependent pathways by forskolin and dibutyryl cAMP rapidly elevates *Dec1* mRNA levels in various cell types *in vitro* (Shen, Kawamoto, Teramoto, Makihira, Fujimoto, Yan, Noshiro, and Kato 2001; Teramoto, Nakamasu, Noshiro, Matsuda, Gotoh, Shen, Tsutsumi, Kawamoto, Iwamoto, and Kato 2001).

A number of CREB-regulated immediate early genes (IEGs), such as *c-fos*, *fosB*, *junB*, *nur77*, *zif268* and *egr-3*, have been shown to be induced in the SCN by nocturnal light exposure (Kornhauser, Nelson, Mayo, and Takahashi 1990; Kornhauser, Nelson, Mayo, and Takahashi 1992; Lin, Kornhauser, Singh, Mayo, and Takahashi 1997; Morris, Viswanathan, Kuhlman, Davis, and Weitz 1998). Photic induction of both *c-fos* and *junB* is phase-restricted to the subjective night, and strictly accompanies behavioural phase shifting even at threshold levels of illumination (Kornhauser et al., 1990, 1992). The Fos and Jun proteins represent members of the AP-1 (activator protein 1) family of bZIP transcription factors, which form functional homo- (Jun/Jun) or hetero- (Fos/Jun) dimers to activate transcription from AP-1 binding sequences (TGA(C/G)TCA) (reviewed in Karin, Liu, and Zandi 1997). Photic stimulation in the subjective night elevates AP-1 binding activity in SCN nuclear extracts (Kornhauser et al., 1992). As an aside, genetic evidence supporting an essential role of c-Fos in photic entrainment is lacking, since light-induced phase delays and advances were only modestly attenuated in *c-fos*-deficient mice (Honrado, Johnson, Golombek, Spiegelman, Papaioannou, and Ralph 1996). However, the possibility that AP-1 complexes formed by alternative dimerization partners may compensate for the absence of a particular subunit, and/or other compensatory mechanisms, which may come into play during development, confounds the interpretation of results using knockout mice. Indeed, acute knockdown of *c-fos* or *junB* expression in adult hamsters using antisense oligonucleotide infusion into the SCN inhibited light-induced phase shifts of behavioural rhythms (Wollnik, Brysch, Uhlmann, Gillardon, Bravo, Zimmermann, Schlingensiepen, and Herdegen 1995). The expression of two other IEGs, *nur77* and *zif268*, is also induced in the SCN following photic stimulation in the subjective night. Unlike *c-fos* induction, induction of *nur77* and *zif268* is extremely sensitive to irradiance levels that are 10-100 fold below the threshold required for behavioural phase shifts (Lin et al., 1997). In the hamster, *egr-3* was induced rapidly (within 30 min) in the SCN following light administration in the early (CT 14) and mid (CT 19) subjective night (Morris et al., 1998). Unlike *c-fos* and *junB,* which are induced in both the dorsal and ventral regions and throughout the rostrocaudal axis of the hamster SCN, light-induced *egr-3* expression is restricted to the ventral core of the central SCN, similar to the distribution of calbindin-D_{28K} immunoreactivity, suggesting that *egr-3* induction may be a direct response to entrainment cues arriving from the RHT (Morris et al., 1998).

16.5 Kinase Signaling: from the Cell Surface to the Nucleus

The activation of key transcriptional events that couple light to clock entrainment requires, firstly, that the photic cue be relayed to the cell nucleus via one (or several) signal transduction pathway(s). Multiple lines of evidence implicate the p42/p44 mitogen-activated protein kinase (MAPK) pathway as a pivotal regulator of photic entrainment of the clock. Signaling within the p42/p44 MAPK cascade involves sequential activation of three protein kinases: the monomeric G protein Ras is activated by a number of cell surface receptors (including G protein-coupled receptors and neurotrophin receptors), recruiting Raf kinases to the cell membrane for subsequent activation. Activated Raf phosphorylates and activates the kinase MEK, which in turn activates the kinases ERK1/2 by phosphorylation at Thr-202 and Tyr-204 (Crews, Alessandrini, and Erikson 1992; Zheng and Guan 1993). Consequently, p-ERK can translocate to other subcellular compartments, including the cell nucleus, and phosphorylate a number of potential cytosolic and nuclear substrates, including, amongst others, the transcription factor Elk-1 (an ETS domain transcription factor), and members of the p90 ribosomal S6 kinase (RSK) and mitogen- and stress-dependent kinase (MSK) families (Chen, Sarnecki, and Blenis 1992; Gille, Kortenjann, Thomae, Moomaw, Slaughter, Cobb, and Shaw 1995; Deak, Clifton, Lucocq, and Alessi 1998).

Obrietan et al., (Obrietan, Impey, and Storm 1998) provided the first evidence in support of a role of the p42/p44 MAPK pathway in entrainment processes in the SCN. A brief light pulse in the early (CT 15) and late (CT 22.5) subjective night robustly increased the levels of the dually phosphorylated forms of p42/p44 in the SCN (Obrietan et al., 1998). Light-induced p-ERK was observed in both the dorsal and ventral regions of the SCN, with highest expression in the central part of the rostrocaudal axis (Obrietan et al., 1998). Light-induced ERK activation was phase restricted, and not observed following light administration in the mid subjective day (CT 6) (Obrietan et al., 1998). Application of glutamate, forskolin, or potassium (to elicit depolarization) directly to SCN slices or cultured SCN neurons *in vitro* triggered an increase in p-ERK expression (Obrietan et al., 1998). Additionally, late-night photic stimulation resulted in co-localization of p-CREB and p-ERK in the cell nucleus. The MEK inhibitor, PD 98095, inhibited inducible p-ERK expression in cultured neurons, as well as glutamate-mediated CREB phosphorylation in SCN slices (Obrietan et al., 1998). *In vivo* disruption of MAPK signaling by infusion of the MEK inhibitor, U0126, into the third ventricle blocked phase delays in behavioural rhythms following light administration at CT 15 (Butcher, Dziema, Collamore, Burgoon, and Obrietan 2002). In addition to light-dependent regulation of the MAPK pathway, p-ERK expression in the SCN is clock-regulated and exhibits a circadian rhythm with peak levels reached at the mid to late subjective day (CT 6-10) (Obrietan et al., 1998).

Other studies by Obrietan et al., have been aimed to address the signaling events that are upstream and downstream of MAPK activation in the SCN. In the study of Butcher et al., (2002), calcium/calmodulin kinase signaling was demonstrated to be upstream of light-induced MAPK activation, as infusion of the broad-spectrum CaMK inhibitor, KN-62, strongly attenuated inducible p-ERK expression. Abrogation of the p42/p44 MAPK pathway by U0126 infusion blocked induction of the immediate early genes (IEGs) *c-fos*, *junB* and *zif268* by a brief light pulse at CT 15, indicating that MAPK activation couples light to transcriptional activation (Dziema et al., 2003).

More recently, Butcher et al., (Butcher, Lee, Hsieh, and Obrietan 2004; Butcher, Lee, Cheng, and Obrietan 2005) examined the role of additional kinases as an intermediate between MAPK and transcriptional activation in clock timing processes. Likely candidates include members of the family of 90 kDa ribosomal S6 kinases (RSKs) and mitogen- and stress-activated kinases (MSKs). Originally identified in *Xenopus* as a kinase for the 31-kDa protein (S6) that is a component of the 40 S ribosomal subunit (Erikson and Maller 1985), four RSK isoforms (RSK1-4) have been reported in mammals to date. RSKs are Ser/Thr kinases that consist of an ERK binding site and an internal linker region bridging two functional kinase domains: a C-terminal kinase (CTK) domain with autophosphorylation capacity and an N-terminal kinase (NTK) domain that phosphorylates other target substrates (Frodin and Gammeltoft 1999).

Butcher et al., (2004) provided the first evidence that RSKs are potential effectors of MAPK signaling within the murine SCN. Phosphorylation within the linker region of RSK1, a requisite step in activation of the kinase, was shown to be induced following photic stimulation in the early (CT 15), and to a lesser extent in the late (CT 22), subjective night (Butcher et al., 2004). However, daytime (CT 6) light exposure had no effect on phospho-RSK1 (pRKS1) expression (Butcher et al., 2004). Expression of PDK1, as well as other isoforms of RSK (RSK2, RSK3), within the SCN was demonstrated (Butcher et al., 2004). Immunofluorescent labeling revealed that ~70% of neurons (NeuN-immunoreactive) in the ventral SCN expressed light-activated RSK1, and that there was strong colocalization of pRSK1 and pERK in this region following a CT 15 light pulse (Butcher et al., 2004). Photic stimulation not only triggered RSK1 phosphorylation, but as determined by a kinase assay using immunoprecipitated RSK1 from SCN tissue, resulted in a three-fold enhancement of kinase activity (Butcher et al., 2004). Finally, U0126 infusion into the third ventricle prior to a light pulse at CT 15 blocked the light-induced increase in pRSK1 levels in the SCN, indicating that RSK1 is a *bona fide* downstream target of MAPK signaling within the SCN (Butcher et al., 2004).

In addition to RSKs, the MAPK pathway also regulates transcription via mitogen- and stress-activated protein kinase 1 and 2 (MSKs). MSKs are structurally and functionally related to RSKs (Deak et al., 1998). There are, however, distinctive differences between the two kinase families. For example, MSKs are nuclear localized, are not dependent on PDK-1 for activation and can be activated by both the ERK/MAPK and p38/MAPK pathways (Deak et al., 1998; Pierrat, Correia, Mary, Tomas-Zuber, and Lesslauer 1998). Furthermore, a number of biochemical studies have shown that MSK1 is a more effective Ser-133 CREB kinase than RSK2 (Pierrat et al., 1998; Deak et al., 1998; Arthur and Cohen 2000). Paralleling the work on RSK in the SCN, Butcher et al., (2005) showed that MSK1 is an ERK/MAPK-regulated, light-responsive kinase in the SCN and, using reporter gene technology, that MSK1 couples to the induction of *mPeriod1* via a CREB-dependent mechanism. Interestingly, MSK1 does not appear to be particularly responsive to glutamatergic stimulation (Arthur, Fong, Dwyer, Davare, Reese, Obrietan, and Impey 2004), thus raising the possibility that light-induced MSK1 activation is mediated by PACAP. Indeed, the infusion of PAC1 receptor antagonist PACAP 6-38 into the 3rd ventricle attenuated light-induced MSK1 activation. Together these data suggest that a signaling cassette formed by PACAP, ERK/MAPK and MSK contributes to the phase shifting effects of light. Additional work will be required to address the roles of both RSKs in MSKs as regulators of clock timing and entrainment.

The diffusible gas nitric oxide (NO) has been implicated in clock entrainment. NO is synthesized as a reaction product from Ca^{2+}/calmodulin-activated NO synthase (NOS) oxidizing arginine to citrulline. Of the three NOS isozymes, neuronal NOS has been the most extensively examined for a role in clock entrainment. NO triggers the synthesis of guanosine 3',5' monophosphate (cGMP), which in turn stimulates the activation of cGMP-dependent protein kinase I and II (PKG). As a Ser/Thr kinase, PKG has been implicated in an array of neuronal processes, including glutamate-induced LTP in the hippocampus and apoptosis (Fiscus 2002). Both PKG activity and cGMP levels are rhythmically regulated in the SCN (Tischkau, Weber, Abbott, Mitchell, and Gillette 2003). Interestingly, this rhythmicity appears to be regulated via two processes: rhythmic expression of PKG II and circadian regulation of cGMP phosphodiesterase activity (Ferreyra and Golombek 2001). A number of studies have shown that NO/PKG affects the clock during the late night. Along these lines, photic stimulation triggers an increase in cGMP levels and PKG activity during the late subjective night (Golombek, Agostino, Plano, and Ferreyra 2004), and the blockade of PKG signaling disrupts light- and glutamate-induced phase advancing of the clock (Weber, Gannon, and Rea 1995; Mathur, Golombek, and Ralph 1996; Ding, Buchanan, Tischkau, Chen, Kuriashkina, Faiman, Alster, McPherson, Campbell, and Gillette 1998). Likewise, PKG activation using the NO donor S-nitroso-N-acetyl penicillamine (SNAP) advanced the peak of the SCN firing rate in brain slice, as well as triggering the phosphorylation of CREB during the late subjective night (Ding, Faiman, Hurst, Kuriashkina, and Gillette 1997). However, PKGII knockout mice exhibit attenuated phase delaying during the early night, whereas phase-advancing was not affected. This is, in some respects counter to data generated using pharmacological approaches showing that NOS/PKG signaling contributes specifically to the phase advancing effects of light during the late night. Interestingly, it was reported (Kriegsfeld, Demas, Lee, Dawson, Dawson, and Nelson 1999) that light entrainment was not affected in nNOS knockout mice. Thus, a rigorous examination of these discrepancies is required to gain a clear picture of how NO signaling regulates the clock.

Calcium/calmodulin kinase II (CaMKII) is a Ser/Thr kinase that has been implicated in clock entrainment. CaMKII is a member of the Ca^{2+}/CaM-dependent kinase family. Through the use of pharmacological approaches and gene deletion studies, CaMKII has been shown to regulate the function of a large number of proteins involved in both cellular homeostasis, as well as key aspects of neuron physiology such as neurotransmitter release, channel conductance, and plasticity-associated gene expression (Malenka, Kauer, Perkel, Mauk, Kelly, Nicoll, and Waxham 1989; Otmakhov, Griffith, and Lisman 1997; Hinds, Tonegawa, and Malinow 1998; Hudmon and Schulman 2002). The CaM-KII α holoenzyme contains 12 subunits that form hexagonally-shaped rings (Soderling, Chang, and Brickey 2001). Each subunit consists of an amino-terminal catalytic region, an autoinhibitory domain, a Ca^{2+}/CaM binding motif and an association domain located at the carboxy terminus (Soderling 1996). CaMKII activity is initiated by an increase in intracellular Ca^{2+}. Elevated Ca^{2+} binds to calmodulin, which in turn associates with CaMKII, thus triggering a conformational change that allows the dissociation of the inhibitory domain from the catalytic domain. In addition, Ca^{2+}/calmodulin association triggers Thr-286 autophosphorylation. This event allows for kinase activity to persist well after the Ca^{2+} rise has subsided. This capacity for lasting enzymatic activity is thought to play a central role in the ability of CaMKII to function as a Ca^{2+} frequency detector (Hudmon and Schulman 2002).

With respect to the circadian clock, the Thr-286 phosphorylated form of CaMKII has been shown to be driven by SCN pacemaker activity and by photic stimulation (Yokota, Yamamoto, Moriya, Akiyama, Fukunaga, Miyamoto, and Shibata 2001; Agostino, Ferreyra, Murad, Watanabe, and Golombek 2004). In addition, the infusion of the broad spectrum CaMK inhibitors KN-62 and KN-93 into the SCN has been shown to abrogate light- and glutamate-induced clock entrainment and clock gene expression (Golombek and Ralph 1994; Fukushima, Shimazoe, Shibata, Watanabe, Ono, Hamada, and Watanabe 1997; Yokota et al., 2001; Agostino et al., 2004). However, given that both KN-62 and KN-93 effectively inhibit other members of the CaMK family, including CaMKIV, the precise mechanistic contribution of CaMKII to clock entrainment cannot be elucidated. Along these lines, unlike CaMKIV, CaMKII appears to be a weak activator of CREB (Sun et al., 1994). Interestingly, as noted above CaMKII can function as an upstream activator of light-induced MAPK activation and nNOS (Butcher et al., 2002; Agostino et al., 2004), thus CaMKII may function as a trigger which elicits activation of pathways which in turn affect the clock via a more direct mechanism. Additional studies will be required to establish where within the complex network of light and clock regulated signaling events CaMKII regulates the SCN timing process.

In this regard, the examination of signaling crosstalk and input integration across multiple kinase networks is key to our understanding of how synaptic activity regulates clock entrainment. Dexras1 is a good example of how signaling pathway activity is regulated to generate context-specific responsiveness of the SCN clock. Dexras1 is a novel member of the Ras family of monomeric G proteins. It has complex signaling qualities that place it within several signal transduction pathways (reviewed in Cheng and Obrietan 2006). Along these lines, Dexras1 has been show to function as a guanine nucleotide exchange factor (GEF) for $G_{i/o}$ heterotrimeric G proteins in the absence of receptor stimulation (Cismowski, Ma, Ribas, Xie, Spruyt, Lizano, Lanier, and Duzic 2000). In addition, Dexras1 interacts with the nNOS adaptor protein CAPON, and can be activated via S-nitrosylation, thus placing Dexras1 downstream of glutamatergic neurotransmission (Fang, Jaffrey, Sawa, Ye, Luo, and Snyder 2000). In response to stimulation of these pathways, Dexras1 regulates ERK/MAPK signaling in a bi-directional manner (Cismowski, Takesono, Ma, Lizano, Xie, Fuernkranz, Lanier, and Duzic 1999, Cismowski et al., 2000, Graham, Prossnitz, and Dorin 2002). Dexras1 initially drew the attention of circadian biologist when it was shown to be a clock-regulated gene in the SCN (Takahashi, Umeda, Tsutsumi, Fukumura, Ohkaze, Sujino, van der Horst, Yasui, Inouye, Fujimori, Ohhata, Araki, and Abe 2003). To address the role of Dexras1 in the SCN, Cheng et al., (Cheng, Obrietan, Cain, Lee, Agostino, Joza, Harrington, Ralph, and Penninger 2004) generated a *dexras1*-null strain of mice. The targeting construct eliminated the GTP binding domain, hydrolysis domain and effector loop of Dexras1. Cheng et al., (2004) demonstrated that Dexras1 is an important modulator of photic and non-photic responsiveness of the SCN clock. *Dexras1*$^{-/-}$ mice exhibited attenuated early night light-induced phase delays and enhanced nonphotic responses to arousal. The attenuated photic responsiveness of *dexras1*$^{-/-}$ mice correlated with a decrease in light-induced MAPK pathway activation, whereas the enhanced nonphotic responses were attributed to derepression of NPYergic neurotransmission from the IGL. Along these lines, electrophysiological experiments found that NPY-mediated signaling was augmented in *dexras1*$^{-/-}$ SCN. Thus, Dexras1 functions as a modulator of photic and nonphotic stimuli: attenuating early night photic responses and enhancing nonphotic response to arousal. Interestingly, circadian behavioral rhythms of *dexras1*$^{-/-}$ mice exhibited instability in constant light, suggesting that Dexras1 plays a role in cellular pacemaker communication within the SCN.

References

Abrahamson, E.E. and Moore, R.Y. (2001) Suprachiasmatic nucleus in the mouse: retinal innervation, intrinsic organization and efferent projections. Brain Res. 916, 172–191.

Agostino, P.V., Ferreyra, G.A., Murad, A.D., Watanabe, Y. and Golombek, D.A. (2004) Diurnal, circadian and photic regulation of calcium/calmodulin kinase II and neuronal nitric oxide synthase in the hamster suprachiasmatic nuclei. Neurochem. Int. 44, 617–625.

Akashi, M. and Takumi, T. (2005) The orphan nuclear receptor RORalpha regulates circadian transcription of the mammalian core-clock Bmal1. Nat. Struct. Mol. Biol. 12, 441–448.

Akashi, M., Tsuchiya, Y., Yoshino, T. and Nishida, E. (2002) Control of intracellular dynamics of mammalian period proteins by casein kinase I epsilon (CKIepsilon) and CKIdelta in cultured cells. Mol. Cell. Biol. 22, 1693–1703.

Akiyama, M., Minami, Y., Nakajima, T., Moriya, T. and Shibata, S. (2001) Calcium and pituitary adenylate cyclase-activating polypeptide induced expression of circadian clock gene mPer1 in the mouse cerebellar granule cell culture. J. Neurochem. 78, 499–508.

Arimura, A., Somogyvari-Vigh, A., Miyata, A., Mizuno, K., Coy, D.H. and Kitada, C. (1991) Tissue distribution of PACAP as determined by RIA; highly abundant in the rat brain and testes. Endocrinology 129, 2787–2789.

Arthur, J.S. and Cohen, P. (2000) MSK1 is required for CREB phosphorylation in response to mitogens in mouse embryonic stem cells. FEBS Lett. 482, 44–48.

Arthur, J.S., Fong, A.L., Dwyer, J.M., Davare, M., Reese, E., Obrietan, K. and Impey, S. (2004) Mitogen- and stress-activated protein kinase 1 mediates cAMP response element-binding protein phosphorylation and activation by neurotrophins. J. Neurosci. 24, 4324–4332.

Bae, K., Jin, X., Maywood, E.S., Hastings, M.H., Reppert, S.M. and Weaver, D.R. (2001) Differential functions of mPer1, mPer2, and mPer3 in the SCN circadian clock. Neuron 30, 525–536.

Barrie, A.P., Clohessy, A.M., Buensucso, C.S., Rogers, M.V. and Allen, J.M. (1997) Pituitary adenylyl cyclase-activating peptide stimulates extracellular signal-regulated kinase 1 or 2 (ERK1/2) activity in a Ras-independent, mitogen-activated protein Kinase/ERK kinase 1 or 2-dependent manner in PC12 cells. J. Biol. Chem. 272, 19666–19671.

Bergstrom, A.L., Hannibal, J., Hindersson, P. and Fahrenkrug, J. (2003) Light-induced phase shift in the Syrian hamster (Mesocricetus auratus) is attenuated by the PACAP receptor antagonist PACAP6-38 or PACAP immunoneutralization. Eur. J. Neurosci. 18, 2552–2562.

Bunger, M.K., Wilsbacher, L.D., Moran, S.M., Clendenin, C., Radcliffe, L.A., Hogenesch, J.B., Simon, M.C., Takahashi, J.S. and Bradfield, C.A. (2000) Mop3 is an essential component of the master circadian pacemaker in mammals. Cell 103, 1009–1017.

Butcher, G.Q., Dziema, H., Collamore, M., Burgoon, P.W. and Obrietan, K. (2002) The p42/p44 mitogen-activated protein kinase pathway couples photic input to circadian clock entrainment. J. Biol. Chem. 277, 29519–29525.

Butcher, G.Q., Lee, B., Cheng, H.Y. and Obrietan, K. (2005) Light stimulates MSK1 activation in the suprachiasmatic nucleus via a PACAP-ERK/MAP kinase-dependent mechanism. J. Neurosci. 25, 5305–5313.

Butcher, G.Q., Lee, B., Hsieh, F. and Obrietan K. (2004) Light- and clock-dependent regulation of ribosomal S6 kinase activity in the suprachiasmatic nucleus. Eur. J. Neurosci. 19, 907–915.

Cermakian, N., Monaco, L., Pando, M.P., Dierich, A. and Sassone-Corsi, P. (2001) Altered behavioral rhythms and clock gene expression in mice with a targeted mutation in the Period1 gene. EMBO J. 20, 3967–3974.

Chen, D., Buchanan, G.F., Ding, J.M., Hannibal, J. and Gillette, M.U. (1999) Pituitary adenylyl cyclase-activating peptide: a pivotal modulator of glutamatergic regulation of the suprachiasmatic circadian clock. Proc. Natl. Acad. Sci. USA 96, 13468–13473.

Chen, R.H., Sarnecki, C. and Blenis, J. (1992) Nuclear localization and regulation of ERK- and RSK-encoded protein kinases. Mol. Cell. Biol. 12, 915–927.

Cheng, H.Y. and Obrietan, K. (2006) Dexras1: shaping the responsiveness of the circadian clock. Semin. Cell. Dev. Biol. 17, 345–351.

Cheng, H.Y., Obrietan, K., Cain, S.W., Lee, B.Y., Agostino, P.V., Joza, N.A., Harrington, M.E., Ralph, M.R. and Penninger, J.M. (2004) Dexras1 potentiates photic and suppresses nonphotic responses of the circadian clock. Neuron 43, 715–728.

Cismowski, M.J., Ma, C., Ribas, C., Xie, X., Spruyt, M., Lizano, J.S., Lanier, S.M. and Duzic, E. (2000) Activation of heterotrimeric G-protein signaling by a ras-related protein. Implications for signal transduction. J. Biol. Chem. 275, 23421–23424.

Cismowski, M.J., Takesono, A., Ma, C., Lizano, J.S., Xie, X., Fuernkranz, H., Lanier, S.M., and Duzic, E. (1999) Genetic screens in yeast to identify mammalian nonreceptor modulators of G-protein signaling. Nat. Biotechnol. 17, 878–883.

Colwell, C.S. (2001) NMDA-evoked calcium transients and currents in the suprachiasmatic nucleus: Gating by the circadian system. Eur. J. Neurosci. 13, 1420–1428.

Colwell, C.S. and Menaker, M. (1992) NMDA as well as non-NMDA receptor antagonists can prevent the phase shifting effects of light on the circadian system of the golden hamster. J. Biol. Rhythms 7, 125–136.

Colwell, C.S., Michel, S., Itri, J., Rodriguez, W., Tam, J., Lelievre, V., Hu, Z. and Waschek, J.A. (2004) Selective deficits in the circadian light response in mice lacking PACAP. Am. J. Physiol. Regul. Integr. Comp. Physiol. 287, R1194–R1201.

Crews, C.M., Alessandrini, A. and Erikson, R.L. (1992) The primary structure of MEK, a protein kinase that phosphorylates the ERK gene product. Science 258, 478–480.

Deak, M., Clifton, A.D., Lucocq, L.M. and Alessi, D.R. (1998) Mitogen- and stress-activated protein kinase-1 (MSK1) is directly activated by MAPK and SAPK2/p38, and may mediate activation of CREB. EMBO J. 17, 4426–4441.

Debruyne, J.P., Noton, E., Lambert, C.M., Maywood, E.S., Weaver, D.R. and Reppert, S.M. (2006) A clock shock: mouse CLOCK is not required for circadian oscillator function. Neuron 50, 465–477.

Ding, J.M., Buchanan, G.F., Tischkau, S.A., Chen, D., Kuriashkina, L., Faiman, L.E., Alster, J.M., McPherson, P.S., Campbell, K.P., and Gillette, M.U. (1998) A neuronal ryanodine receptor mediates light-induced phase delays of the circadian clock. Nature 394, 381–384.

Ding, J.M., Chen, D., Weber, E.T., Faiman, L.E., Rea, M.A. and Gillette, M.U. (1994) Resetting the biological clock: mediation of nocturnal circadian shifts by glutamate and NO. Science 266, 1713–1717.

Ding, J.M., Faiman, L.E., Hurst, W.J., Kuriashkina, L.R., and Gillette, M.U. (1997) Resetting the biological clock: mediation of nocturnal CREB phosphorylation via light, glutamate, and nitric oxide. J. Neurosci. 17, 667–675.

Dioum, E.M., Rutter, J., Tuckerman, J.R., Gonzalez, G., Gilles-Gonzalez, M.A. and McKnight, S.L. (2002) NPAS2: a gas-responsive transcription factor. Science 298, 2385–2387.

Drolet, D.W., Scully, K.M., Simmons, D.M., Wegner, M., Chu, K.T., Swanson, L.W. and Rosenfeld, M.G. (1991) TEF, a transcription factor expressed specifically in the anterior pituitary during embryogenesis, defines a new class of leucine zipper proteins. Genes & Dev. 5, 1739–1753.

Dziema, H., Oatis, B., Butcher, G.Q., Yates, R., Hoyt, K.R. and Obrietan, K. (2003) The ERK/MAP kinase pathway couples light to immediate-early gene expression in the suprachiasmatic nucleus. Eur. J. Neurosci. 17, 1617–1627.

Dziema, H. and Obrietan, K. (2002) PACAP potentiates L-type calcium channel conductance in suprachiasmatic nucleus neurons by activating the MAPK pathway. J. Neurophysiol. 88, 1374–1386.

Edelstein, K. and Amir, S. (1999) The role of the intergeniculate leaflet in entrainment of circadian rhythms to a skeleton photoperiod. J. Neurosci. 19, 372–380.

Erikson, E. and Maller, J.L. (1985) A protein kinase from Xenopus eggs specific for ribosomal protein S6. Proc. Natl. Acad. Sci. USA 82, 742–746.

Etchegaray, J.P., Lee, C., Wade, P.A. and Reppert, S.M. (2003) Rhythmic histone acetylation underlies transcription in the mammalian circadian clock. Nature 421, 177–182.

Fang, M., Jaffrey, S.R., Sawa, A., Ye, K., Luo, X. and Snyder, S.H. (2000) Dexras1: a G protein specifically coupled to neuronal nitric oxid synthase via CAPON. Neuron 28, 183–193.

Ferreyra, G.A. and Golombek, D.A. (2001) Rhythmicity of the cGMP-related signal transduction pathway in the mammalian circadian system. Am. J. Physiol. Regul. Integr. Comp. Physiol. 280, R1348-R1355.

Fiscus, R.R. (2002) Involvement of cyclic GMP and protein kinase G in the regulation of apoptosis and survival in neural cells. Neurosignals 11, 175–190.

Frodin, M. and Gammeltoft, S. (1999) Role and regulation of 90 kDa ribosomal S6 kinase (RSK) in signal transduction. Mol. Cell. Endocrinol. 151, 65–77.

Fukushima, T., Shimazoe, T., Shibata, S., Watanabe, A., Ono, M., Hamada, T. and Watanabe, S. (1997) The involvement of calmodulin and Ca2+/calmodulin-dependent protein kinase II in the circadian rhythms controlled by the suprachiasmatic nucleus. Neurosci. Lett. 227, 45–48.

Gau, D., Lemberger, T., von Gall, C., Kretz, O., Le Minh, N., Gass, P., Schmid, W., Schibler, U., Korf, H.W. and Schutz, G. (2002) Phosphorylation of CREB Ser142 regulates light-induced phase shifts of the circadian clock. Neuron 34, 245–253.

Gekakis, N., Staknis, D., Nguyen, H.B., Davis, F.C., Wilsbacher, L.D., King, D.P., Takahashi, J.S. and Weitz, C.J. (1998) Role of the CLOCK protein in the mammalian circadian mechanism. Science 280, 1564–1569.

Ghatei, M.A., Takahashi, K., Suzuki, Y., Gardiner, J., Jones, P.M. and Bloom, S.R. (1993) Distribution, molecular characterization of pituitary adenylate cyclase-activating polypeptide and its precursor encoding messenger RNA in human and rat tissues. J. Endocrinol. 136, 159–166.

Gille, H., Kortenjann, M., Thomae, O., Moomaw, C., Slaughter, C., Cobb, M.H. and Shaw, P.E. (1995) ERK phosphorylation potentiates Elk-1-mediated ternary complex formation and transactivation. EMBO J. 14, 951–962.

Ginty, D.D., Kornhauser, J.M., Thompson, M.A., Bading, H., Mayo, K.E., Takahashi, J.S. and Greenberg, M.E. (1993) Regulation of CREB phosphorylation in the suprachiasmatic nucleus by light and a circadian clock. Science 260, 238–241.

Golombek, D.A., Agostino, P.V., Plano, S.A., Ferreyra, G.A. (2004) Signaling in the mammalian circadian clock: the NO/cGMP pathway. Neurochem. Int. 45, 929–936.

Golombek, D.A. and Ralph, M.R. (1994) KN-62, an inhibitor of Ca2+/calmodulin kinase II, attenuates circadian responses to light. Neuroreport 5, 1638–1640.

Graham, T.E., Prossnitz, E.R., and Dorin, R.I. (2002) Dexras1/AGS-1 inhibits signal transduction from the Gi-coupled formyl peptide receptor to Erk-1/2 MAP kinases. J. Biol. Chem. 277, 10876–10882.

Griffin, E.A. Jr., Staknis, D. and Weitz, C.J. (1999) Light-independent role of CRY1 and CRY2 in the mammalian circadian clock. Science 286, 768–771.

Haak, L.L. (1999) Metabotropic glutamate receptor modulation of glutamate responses in the suprachiasmatic nucleus. J. Neurophysiol. 81, 1308–1317.

Hamaguchi, H., Fujimoto, K., Kawamoto, T., Noshiro, M., Maemura, K., Takeda, N., Nagai, R., Furukawa, M., Honma, S., Honma, K., Kurihara, H. and Kato, Y. (2004) Expression of the gene for Dec2, a basic helix-loop-helix transcription factor, is regulated by a molecular clock system. Biochem. J. 382, 43–50.

Hannibal, J. (2002) Neurotransmitters of the retino-hypothalamic tract. Cell Tissue Res. 309, 73–88.

Hannibal, J., Ding, J.M., Chen, D., Fahrenkrug, J., Larsen, P.J., Gillette, M.U. and Mikkelsen, J.D. (1997) Pituitary adenylate cyclase-activating peptide (PACAP) in the retinohypothalamic tract: a potential daytime regulator of the biological clock. J. Neurosci. 17, 2637–2644.

Hannibal, J., Ding, J.M., Chen, D., Fahrenkrug, J., Larsen, P.J., Gillette, M.U. and Mikkelsen, J.D. (1998) Pituitary adenylate cyclase activating peptide (PACAP) in the retinohypothalamic tract: a daytime regulator of the biological clock. Ann. N. Y. Acad. Sci. 865, 197–206.

Hannibal, J., Hindersson, P., Knudsen, S.M., Georg, B. and Fahrenkrug, J. (2002) The Photopigment Melanopsin Is Exclusively Present in Pituitary Adenylate Cyclase-Activating Polypeptide-Containing Retinal Ganglion Cells of the Retinohypothalamic Tract. J. Neurosci. 22, RC191.

Hannibal, J., Jamen, F., Nielsen, H.S., Journot, L., Brabet, P. and Fahrenkrug, J. (2001) Dissociation between light-induced phase shift of the circadian rhythm and clock gene expression in mice lacking the pituitary adenylate cyclase activating polypeptide type 1 receptor. J. Neurosci. 21, 4883–4890.

Hannibal, J., Mikkelsen, J.D., Clausen, H., Holst, J.J., Wulff, B.S. and Fahrenkrug, J. (1995) Gene expression of pituitary adenylate cyclase activating polypeptide (PACAP) in the rat hypothalamus. Regul. Pept. 55, 133–148.

Hannibal, J., Moller, M., Ottersen, O.P. and Fahrenkrug, J. (2000) PACAP and glutamate are co-stored in the retinohypothalamic tract. J. Comp. Neurol. 418, 147–155.

Harmar, A.J., Arimura, A., Gozes, I., Journot, L., Laburthe, M., Pisegna, J.R., Rawlings, S.R., Robbrecht, P., Said, S.I., Sreedharan, S.P., Wank, S.A. and Waschek, J.A. (1998) International Union of Pharmacology. XVIII. Nomenclature of receptors for vasoactive intestinal peptide and pituitary adenylate cyclase-activating polypeptide. Pharmacol. Rev. 50, 265–270.

Harrington, M.E., Hoque, S., Hall, A., Golombek, D. and Biello, S. (1999) Pituitary adenylate cyclase activating peptide phase shifts circadian rhythms in a manner similar to light. J. Neurosci. 19, 6637–6642.

Hinds, H.L., Tonegawa, S. and Malinow, R. (1998) CA1 long-term potentiation is diminished but present in hippocampal slices from alpha-CaMKII mutant mice. Learn. Mem. 5, 344–354.

Hogenesch, J.B., Gu, Y.Z., Jain, S. and Bradfield, C.A. (1998) The basic-helix-loop-helix-PAS orphan MOP3 forms transcriptionally active complexes with circadian and hypoxia factors. Proc. Natl. Acad. Sci. USA 95, 547–549.

Honma, S., Kawamoto, T., Takagi, Y., Fujimoto, K., Sato, F., Noshiro, M., Kato, Y. and Honma, K. (2002) Dec1 and Dec2 are regulators of the mammalian molecular clock. Nature 419, 841–844.

Honrado, G.I., Johnson, R.S., Golombek, D.A., Spiegelman, B.M., Papaioannou, V.E. and Ralph, M.R. (1996) The circadian system of c-fos deficient mice. J. Comp. Physiol. 178, 563–570.

Hsu, D.S., Zhao, X., Zhao, S., Kazantsev, A., Wang, R.P., Todo, T., Wei, Y.F. and Sancar, A. (1996) Putative human blue-light photoreceptors hCRY1 and hCRY2 are flavoproteins. Biochemistry 35, 13871–13877.

Hudmon, A. and Schulman, H. (2002) Neuronal CA2+/calmodulin-dependent protein kinase II: the role of structure and autoregulation in cellular function. Annu. Rev. Biochem. 71, 473–510.

Inaba, T., Roberts, W.M., Shapiro, L.H., Jolly, K.W., Raimondi, S.C., Smith, S.D. and Look, A.T. (1992) Fusion of the leucine zipper gene HLF to the E2A gene in human acute B-lineage leukemia. Science **257**, 531–534.

Janik, D., Mikkelsen, J.D. and Mrosovsky, N. (1995) Cellular colocalization of Fos and neuropeptide Y in the intergeniculate leaflet after nonphotic phase-shifting events. Brain Res. 698, 137–145.

Johannessen, M., Delghandi, M.P. and Moens, U. (2004) What turns CREB on? Cell Signal. 16, 1211–1227.

Karin, M., Liu, Z. and Zandi, E. (1997) AP-1 function and regulation. Curr. Opin. Cell. Biol. 9, 240–246.

Kawaguchi, C., Tanaka, K., Isojima, Y., Shintani, N., Hashimoto, H., Baba, A. and Nagai, K. (2003) Changes in light-induced phase shift of circadian rhythm in mice lacking PACAP. Biochem. Biophys. Res. Commun. 310, 169–175.

Kawamoto, T., Noshiro, M., Sato, F., Maemura, K., Takeda, N., Nagai, R., Iwata, T., Fujimoto, K., Furukawa, M., Miyazaki, K., Honma, S., Honma, K. and Kato, Y. (2004) A novel autofeedback loop of Dec1 transcription involved in circadian rhythm regulation. Biochem. Biophys. Res. Commun. 313, 117–124.

King, D.P., Zhao, Y., Sangoram, A.M., Wilsbacher, L.D., Tanaka, M., Antoch, M.P., Steeves, T.D., Vitaterna, M.H., Kornhauser, J.M., Lowrey, P.L., Turek, F.W. and Takahashi, J.S. (1997) Positional cloning of the mouse circadian clock gene. Cell 89, 641–653.

Kornhauser, J.M., Nelson, D.E., Mayo, K.E. and Takahashi, J.S. (1990) Photic and circadian regulation of c-fos gene expression in the hamster suprachiasmatic nucleus. Neuron 5, 127–134.

Kornhauser, J.M., Nelson, D.E., Mayo, K.E. and Takahashi, J.S. (1992) Regulation of jun-B messenger RNA and AP-1 activity by light and a circadian clock. Science 255, 1581–1584.

Kriegsfeld, L.J., M.J., Demas, G.E., Lee, S.E. Jr., Dawson, T.M., Dawson, V.L., and Nelson, R.J. (1999) Circadian locomotor analysis of male mice lacking the gene for neuronal nitric oxide synthase (nNOS-/-) J. Biol. Rhythms 14, 20–27.

Lin, J.T., Kornhauser, J.M., Singh, N.P., Mayo, K.E. and Takahashi, J.S. (1997) Visual sensitivities of nur77 (NGFI-B) and zif268 (NGFI-A) induction in the suprachiasmatic nucleus are dissociated from c-fos induction and behavioral phase-shifting responses. Brain Res. Mol. Brain Res. 46, 303–310.

Lopez-Molina, L., Conquet, F., Dubois-Dauphin, M. and Schibler, U. (1997) The DBP gene is expressed according to a circadian rhythm in the suprachiasmatic nucleus and influences circadian behavior. EMBO J. 16, 6762–6771.

Lowrey, P.L., Shimomura, K., Antoch, M.P., Yamazaki, S., Zemenides, P.D., Ralph, M.R., Menaker, M. and Takahashi, J.S. (2000) Positional Syntenic Cloning and Functional Characterization of the Mammalian Circadian Mutation *tau*. Science 288, 483–492.

Low-Zeddies, S.S. and Takahashi, J.S. (2001) Chimera analysis of the Clock mutation in mice shows that complex cellular integration determines circadian behavior. Cell 105, 25–42.

Malenka, R.C., Kauer, J.A., Perkel, D.J., Mauk, M.D., Kelly, P.T., Nicoll, R.A. and Waxham, M.N. (1989) An essential role for postsynaptic calmodulin and protein kinase activity in long-term potentiation. Nature 340, 554–557.

Martinek, S., Inonog, S., Manoukian, A.S. and Young, M.W. (2001) A role for the segment polarity gene shaggy/GSK 3 in the Drosophila circadian clock. Cell 105, 769–779

Mathur, A., Golombek, D.A. and Ralph, M.R. (1996) cGMP-dependent protein kinase inhibitors block light-induced phase advances of circadian rhythms in vivo. Am. J. Physiol. 270, R1031-R1036.

Michel, S., Itri, J. and Colwell, C.S. (2002) Excitatory mechanisms in the suprachiasmatic nucleus: the role of AMPA/KA glutamate receptors. J. Neurophysiol. 88, 817–828.

Mintz, E.M., Marvel, C.L., Gillespie, C.F., Price, K.M. and Albers, H.E. (1999) Activation of NMDA receptors in the suprachiasmatic nucleus produces light-like phase shifts of the circadian clock in vivo. J. Neurosci. 19, 5124–5130.

Mitsui, S., Yamaguchi, S., Matsuo, T., Ishida, Y. and Okamura, H. (2001) Antagonistic role of E4BP4 and PAR proteins in the circadian oscillatory mechanism. Genes Dev. 15, 995–1006.

Miyata, A., Arimura, A., Dahl, R.R., Minamino, N. Uehara, A., Jiang, L., Culler, M.D. and Coy, D.H. (1989) Isolation of a novel 38 residue-hypothalamic polypeptide which stimulates adenylate cyclase in pituitary cells. Biochem. Biophys. Res. Commun. 164, 567–574.

Miyata, A., Jiang, L., Dahl, R.D., Kitada, C., Kubo, K., Fujino, M., Minamino, N. and Arimura, A. (1990) Isolation of a neuropeptide corresponding to the N-terminal 27 residues of the pituitary adenylate cyclase activating polypeptide with 38 residues (PACAP38). Biochem. Biophys. Res. Commun. 170, 643–648.

Moore, R.Y. (1996) Entrainment pathways and the functional organization of the circadian system. Prog. Brain Res. 111, 103–119.

Morin, L.P. and Blanchard, J.H. (2001) Neuromodulator content of hamster intergeniculate leaflet neurons and their projection to the suprachiasmatic nucleus or visual midbrain. J. Comp. Neurol. 437, 79–90.

Morris, M.E., Viswanathan, N. Kuhlman, S., Davis, F.C. and Weitz, C.J. (1998) A screen for genes induced in the suprachiasmatic nucleus by light. Science 279, 1544–1547.

Mueller, C.R., Maire, P. and Schibler, U. (1990) DBP, a liver-enriched transcriptional act-ivator, is expressed late in ontogeny and its tissue specificity is determined posttranscriptionally. Cell 61, 279–291.

Nielsen, H.S., Hannibal, J., Knudsen, S.M. and Fahrenkrug, J. (2001) Pituitary adenylate cyclase-activating polypeptide induces period1 and period2 gene expression in the rat suprachiasmatic nucleus during late night. Neuroscience 103, 433–441.

Obrietan, K., Impey, S., Smith, D., Athos, J. and Storm, D.R. (1999) Circadian regulation of cAMP response element-mediated gene expression in the suprachiasmatic nuclei. J. Biol. Chem. 274, 17748–17756.

Obrietan, K., Impey, S. and Storm, D.R. (1998) Light and circadian rhythmicity regulate MAP kinase activation in the suprachiasmatic nuclei. Nat. Neurosci. 1, 693–700.

Otmakhov, N., Griffith, L.C. and Lisman, J.E. (1997) Postsynaptic inhibitors of calcium/calmodulin-dependent protein kinase type II block induction but not maintenance of pairing-induced long-term potentiation. J. Neurosci. 17, 5357–5365.

Pennartz, C.M., Hamstra, R. and Geurtsen, A.M. (2001) Enhanced NMDA receptor activity in retinal inputs to the rat suprachiasmatic nucleus during the subjective night. J. Physiol. 532, 181–194.

Pickard, G.E., Ralph, M.R. and Menaker, M. (1987) The intergeniculate leaflet partially mediates effects of light on circadian rhythms. J. Biol. Rhythms 2, 35–56.

Pierrat, B., Correia, J.S., Mary, J.L., Tomas-Zuber, M. and Lesslauer, W. (1998) RSK-B, a novel ribosomal S6 kinase family member, is a CREB kinase under dominant control of p38alpha mitogen-activated protein kinase (p38alphaMAPK). J. Biol. Chem. 273, 29661–29671.

Preitner, N., Damiola, F., Lopez-Molina, L., Zakany, J., Duboule, D., Albrecht, U. and Schibler, U. (2002) The orphan nuclear receptor REV-ERBalpha controls circadian transcription within the positive limb of the mammalian circadian oscillator. Cell 110, 251–260.

Reick, M., Garcia, J.A., Dudley, C. and McKnight, S.L. (2001) NPAS2: an analog of clock operative in the mammalian forebrain. Science 293, 506–509.

Ripperger, J.A., Shearman, L.P., Reppert, S.M. and Schibler, U. (2000) CLOCK, an essential pacemaker component, controls expression of the circadian transcription factor DBP. Genes Dev. 14, 679–689.

Roberto, M. and Brunelli, M. (2000) PACAP-38 enhances excitatory synaptic transmission in the rat hippocampal CA1 region. Learn Mem. 7, 303–311.

Romano, C., Sesma, M.A., McDonald, C.T., O'Malley, K., Van den Pol, A.N. and Olney, J.W. (1995) Distribution of metabotropic glutamate receptor mGluR5 immunoreactivity in rat brain. J. Comp. Neurol. 355, 455–469.

Shearman, L.P., Jin, X., Lee, C., Reppert, S.M. and Weaver, D.R. (2000) Targeted disruption of the mPer3 gene: subtle effects on circadian clock function. Mol. Cell. Biol. 20, 6269–6275.

Shen, M., Kawamoto, T., Teramoto, M., Makihira, S., Fujimoto, K., Yan, W., Noshiro, M. and Kato, Y. (2001) Induction of basic helix-loop-helix protein DEC1 (BHLHB2)/Stra13/Sharp2 in response to the cyclic adenosine monophosphate pathway. Eur. J. Cell. Biol. 80, 329–334.

Shibata, S., Watanabe, A., Hamada, T., Ono, M. and Watanabe, S. (1994) N-methyl-D-aspartate induces phase shifts in circadian rhythm of neuronal activity of rat SCN in vitro. Am. J. Physiol. 267, R360-R364.

Soderling, T. R. (1996) Structure and regulation of calcium/calmodulin-dependent protein kinases II and IV. Biochim. Biophys. Acta 1297, 131–138.

Soderling, T. R., Chang, B. and Brickey, D. (2001) Cellular signaling through multifunctional Ca2+/calmodulin-dependent protein kinase II. J. Biol. Chem. 276, 3719–3722.

Spengler, D., Waeber, C., Pantaloni, C., Holsboer, F., Bockaert, J., Seeburg, P.H. and Journot, L. (1993) Differential signal transduction by five splice variants of the PACAP receptor. Nature 365, 170–175.

Sun, P., Enslen, H., Myung, P.S. and Maurer, R.A. (1994) Differential activation of CREB by Ca2+/calmodulin-dependent protein kinases type II and type IV involves phosphorylation of a site that negatively regulates activity. Genes Dev. 8, 2527–2539.

Sun, P. and Maurer, R.A. (1995) An inactivating point mutation demonstrates that interaction of cAMP response element binding protein (CREB) with the CREB binding protein is not sufficient for transcriptional activation. J. Biol. Chem. 270, 7041–7044.

Takahashi, H., Umeda, N., Tsutsumi, Y., Fukumura, R., Ohkaze, H., Sujino, M., van der Horst, G., Yasui, A., Inouye, S.T., Fujimori, A., Ohhata, T., Araki, R., and Abe, M. (2003) Mouse dexamethasone-induced RAS protein 1 gene is expressed in a circadian rhythmic manner in the suprachiasmatic nucleus. Brain Res. Mol. Brain Res. 110, 1–6.

Tanaka, K., Shibuya, I., Harayama, N., Nomura, M., Kabashima, N., Ueta, Y. and Yamashita, H. (1997) Pituitary adenylate cyclase-activating polypeptide potentiation of Ca2+ entry via protein kinase C and A pathways in melanotrophs of the pituitary pars intermedia of rats. Endocrinology. 138, 4086–4095.

Tanaka, K., Shibuya, I., Nagamoto, T., Yamashita, H. and Kanno, T. (1996) Pituitary adenylate cyclase-activating polypeptide causes rapid Ca2+ release from intracellular stores and long lasting Ca2+ influx mediated by Na+ influx-dependent membrane depolarization in bovine adrenal chromaffin cells. Endocrinology. 137, 956–966.

Taylor, B.L. and Zhulin, I.B. (1999) PAS domains: internal sensors of oxygen, redox potential, and light. Microbiol. Mol. Biol. Rev. 63, 479–506.

Teramoto, M., Nakamasu, K., Noshiro, M., Matsuda, Y., Gotoh, O., Shen, M., Tsutsumi, S., Kawamoto, T., Iwamoto, Y. and Kato, Y. (2001) Gene structure and chromosomal location of a human bHLH transcriptional factor DEC1 x Stra13 x SHARP-2/BHLHB2. J. Biochem (Tokyo) 129, 391–396.

Thankachan, S. and Rusak, B. (2005) Juxtacellular recording/labeling analysis of physiological and anatomical characteristics of rat intergeniculate leaflet neurons. J. Neurosci. 25, 9195–9204.

Tischkau, S.A., Mitchell, J.W., Tyan, S.H., Buchanan, G.F. and Gillette, M.U. (2003) Ca2+/cAMP response element-binding protein (CREB)-dependent activation of Per1 is required for light-induced signaling in the suprachiasmatic nucleus circadian clock. J. Biol. Chem. 278, 718–723.

Tischkau, S.A., Weber, E.T., Abbott, S.M., Mitchell, J.W. and Gillette, M.U. (2003) Circadian clock-controlled regulation of cGMP-protein kinase G in the nocturnal domain. J. Neurosci. 23, 7543–7550.

Travnickova-Bendova, Z., Cermakian, N., Reppert, S.M. and Sassone-Corsi, P. (2002) Bimodal regulation of mPeriod promoters by CREB-dependent signaling and CLOCK/BMAL1 activity. Proc. Natl. Acad. Sci. USA 99, 7728–7733.

Van den Pol, A.N., Kogelman, L., Ghosh, P., Liljelund, P. and Blackstone, C. (1994) Developmental regulation of the hypothalamic metabotropic glutamate receptor mGluR1. J. Neurosci. 14, 3816–3834.

Van der Horst, G.T., Muijtjens, M., Kobayashi, K., Takano, R., Kanno, S., Takao, M., de Wit, J., Verkerk, A., Eker, A.P., van Leenen, D., Buijs, R., Bootsma, D., Hoeijmakers, J.H. and Yasui, A. (1999) Mammalian Cry1 and Cry2 are essential for maintenance of circadian rhythms. Nature 398, 627–630.

Vitaterna, M.H., King, D.P., Chang, A.M., Kornhauser, J.M., Lowrey, P.L., McDonald, J.D., Dove, W.F., Pinto, L.H., Turek, F.W. and Takahashi, J.S. (1994) Mutagenesis and mapping of a mouse gene, Clock, essential for circadian behavior. Science 264, 719–725.

Von Gall, C., Duffield, G.E., Hastings, M.H., Kopp, M.D., Dhghani, F., Korf, H.W. and Stehle, J.H. (1998) CREB in the mouse SCN: a molecular interface coding the phase-adjusting stimuli light, glutamate, PACAP, and melatonin for clockwork access. J. Neurosci. 18, 10389–10397.

Vrang, N., Mrosovsky, N. and Mikkelsen, J.D. (2003) Afferent projections to the hamster intergeniculate leaflet demonstrated by retrograde and anterograde tracing. Brain Res. Bull. 59, 267–288.

Weber, E.T., Gannon, R.L. and Rea, M.A. (1995) cGMP-dependent protein kinase inhibitor blocks light-induced phase advances of circadian rhythms in vivo. Neurosci. Lett. 197, 227–230.

Wickland, C. and Turek, F.W. (1994) Lesions of the thalamic intergeniculate leaflet block activity-induced phase shifts in the circadian activity rhythm of the golden hamster. Brain Res. 660, 293–300.

Wollnik, F., Brysch, W., Uhlmann, E., Gillardon, F., Bravo, R., Zimmermann, M., Schlingensiepen, K.H. and Herdegen, T. (1995) Block of c-Fos and JunB expression by anti-sense oligonucleotides inhibits light-induced phase shifts of the mammalian circadian clock. Eur. J. Neurosci. 7, 388–393.

Yamaguchi, S., Mitsui, S., Yan, L., Yagita, K., Miyake, S. and Okamura, H. (2000) Role of DBP in the circadian oscillatory mechanism. Mol. Cell. Biol. 20, 4773–4781.

Yin, L., Wang, J., Klein, P.S. and Lazar, M.A. (2006) Nuclear receptor Rev-erbalpha is a critical lithium-sensitive component of the circadian clock. Science 311, 1002–1005.

Yokota, S., Yamamoto, M., Moriya, T., Akiyama, M., Fukunaga, K., Miyamoto, E. and Shibata, S. (2001) Involvement of calcium-calmodulin protein kinase but not mitogen-activated protein kinase in light-induced phase delays and Per gene expression in the suprachiasmatic nucleus of the hamster. J. Neurochem. 77, 618–627.

Yujnovsky, I., Hirayama, J.., Doi, M., Borrelli, E. and Sassone-Corsi, P. (2006) Signaling mediated by the dopamine D2 receptor potentiates circadian regulation by CLOCK:BMAL1. Proc. Natl. Acad. Sci. USA. 103, 6386–6391.

Zheng, C.F. and Guan, K.L. (1993) Cloning and characterization of two distinct human extracel-lular signal-regulated kinase activator kinases, MEK1 and MEK2. J. Biol. Chem. 268, 11435–11439.

Zheng, B., Larkin, D.W., Albrecht, U., Sun, Z.S., Sage, M., Eichele, G., Lee, C.C. and Bradley, A. (1999) The mPer2 gene encodes a functional component of the mammalian circadian clock. Nature 400, 169–173.

Zheng, B., Albrecht, U., Kaasik, K., Sage, M., Lu, W., Vaishnav, S., Li, Q., Sun, Z.S., Eichele, G., Bradley, A. and Lee, C.C. (2001) Nonredundant roles of the mPer1 and mPer2 genes in the mammalian circadian clock. Cell 105, 683–694.

Chapter 17
Intersecting Genetics with Lifestyle: the Role of Exercise and Diet in Synaptic Plasticity and Cognitive Enhancement

Fernando Gomez-Pinilla and Shoshanna Vaynman

Abstract Science in the last two decades has revealed the plastic nature of the brain, displaying its remarkable ability to undergo dynamic structural and functional changes. The brain is a dynamic system that derives its structural and functional capacity from its interaction with environmental factors and synaptic plasticity is a property of the nervous system that utilizes changes at the transcriptional level to enable neurons to shape the efficacy of their connections with activity and experience. This chapter discusses the ability of specific aspects of lifestyle such as exercise and diet to promote synaptic plasticity and to modulate cognitive function by the activity-dependent regulation of brain-derived neurotrophic factor (BDNF). We will discuss new evidence suggesting that the supporting role of BDNF on synaptic plasticity and learning and memory may be partially achieved by interacting with mechanisms that modulate cell energy metabolism. We will also discuss provocative evidence suggesting that the involvement of BDNF with synaptic plasticity and energy metabolism may underlie even more profound biological processes such as those related to the epigenetic inheritance of cognitive traits.

17.1 Experience-dependent Modulation of Neurotrophins

To understand how behaviors use the action of BDNF to induce synaptic plasticity in the brain, it is essential to take an introductory tour of the neurotrophin family. The neurotrophin family comprises of homodimeric proteins, including, nerve growth factor (NGF), brain-derived neurotrophic factor (BDNF), neurotrophin-3 (NT-3), neurotrophin-4/5 (NT-4/5), and neurotrophin-6 (NT-6). Neurotrophins exert their effects with a degree of specificity by their binding to their cognate tyrosine kinase receptors (Trks; NGF to TrkA, BDNF and NT-4/5 to TrkB, and NT-3 to Trk-C (Barbacid, 1994). In addition, the different neurotrophins share a similar affinity for the low-affinity p75 neurotrophin receptor.

Initiatory discoveries regarding the role of neurotrophins produced the conclusion that they acted as primary factors in the regulation of neuronal survival and differentiation in the developmental organism (Barde, 1994). A visual demonstration of this capacity is seen from a study using EM analysis in TrkB and TrkC knockout mice. Eliminating the receptors through which neurotrophic

S. M. Dudek (ed.), *Transcriptional Regulation by Neuronal Activity.*
© Springer 2008

factors mediate their action resulted in a reduction in axonal arborization and the number of synapses in the postnatal hippocampus (Martinez et al., 1998). The time frame of these experiments, the postnatal developmental stage (P12-P13) when synaptogenesis is normally taking place, illustrate and reinforce the importance for neurotrophic involvement in the regulation of synapse formation during development.

In the last decade, their role has greatly expanded with findings that neurotrophins are prominent in mediating activity-dependent functional and structural plasticity in both the embryonic and mature CNS (Poo, 2001, Snider et al., 1996, McAllister et al., 1999, Prakash et al., 1996). The much publicized neurotrophin factor, BDNF, provides insight into how activity-dependent regulation of neurotrophic factors and related factors can alter neuronal connectivity and modulate the functional complexity of neuronal circuits in the hippocampus. This can been seen from that findings that BDNF participates in the regulation of axonal and dendritic branching and remodeling (Lom et al., 1999, McAllister et al., 1996, Yacoubian et al., 2000, Shimada et al., 1998), augments the efficacy of synaptic transmission (Boulanger et al., 1999, Kafitz et al., 1999, Kang & Schuman 1996, Lohof et al., 1993), and participates to modulate the functional maturation of excitatory and inhibitory synapses (Seil et al., 2000, Vicario-Abejon et al., 1998, Rutherford et al., 1998).

Regulation by activity is an integral and fundamental property of neurotrophins such as BDNF. Neuronal activity enhances the expression, secretion, and/or actions of BDNF at the synapse to result in the modification of synaptic transmission and connectivity. The literature shows that multiple experimental paradigms, structured around the concept of increasing neuronal activity, effectively augment neurotrophin expression. Early on it was discovered that seizures dramatically increase the expression of BDNF mRNA (Zafra et al., 1990, Ernfors et al., 1991) as well as the mRNA expression of another member of the neurotrophin family, NGF, in the hippocampus (Gall et al., 1989). In a similar vein, it was discovered that sensory stimulation regulates BDNF with visual input in the visual cortex (Castren et al., 1992), and whisker stimulation in the barrel cortex (Rocamora et al., 1996). Additionally, physiological activity such as exercise (Neeper et al., 1996, Vaynman et al., 2003), learning (Kesslak et al., 1998) and sleep and circadian rhythm (Bova et al., 1998, Liang et al., 1998) increase BDNF. There is evidence to suggest that BDNF may be more sensitive to regulation by activity than other members of its family. A particular study demonstrated that when the mRNAs for the precursor proteins pro-BDNF and pro-NGF were over-expressed in cultured hippocampal neurons, that secretion by activity was specifically a property of BDNF and not NGF, i.e., the application off a depolarizing stimulus was selective to triggering BDNF release (Mowla et al., 1999).

Several experiments employing vaccina virus expression system have demonstrated that, at least in the hippocampus, BDNF is sorted into the activity-regulated pathway whereas other neurotrophins are mainly sorted into the constitutive pathway (Mowla et al., 1999, Farhadi et al., 2000). Transfection experiments employing BDNF-GFP (green fluorescent protein) fusion constructs have enabled the actual visualization of BDNF in hippocampal and cortical neurons. Accordingly, these studies have revealed that BDNF is packaged in secretory vesicles (Haubensak et al., 1998, Kojima et al., 2001). Colocalization of BDNF with specific markers, the presynaptic secretory protein synapsin I and the postsynaptic scaffolding protein PSD95, revealed that the BDNF-GFP fluorescence was found to be concentrated at synaptic junctions (Haubensak et al., 1998, Kojima et al., 2001). The BDNF-GFP fluorescence spots were found to quickly disappear when depolarization or high frequency stimulation was applied, therefore, suggesting that BDNF was secreted from these synaptically localized secretory vesicles (Hartmann et al., 2001, Kojima et al., 2001). Thus BDNF, coupled with its prominence in the hippocampus, seems to exhibit a property that makes its particularly capable of mediating the benefits of behaviors such as exercise and diet on neuronal and cognitive plasticity.

An interesting finding regarding the role of activity-dependent BDNF gene regulation comes form a study conducted by Chen et al., (2003). Using a chromatin immunoprecipitation technique, they found that the transcriptional repressor MeCP2 is bound to the rat BDNF promoter III (homologous to mouse BDNF IV promoter) in resting cortical neurons. However, upon the application of activity, i.e., membrane depolarization and subsequent calcium influx, BDNF transcription occurs concurrent with the dissociation of Mecp2 repression from the BDNF promoter. In conclusion, BDNF's activity dependence, such that both it's expression and release of BDNF are regulated by activity (Lu & Chow, 1999, Schinder & Poo, 2000), enable it to be responsive to stimulation from behavioral interventions such as diet and exercise.

17.1.1 From LTP to Learning and Memory

The role that BDNF plays in implementing structural and functional changes at the synapse, particularly those enabled by activity-dependent transcriptional regulation, reinforces the importance of BDNF to cognitive function. Prime examples are the findings that BDNF modulates the efficacy of neurotransmitter release (Kang & Schuman, 1995; Bolton et al., 2000), stimulates the synthesis of vesicle-associated proteins (Lu & Chow, 1999; Schinder & Poo, 2000), and contributes to long lasting neuronal changes by regulating transcriptional factors in the brain (Finkbeiner et al., 1997; Tully, 1997). Studies have found that inducing long-term potentiation (LTP), by direct electrical stimulation to the hippocampus, is highly effective at increasing the mRNA expression of BDNF and modulating the expression of NT-3 and NGF in specific regions of the hippocampus (Patterson et al., 1992; Castren et al., 1993). As LTP has been regarded as the transcriptional-dependent electrophysiological correlate of learning and memory (Nyguyen & Kandel, 1996), this finding was seminal for later studies showing that cognitive tasks modulate BDNF expression in the hippocampus.

Numerous studies have documented the role of BDNF in supporting learning and memory, from findings that BDNF expression is increased during learning tasks in the hippocampus (Kesslak et al., 1998; Hall et al., 2000) to transgenic studies, which show that knocking out the BDNF gene impairs memory formation (Mizuno et al., 2000; Linnarson et al., 1997; Ma, 1998). This aptitude of BDNF holds particular relevance to exercise, as it has been shown that BDNF mediates the ability of exercise to enhance learning and memory (Vaynman et al., 2004). BDNF expression is higher in the hippocampi of rats that underwent hippocampal-dependent learning paradigms such as the Morris water maze (MWM) task or contextual fear conditioning (Kesslak et al., 1998; Hall et al., 2000). For example BDNF mRNA levels have been found to be increased in the hippocampi of rats undergoing 3 or 6 days of MWM training (Kesslak et al., 1998). For the case of contextual fear conditioning, another hippocampal dependent learning paradigms, BDNF mRNA levels have been found to be significantly increased in the CA1 region of the hippocampus (Hall et al., 2000). An association between BDNF and learning and memory was found to exist when measuring the performance of rats on the MWM task (Molteni et al., 2002). Both performance on the acquisition phase, which is measure by the latency to find the platform (A), and the performance on the retention phase, measured as the distance swam in the platform quadrant after the platform is removed (B), were correlated with the levels of BDNF for individual animals. The results of this study suggest that hippocampal levels of BDNF may be directly related to learning efficiency and memory stability (Fig. 1). More direct approaches in evaluating the role BDNF in learning and memory have shown that BDNF gene deletion or BDNF function-blocking precludes BDNF ability to induce LTP and produces an impairment in learning and memory on both the water maze and the inhibitory avoidance task (Patterson et al., 1996; Mu et al., 1999; Linnarsson et al., 1997). In the study conducted by Mu, and colleagues (1999), it was found that rats who received 7 days of intracerebroventricular infusions of anti-BDNF antibodies had impaired performance on the MWM task, specifically these animals exhibited longer escape latencies during the training trials and poorer task completion during the probe trial when compared to controls. Another study that performed a more localized inhibition of BDNF found similar results. Injecting BDNF antisense oligonucleotides specifically into the dentate gyrus of the hippocampus, Ma, and colleagues (1998) showed that blocking BDNF in a specific sub-field of the hip-

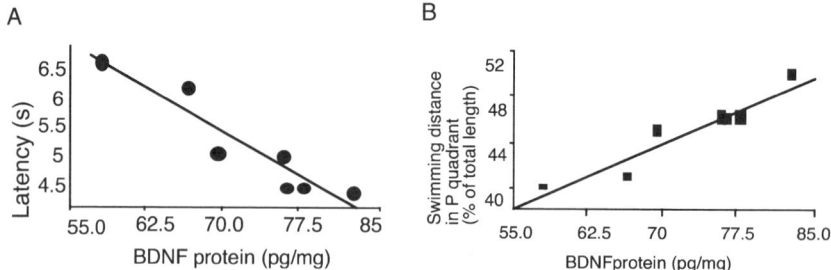

Fig. 1 Performance on the MWM task illustrating that learning acquisition (the latency to find the platform) (A), and memory recall (the distance swam in the platform quadrant after the platform is removed) (B), are correlated with the levels of BDNF for individual animals. Adopted from Molteni et al., 2002

pocampus, was sufficient to decrease hippocampal BDNF mRNA expression and impair memory consolidation on an inhibitory avoidance task. Moreover, this study found a connection with BDNF in the dentate gyrus with LTP. Blocking BDNF in the dentate gyrus of the hippocampus significantly reduced long-term-potentiation (LTP) in perforant path-dentate gyrus synapses (Ma et al., 1998).

In order to properly evaluate the role of BDNF in learning and memory, it is important to consider the imperfect association between memory and LTP, or rather the existence of a dissociation between spatial learning and hippocampal LTP. Several studies have demonstrated that spatial learning does not necessarily require hippocampal CA1 LTP (Bannerman et al., 1995, Huang et al., 1995, Okabe et al., 1998). In fact, in TrkB.T1 mutants, animals exhibited learning deficits the MWM maze while retaining normal CA1 LTP (Saarelainen et al., 2000). On the other hand, as van Praag et al., (1999) found that mice showing poor performance on the MWM task also decreased BDNF levels and impaired dentate gyrus LTP, hippocampal sub-field specificity may contribute to deciphering the relationship between hippocampal dependent learning and LTP.

Interesting findings concerning the role of BDNF and other trophic factors in memory consolidation have shown that BDNF, but not NGF or NT-3, play a role in consolidating short-term memories into long-term memories (Johnston & Rose, 2001). Antisense oligonucleotides for BDNF, NGF, and NT-3 were administered in day old chicks and their performance was tested on the one-trial avoidance task with each respective administration. Although, each of the treatments inhibited both the mRNA and protein levels of the respective trophic factor during the time of training, only inhibition of BDNF produced amnesia for the avoidance behavior 3 to 24 hrs after training. Interestingly, BDNF inhibition did not produce amnesia 1 hr following the avoidance training, indicating that BDNF may not be necessary for long-term memory of this task. Unfortunately, there are inconsistencies in the findings regarding neurotrophic factors effects on learning and memory. For instance another study found that NT-3 overexpression prevented age-related memory deficits in mice (Kaisho et al., 1999). Others found that high levels of hippocampal NGF mRNA levels are correlated with poorer MWM performance (Sugaya et al., 1998). Therefore, it is important to consider that these discrepancies may be due to the differences in protocol, species, memory task, age, in vitro vs. in vivo conditions, amongst others.

17.1.2 The BDNF Genotype in Human Cognitive Function

Going beyond the findings from animals, clinical studies reveal that the BDNF genotype may be important for influencing cognitive function in humans. The Val66Met BDNF polymorphism, is a common single nucleotide polymorphism, consisting of a missense change (G196A) that produces a non-conservative amino acid substitution of valine to methionine at codon coding exon of the BDNF-gene at position 66 (Val66Met), has been implicated in abnormal hippocampal functioning and memory processing in humans (Egan et al., 2003; Hariri et al., 2003). Egan et al., (2003) examined the effect of the BDNF val66met polymorphism using a cohort of normal subjects, patients with schizophrenia and their unaffected patients. Study subjects were tested using the Wechsler Memory Scale, revised version (WMS-R), and the California Verbal Learning Memory Test (CVLT), tests dependent on episodic memory as well as standard tests of medial temporal lobe function. To control for general intelligence, IQ and reading comprehension (using the WRAT) were included. Subjects homozygous for the Met allele (met/met) scored significantly lower on the WMS-R delayed recall test than subjects who were either heterozygous for the Met allele (val/met) or homozygous for the Val allele (val/val). In contrast, no such association was detected on the other test of episodic memory, the CVLT. Thus, the BDNF polymorphism, i.e. the Met allele, seems to be linked with specific deficits in episodic memory such that it may affect cognitive function on a modular basis.

In the Hariri et al., study (2003) the contribution of the BDNF val66met substitution to memory-related hippocampal activity was studied by using Blood oxygenation level-dependent functional magnetic resonance imaging (BOLD fMRI) to monitor hippocampal activity in healthy volunteers exposed to a declarative memory task. Hariri et al., 2003 picked a task that has been reproducibly shown to result in hippocampal activation in neuroimaging studies (Stern et al., 1996; Gabrieli et al., 1997; Zeineh et al., 2000). The declarative memory task consisted of the encoding and subsequent retrieval of novel scenes. During the encoding blocks, subjects were asked to view 6 images, presented serially for 3 sec each, to determine whether each image represented an "indoor" or "outdoor" scene. In the subsequent retrieval blocks, subjects were asked to determine whether presented scenes (6 images presented serially at 3 sec each) were "old" (presented during the encoding blocks) or "new" (not presented during the encoding blocks). Findings from this study showed that the val66met substitution had an effect on hippocampal activity during the encoding phase. Specifically, Met carriers, whether heterozygous or homozygous, exhibited diminished hippocampal activity as compared to Val homozygotes during both the encoding and retrieval parts of the declarative memory task. Another important finding from this study illustrated that the interaction of the BDNF val66met genotype with the hippocampal response during encoding accounted for 25% of the variance in the recognition/memory phase of the task. This last finding indicates that the BDNF val66met polymorphism may not only be related to impaired memory functioning, but may also impact hippocampal brain activity to contribute to the production of the variance in memory capabilities or what can be called "the spectrum of memory abilities" in humans.

Another study used a cohort of 114 healthy young Chinese females to test the functional implication of the BDNF Val66Met variants on intelligence (Tsai et al., 2004). They found that Val homozygotes scored significantly higher than heterozygotes on the performance IQ subtest of the Wechsler Adult Intelligence Scale-revised IQ test (Tsai et al., 2004). Similar to the conclusion derived from Hariri et al., 2003, this study provides evidence that BDNF genetic variants may contribute to specific cognitive functions, but not necessarily overall intelligence.

17.2 Genetically Driven Variation in BDNF Function may Influence Learning and Memory

Evidence suggests that the BDNF val66met polymorphism affects the intracellular trafficking and the regulated secretion of BDNF in neuronal cells (Egan et al., 2003; Chen et al., 2004). Moreover, the variant BDNF may additionally affect the intracellular trafficking and secretion of the wild-type BDNF. A study conducted by Chen et al., 2004 showed that coexpressing the variant BDNF with the wild type, results in neuronal cells with > 70% of the BDNF met bound up in the BDNF val BDNF met heterodimer form (Chen et al., 2004). This finding may have reverberating implications for the activity-dependent regulation of BDNF as well as behavioral consequences. Genetically driven variation in BDNF function, such as its secretion, may influence learning and memory (Egan et al., 2003; Hariri et al., 2003). The secretion of neurotrophins is either regulated or constitutive. Constitutive secretion enables neurotrophins to be continuously available to cells that need it by being spontaneously released shortly after being synthesized. The antipode of this is the regulated pathway, in which neurotrophins are synthesized and then stored in secretory granules to be released in response to extracellular cues (Farhadi et al., 2000). Several seminal experiments have shown that BDNF is sorted into the regulated pathway whereas other neurotrophins are mainly sorted into the constitutive pathway (Mowla et al., 1999, Farhadi et al., 2000). Moreover, the majority of BDNF secreted from neuronal cells seems to be a product of activity-dependent secretion from the regulatory secretory pathway (Lu, 2003). The trafficking of BDNF is a highly regulated process BDNF is sorted into the regulated pathway whereas other neurotrophins are mainly sorted into the constitutive pathway (Mowla et al., 1999, Farhadi et al., 2000). The expression of one copy of the BDNF met gene, by engaging in heterodimer formation with the wild-type BDNF val form, may decrease the efficiency of BDNF trafficking into the regulated pathway. This could compromise the ability of BDNF to respond to activity-dependent secretion in neuronal cells. Given that the val66met polymorphism comprises the pro region of the BDNF protein, the mature BDNF met product secreted is not different from the wild-type BDNF product (Chen et al., 2004). Thus, any defect in BDNFmet function seems to be the 'product' of faulty activity-dependent BDNF regulation. This finding becomes especially relevant when considering that an estimated 20-30% of the population is heterozygous for the met polymorphism (Neves-Pereira et al., 2002; Egan et al., 2003; Hariri et al., 2003; Sen et al., 2003).

17.3 Behaviorally-induced Neurotrophins Improve Neural Function; Exercise, Synaptic Plasticity, and Learning and Memory

Exercise has the ability to improve and maintain cognition, even in the ageing organism. Besides raising alertness and responsiveness to various stimuli and situations, human studies have found that adherence to an exercise regimen reduces the normal decay of cognitive function observed during aging (Kramer et al., 1999), and may reduce the risk of getting Alzheimer's disease. Multiple clinical studies (Kramer et al., 1999, Suominen-Troyer et al., 1986, Rogers et al., 1980) and studies in rodents (van Praag et al., 1999, Fordyce et al., 1991, Somarajski et al., 1985), demonstrate the beneficial effects of exercise on cognitive function. The findings report that exercise has the capacity to enhance learning and memory (Rogers et al., 1990; Suominen-Troyer et al., 1986; van Praag et al., 1999) under a variety of conditions, from counteracting the mental decline associated with ageing (Kramer et al., 1999; Laurin et al., 2001) to facilitating functional recovery in patients suffering from brain injury or disease (Bohannon et al., 1993; Grealy et al., 1999; Lindvall et al., 1992). A metaanalysis of clinical studies about the effects of exercise conducted by Colcombe & Kramer (2003) supported the conclusion that exercise has a positive effect on cognitive function in humans. The study analyzed 18 longitudinal fitness-training studies that revealed cardiovascular fitness training improves overall cognitive function regardless of task type.

Exercise, as other types of behavioral stimuli, may activate specific neural circuits, to modify the way that information is transmitted across cells at the synapse and initiate long-term changes through transcriptional regulation. The mechanism of how exercise affects neuronal and behavioral plasticity seems to center around the finding that exercise induces an increase in BDNF levels in the area vital for learning and memory formation, the hippocampus (Neeper et al., 1996, Gomez-Pinilla et al., 2002; Vaynman et al., 2003), such that blocking BDNF action using specific immuno adhesive chimeres abolished the ability of exercise to augment learning and memory in the rat (Vaynman et al., 2004; Fig. 2).

The evaluation of an array of genes activated by exercise has showed that although exercise affects the expression of other neurotrophic factors, BDNF is the only neurotrophic factor consistently elevated after a few weeks of continuous exercise (Molteni et al., 2002). The actions of BDNF upon synaptic plasticity occur through increased BDNF gene expression, processing and release, to ultimately result in the binding and activation of its cognate tyrosine kinase receptor (TrkB) at both pre- and postsynaptic sites (Poo, 2001, Lu, 2003). Exercise may potentiate the effects of BDNF on synaptic-plasticity, by employing a positive feedback loop through transcriptional regulation such that BDNF action concurrently increases the mRNA levels of both itself and its TrkB receptor (Vaynman et al., 2003). Additionally, BDNF interacts with a variety of mechanisms and downstream signaling events that induce structural and functional changes at the synapse.

Blocking experiments have demonstrated that different pathways contribute to the exercise-induced increases in the mRNA levels of BDNF, TrkB, and molecules involved in transcription and synaptic transmission, cAMP response element binding protein (CREB), and synapsin I, respectively (Vaynman et al., 2003). In addition to BDNF, exercise employs several different conduits of signal transduction, such as mitogen-activated protein kinase (MAPK), calcium/calmodulin protein kinase II (CAMKII), and the N-methyl–D-aspartate (NMDA) receptor, to mediate its effects on hippocampal synaptic plasticity (Vaynman et al., 2003). Given the interaction of BDNF with these molecules, a basic mechanism has been proposed through which exercise may promote synaptic-plasticity in the adult brain.

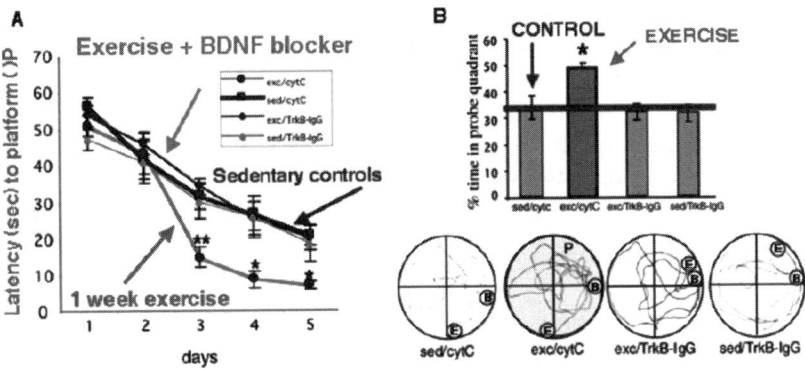

Fig. 2 Exercise augments learning and memory on the Morris water maze task through the action of BDNF: (A) Animals who underwent a two week voluntary running wheel regimen had located the hidden platform in a significantly shorter time frame than sedentary control animals (exc/cutC demonstrated shorter escape latencies compared to sed/cytC controls). Blocking the action of BDNF with TrkBIgG during the exercise period successfully abolished the exercise-induced enhancement in learning ability (exc/TrkBIgG had escape latencies comparable to sed/cytC controls), to support the role of BDNF in advancing exercise-induced learning ability. Data are expressed as mean ± SEM (ANOVA, Fischer test, Scheffe F-test, HP < 0.05, HHP < 0.01). (B) Exercise enhanced memory retention on the MWM. Exercise animals performed significantly better on the probe trial of the MWM task as indicated by their increased time spent in the p quadrant where the platform used to be located (exc/cytC vs. sed/cytC). Blocking BDNF abolished the preference of exercise animals for the quadrant, illustrating the importance of BDNF mechanisms in mediating the effect of exercise on memory recall (time spent in the p quadrant of exc/TrkBIgG animals was not significantly different form sed/cytC controls). Blocking BDNF action in sedentary animals had no effect for p quadrant preference, thereby reinforcing the activity dependence of BDNF (sed/TrkBIgG vs. sed/cytC). Data are expressed as mean ± SEM (ANOVA, Fischer test, HP < 0.05). Representative samples of the times spent in the quadrants, illustrate the marked preference of the exercise animals for the P quadrant as compared to all other groups. (B, begin, E, end, P quadrant that previously housed the platform). Adopted from Vaynman et al., 2004. (*See* Color Plate 9).

Evaluation of the pathways activated downstream to BDNF induction provide further insight into how exercise is capable of orchestrating its beneficial effects on brain health and learning and memory. The two main intracellular signaling cascades found to be activated by BDNF, CAMKII and MAPKII, have important roles in neuronal and behavioral plasticity. BDNF activation of its TrkB receptor has been shown to lead to the activation of the MAPK cascade (Stephens et al., 1994), which serves as an intracellular signaling mechanism that integrates multiple signals (Sweatt et al., 2001), and leads to the activation of a diverse set of targets, such as CREB mediated transcription, protein synthesis, and voltage/ion gated channels. BDNF also activates CAMKII (Blanquet & Lamour, 1997), which can integrate with the BDNF-mediated MAPK activation as CAMKII has been shown to converge on the MAPK cascade (Blanquet et al., 2003).

MAPK is integral to the regulation of nuclear signaling (Adams et al, 2000), LTP (English & Sweatt, 1997), and may be especially necessary for learning and memory formation (Atkins et al., 1998, Blum et al., 1999; Sweat, 2001; Impey et al., 1999). It has been proposed that the ability of MAPK to induce long-lasting changes in synaptic-plasticity is connected with its ability to regulate (Finkbeiner et al., 1997), and prolong the transcriptionally active state of CREB (Hardingham et al., 2001), a consummate end-product of BDNF action that is important for various forms of learning and memory (Ying et al., 2002, Silva et al., 1998). In a similar fashion, CAMKII is important in mechanisms underlying learning and memory (Yin & Tully, 1996). Gene deletions of αCAMKII in mice result in their impaired performance on spatial learning tasks (Silva et al., 1992). Like MAPK, CAMKII have been repeatedly described as conserved signaling pathways that leads to CREB mediated gene transcription (Finkbeiner et al., 1997; Ying et al., 2002).

CREB has a well-described role in activity-dependent long-term neuronal plasticity, namely, it has been documented as evolutionarily conserved molecule requisite for the formation of long-term memory (LTM; Dash et al., 1990; Bourtchouladze et al., 1994; Yin et al., 1995). The importance of CREB can be seen across species, as disrupting CREB function with a dominant negative CREB protein impairs odor memory in Drosophila (Yin et al., 1994) while a targeted disruption of CREB isoforms results in LTM deficiency in mice (Bourtchouladze et al., 1994). CREB is a critical aspect of the BDNF-mediated machinery responsible for the potentiating effects of exercise on learning and memory. It has been found that blocking BDNF action during exercise is sufficient to abolish the exercise-induced enhancement in learning and memory and prevent the ability of exercise to increase CREB mRNA levels and the phoshorylated active form of CREB in the hippocampus (p-CREB; Vaynman et al., 2004). During exercise, BDNF and CREB mRNA levels have been found to be significantly and positively associated with each other as well as with memory performance on the MWM task; animals with the highest BDNF expression also had the highest CREB expression and the best memory recall. Additional findings suggest that CREB may provide a self-perpetuating loop for BDNF action. Not only is CREB regulated by BDNF (Finkbeiner et al., 1997; Tully, 1997), but CREB has been found to regulate BDNF transcription (Tao et al., 1998).

Another important way that exercise and BDNF impact synaptic plasticity is by modulating the transmission properties at the synapse. Synaptic remodeling and synaptogenesis are mechanisms that have been found to be largely mediated by neurotrophic factors, such as BDNF (Patapoutian & Reichardt, 2001). Studies have shown that BDNF regulates synapsin I, a phospho-protein localized to the presynaptic membrane (Vaynman et al., 2006c) that is important for formation and maintenance of the presynaptic structure (Melloni et al., 1994, Takei et al., 1995) and axonal elongation (Akagi et al., 1996). Synapsin I tethers synaptic vesicles to the actin cytoskeleton, creating a synaptic vesicle reserve pool that is important for modulating vesicular release (Greengard et al., 1993). Inhibiting synapsin I has been found to reduce both the SV reserve pool and neurotransmitter release (Hilfiker et al., 1999). BDNF also regulates other synaptic proteins, as BDNF has been found to regulate synaptophysin, a major integral protein located on synaptic vesicles (Vaynman et al., 2006c). BDNF gene in mice results in a reduction in synaptic proteins, sparsely docked vesicles, and impaired neurotransmitter release and decreases synaptophysin levels (Pozzo-Miller et al., 1999). Synaptophysin seems to be a key protein in the biogenesis of synaptic vesicles from cholesterol, by promoting membrane curvature to facilitate vesicular budding and membrane retrieval (Thiele et al., 2000). Studies of the giant squid axon provided evidence that synaptophysin has a role in rapid clathrin-independent vesicle endocytosis at the active zone (Daly et al., 2000). The ability to retrieve synaptic vesicle proteins through endocytosis is essential for generating fusion competent vesicles. Thus exercise may facilitate rapid neurotransmission by affecting vesicular proteins involved in synaptic release recycling, such as synapsin I and synaptophysin.

BDNF may be integral to the ability of synapsin to modulate synaptic vesicular release. This is supported by the finding that presynaptic neurotransmitter release via MAPK phosphorylation of synapsin I is coupled with the ability of BDNF to enhance communication between neurons by activating its TrkB receptor (Jovanovic et al., 1996). A possible mechanism by which BDNF regulates synapsin I levels is by activation of CAMKII through phospho-lipase C (PLC), which is anchored to its TrkB receptor (Blanquet & Lamour, 1997). During exercise, CAMKII, but not MAPKII, has been shown to contribute to the BDNF regulation of synapsin I expression (Vaynman et al., 2003). However, MAPK may still regulate the phosphylated form during exercise to influence synaptic vesicle release. A clinical study on familial epilepsy showed that a genetic mutation in the synapsin I gene seems to be associated with learning difficulties (Garcia et al., 2004). An adequate vesicular release pool and adequate and sustainable transmitter release provided by functional levels of synapsin I may afford the level of synaptic communication that may be required to successful learning and memory formation.

17.4 From Gut Feelings to BDNF-mediated Synaptic Plasticity

The phrase "A gut feeling" may be recognized as more than a poetic maxim. While the brain influences the digestive system, the opposite scenario is also true. Gastrointestinal disorders can affect the central nervous system, to trigger stress, anxiety, and other psychological symptoms. Symptoms of a long-standing or recurring gastrointestinal disorder intrude on a person's ability to concentrate on their normal activities, and even contribute to the etiology of headaches, fatigue, and depression. The interconnection of the gut with the brain is achieved by the autonomic nervous system composed by the sympathetic and parasympathetic nervous systems and their interconnections with the "enteric nervous system" (ENS; The word enteric is a Greek term for "intestine"). The enteric nervous system consists of neurons, neurotransmitters, and messenger proteins embedded in the layers or coverings of tissue that line the esophagus, stomach, small intestine, and colon as the "second brain" or "little brain."

Ancient and medieval anatomists and philosophers recognized the importance of the autonomic or visceral nervous system to maintain the harmony between viscera and the brain. Containing a fairly accurate gross physiological knowledge of the structure of the stomach, colon, and intestines, dividing the later into six sections whose names are still retained today by modern anatomist, the importance of digestion was recognized as a key element to maintaining the humoral balance of the body. This view maintained that the gut was an active, almost thinking agent in the body, which functioned as a source of emotional input crucial to the decision-making process. Interestingly, the early concept that the gut plays an important role on emotions has not been completely overridden in the modern age. Indeed, as discussed below the importance of the viscera to drive emotions, feelings, and sorrows is emphasized in current psychiatry. The study of the molecular mechanisms involved with the effects of diet and exercise on the CNS provides a new window to interpret the contribution of the gut to emotions and cognition in the context of modern neuroscience. Modern neuroscience is still discovering ways the enteric nervous system mirrors and interacts with the central nervous system. Nearly every substance that helps run and control the brain has turned up in the "little brain." Major neurotransmitters associated with the brain, such as serotonin, dopamine, glutamate, norepinephrine, and nitric oxide, have been found in plentiful amounts in the gut.

The physiological reactions elicited by the action of exercise and diet integrates molecular mechanisms proper of energy metabolism to influence brain function. Inherently, exercise leads to energy expenditure while diet enables energy acquisition through food consumption. The brain, an organ that comprises 2% of the total body mass, relies on harnessing aspects of metabolism to account for 50% of total body glucose utilization (Fehm et al., 2006). As the brain does not house its own energy stores, fluctuations in bodily energy resources serve to provide signals to the animal when to initiate reactions that modify energy metabolism, i.e., feeding behaviors, food breakdown, energy acquisition, expenditure, utilization, storage, and transformation. In fact, the nervous system possesses the capacity to integrate signals about energy status from the periphery with the brain. The gastrointestinal (GI) tract contains a rich sensory innervation that it uses to monitor and relay information about various aspects of digestion such as motility, secretion, and blood flow, to the central nervous system. Vagal afferents are both chemosensitive to chemicals absorbed by the epithelium and/or released from endocrine cells that sample luminal components and mechanosensitive to the levels of stomach distension. The information signaled by the vagus nerve may serve to influence higher order cognitive processing. It has been found that vagus stimulation enhances memory in animal and human subjects (Clarke et al., 1995, 1999) and that that this memory enhancement is due to the activation of vagal afferents (Clark et al., 1998). Thus, behaviors, such as exercise and feeding, which hold the ability to alter energy metabolism, also inferentially impact energy homeostasis in the brain. Resultant, exercise and diet append signals from the gut with synaptic plasticity mechanisms in the brain and modify neuronal processes of energy production (Vaynman et al., 2006). Recent findings show that neurotrophic factors may comprise key molecular components of a system that engages brain cellular and whole body energy metabolism to impact gene expression and interface with learning and memory mechanisms. These advances further our understanding of how the body interacts with and modifies brain functions, even key cognitive features, such as learning and memory, believed to be strictly under the hospice of the brain.

An integral feature of exercise and diet is their capacity to impact the central nervous system through the transcriptional regulation of neurotrophic systems. Conspicuously, studies have reproducibly found that exercise and diet can regulate the expression of BDNF in the hippocampus, a brain structure long held as a critical substrate for learning and memory. Exercise upregulates BDNF levels in the hippocampus (Neeper et al., 1996, Gomez-Pinilla et al., 2002; Vaynman et al., 2003), while a high fat diet decreases BDNF levels in the hippocampus and associated learning and memory (Molteni et al., 2002) and increases oxidative stress (Wu et al., 2004a). Conversely, supplementing the diet with omega-3 fatty acids, the primary constituents of fish oils, has been found to increase BDNF and enhance cognitive function while reducing oxidative stress in the traumatic injury model (Wu et al., 2004b).

Early studies for the action of BDNF, proposed that like other members of its neurotrophin family, it was delimited to promoting neuronal survival and differentiation during development (Barde, 1994). During the process of realization of the action of BDNF on modulating synapse development and function (Poo, 2001), it became apparent that BDNF mediates activity-dependent functional and structural plasticity in both the embryonic and adult CNS (Poo, 2001, Snider & Lichtman, 1996, McAllister et al., 1999, Prakash et al., 1996). As a varied assortment of behaviors has been shown to modulate neurotrophic levels, it is becoming clear that the activity-dependence capacity of BDNF is instrumental for transducing the effects of experience on brain structure and function (Branchi et al., 2004).

Moving beyond the role of mediator for synaptic and behavioral plasticity, BDNF is intimately connected with energy metabolism. In the mature CNS, the BDNF protein is most abundant in brain areas foremost associated with cognitive and neuroendocrine regulation, the hippocampus and hypothalamus, respectively (Nawa et al., 1995). BDNF function has been shown to regulate obesity (Rios et al., 2001; Kernie et al., 2000; Lyons et al., 1999), insulin sensitivity (Pelleymounter et al., 1995; Nakagawa et al., 2002) glucose (Tonra et al., 1999) and lipid metabolism (Tsuchida et al., 2002), and oxidative stress (OS) levels, the harmful by-products of metabolism (Lindvall et al., 1992; Lee et al., 2002; Wu et al., 2004a). A prime example is the effect of BDNF on multiple parameters of energy metabolism in a rodent model of diabetes. Central administration of BDNF to diabetic mice lowered blood glucose levels and simultaneously increased insulin levels, enhanced thermogenesis, and upregulated the mRNA expression of the uncoupling protein 1(UCP-1) in brown adipose tissue (Nonomura et al., 2001).

In light of the aforementioned examples, the BDNF system, from the level of gene expression to circuitry development, can be regarded as an archetype for biological systems that interface energy metabolism with synaptic plasticity mechanisms underlying cognitive function. Recent findings from our own laboratory illustrate the interdependency of metabolic processes with synaptic plasticity in the brain (Ding et al., 2006). We found that in the hippocampus, exercise modified aspects of energy metabolism by decreasing oxidative stress and increasing the levels of cytochrome c oxidase-II, a specific component of the mitochondrial machinery. Infusion of 1,25-dihydroxyvitamin D3, a modulator of energy metabolism, directly into the hippocampus during 3 days of voluntary wheel running decreased exercise-induced BDNF and abolished the effects of exercise on the consummate end-products of BDNF action, i.e. cyclic AMP response element-binding protein and synapsin I, and modulated phosphorylated calmodulin protein kinase II, a signal transduction cascade downstream to BDNF action that is important for learning and memory. Exercise also significantly increased the expression of the mitochondrial uncoupling protein 2 (UCP2), an energy-balancing factor concerned with ATP production and free radical management (Kramer et al., 1999; Laurin et al., 2001), to suggest that mitochondrial cellular energy metabolism interacts with the BDNF-mediated system (Ding et al., 2006). In this vein, BDNF has been recognized as an arbitrator of metabolic efficiency, synaptic plasticity, and learning and memory. Applied to the human condition, a study by Yeo et al., (2004) documented a case conjoining the importance of BDNF to energy metabolism and cognitive function. A patient exhibiting a denovo mutation that affects the consort receptor of BDNF, TrKB, not only developed hyperphagia and obesity but also suffered from developmental delays and other deficits in higher order neurological functions.

17.4.1 The Energy Metabolism – Mind Connection

To understand how energy metabolism relates to brain function, it is important to consider the evolutionary underpinning of the gut and brain connection, in light of the current prevalence of metabolic disorders in the western world. It has been postulated that as a result of the environmental pressures of 'feast and famine', adaptive mechanisms of metabolic systems evolved to maximize survival rates during those times of food shortage (Holliday, 1999). As part of these adaptive metabolic mechanisms, it has been suggested that adipocytes evolved to store energy as triglycerides (Friedman & Halaas, 1998), thereby producing an energy source to accommodate energy requirements during times of dearth. Thus, not only was a genotype that would support physical activity selected for, but the constraints of an environment consisting of 'feast and famine' may have been a major force in the evolution.

Basically, individuals who could outrun and outplan their peers in the pursuit of food and then utilize the gained food resources more efficiently became the fittest, those who could store fat better to survive the time of famine. Conjoining with these developments, hypothalamic neuro-metabolic mechanisms adapted to interact with peripheral signals carrying information about energy status. Thus, mechanisms of energy metabolism occurring in the periphery evolved in conjunction with brain function. Indeed, new research shows that metabolic signals interface with the hippocampus, to affect the mechanisms of synaptic plasticity underlying cognitive function (Vaynman et al., 2006c; Fig. 3). Learning and memory are central to the ability of animals to acquire energy sources and ultimately to survive. Mechanisms of learning and memory processing come into play during different aspects of food allocation, in actively exploring and orientating in the environment, locating and recalling food sources, differentiating between high vs. low density resources, alluding predators, and remembering whether food sources are noxious or innocuous. It is interesting to note that only recently has scientific research has begun to unearth that energy metabolism and cognitive function may be part and parcel of converging molecular pathways.

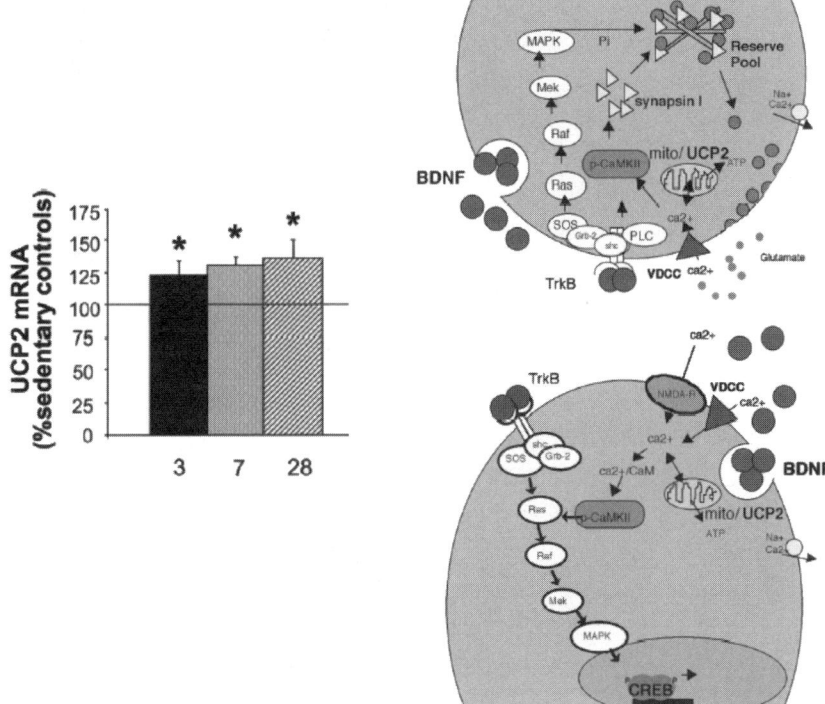

Fig. 3 (A) Exercise increases UCP2 expression in the hippocampus. UCP2 expressions was significantly elevated after three days of exercise and remained significantly above sedentary after 7 and 28 days of exercise. Result as displayed as percentage of sedentary controls (represented by the 100% line). Each value represent the mean± SEM (ANOVA, Fischer test, * P < 0.05). (B) Illustration of the potential mechanism through which elements central to energy metabolism, such as mitochondria and UCP2, interface with BDNF-mediated synaptic plasticity in the hippocampus during exercise. BDNF activates intracellular signal transduction cascades (MAPK and CAMKII) that recruit CREB and synapsin I. The uncoupling protein, UCP2 can interface with these signal transduction cascades to impact synaptic transmission and transcription, by modulating calcium homeostasis, OS production, and ATP production. Adopted from Ding et al., 2006. (*See* Color Plate 10).

The discordance between our genes and the environment manifests on the level of diseases related to higher order cognitive function. Numerous studies have found that there may be a link between abnormal glucose metabolism, in particularly an increased risk for diabetes type II, and psychiatric disorders. Psychiatric disorders such as depression, bipolar, and schizophrenia are associated with cognitive deficits and in many instances in sever cognitive impairment (O'Brien, 2005).

Since the industrial revolution, studies have provided evidence of a connection between abnormal glucose metabolism and psychiatric disorders (Kooy, 1919). Studies conducted over the past century have reported increased rates of impaired glucose tolerance, insulin resistance, and diabetes mellitus in the psychiatric population (Braceland et al., 1945; Waitzkin et al., 1966; Meuller et al., 1969; Keskiner et al., 1973, Brambilla et al., 1976; Winokur et al., 1988). Major depression has been found to be associated with a 2.23 relative risk factor of diabetes onset over the 13 years since the onset of diagnosis (Eaton et al., 1996). In the reverse scenario, it has been reported that the prevalence of depression is 2-3 times higher in diabetic as compared to the non-diabetic population (Gavard et al., 1993; Anderson et al., 2001). Similar findings have been reported for other psychiatric illnesses, such as manic depression. In a study of 203 inpatient manic-depressive subjects, Lilliker (1980) found a three fold increased rate of diabetes as compared to other psychiatric inpatients and the general US population. Schizophrenia shows the same increasing rates of diabetes as compared to controls (Tabata et al., 1987; Mukherjee et al., 1996; Dixon et al., 2000). Controlling for the confounding factor of psychotropic medication, many of which have cause a disturbance in glucose metabolism, cause weight gain and are associated with the onset of diabetes, a study by Regenold et al., 2002 determined that, there is an intrinsic relationship between abnormal glucose metabolism and bipolar disorder type I as well as schizoaffective disorders.

The specific BDNF genotype polymorphism has been identified in the etiology of psychiatric disorders. The Val66Met BDNF gene polymorphism has recently been linked with cognitive impairment and brain morphometric correlates in schizophrenia (Ho et al., 2006). Specifically this study found that schizophrenic subjects exhibited an impairment in medial temporal lobe-related memory performances which were associated with the specific BDNF genotype effects on gray matter volumes within brain regions known to subserve two cognitive domains. Specifically, Met allele carriers exhibited smaller temporal and occipital lobar gray matter volumes (Ho et al., 2006). The Val66Met BDNF gene polymorphism has also been linked with geriatric depression and cognitive performance (Hwang et al., 2005). From the standpoint of affective disorders, BDNF has been identified as the most important neurotrophin contributing to the pathogenesis of the depressive disorders. Preclinical and clinical studies demonstrate altered BDNF expression during chronic stress and increased BDNF activity during antidepressant treatment (Filus & Rybakowski, 2005). Stress models of depression have proposed that stress-induced BDNF downregulation is a result of a repression in the transcription of the BDNF gene promoter by the activated corticosteroid receptor in the hippocampus (Schaaf et al., 2000).

The clustering effect of psychiatric disorders with metabolic dysfunction, especially type 2 diabetes, may be related to decreased BDNF expression (Krabbe et al., 2007). Perhaps the best example of the relationship between metabolism and genetics is the finding that the BDNF polymorphism contributes to the genetic vulnerability to the development of eating disorders such as bulimia nervosa and binge eating disorder (Monteloeone et al., 2006)

17.4.2 From Hypothalamus to Hippocampus

The mechanisms intrinsic to energy metabolism interface with neuronal plasticity, such as provided by the cases of insulin and BDNF, who feed into signaling pathways that regulate food intake, glucose metabolism, and learning and memory

(Mattson et al., 2002). An impairment in insulin signaling in humans is a casual factor responsible for the "metabolic syndrome," a condition characterized by insulin resistance, hypertension, dyslipidemia, and disturbances in energy metabolism and autonomic function (Mattson et al., 2004) and is a major risk factor for cardiovascular disease, diabetes, and premature death (Reaven, 2004). In addition to the impairment in insulin signaling in peripheral tissues the action of insulin in the CNS is a contributing factor. The intra-cerebroventicular infusion of insulin into sheep results in decreased plasma insulin levels, increased insulin sensitivity, and reduced food intake and body weight (Foster et al., 1991). The finding that decreasing the amount of insulin receptors in the hypothalamus of rats creates animals that are both insulin resistant and hyperphagic (Obici et al., 2002), reinforces the understanding that insulin regulates whole body energy metabolism and food intake. However, the impact of insulin in the brain extends to structures critical for learning and memory. Insulin receptors are expressed in high concentrations in areas involved in learning and memory and synaptic plasticity, such as the hippocampus (Zhao et al., 2004). Insulin resistance is linked with cognitive impairment and an increased risk for Alzheimer's disease (Aravanitakis et al., 2004).

17.4.3 BDNF: The Neurotrophin Who Was Tantalus

Energy Expenditure, Metabolism, and BDNF

BDNF provides the classic example of a signaling mechanism, which is both intimately connected with cognitive function and energy metabolism. BDNF is a recognized arbitrator of metabolic efficiency, eating behavior, synaptic plasticity, and learning and memory. The ability of BDNF to mediate critical aspects of energy metabolism has entitled it to be described as a "metabotrophin". Studies of transgenic mice heterozygous for BDNF, show that BDNF insufficiency results in hyperphagia, obesity, and hyperinsulinemia (Lyons et al., 1999; Kernie et al., 2000). The peripheral or central administration of BDNF reduces body weight, normalize glucose levels (Tonra et al., 1999), ameliorate lipid metabolism in diabetic rodents (Tsuchida et al., 2002), and increase insulin sensitivity (Pelleymounter et al., 1995; Nakagawa et al., 2002). Hyperphagia and high oxidative stress (OS) levels, the harmful by-products of energy metabolism, decrease BDNF levels, while hypoglycemia and intermittent fasting both increase BDNF levels (Lindvall et al., 1992; Lee et al., 2002; Wu et al., 2004a).

The action of BDNF seems to be conducive to energy expenditure, as BDNF treatments have been demonstrated to prevent body temperature reduction during cold exposure or food deprivation (Tsuchida et al., 2001). In addition, it is known that BDNF increases mitochondrial activity (El Idrissi and Trenker, 1999) and glucose utilization (Birkhalter et al., 2003). The relationship between energy metabolism and higher cognitive function involving BDNF can bee seen from the finding that a high fat diet, whose composition was formulated to closely parallel "the all American diet," decreases BDNF levels in the hippocampus and impairs learning and memory in rats (Molteni et al., 2002). Findings from humans support the same conclusion. In humans, a de novo mutation affecting TrKB, the consort receptor to BDNF, is linked with both hyperphagia and obesity and developmental delays and other defects in higher order neurological functions (Yeo et al., 2004).

In the mature CNS, the BDNF protein is found to be most abundant in brain areas foremost associated with cognitive and neuroendocrine regulation, the hippocampus and hypothalamus, respectively (Nawa et al., 1995). In fact, BDNF's cognate receptor, TrKB, mediates signaling mechanisms coupled with the melanocortin-4 receptor (MC4R), a critical receptor involved in energy balance. MC4R has been shown to regulate the expression of BDNF in the ventral medial hypothalamus (Xu et al., 2003). Using an exercise paradigm, revealed that changes in cellular energy metabolism can modulate BDNF-mediated synaptic plasticity in the hippocampus. Infusing 1, 25-dihydroxyvitamin D3 (D3), a modulator of energy metabolism that acts on the mitochondria, directly into the hippocampus during 3 days of voluntary wheel running, reduced the levels of BDNF and its consummate end-products, synapsin I, and CREB (Vaynman et al., 2006b). Additionally, disrupting energy metabolism in the hippocampus reduced the expression p-CAMKII, the signal transduction cascade downstream to BDNF action. Exercise also increased the expression of the uncoupling protein 2 (UCP2), a mitochondrial protein, which uncouples substrate oxidation from ATP synthesis (Bouillaud et al., 1985; Boss et al., 1997; Vidal-Puig et al., 1997; Mao et al., 1999; Sanchis et al., 1998). It has been suggested that the aptitude of UCP2 to decrease OS, generate ATP, and buffer calcium may contribute to the ability of the mitochondria to modulate synaptic release and gene expression. Fig. 3 provides a potential mechanism by which cellular energy metabolism may interface with BDNF-mediated synaptic plasticity in the hippocampus during exercise.

17.4.4 Other Signals Reinforcing the Connection of the Gut with the Brain

Recent findings have revealed that peripheral metabolic signals such as ghrelin have a profound influence on hippocampal architecture and higher cognitive functions (Diano et al., 2006). Ghrelin is an endogenous ligand to the growth hormone secretagogue (GHS) receptor, which is both a hormone and neuropeptide. It is synthesized primarily in the stomach and upper intestine (Kojima et al., 1999), but can also be produced centrally (Cowley et al., 2003). Ghrelin is defined as an adipogenic hormone that is secreted from the stomach when the stomach is empty (van der Lely et al., 2004). Peripheral and central ghrelin administration increases food consumption (Faulconbridge et al., 2003; Wren et al., 2001). Ghrelin affects cognitive functions, in addition to its involvement in endocrine and metabolic regulation. This is notably demonstrated by the finding that injections of ghrelin into the hippocampus increase memory retention in rats (Carlini et al., 2004) and is consistent with the earlier finding that the receptors for ghrelin, growth hormone secretagogue receptor, have been identified in the hippocampus (Guan et al., 1997). Indeed, a recent study shows that ghrelin may have a profound action on hippocampal synaptic plasticity, altering morphology and electrophysiological parameters such as long-term potentiation (LTP), and hippocampal-dependent behavioral functions, enhancing learning and memory (Diano et al., 2006). Ghrelin may be one of a set of factors that serve as molecular interfaces between energy metabolism and neuronal and cognitive function (Fig. 4).

There may be a connection between ghrelin secretion and dietary modifications designed around restricting the amount of food intake, given that ghrelin is secreted from the stomach when it is empty (van der Lely et al., 2004) and circulating ghrelin is increased by energy restriction and decreased by food intake (Ariyasu et al., 2001; Tschop et al., 2001). Different dietary restriction models have found that dietary constraint have an enhancing effect on learning and memory. For example, either reducing the amount of calories per meal (CR) or every-other-day-fasting (EODF) both demonstrated an affect on mental health. Maintaining mice or rats on a 30–40% CR or EODF arrested or delayed the deficits in motor and cognitive function associated with ageing (Ingram et al., 1987; Means et al., 1993). Ageing is associated with decreased ghrelin levels (Rigamonti et al., 2002). Additionally, both these forms of

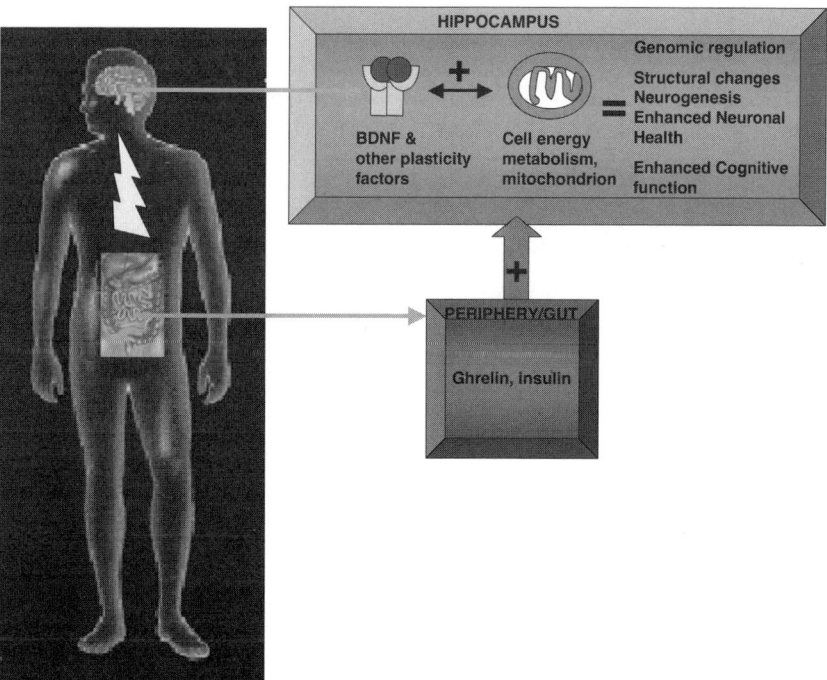

Fig. 4 Metabolism: The interface between the gut and the brain: Factors produced in the periphery such as ghrelin and insulin are shown to impact BDNF-mediated plasticity and energy metabolism in the hippocampus. As a result programs orchestrating genomic regulation, structural changes, and neurogenesis are activated to promote neuronal and cognitive plasticity. (*See* Color Plate 11).

dietary restriction models, CR and EODF, seem to protect hippocampal and basal cholinergic neurons against excitotoxicity induced death (Bruce-Keller et al., 1999; Contestabile et al., 2004). Hippocampal neurons became much resistant to degeneration induced by kainic acid, an amnesic excitotoxin in rats maintained on EODF. In parallel to this morphological finding, these EODF rats exhibited a greater preserved memory than rats fed ad lib and their degree of hippocampal neuronal resistance correlated with learning and memory on a water maze task (Bruce-Keller et al., 1999). Excessive energy intake may are associated with an enhanced risk for Alzheimer's and Parkinson's disease. A cohort study showed that people who ate a low-calorie or low-fat diet had a significantly lower risk for acquiring these neurodegenerative diseases than those who maintained a high-caloric intake. It is interesting to note that this increased risk factor was more strongly correlated with caloric intake than with weight or body mass index (Logroscino et al., 1996; Luchsinger et al., 2002). These findings support the evolutionary contention that energy balance is related to learning ability, as animals may be more determined to engage in learning and memory activities that may led to food procurement.

17.5 Exercise: From parents to offspring

Physical exercise can improve the learning and memory in both animals and humans. However, novel findings suggest that exercise in a pregnant mother can also have a positive effect on the brain and spatial learning ability of the offspring. Parnipiansil et al., (2003) found that the pups of pregnant rats who were exercised by running on a treadmill regimen (20m/min for 30 min/day for 5 consecutive days a week) had increased hippocampal BDNF mRNA expression and performed better on tests of spatial learning than control pups from unexercised mothers.

The method of action by which exercise in the pregnant mother reaches to benefit the newborn pup remains elusive. Studies have shown that exercise during pregnancy maintains the aerobic fitness of the mother and reduces pregnancy-associated discomforts (Heffernan, 2000; Uzenndoski et al., 1990), all potential factors, which can indirectly influence the fetus. In fact regular exercise during pregnancy has been shown to improve placental and fetal growth and result in an increased birth weight (Clapp et al., 2002). However, more direct methods of action have been suggested. The placenta and amniotic fluid may be sources of neurotrophic factors for the developing fetus (Gilmore, Jarskog, & Vadlamudi, 2003; Uchida et al., 2002) that may be accessed by the fetus by the initiation of exercise the mother. Maternally derived neurotrophic factors may cross the placenta to influence the health and development of the fetus (Gilmore, Jarskog, & Vadlamudi, 2003; Uchida et al., 2002). A more recent study on the effect of exercise during pregnancy on the offspring has found that maternal swimming exercise increases neurogenesis in the offspring (Lee et al., 2006). As hippocampal neurogenesis is very well correlated with learning and memory abilities (Kempermann, Kuhn, and Gage. 1998; Gould et al., 1999; Shors et al., 2001), it may well play a significant role in the exercise-induced enhancement in cognitive function in the offspring. Already exercise is known for its effective ability to induce neurogenesis in the active animal (van Praag et al., 1999). Described in the following section, alternate program through which exercise may influence the new generation is by exerting changes at the epigenetic level.

17.6 Epigenetics, Chromatin, and Gene Expression

What is epigenetics? Epigenetics literally means "in addition to changes in the genetic sequence," and is a combination of Greek prefix $\varepsilon\pi\iota$ (epi-) " in addition" and genetics. Epigenetic mechanisms include any processes that alter gene activity without changing the DNA sequence itself. For a long time, it was believed that epigenetic changes acquired during the lifetime of an organism were not transferable to the next generation. However, we are beginning to discover the existence and importance of the transgenerational inheritance of the epigenetic state.

The types of epigenetic mechanisms include DNA methylation and chromatin modification, such as methylation, acetylation, phosphorylation, and ubiquitylation. Epigenetic inheritance allows cells of a different phenotype but identical genotype to transfer that phenotype to their offspring. The potential of epigenetic mechanisms, that gene-regulatory information that is not expressed in DNA sequence itself can be transmitted from one generation (of cells or organisms) to the next, has many implications for evolution and the etiology of diseases, including coping behaviors and other psychological disturbances. Recent findings regarding the heritability of epigenetics mechanisms may provide startling insight into the effect of environment on heredity.

Chromatin refers to the complex of histones (proteins) and DNA that is tightly wound up in the nucleus. The expression of genes can be regulated by the way that chromatin is structured. When chromatin is tightly wound up it is expressionally silent, while open chromatin tends to be more functional, i.e., conducive for gene expression (Fig. 5). Epigenetic mechanisms play an important role in the silencing of genes, primarily DNA methylation and deacetylation of histones (Burzynski, 2003). Changes in the way that chromatin is structured and the modification status of histone tails governs the access of transcription and regulator factors to the DNA (Song, Taniura, & Yoneda, 2006). The acetylation of histones (H3 and H4) has been shown to loosen the DNA-histone interactions, thereby allowing transcriptional machinery to bind to DNA and increase transcription (Felsenfeld & Groudine, 2003; Jenuwein & Allis 2001). On the other hand, histone methylation has been shown to correlate with either transcriptional activation or repression, (Lachner & Jenuwein, 2002). Moreover, histone methylation can facilitate DNA methylation, which causes the repression of transcription (Lachner, O'Sullivan, & Jenuwein, 2003; Lachner & Jenuwein, 2002). One model for transcriptional repression shows that the methylation of CpG pairs recruits protein complexes that cause histone deacetylation and the repression of transcription (Fig. 5). A study was

A mechanism of epigenetic transcriptional repression

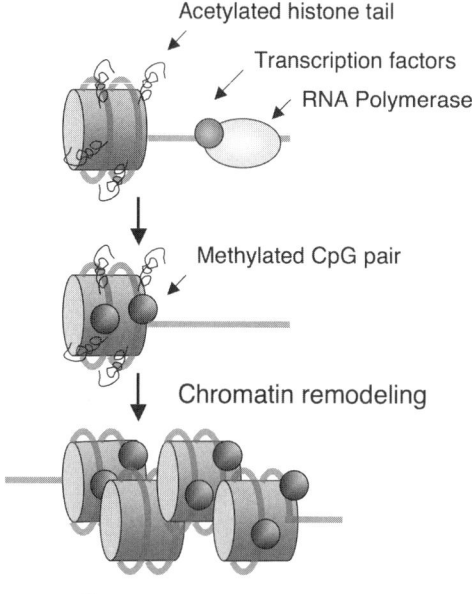

Fig. 5 Model for transcriptional repression showing that the methylation of CpG pairs recruits protein complexes that cause histone deacytylastion and the repression of transcription. (*See* Color Plate 12).

conducted on the transgenerational effects of the environmental toxins, the insecticide methyoxychlor and the fungicide vinclozolin (Anway et al., 2005). The methylation state of the germ line is reprogrammed during mammalian germ cell development (Reik & Walter, 2001a,b). Gonadal sex determination and testis development occur between embryonic days 12 and 15 in the rat, a time frame that corresponds to the period after midgestation in humans. Although steroids are not produced by the testis at this stage of development, during this stage of gonadal sex determination, endocrine agents have the ability to target the fetal testis as it contains steroid receptors. Exposure of pregnant rats to a high level of these toxins, which are also endocrine agents, decreased sperm production and fertility in the offspring, with the effects being prominent in 90% of the males in four subsequent generations. The effects of the reproduction correlated with an altered pattern of DNA methylation in the germ line (Reik & Walter, 2001a,b). The ramifications from this finding go beyond the concept that environmentally derived agents have the potential to reprogram the germ line through epigenetic mechanisms to produce the transgenerational transmission of a phenotype, to enter the fields of disease etiology and evolution.

17.6.1 What About the Role of Epigenetics in the Brain?

Epigenetic mechanism can affect the brain as well. There is growing interest in determining the role that chromatin remodeling plays in the gene regulation of neurons (Tsankova et al., 2004; Huang et al., 2002). A recent study (Weaver et al., 2004) found that the maternal behavior of mother rats affects the long-term stress response behavior of their offspring that is tied into the pattern of DNA methylation and histone acetylation at the glucocorticoid receptor gene in the offsprings' hippocampi.

Past studies have shown that the offspring of high maternal care mothers have a better hypothalamic-pituitary-adrenal (HPA) response to stress as adults (Liu et al., 1997). The magnitude of the HPA response is measured as a function of hypothalamic corticotropin-releasing factor (CRF) release, which in turn moves into the anterior pituitary via the portal circulation, where it induces the production of adrenocorticotrophin hormone (ACTH). ACTH is transported through the circulation to the adrenal gland cortex, which secretes glucocorticoids (cortisol). The HPA axis is activated in response to stress, initiating a cascade in which prolonged stress leads to the secretion of the adrenal hormone cortisol, which elevates blood sugar and increases metabolism. These changes help the body sustain prolonged activity but at the expense of decreased immune system activity. Therefore, there needs to be in intrinsic negative feedback switch installed into the HPA axis. Indeed, this is the role that glucocorticoids play. Glucocorticoids initiate a process of negative feedback inhibition over CRF synthesis and release, thereby dampening the HPA axis response to stress (De Kloet, et al., 1998). This glucocorticoids negative feedback response in part depends on the ability of glucocorticoids to bind to their cognate glucocorticoids receptors (GR) in brain regions such as the hippocampus.

The offspring that received increased maternal care (licking, grooming, arched back nursing behavior) exhibited increased expression of glucocorticoids receptors (GR) in their hippocampi, which could accommodate better detection of glucocorticoid secretion from the HPA axis, and therefore initiate a negative feedback response. Indeed, this was the case, as these animals also showed enhanced glucocorticoid feedback sensitivity compared to offspring that had poor maternal care (Liu et al., 1997; Francis et al., 1999; Weaver et al., 2004). In the rat increased fearful behavior in response to stress has been found to be associated with a decreased in hippocampal neurogenesis (the birth of new neuronal cells) and the density of synaptic connections (Bredy et al., 2003; Caldji et al., 1998). As early maternal care seems to have such a profound effect on the hippocampus, a brain region that is critical for learning and memory and in the way that an organism learns behaviors to adapt with the environment, it is not surprising that animals exposed to increased maternal behavior during the first week of life also demonstrate an increase in hippocampal cell survival, synaptogenesis (the creation of synaptic connections) and improved cognitive performance under stressful situations (Liu et al., 1997; Weaver et al., 2002; Liu et al., 2000). New evidence shows that in addition to these plastic changes in the brain, the level of maternal care can "program" the genome through epigenetic mechanisms. Specifically, animals that receive high maternal care show increased levels of histone acetylation and decreased DNA methylation on a specific sequence of the brain specific GR promoter in the hippocampus (specifically at the nerve growth factor inducible protein A (NGFI-A) transcription factor binding site), increased NGFI-A binding and increased GR expression in the hippocampus (Weaver et al., 2004; McCormick et al., 2000).

Increased pup licking, grooming, and arched back nursing behavior by mothers affected the methylation state of the offspring. For example, the cytosine residue within the 5'CpG dinucleotide site of the consensus sequence for the transcription factor, nerve growth factor inducible protein A (NGFI-A) at the glucocorticoid promotor, was regularly methylated in the offspring that received poor maternal care. In contrast, this site was rarely methylated in the offspring that received a high state of maternal care, lot of pup licking, grooming, and arched back nursing (Weaver et al., 2004). Hypomethylation of CpG dinucleotides of regulatory region on genes, as found in the high maternal care offspring, is associated with an expressionally active chromatin structure, i.e., a chromatin structure that is conducive for transcriptional activity (Razin, 1998; Razin & Cedar, 1977), therefore explaining the increase in GR expression in the hippocampus and subsequently the enhanced stress responsivity (ability to inhibit the HPA axis and decrease the amount of CRF released). Remember the magnitude of the HPA response is measured as a function of hypothalamic CRF release and the GR in the hippocampus is critical to dampening/inhibiting the HPA response to stress. These studies provide evidence for a mechanism by which stable inter individual variation in gene expression can be initiated and maintained in the brain, and specifically in the hippocampus, the brain region that is necessary for the ability of an organism to learn and develop coping mechanisms in response to the environment.

17.6.2 Are Epigenetic Changes Reversible?

Central infusion of the histone deacetylase (HDAC) inhibitor trichostatin into the low maternal care offspring as adults not only induced hypomethylation of the CpG dinucleutide sequence but also reversed other epigenetic mechanisms associated with poor maternal care, to increase histone H3-K9 acetylation of the exon I_7 glucocorticoid promoter. Moreover, the inhibitor dampened the increased stress response of the poor maternal care offspring by decreasing plasma cortisone levels to those comparable to the high maternal care offspring (Weaver et al., 2004). The results from the HDAC inhibitor causally link epigenetic mechanisms, the state of maternal care of offspring, and stress responsivity. In turn, these results implicate that that environmental factors, such as maternal care, could transmit adaptive responses across generations, a behavioral alteration in phenotype, by imprinting environmental experiences though epigenetic mechanisms onto the genome. However, these findings also suggest that it is possible to reverse the effects of early maternal care on epigenetic mechanisms and behavior with the proper intervention in adulthood.

17.6.3 Epigenetics and memory

Changing the structure of chromatin is a dynamic way to regulate gene expression. Histone acetylation at a region of ongoing transcription may be an essential mechanism to maintain a high level of gene expression (Fischle et al., 2003; Neely & Workman, 2002). Besides inducing transcription factors and removing promoter-specific repressors, chromatin remodeling is necessary to promote the transition of silent to transcriptionally permissive chromatin (Fischle et al., 2003). One histone acetyltransferase, CREB binding protein (CBP) reveals the importance of epigenetic processes to learning and memory.

In addition to functioning as a molecular scaffold for recruiting components of the transcription machinery, CBP has the ability to regulate gene expression by using its histone acetyltransferase activity to induce chromatin remodeling. Interestingly, each of these functions seems to play a different role in memory. To dissect out the importance of these two components, transgenic mice were developed in which HAT activity of CBP (CBP{HAT$^-$} mice) was eliminated in the adult stage to preclude any developmental impairments. A visual-paired comparison task (VPC), which depends on the integrity of the hippocampal formation and has been successfully used to assess declarative memory formation in humans, monkey, and rodents (Bachevalier, 1990; Ennaceur & Delacour 1988; Fagan, 1970) was used to test the effect of inhibiting CBP HAT activity in the hippocampus on short term and long-term memory. Amnesic patients suffer from damage to the temporal lobe perform poorly on the VPC task (McKee & Squire, 1993). In this task, the subject's innate ability to explore a novel object was used to assess recognition memory at a 30 min and 24 hr delay. The acquisition of new information, tested at the 30 min interval, was spared in these mice. However, the lack of HAT activity caused these animals to suffer from deficit in long-term memory, when tested 24 hrs later. Testing on the Morris water maze (Morris et al., 1982) also showed that the transgene animals performed just as well on the acquisition phase of the task as controls (2 trials per day for 12 days). However, there was an evident deficit in memory recall when they were tested on the probe trial one day after. This phenotype was reversible as either turning off the transgene expression or administering a histone deacetylase inhibitor (Trichostatin A) during training restored long-term memory formation. Combined, these results indicate that the HAT activity of CBP plays an important role in the transfer of short-term memory in to long-term memory. These findings also bring up the importance of epigenetic mechanisms in regulating transcriptional changes necessary for neuronal and behavioral plasticity.

It should not be left for granted that CBP is so well connected to CREB, the well-described evolutionarily conserved transcription factor necessary for the formation of LTM (Dash et al., 1990, Bourtchouladze et al., 1994, Yin et al., 1995; Josselyn et al., 2001; Kida et al., 2002;Pittenger et al., 2002). The importance of CREB for LTM formation across species is well documented. For example, odor memory is impaired in Drosophilia in which CREB function has been disrupted by a dominant negative CREB protein (Yin et al., 1995), while in mice, the targeted disruption of CREB isoforms results in LTM deficiency (Bourtchouladze et al., 1994). CREB has been described as a "molecular memory switch" necessary to activate the transcription necessary for LTM formation (Yin & Tully, 1996). One study in particular found that an overexpression of CREB has been shown to enhance LTM formation by reducing the usual requirements for repetition and rest during training (Yin & Tully, 1996).

However, the new findings concerning CBP and chromatin remodeling indicate the multi level of regulation that goes into transcriptional regulation, neuronal and behavioral plasticity. Although CREB phosphorylation at serine 133 is necessary to recruit CBP and subsequent transcriptional activation (Chrivia et al., 1993), but serine 133 phosphorylation of CREB alone is not sufficient to induce transcription (Bading et al., 1993; Bito et al., 1996). It seems that additional signaling pathways are needed to stimulate CBP-dependent gene expression, such as the requirement for NMDA-dependent phosphorylation of serine 301 on CBP (Impey et al., 2002). Furthermore, phosphorylation of CBP by other cellular signaling pathways such as PKA and mitogen-activated protein kinase may be required to enhance the trans-activation potential of CBP (Janknecht & Hunter, 1996). The contribution of functional histone acetyltransferases activity on CBP puts another level of complexity on activity-regulated transcription.

The epigenetic platform integrates with BDNF–mediated plasticity. The strongest connection to date is methyl-CpG-binding protein (MeCP2). MeCP2 belongs to the family of methylcytosine-binding proteins that are abundantly expressed in the central nervous system and contribute to the gene silencing effect of DNA methylation (Lewis et al., 1992; Ng & Bird, 1999). The dynamic regulation of DNA methylation and MeCP2 modification contribute to the activity driven regulation of the BDNF gene. Findings indicate that MeCP2 occupies a site on the BDNF promoter in the absence of stimulation (Martinovitch et al., 2003). When membrane depolarization is applied, MeCP2 dissociates from the BDNF exon IV promoter and methylation of several cysteine residues within the core promoter that resulted in the transcriptional repression of BDNF (Chen et al., 2003). MeCP2 extends the importance of BDNF to neural and cognitive plasticity. Mutations in the MeCP2 gene have been linked to a neuro-developmental disorder, Rett syndrome (Amir et al., 1999). Multiple studies demonstrating that MeCP2 deficiency in mice results in Rett syndrome–like abnormalities substantiate the role of MeCP2 in neuronal function (Amir et al., 1999; Chen et al., 2001; Guy et al., 2001; Habazianet et al., 2002). New findings point out that the activity dependent BDNF transcription is also regulated by the phosphorylation of MePC2 at serine 421 (S421; Zhou et al., 2006). Neuronal activity and subsequent calcium influx was found to selectively induce a CAMKII dependent mechanism of MeCP2 phosphorylation at S421 in the brain, that was required for dendritic patterning, spine morphogenesis, and activity-dependent gene expression. As MeCP2 functions as a BDNF transcriptional repressor, MeCP2 S421 phosphorylation relieves its transcriptional repressor function on the BDNF promoter IV (Zhou et al., 2006).

17.7 Conclusions and Research Demands

After mapping the human genome, there is growing emphasis to study the way that epigenetic processes alter non-Mendelian processes in disease such as schizophrenia, multiple sclerosis, diabetes, and especially affective disorders such as depression. More than just the interaction of genes with environment, science has enabled us to realize that the remarkable capacity of the brain for synaptic and cognitive plasticity requires the working connection between the genome and epigenome and the gut with the central nervous system. Both elements will be central to developing an understanding of and treating disorders of metabolic and cognitive function. The example of BDNF-mediated plasticity shows that, indeed, dynamic changes occurring at the level of neuron to system are influenced by the processes of energy metabolism and rely on an activity dependent regulation that has crosstalk between the genes and its chromatin. The contribution of behaviors that intrinsically rely on altering energy metabolism, such as exercise and diet, have provided evidence to support a model of this "collaborative nervous system " and will be resourceful for defining the interlinking role of the genome and epigenome.

Acknowledgments Supported by NIH awards NS45804 and NS50465

References

Adams J., Roberson E.D., English J.D., Selcher J.C., Sweatt J.D. (2000) MAPK regulation of gene expression in the central nervous system. Acta. Neurobiol. Exp. 60, 377–394.

Allison D.B., Fontaine K.R., Manson J.E., Stevens J., VanItallie T.B. (1999) Annual deaths attributable to obesity in the United States. JAMA 282, 1530–1538.

American Diabetes Association. Diabetes (1996) *Vital Statistics*. American Diabetes, Alexandria.

Association.American Heart Association. (1999) *Heart and Stroke Statistical Update*. American Heart Association. Dallas.

Anderson, R.J., Freedland, K.E., Clouse, R.E. and Lustman, P.J. (2001) The prevalence of comorbid depression in adults with diabetes: a meta-analysis. Diabetes Care 24, 1069–1078.

Anway, M.D., Cupp, A.S., Uzumcu, M. and Skinner, M.K. (2005) Epigenetic transgenerational actions of endocrine disruptors and male fertility. Science 308, 1466–1469.

Ariyasu, H., Takaya, K., Tagami, T., Ogawa, Y., Hosoda, K., Akamizu, T., Suda, M., Koh, T., Natsui, K., Toyooka, S., Shirakami, G., Usui, T., Shimatsu, A., Doi, K., Hosoda, H., Kojima, M., Kangawa, K. and Nakao, K. (2001) Stomach is a major source of circulating ghrelin, and feeding state determines plasma ghrelin-like im munoreactivity levels in humans. J. Clin. Endocrinol. Metab. 86, 4753–4758.

Arvanitakis, Z., Wilson, R.S., Bienias, J.L., Evans, D.A. and Bennett, D.A. (2004) Diabetes mellitus and risk of Alzheimer disease and decline in cognitive function. Arch. Neurol. 61, 661–666.

Astrand, P.O. and Rodahl, K. (1986) *Textbook of Work Physiology*. McGraw Hill, New York, pp 1–11.

Atkins, C.M., Selcher, J.C., Petraitis, J.J., Trzaskos, J.M. and Sweatt, J.D. (1998) The MAPK cascade is required for mammalian associative learning. Nat. Neurosci. 1, 602–609.

Bachevalier J. (1990) Ontogenetic development of habit and memory formation in primates. Ann. N. Y. Acad. Sci. 608, 457-77 discussion 477–484.

Bading, H., Ginty, D.D., Greenberg, M.E. (1993) Regulation of gene expression in hippo campal neurons by distinct calcium signaling pathways. Science 260, 181–186.

Bannerman, D.M., Good, M.A., Butcher, S.P., Ramsay, M. and Morris, R.G. (1995) Distinct components of spatial learning revealed by prior training and NMDA receptor blockade. Nature 378, 182–186.

Barbacid, M. (1994) The Trk family of neurotrophin receptors. J. Neurobiol. 25, 1386–1403.

Barde, Y.A. (1994) Neurotrophins: A family of proteins supporting the survival of neurons. Prog. Clin. Biol. Res. 390, 45–56.

Bar-Yosef O, Belfer-Cohen (1992) From foraging to farming in the Mediterranean Levant. In: A.B. Gebauer and T.D. Price, (Eds.)*Transitions to agriculture in prehis tory*. Prehistory Press. pp. 21–48.

Bito, H., Deisseroth, K. and Tsien, R.W. (1996) CREB phosphorylation and dephosphorylation: a Ca(2+)- and stimulus duration-dependent switch for hippocampal gene expression. Cell 87, 1203–1214.

Blanquet, P.R., Mariani, J. and Derer, P. (2003) A calcium/calmodulin kinase pathwayconnects brain-derived neurotrophic factor to the cyclic AMP-responsive transcrip tion factor in the rat hippocampus. Neurosci. 118, 477–490.

Blanquet, P.R., and Lamour, Y. (1997) Brain-derived neurotrophic factor increases Ca2+/calmodulin-dependent protein kinase 2 activity in hippocampus. J. Biol . Chem. 272, 24133–24136.

Blum, S., Moore, A.N., Adams, F. and Dash, P.K. (1999) A mitogen-activated protein kinase cascade in the CA1/CA2 subfield of the dorsal hippocampus is essential for long-term spatial memory. J. Neurosci. 19, 3535–3544.

Boaz, N.T. (2002) *Evolving health: The origins of illness and how the modern world is making us sick*. Wiley & Sons, Inc. New York.

Bohannon, R.W. (1993) Physical rehabilitation in neurologic diseases. Curr. Opin. Neu rol. 6, 765–772.

Bolton, M.M., Lo, D.C. and Sherwood, N.T. (2000) Long-term regulation of excitatory and inhibitory synaptic transmission in hippocampal cultures by brain-derived neurotrophic factor. Prog. Brain Res. 128, 203–18.

Bolton, M.M., Pittman, A.J. and Lo, D.C. (2000) Brain-Derived Neurotrophic Factor Differentially Regulates Excitatory and Inhibitory Synaptic Transmission in Hippocam pal Cultures. J. Neurosci.. 20, 3221–3232.

Booth, F.W., Chakravarthy, M.V., Gordon, S.E. and Spangenburg, E.E. (2002) Waging war on physical inactivity: using modern molecular ammunition against an ancient enemy. J. Appl. Physiol. 93, 3–30.

Boss, O., Samec, S., Dulloo, A., Seydoux, J., Muzzin, P. and Giacobino, J.P. (1997) Tissue-dependent upregulation of rat uncoupling protein-2 expression in response to fasting or cold. FEBS 412, 111–114.

Bouillaud, F., Ricquier, D., Thibault, J. and Weissenbach, J. (1985) Molecular approach to thermogenesis in brown adipose tissue: cDNA cloning of the mitochondrial un coupling protein. Proc. Natl. Acad. Sci. USA 82, 445–448.

Boulanger, L. and Poo, M.M. (1999) Gating of BDNF-induced synaptic potentiation by cAMP. Science 284,1982–1984.

Bourtchouladze, R., Frenguelli, B., Blendy, J., Cioffi, D., Schutz, G. and Silva, A.J. (1994) Deficient long-term memory in mice with a targeted mutation of the cAMP- responsive element-binding protein. Cell 79, 59–68.

Bourtchouladze, R., Frenguelli, B., Blendy, J., Cioffi, D., Schutz, G., Silva, A.J. (1994) Deficient long-term memory in mice with a targeted mutation of the cAMP- re sponsive element-binding protein. Cell, 79, 59–68.

Bova, R., Micheli, M.R., Qualadrucci, P. and Zucconi, G.G. (1998) BDNF and trkB mRNAs oscillate in rat brain during the light-dark cycle. Brain Res. Mol. Brain Res. 57, 321–324.

Braceland, F.J., Meduna L.J. and Vaichulis J.A. (1945) Delayed action of insulin in schizophrenia. Am. J. Psychiatry 102, 108–110.

Brambilla, F., Guastalla, A., Guerrini, A., Riggi, F., Rovere, C., Zanoboni A. and Zanoboni-Muciaccia W. (1976) Glucose–insulin metabolism in chronic schizophrenia. Dis. Nerv. Syst. 37, 98–103.

Branchi, I., Francia, N. and Alleva, E. (2004) Epigenetic control of neurobehavioral plasticity: the role of neurotrophins. Behav. Pharm.. 15, 353–362.

Bredy, T.W., Grant, R.J., Champagne, D.L. and Meaney, M.J. (2003) Maternal care influences neuronal survival in the hippocampus of the rat. Eur. J. Neurosci. 18, 2903–2909.

Bruce-Keller, A.J., Umberger, G., McFall, R. and Mattson, M.P. (1999) Food restriction reduces brain damage and improves behavioral outcome following excitotoxic and metabolic insults. Ann. Neurol. 45, 8–15.

Burzynski, S.R. (2003) Gene silencing–a new theory of aging. Med. Hypotheses. 60, 578–583.

Caldji, C., Tannenbaum, B., Sharma, S., Francis, D., Plotsky, P.M. and Meaney, M.J. (1998) Maternal care during infancy regulates the development of neural systems mediating the expression of fearfulness in the rat. Proc. Natl. Acad. Sci. USA 95, 5335–5340.

Carlini, V.P., Varas, M.M., Cragnolini, A.B., Schioth, H.B., Scimonelli, T.N. and de Barioglio, S.R. (2004) Differential role of the hippocampus, amygdala, and dorsal raphe nucleus in regulating feeding, memory, and anxiety-like behavioral responses to ghrelin. Biochem. Biophys. Res. Commun. 313, 635–641.

Castren, E., Pitkanen, M., Sirvio, J., Parsadanian, A., Lindholm, D., Thoenen, H. and Riekkinen, P.J. (1993) The induction of LTP increases BDNF and NGF mRNA but decreases NT-3 mRNA in the dentate gyrus. Neuroreport 4, 895–898.

Castren, E., Zafra, F., Thoenen, H., Lindholm, D. (1992) Light regulates expression of brain-derived neurotrophic factor mRNA in rat visual cortex. Proc. Natl. Acad. Sci. USA 89, 9444–9448.

Center for Disease Control and Prevention (CDC) (2001) *Behavioral Risk factor Surveil lance System survey data.* US Department of Health and Human Services, Center for Disease Control and Prevention, Atlanta.

Chen, W.G., Chang, Q., Lin, Y., Meissner, A., West, A.E., Griffith, E.C., Jaenisch, R. and Greenberg, M.E. (2003) Derepression of BDNF transcription involves calcium- dependent phosphorylation of MeCP2. Science 302, 885–889.

Chen, W.P., Chang, Y.C. and Hsieh, S.T. (1999) Trophic interactions between sensory nerves and their targets. J. Biomed. Sci. 6, 79–85.

Chen, Z.Y., Patel, P.D., Sant, G., Meng, C.X., Teng, K.K., Hempstead, B.L. and Lee, F.S. (2004) Variant brain-derived neurotrophic factor (BDNF) (Met66) alters the in tracellular trafficking and activity-dependent secretion of wild-type BDNF in neu rosecretory cells and cortical neurons. J. Neurosci. 24, 4401–4411.

Chrivia, J.C., Kwok, R.P., Lamb, N., Hagiwara, M., Montminy, M.R. and Goodman, R.H. (1993) Phosphorylated CREB binds specifically to the nuclear protein CBP. Nature 365, 855–859.

Clapp, J.F. 3rd, Kim, H., Burciu, B., Schmidt, S., Petry, K. and Lopez, B. (2002) Continuing regular exercise during pregnancy: effect of exercise volume on fetoplacental growth. Am. J. Obstet. Gynecol. 186, 142–147.

Clark, K.B., Krahl, S.E., Smith, D.C. and Jensen, R.A. (1995) Post-training unilateral vagal stimulation enhances retention performance in the rat. Neurobiol. Learn. Mem. 63, 213–6.

Clark, K.B., Naritoku, D.K., Smith, D.C., Browning, R.A. and Jensen, R.A. (1999) Enhanced recognition memory following vagus nerve stimulation in human subjects. Nat. Neurosci. 2, 94–8.

Clark, K.B., Smith, D.C., Hassert, D.L., Browning, R.A., Naritoku, D.K. and Jensen, R.A. (1998) Posttraining electrical stimulation of vagal afferents with concomitant vagal efferent inactivation enhances memory storage processes in the rat. Neurobiol. Learn. Mem. 70, 364–73.

Colcombe, S. and Kramer, A.F. (2003) Fitness effects on the cognitive function of older adults: a meta-analytic study. Psychol. Sci. 14, 125–130.

Contestabile, A., Ciani, E. and Contestabile A. (2004) Dietary restriction differentially protects from neurodegeneration in animal models of excitotoxicity. Brain Res. 1002, 162–166.

Cordain, L., Eaton, S.B., Sebastian, A., Mann, N., Lindeberg, S., Watkins, B.A., O'Keefe, J.H. and Brand-Miller, J. (2005) Origins and evolution of the Western diet: health implications for the 21st century. Am. J. Clin. Nutr. 81, 341–354.

Cordain, L., Gotshall, R.W., Eaton, S.B., Eaton, S.B. 3rd. (1998) Physical activity, energy expenditure and fitness: an evolutionary perspective. Int. J. Sports Med. 19, 328–335.

Cordain, L., Miller, J.B., Eaton, S.B., Mann, N., Holt, S.H. and Speth, J.D. (2000) Plant- animal subsistence ratios and macronutrient energy estimations in worldwide hunter- gatherer diets. Am. J. Clin. Nutr. 71, 682–692.

Cordain, L. (1999) Cereal grains: humanity's double-edged sword. World Rev. Nutr. Diet. 84, 19–73

Cowley, M.A. et al. (2003) The distribution and mechanism of action of ghrelin in the CNS demonstrates a novel hypothalamic circuit regulating energy homeostasis. Neuron 37, 649–661

Cunnane, S.C. (2005) Origins and evolution of the Western diet: health implications for the 21st century. Am. J. Clin. Nutr. 81, 341–354.

Daly, C., Sugimori, M., Moreira, J.E., Ziff, E.B. and Llinas, R. (2000) Synaptophysin regulates clathrin-independent endocytosis of synaptic vesicles, Proc. Natl. Acad. Sci. USA 97, 6120–6125.

Dash, P.K., Hochner, B. and Kandel, E.R. (1990) Injection of the cAMP-responsive element into the nucleus of *Aplysia* sensory neurons blocks long-term facilitation. Nature 345, 718–721.

De Kloet, E.R., Vreugdenhil, E., Oitzl, M.S. and Joels, M. (1998) Brain corticosteroid receptor balance in health and disease. Endocr. Rev. 19, 269–301.

Diano, S., Farr, S.A., Benoit, S.C., McNay, E.C., da Silva, I., Horvath, B., Gaskin, F.S., Nonaka, N., Jaeger, L.B., Banks, W.A., Morley, J.E., Pinto, S., Sherwin, R.S., Xu, L., Yamada, K.A., Sleeman, M.W., Tschop, M.H. and Horvath TL. (2006) Ghrelin controls hippocampal spine synapse density and memory performance. Nat. Neurosci. 9, 381–388.

Ding, Q., Vaynman, S., Souda, P., Whitelegge, J.P. and Gomez-Pinilla, F. (2006) Exercise affects energy metabolism and neural plasticity-related proteins in the hippo campus as revealed by proteomic analysis. Eur. J. Neurosci. 24, 1265–76.

Dixon, L., Weiden, P., Delahanty, J., Goldberg, R., Postrado, L., Lucksted, A. and Lehman, A. (2000) Prevalence and correlates of diabetes in national schizophrenia sam ples. Schizophr. Bull. 26, 903–912.

Eades, M.R. and Eades, M.D. (2000) *The protein power lifeplan.* Warner Books, New York.

Eaton, W.W., Armenian, H., Gallo, J., Pratt, L. and Ford, D.E. (1996) Depression and risk for onset of type II diabetes. A prospective population-based study. Diabetes Care 19, 1097–2102.

Eaton, S.B., Eaton, S.B. 3rd and Konner, M.J. (1997) Paleolithic nutrition revisited: a twelve-year retrospective on its nature and implications. Eur. J. Clin. Nutr. 51, 207–216.

Eaton, S.B. and Konner, M. (1985) Paleolithic nutrition. A consideration of its nature and current implications. N. Engl. J. Med. 312, 283–289.

Egan, M.F., Kojima, M., Callicott, J.H., Goldberg, T.E., Kolachana, B.S., Bertolino, A., Zaitsev, E., Gold, B., Goldman, D., Dean, M., Lu. B. and Weinberger, D.R. (2003) The BDNF val66met polymorphism affects activity-dependent secretion of BDNF and human memory and hippocampal function. Cell 112 , 257–269.

English, J.D. and Sweatt, J.D. (1997) A requirement for the mitogen-activated protein kinase cascade in hippocampal long term potentiation. J. Biol. Chem. 272, 19103–19106.

Ennaceur, A, Delacour J. (1988) A new one-trial test for neurobiological studies of memory in rats. I: Behavioral data. Behav. Brain. Res. 31, 47–59.

Ernfors, P., Bengzon, J., Kokaia, Z., Persson, H. and Lindvall, O. (1991) Increased levels of messenger RNAs for neurotrophic factors in the brain during kindling epilepto genesis. Neuron 7, 165–176.

Fagan JF 3rd. (1970) Memory in the infant. J. Exp. Child Psychol. 9, 217–226.

Farhadi, H.F., Mowla, S.J., Petrecca, K., Morris, S.J., Seidah, N.G. and Murphy, R.A. (2000) Neurotrophin-3 sorts to the constitutive secretory pathway of hippocampal neurons and is diverted to the regulated secretory pathway by coexpression with brain-derived neurotrophic factor. J. Neurosci. 20, 4059–4068.

Faulconbridge, L.F., Cummings, D.E., Kaplan, J.M. and Grill, H.J.(2003) Hyperphagic effects of brainstem ghrelin administration. Diabetes 52, 2260–2265.

Fehm, H.L., Kern, W. and Peters, A. (2006) The selfish brain: competition for energy resources. Prog. Brain Res. 153, 129–40.

Felsenfeld, G. and Groudine M. (2003) Controlling the double helix. Nature 421, 448–453.

Filus, J.F. and Rybakowski, J. (2005) Neurotrophic factors and their role in the pathogenesis of affective disorders. Psychiatr. Pol. 39, 883–97.

Finkbeiner, S., Tavazoie, S.F., Maloratsky, A., Jacobs, K.M., Harris, K.M. and Greenberg, M.E. (1997) CREB: a major mediator of neuronal neurotrophin responses. Neuron 19, 1031–1047.

Fischle, W., Wang, Y. and Allis, C.D. (2003) Histone and chromatin cross-talk. Curr Opin. Cell Biol. 15, 172–183.

Flegal, K.M., Carrol, M.D., Kuczmarski, R.J. and Johnson, C.L. (1998) Overweight and obesity in the United States: prevalence and trends, 1960–1994. Int. J. Obesity 22, 39–47.

Fordyce, D.E. and Farrar, R.P. (1991) Enhancement of spatial learning in F344 rats by physical activity and related learning-associated alterations in hippocampal and cor tical cholinergic functioning. Behav. Brain Res. 46, 123–33.

Foster, L.A., Ames, N.K. and Emery, R.S. (1991) Food intake and serum insulin responses to intraventricular infusions of insulin and IGF-I. Physiol. Behav. 50, 745–749.

Francis, D., Diorio, J., Liu, D. and Meaney, M.J. (1999a) Nongenomic transmission across generations of maternal behavior and stress responses in the rat. Science 286, 1155–1158.

Francis, D.D., Champagne, F.A., Liu, D. and Meaney, M.J. (1999b) Maternal care, gene expression, and the development of individual differences in stress reactivity. Ann. N.Y. Acad. Sci. 896, 66–84.

Friedman, J.M. and Halaas, J.L. (1998) Leptin and the regulation of body weight in mammals. Nature 395, 763–770.

Gabrieli, J.D., Brewer, J.B., Desmond, J.E. and Glover, G.H. (1997) Separate neural bases of two fundamental memory processes in the human medial temporal lobe. Science 276, 264–266.

Gall, C.M. and Isackson, P.J. (1989) Limbic seizures increase neuronal production of messenger RNA for nerve growth factor. Science 245, 758–761.

Garcia, C.C., Blair, H.J., Seager, M., Coulthard, A., Tennant, S., Buddles, M., Curtis, A. and Goodship, J.A. (2004) Identification of a mutation in synapsin I, a synaptic vesicle protein, in a family with epilepsy. J. Med. Genet. 41, 183–186.

Gavard, J.A., Lustman, P.J. and Clouse, R.E. (1993) Prevalence of depression in adults with diabetes. An epidemiological evaluation. Diabetes Care 16, 1167–1178.

Gilmore, J.H., Jarskog, L.F. and Vadlamudi, S. (2003) Maternal infection regulates BDNF and NGF expression in fetal and neonatal brain and maternal-fetal unit of the rat. J. Neuroimmunol. 138, 49–55.

Gomez-Pinilla, F., Ying, Z., Roy, R.R., Molteni, R. and Edgerton, R. (2002) Voluntary exercise induces a BDNF-mediated mechanism that promotes neuroplasticity. J. Neu rophysiol. 88, 2187–2195.

Gould, E., Tanapat, P., Rydel, T. and Hastings, N. (2000) Regulation of hippocampal neurogenesis in adulthood. Biol. Psychiatry 48, 715–720.

Gould, S.J. (2002) *The structure evolutionary theory*. Harvard University Press, Cambridge.

Grealy, M.A., Johnson, D.A. and Rushton, S.K. (1999) Improving cognitive function after brain injury: the use of exercise and virtual reality. Arch. Phys. Med. Rehabil. 80, 661–667.

Greengard, P. and Czernik, A.J. (1996) Neurotrophins stimulate phosphorylation of synapsin I by MAP kinase and regulate synapsin I-actin interactions. Proc. Natl. Acad. Sci. USA 93, 3679–83.

Guan, X.M. et al. (1997) Distribution of mRNA encoding the growth hormone secretagogue receptor in brain and peripheral tissues. Brain Res. Mol. Brain Res. 48, 23–29

Hall, J., Thomas, K.L. and Everitt, B.J. (2000) Rapid and selective induction of BDNF expression in the hippocampus during contextual learning. Nat. Neurosci. 3, 533–535.

Hardingham, G., Arnold, F.J. and Bading, H. (2001) Nuclear calcium signaling controls CREB-mediated gene expression triggered by synaptic activity. Nat. Neurosci. 4, 261–267.

Hariri, A.R., Goldberg, T.E., Mattay, V.S., Kolachana, B.S., Callicott, J.H., Egan, M.F. and Weinberger, D.R. (2003) Brain-derived neurotrophic factor val66met polymor phism affects human memory-related hippocampal activity and predicts memory performance. J. Neurosci. 23(17) 6690–6694.

Hartmann, H., Heumann, R. and Lessmann, V. (2001) Synaptic secretion of BDNF after high frequency stimulation of glutamatergic synapses, EMBO J. 20, 5887–5897.

Haubensak, W., Narz, F., Heumann, R. and Lessmann, V. (1998) BDNF-GFP containing secretory granules are localized in the vicinity of synaptic junctions of cultured cortical neurons. J. Cell Sci. 111, 1483–1493.

Heffernan, A.E. (2000) Exercise and pregnancy in primary care. Nurse Pract. 25, 42–49.

Ho, B.C., Milev, P., O'Leary, D.S., Librant, A. Andreasen, N.C. and Wassink, T. H. (2006) Cognitive and magnetic resonance imaging brain morphometric correlates of brain-derived neurotrophic factor Val66Met gene polymorphism in patients with schizophrenia and healthy volunteers. Arch. Gen. Psychiatry. 63, 731–740.

Holliday, R. (1998) Ageing in the 21st century. Lancet. 354 Suppl:SIV4.

Huang, Y.Y., Kandel, E.R., Varshavsky, L., Brandon, E.P., Qi, M., Idzerda, R.L., McKnight, G.S. and Bourtchouladze, R. (1995) A genetic test of the effects of mutations in PKA on mossy fiber LTP and its relation to spatial and contextual learning. Cell 83, 1211–1222.

Hwang, J.P., Tsai, S.J., Hong, C.J., Yang, C.H., Lirng, J.F. and Yang, Y.M. (2005) The Val66Met polymorphism of the brain-derived neurotrophic-factor gene is associated with geriatric depression. Neurobiol. Aging. 27, 1834–1837.

Impey, S., Fong, A.L., Wang, Y., Cardinaux, J.R., Fass, D.M., Obrietan, K., Wayman, G.A., Storm, D.R., Soderling, T.R. and Goodman, R.H. (2002) Phosphorylation of CBP mediates transcriptional activation by neural activity and CaM kinase IV. Neuron 34, 235–244.

Impey, S., Obrietan, K. and Storm, D.R. (1999) Making new connections: role of ERK/MAP kinase signaling in neuronal plasticity. Neuron 23, 11–14.

Ingram, D.K., Weindruch, R., Spangler, E.L., Freeman, J.R., Walford, R.L. (1987) Dietary restriction benefits learning and motor performance of aged mice. J. Gerontol. 42, 78–81.

Janknecht, R. and Hunter, T. (1996) Versatile molecular glue. Transcriptional control. Curr. Biol. 6, 951–954.

Jenuwein, T. and Allis, C.D. (2001) Translating the histone code. Science 293, 1074–1080.

Johnston, A.N. and Rose, S.P. (2001) Memory consolidation in day-old chicks requires BDNF but not NGF or NT-3: and anti-sense study. Mol. Brain Res. 88, 26–36.

Josselyn, S.A., Shi, C., Carlezon, W.A. Jr., Neve, R.L., Nestler, E.J. and Davis, M. (2001) Long-term memory is facilitated by cAMP response element binding protein overexpression in the amygdala. J. Neurosci. 21, 2404–2412.

Jovanovic, J.N., Benfenati, F., Siow, Y.L., Sihra, T.S., Sanghera, J.S., Pelech, S.I., Greengard, P. and Czernik, A.J. (1996) Neurotrophins stimulate phosphorylation of synapsin I by MAP kinase and regulate synapsin I-actin interactions, Proc. Natl. Acad. Sci. USA 93, 3679–3683.

Jung, R.T. (1997) Obesity as a disease. Br. Med. Bull. 53, 307–321.

Kafitz, K.W., Rose, C.R., Thoenen, H. and Konnerth, A. (1999) Neurotrophin-evoked rapid excitation through TrkB receptors. Nature 401, 918–921

Kaisho, Y., Ohta, H., Miyamato, M. and Igarashi, K. (1999) Nerve growth factor promotor driven neurotrophin-3 overexpression in the mouse and the neuroprotective effects of transgene on age related behavioral deficits. Neurosci. Lett. 277, 181–184.

Kang, H. and Schuman, E.M. (1996) Long-lasting neurotrophin-induced enhancement of synaptic transmisssion in the adult hippocampus. Science 273, 1402–1406.

Kempermann, G., Kuhn, H.G. and Gage, F.H. (1998) Experience-induced neurogenesis in the senescent dentate gyrus. J. Nuerosci. 18, 3206–3212.

Kernie, S.G., Liebl, D.J. and Parada, L.F. (2000) BDNF regulates eating behavior and locomotor activity in mice. EMBO J. 19, 1290–1300.

Keskiner, A., el-Toumi, A. and Bousquet, T. (1973) Psychotropic drugs, diabetes and chronic mental illness. Psychosomatics 14, 176–181.

Kesslak, J.P., So, V., Choi, J., Cotman, C.W. and Gomez-Pinilla, F. (1998) Learning upregulates brain-derived neurotrophic factor messenger ribonucleic acid: a mechanism to facilitate encoding of circuit maintenance. Beh. Neurosci. 112, 1012–1019.

Kida, S., Josselyn, S.A., de Ortiz, S.P., Kogan, J.H., Chevere, I., Masushige, S. and Silva, A.J. (2002) CREB is required for the stability of new and reactivated fear memories. Nat. Neurosci. 5, 348–355.

Kojima, M., Takei, N., Numakawa, T., Ishikawa, Y., Suzuki, S., Matsumoto, T., Katoh-Semba, R., Nawa, H. and Hatanaka, H. (2001) Biological characterization and optical imaging of brain-derived neurotrophic factor-green fluorescent protein suggest an activity-dependent local release of brain-derived neurotrophic factor in neurites of cultured hippocampal neurons. J. Neurosci. Res. 64, 1–10.

Kojima, M., Hosoda, H., Date, Y., Nakazato, M. and Kangawa, K. (1999) Ghrelin is a growth-hormone-releasing acylated peptide from stomach. Nature 402, 656–660.

Kooy, F.H. (1919) Hyperglycemia in mental disorders. Brain 42, 214–289.

Krabbe, K.S., Nielsen, A.R., Krogh-Madsen, R., Plomgaard, P., Rasmussen, P., Erikstrup, C., Fischer, C.P., Lindegaard, B., Petersen, A.M., Taudorf, S., Secher, N.H., Pilegaard, H., Bruunsgaard, H. and Pedersen, B.K. (2007) Brain-derived neurotrophic factor (BDNF) and type 2 diabetes. Diabetologia. 50, 431–438.

Kramer, A.F., Hahn, S., Cohen, N.J., Banich, M.T., McAuley, E., Harrison, C.R., Chason, J., Vakil, E., Bardell, L., Boileau, R.A., Colcombe, A. (1999) Ageing, fitness and neurocognitive function. Nature 400, 418–419.

Lachner, M. and Jenuwein, T. (2002) The many faces of histonelysine methylation. Curr. Opin. Cell Biol. 14, 286–298.

Lachner, M., O'Sullivan, R.J. and Jenuwein, T. (2003) An epigenetic road map for histone lysine methylation. J. Cell Sci. 116, 2117–2124.

Laurin, D., Verreault, R., Lindsay, J., MacPherson, K. and Rockwood, K. (2001) Physical activity and risk of cognitive impairment and dementia in elderly persons. Arch. Neurol. 58, 498–504.

Lee, H.H., Kim, H., Lee, J.W., Kim, Y.S., Yang, H.Y., Chang, H.K., Lee, T.H., Shin, M.C., Lee, M.H., Shin, M.S., Park, S., Baek, S. and Kim, C.J. (2006) Maternal swimming during pregnancy enhances short-term memory and neurogenesis in the hippocampus of rat pups. Brain. Dev. 28, 147–154.

Lee, J., Duan, W. and Mattson, M.P. (2002) Evidence that brain-derived neurotrophic factor is required for basal neurogenesis and mediates, in part, the enhancement of neurogenesis by dietary restriction in the hippocampus of adult mice. J. Neurochem. 82, 1367–1375.

Liang, F.Q., Walline, R., Earnest, D.J. (1998) Circadian rhythm of brain-derived neurotrophic factor in the rat suprachiasmatic nucleus. Neurosci. Lett. 242, 89–92.

Libman, I. and Arslanian, S.A. (1999) Type 2 diabetes mellitus: no longer just adults. Pediatr. Ann. 28, 589–593.

Lilliker, S.L. (1980) Prevalence of diabetes in a manic-depressive population. Compr. Psychiatry 21, 270–275.

Lindvall, O., Ernfors, P., Bengzon, J., Kokaia, Z., Smith, M.L., Siesjo, B.K. and Persson, H. (1992) Differential regulation of mRNAs for nerve growth factor, brain-derived neurotrophic factor, and neurotrophin 3 in the adult rat brain following cerebral ischemia and hypoglycemic coma. Proc. Natl. Acad. Sci. USA 89, 648–52.

Linnarsson, S., Bjorklund, A. and Ernfors, P. (1997) Learning deficit in BDNF mutant mice. Eur. J. Neurosci. 9, 2581–2587.

Liu, D., Diorio, J., Day, J.C., Francis, D.D. and Meaney, M.J. (2000) Maternal care, hippocampal synaptogenesis and cognitive development in rats. Nat. Neurosci. 3, 799–806.

Liu, D., Diorio, J., Tannenbaum, B., Caldji, C., Francis, D., Freedman, A., Sharma, S., Pearson, D., Plotsky, P.M. and Meaney, M.J. (1997) Maternal care, hippocampal glucocorticoid receptors, and hypothalamic-pituitary-adrenal responses to stress. Science 277, 1659–1662.

Logroscino, G., Marder, K., Cote, L., Tang, M.X., Shea, S. and Mayeux, R. (1996) Dietary lipids and antioxidants in Parkinson's disease: a population-based, case-control study. Ann. Neurol. 39, 89–94.

Lohof, A.M., Ip, N.Y. and Poo, M.M. (1993) Potentiation of developing neuromuscular synapses by the neurotrophins NT-3 and BDNF, Nature 363, 350–353.

Lom, B. and Cohen-Cory, S. (1999) Brain-derived neurotrophic factor differentially regulates retinal ganglion cell dendritic and axonal arborization in vivo. J. Neurosci. 19, 9928–9938.

Lu, B. and Chow, A. (1999) Neurotrophins and hippocampal synaptic transmission and plasticity. J. Neurosci. Res. 58, 76–87.

Lu, B. (2003) BDNF and activity-dependent synaptic modulation. Learn. Mem. 10, 86–98.

Luchsinger, J.A., Tang, M.X., Shea, S. and Mayeux, R. (2002) Caloric intake and the risk of Alzheimer disease. Arch. Neurol. 59, 1258–1263.

Lyons, W.E., Mamounas, L.A., Ricaurte, G.A., Coppola, V., Reid, S.W., Bora, S.H., Wihler, C., Koliatsos, V.E. and Tessarollo, L. (1999) Brain-derived neurotrophic factor-deficient mice develop aggressiveness and hyperphagia in conjunction with brain serotonergic abnormalities. Proc. Natl. Acad. Sci. USA 96, 15239–15244.

Ma, Y.L., Wang, H.L., Wu, H.C., Wei, C.L. and Lee, E.H. (1998) Brain-derived neurotrophic factor antisense oligonucleotide impairs memory retention and inhibits long-term potentiation in rats. Neurosci. 82, 957–967.

Mao, W., Yu, X.X., Zhong, A., Li, W., Brush, J., Sherwood, S.W., Adams, S.H., Pan, G. (1999) UCP4, a novel brain-specific mitochondrial protein that reduces membrane potential in mammalian cells. FEBS Lett. 443, 326–330.

Martinez, A., Alcantara, S., Borrell, V., Del Rio, J.A., Blasi, J., Otal, R., Campos, N., Boronat, A., Barbacid, M., Silos-Santiago, I. and Soriano, E. (1998) TrkB and TrkC signaling are required for maturation and synaptogenesis of hippocampal connections. J. Neurosci. 19, 7336–7350.

Mattson, M.P., Maudsley, S. and Martin, B. (2004) A neural signaling triumvirate that influences ageing and age-related disease: insulin/IGF-1, BDNF and serotonin. Ageing Res. Rev. 3, 445–464.

Mattson, M.P. (2002) Brain evolution and lifespan regulation: conservation of signal transduction pathways that regulate energy metabolism. Mech. Ageing Dev. 123, 947–953.

McAllister, A.K., Katz, L.C. and Lo, D.C. (1999) Neurotrophins and synaptic plasticity. Annu. Rev. Neurosci. 22, 295–318.

McAllister, A.K., Katz, L.C. and Lo, D.C. (1996) Neurotrophin regulation of cortical dendritic growth requires activity. Neuron 17, 1057–1064.

McCormick, J.A., Lyons, V., Jacobson, M.D., Noble, J., Diorio, J., Nyirenda, M., Weaver, S., Ester, W., Yau, J.L., Meaney, M.J., Seckl, J.R. and Chapman, K.E. (2000) 5'-heterogeneity of glucocorticoid receptor messenger RNA is tissue specific: differential regulation of variant transcripts by early-life events. Mol. Endocrinol. 14, 506–517.

McKee, R.D. and Squire, L.R. (1993) On the development of declarative memory. J. Exp. Psychol. Learn. Mem. Cogn. 19, 397–404.

Means, L.W., Higgins, J.L., Fernandez, T.J. (1993) Mid-life onset of dietary restriction extends life and prolongs cognitive functioning. Physiol. Behav. 54, 503–508.

Melloni, R.H. Jr., Apostolides, P.J., Hamos, J.E. and DeGennaro, L.J. (1994) Dynamics of synapsin I gene expression during the establishment and restoration of functional synapses in the rat hippocampus. Neurosci. 58, 683–703.

Mizuno, M., Yamada, K., Olariu, A., Nawa, H. and Nabeshima T. (2000) Involvement of brain-derived neurotrophic factor in spatial memory formation and maintenance in a radial arm maze test in rats. J. Neurosci. 20, 7116–7121.

Mokdad, A.H., Bowman, B.A., Ford, E.S., Vinicor, F., Marks, J.S. and Koplan, J.P. (2001) The continuing epidemics of obesity and diabetes in the United States. JAMA 286, 1195–1200.

Molteni, R., Ying, Z. and Gomez-Pinilla, F. (2002) Differential effects of acute and chronic exercise on plasticity-related genes in the rat hippocampus revealed by microarray. Eur. J. Neurosci. 16, 1107–1116.

Molteni, R., Wu, A., Vaynman, S., Ying, Z., Barnard, R.J. and Gomez-Pinilla, F. (2004) Exercise reverses the effects of consumption of a high-fat diet on synaptic and behavioral plasticity associated to the action of Brain-derived neurotrophic factor. Neurosci. 123, 429–440.

Monteleone, P., Zanardini, R., Tortorella, A., Gennarelli, M., Castaldo, E., Canestrelli, B. and Maj, M. (2006) The 196G/A (val66met) polymorphism of the BDNF gene is significantly associated with binge eating behavior in women with bulimia nervosa or binge eating disorder. Neurosci. Lett. 406, 133–137.

Morris, R.G.M., Garrud, P., Rawlins, J.N.P. and O'Keefe, J. (1982) Place navigation impaired in rats with hippocampal lesions. Nature 297, 681–683.

Mowla, S.J., Pareek, S., Farhadi, H.F., Petrecca, K., Fawcett, J.P., Seidah, N.G., Morris, S.J., Sossin, W.S. and Murphy, R.A. (1999) Differential sorting of nerve growth factor and brain-derived neurotrophic factor in hippocampal neurons. J. Neurosci. 19, 2069–2080.

Mu, J., Li, W., Yao, Z. and Zhou, X. (1999) Deprivation of endogenous brain-derived neurotrophic factor results in impairment of spatial learning and memory in adult rats, Brain Res. 835, 259–265.

Mueller, P.S., Heninger, G.R. and McDonald, R.K. (1969) Intravenous glucose tolerance test in depression. Arch. Gen. Psychiatry 21, 470–477.

Mukherjee, S., Decina, P., Bocola, V., Saraceni, F. and Scapicchio, P.L. (1996) Diabetes mellitus in schizophrenic patients. Comp. Psychiatry 37, 68–73.

Must, A., Spadano, J., Coakley, E.H., Field, A.E., Colditz, G. and Dietz, W.H. (1999) The disease burden associated with overweight and obesity. JAMA 282, 1523–1529.

Nakagawa, T., Tsuchida, A., Itakura, Y., Nomonura, T., Ono, M., Hirota, F., Inoue, T., Nakayama, C., Taiji, M. and Noguchi, H. (2000) Brain-derived neurotrophic factor regulates glucose metabolism by modulating energy balance in diabetic mice. Diabetes 49, 436–444.

Nawa, H., Carnahan, J. and Gall, C. (1995) BDNF protein measured by a novel enzyme immunoassay in normal brain and after seizure: partial disagreement with mRNA levels. Eur. J. Neurosci. 7, 1527–35.

Neely, K.E. and Workman, J.L. (2002) Histone acetylation and chromatin remodeling: which comes first? Mol. Genet. Metab. 76, 1–5.

Neeper, S.A., Gomez-Pinilla, F., Choi, J. and Cotman, C.W. (1996) Physical activity increases mRNA for brain-derived neurotrophic factor and nerve growth factor in rat brain. Brain Res. 726 49–56.

Neese, R.M. and Williams, G.C. (1994) *Why we get sick. The new science of Darwinian Medicine.* Times Books, New York.

Neves-Pereira, M., Mundo, E., Muglia, P., King, N., Macciardi, F. and Kennedy, J.L. (2002) The brain-derived neurotrophic factor gene confers susceptibility to bipolar disorder: evidence from a family-based association study. Am. J. Hum. Genet. 71, 651–655.

Nguyen, P.V. and Kandel, E.R. (1996) A macromolecular synthesis-dependent late phase of long-term potentiation requiring cAMP in the medial perforant pathway of rat hippocampal slices. J. Neurosci. 16, 3189–3198.

Nonomura, T., Tsuchida, A., Ono-Kishino, M., Nakagawa, T., Taiji, M. and Noguchi, H. (2001) Brain-derived neurotrophic factor regulates energy expenditure through the central nervous system in obese diabetic mice. Int. J. Exp. Diabetes Res. 2, 201–209.

Obici, S., Feng, Z., Karkanias, G., Baskin, D.G., Rossetti, L. (2002) Decreasing hypothalamic insulin receptors causes hyperphagia and insulin resistance in rats. Nat. Neurosci. 5, 566–572.

O'Brien, J. (2005) Dementia associated with psychiatric disorders. Int. Psychogeriatr. 17 Suppl. 1, S207–21.

Ogden, C.L., Flegal, K.M., Carroll, M.D. and Johnson, C.L. (2002) Prevalence and trends in overweight among US children and adolescents, 1999–2000. JAMA 288, 1728–1732.

Okabe, S., Collin, C., Auerbach, J.M., Meiri, N., Bengzon, J., Kennedy, M.B., Segal, M., McKay, R.D. (1998) Hippocampal synaptic plasticity in mice overexpressing an embryonic subunit of the NMDA receptor. J. Neurosci. 18, 4177–4188.

Parnipiansil, P., Jutapakdeegul, N., Chentanez, T. and Kotchabhakdi, N. (2003) Exercise during pregnancy increases hippocampal brain-derived neurotrophic factor mRNA expression and spatial learning in neonatal rat pup. Neurosci. Lett. 352, 45–48.

Patapoutian, A. and Reichardt, L.F. (2001) Trk receptors: mediators of neurotrophin action. Curr. Opin. Neurobiol. 11, 272–80.

Patterson, S.L., Abel, T., Deuel, T.A.S., Martin, K.C., Rose, J.C., Kandel, E.R. (1996) Recombinant BDNF rescues deficits in basal synaptic transmission and hippocampal LTP in BDNF knockout mice. Neuron 16, 1137–1145.

Patterson, S.L., Grover, L.M., Schwartzkroin, P.A. and Bothwell, M. (1992) Neurotrophin expression in rat hippocampal slices: a stimulus paradigm inducing LTP in CA1 evokes increases in BDNF and NT-3 mRNAs. Neuron 9, 1081–88.

Pelleymounter, M.A., Cullen, M.J. and Wellman, C.L. (1995) Characteristics of BDNF-induced weight loss. Neurol. 131, 229–238.

Pinhas-Hamiel, O., Dolan, L.M., Daniels, S.R., Standiford, D., Khoury, P.R. and Zeitler, P. (1996) Increased incidence of non-insulin-dependent diabetes mellitus among adolescents. J. Pediatr. 128, 608–615.

Pittenger, C., Huang, Y.Y., Paletzki, R.F., Bourtchouladze, R., Scanlin, H., Vronskaya, S. and Kandel, E.R. (2002) Reversible inhibition of CREB/ATF transcription factors in region CA1 of the dorsal hippocampus disrupts hippocampus-dependent spatial memory. Neuron 34, 447–462.

Poo, M.M., (2001) Neurotrophins as synaptic modulators. Nat. Rev. Neurosci. 2, 24–32.

Powell, K.E. and Blair, S.N. (1994) The public health burdens of sedentary living habits: theoretical but realistic estimates. Med. Sci. Sports Exerc. 26, 851–856.

Pozzo-Miller, L.D., Gottschalk, W., Zhang, L., McDermott, K., Du, J., Gopalakrishnan, R., Oho, C., Sheng, Z.H. and Lu, B. (1999) Impairments in high-frequency transmission, synaptic vesicle docking, and synaptic protein distribution in the hippocampus of BDNF knockout mice. J. Neurosci. 19, 4972–4983.

Prakash, N., Cohen-Cory, S. and Frostig, R.D. (1996) Rapid and opposite effects of BDNF and NGF on the functional organization of the adult cortex in vivo. Nature 381, 702–706.

Razin, A. and Cedar, H. (1977) Distribution of 5-methylcytosine in chromatin. Proc. Natl. Acad. Sci. USA 74, 2725.

Razin, A. (1998) CpG methylation, chromatin structure and gene silencing-a three-way connection. EMBO J. 17, 4905–4908.

Reaven, P. (2004) Metabolic syndrome. J. Insur. Med. 36, 132–142.

Regenold, W.T., Thapar, R.K., Marano, C., Gavirneni, S., Kondapavuluru, P.V. (2002) Increased prevalence of type 2 diabetes mellitus among psychiatric inpatients with bipolar I affective and schizoaffective disorders independent of psychotropic drug use. J. Affect. Disord. 70, 19–26.

Reik, W. and Walter, J. (2001a) Genomic imprinting: parental influence on the genome. Nat. Rev. Genet. 2, 21–32.

Reik, W. and Walter, J. (2001b) Evolution of imprinting mechanisms: the battle of the sexes begins in the zygote. Nat. Genet. 27, 255–256.

Rigamonti, A.E., Pincelli, A.I., Corra, B., Viarengo, R., Bonomo, S.M., Galimberti, D., Scacchi. M., Scarpini, E., Cavagnini, F., Muller, E.E. (2002) Plasma ghrelin concentrations in elderly subjects: comparison with anorexic and obese patients. J. Endocrinol. 175, R1–5.

Rios, M., Fan, G., Fekete, C., Kelly, J., Bates, B., Kuehn, R., Lechan, R.M., Jaenisch, R. (2001) Conditional deletion of brain-derived neurotrophic factor in the postnatal brain leads to obesity and hyperactivity. Mol. Endocrinol. 15, 1748–1757.

Rocamora, N., Pascual, M., Acsady, L., de Lecea, L., Freund, T.F. and Soriano, E. (1996) Expression of NGF and NT3 mRNAs in hippocampal interneurons innervated by the GABAergic septohippocampal pathway. J. Neurosci. 16, 3991–4004.

Rogers, R.L., Meyer, J.S. and Mortel, K.F. (1990) After reaching retirement age physical activity sustains cerebral perfusion and cognition. J. Am. Geriatr. Soc. 38 ,123–128.

Rutherford, L.C., Nelson, S.B. and Turrigiano, G.G. (1998) BDNF has opposite effects on the quantal amplitude of pyramidal neuron and interneuron excitatory synapses. Neuron 21, 521–530.

Saarelainen, T., Pussinen, R., Koponen, E., Alhonen, L., Wong, G., Sirvio, J., Castren, E. (2000) Transgenic mice overexpressing truncated trkB neurotrophin receptors in neurons have impaired long-term spatial memory but normal hippocampal LTP. Synapse. 38, 102–104.

Samorajski, T., Delaney, C., Durham, L., Ordy, J.M., Johnson, J.A. and Dunlap, W.P. (1985) Effect of exercise on longevity, body weight, locomotor performance, and passive-avoidance memory of C57BL/6J mice. Neurobiol. Aging 6, 17–24.

Sanchis, D., Fleury, C., Chomiki, N., Goubern, M., Huang, Q., Neverova, M., Gregoire, F., Easlick, J., Raimbault, S., Levi-Meyrueis, C., Miroux, B., Collins, S., Seldin, M., Richard, D., Warden, C., Bouillaud, F. and Ricquier, D. (1998) BMCP1, a novel mitochondrial carrier with high expression in the central nervous system of humans and rodents, and respiration uncoupling activity in recombinant yeast. J. Biol. Chem. 273, 34611–34615.

Schaaf, M.J., DeKloet, E.R. and Vreugdenhil, E. (2000) Corticosterone effects on BDNF expression in the hippocampus: Implications for memory formation. Stress 3, 201–208.

Schinder, A.F. and Poo, M. (2000) The neurotrophin hypothesis for synaptic plasticity. Trends Neurosci. 23, 639–645.

Seil, F.J. and Drake-Baumann, R. (2000) TrkB receptor ligands promote activity-dependent inhibitory synaptogenesis. J. Neurosci. 20 , 5367–5373.

Sen, S., Nesse, R.M., Stoltenberg, S.F., Li, S., Gleiberman, L., Chakravarti, A., Weder, A.B. and Burmeister, M. (2003) A BDNF coding variant is associated with the NEO personality inventory domain neuroticism, a risk factor for depression. Neuropsychopharmacol. 28, 397–401.

Shimada, A., Mason, C.A., Morrison, M.E. (1998) TrkB signaling modulates spine density and morphology independent of dendrite structure in cultured neonatal Purkinje cells. J. Neurosci. 18, 8559–8570

Shors, T.J., Miesegaes, G., Beylin, A., Zhao, M., Rydel, T. and Gould, E. (2001) Neurogenesis in the adult is involved in the formation of trace memories. Nature 410, 372–376.

Silva, A.J., Kogan, J.H., Frankland, P.W., Kida, S. (1998) CREB and memory, Annu. Rev. Neurosci. 21, 127–148.

Silva, A.J., Paylor, R., Wehnet, J.M. and Tonegawa, S. (1992) Impaired spatial learning in α-calcium-calmodulin kinase II mutant mice. Science 257, 206–211.

Sng, J.C., Taniura, H. and Yoneda, Y. (2006) Histone modifications in kainate-induced status epilepticus. Eur. J. Neurosci. 23, 1269–1282.

Snider, W.D. and Lichtman, J.W. (1996) Are neurotrophins synaptotrophins? Mol. Cell. Neurosci. 7, 433–442.

Sothern, M.S., Hunter, S., Suskind, R.M., Brown, R., Udall, J.N. Jr. and Blecker, U. (1999) Motivating the obese child to move: the role of structured exercise in pediatric weight management. South. Med. J. 92, 577–584.

Stephens, L., Smrcka, A., Cooke, F.T., Jackson, T.R., Sternweis, P.C., Hawkins, P.T. (1994) A novel phosphoinositide 3 kinase activity in myeloid-derived cells is activated by G protein beta gamma subunits. Cell. 77, 83–93.

Stern, C.E., Corkin, S., Gonzalez, R.G., Guimaraues, A.R., Baker, J.R., Jennings, P.J., Carr, C.A., Sugiura, R.M., Vedantham. V., Rosen, B.R. (1996) The hippocampal formation participates in novel picture encoding: evidence from functional magnetic resonance imaging. Proc. Natl. Acad. Sci. USA 93, 8660–8665.

Sugaya, K., Greene, R., Personett, D., Robbins, M., Kent, C., Bryan, D., Skiba, E., Gallagher, M., McKinney, M. (1998) Septo-hippocampal cholinergic and neurotrophin markers in age-induced cognitive decline. Neurobiol. Aging. 19, 351–361.

Suominen-Troyer, S., Davis, K.J., Ismail, A.H. and Salvendy, G. (1986) Impact of physical fitness on strategy development in decision-making tasks. Percept. Mot. Skills 62, 71–77.

Sweatt, J.D. (2001) The neuronal MAP kinase cascade: a biochemical signal integration system subserving synaptic plasticity and memory. J. Neurochem. 76, 1–10.

Tabata, H., Kikuoka, M., Kikuoka, H., Bessho, H., Hirayama, J., Hanabusa, T., Kubo, K., Momotani, Y., Sanke, T., Nanjo, K., Higashi, Y. and Miyamura, K. (1987) Characteristics of diabetes mellitus in schizophrenic patients. J. Med. Assoc. Thai. 70, Suppl. 2, 90–93.

Takei, Y., Harada, A., Takeda, S., Kobayashi, K., Terada, S., Noda, T., Takahashi, T. and Hirokawa N. (1995) Synapsin I deficiency results in the structural change in the presynaptic terminals in the murine nervous system. J. Cell. Biol. 131, 1789–1800.

Tao, X., Finkbeiner, S., Arnold, D.B., Shaywitz, A. and Greenberg, M.E. (1998) Ca2+ influx regulates BDNF expression by a CREB dependent mechanism. Neuron 20, 709–726.

Thiele, C., Hannah, M., Fahrenholz, F. and Huttner, W.B. (2000) Cholesterol binds to synaptophysin and is required for biogenesis of synaptic vesicles. Nat. Cell Biol. 2 42–49.

Tonra, J.R., Ono, M., Liu, X., Garcia, K., Jackson, C., Yancopoulos, G.D., Wiegand, S.J. and Wong, V. (1999) Brain-derived neurotrophic factor improves blood glucose control and alleviates fasting hyperglycemia in C57BLKS-Lepr(db)/lepr(db) mice. Diabetes 48, 588–594.

Tsai, S.J., Hong, C.J., Yu, Y.W. and Chen, T.J. (2004) Association study of a brain-derived neu-rotrophic factor (BDNF) Val66Met polymorphism and personality trait and intelligence in healthy young females. Neuropsychobiol. 49, 13–6.

Tsankova, N.M., Kumar, A. and Nestler, E.J. (2004) Histone modifications at gene promoter regions in rat hippocampus after acute and chronic electroconvulsive seizures. J. Neurosci. 24, 5603–5610.

Tschop, M., Wawarta, R., Riepl, R.L., Friedrich, S., Bidlingmaier, M., Landgraf, R., Folwaczny, C. (2001) Postprandial decrease of circulating human ghrelin levels. J. Endocrinol. Invest. 24, RC19–21.

Tsuchida, A., Nonomura, T., Nakagawa, T., Itakura, Y., Ono-Kishino, M., Yamanaka, M., Sugaru, E., Taiji, M. and Noguchi, H. (2002) Brain-derived neurotrophic factor ameliorates lipid metabolism in diabetic mice. Diabetes Obes. Metab. 4, 262–269.

Tsuchida, A., Nonomura, T., Ono-Kishino, M., Nakagawa, T., Taiji, M., Noguchi, H. (2001) Acute effects of brain-derived neurotrophic factor on energy expenditure in obese diabetic mice. Int. J. Obes. Relat. Metab. Disord 25, 1286–1293.

Tully, T. (1997) Regulation of gene expression and its role in long-term memory and synaptic-plasticity. Proc. Nat. Acad. Sci. USA 94, 4239–4241.

Uchida, S., Inanaga, Y., Kobayashi, M., Hurukawa, S., Araie, M. and Sakuragawa, N. (2002) Neu-rotrophic function of conditioned medium from human amniotic epithelial cells. J. Neurosci. Res. 62, 585–590.

US Department of Health and Human Services. (1996) *Physical Activity and Health: A report of the Surgeon General*. US Dept. Of Health and Human Servs., Center for Disease Control and Prevention, National Center for Chronic Prevention and Health Promotion, Atlanta.

Uzendoski, A.M., Latin, R.W., Berg, K.E. and Moshier, S. (1990) Physiological responses to aer-obic exercise during pregnancy and post-partum. J. Sports Med. Phys. Fitness. 30, 77–82.

van der Lely, A.J., Tschop, M., Heiman, M.L. and Ghigo, E. (2004) Biological, physiological, pathophysiological, and pharmacological aspects of ghrelin. Endocr. Rev. 25, 426–457

van Praag, H., Christie, B.R., Sejnowski, T.J., and Gage, F.H. (1999) Running enhances neurogen-esis, learning, and long-term potentiation in mice. Proc. Nat. Acad. Sci. USA 96,13427–13434.

van Praag, H., Kempermann, G. and Gage, F.H. (1999) Running increases cell proliferation and neurogenesis in the adult mouse dentate gyrus. Nat. Neurosci. 2, 266–70.

Vaynman, S., Ying, Z., Wu, A., Gomez-Pinilla, F. (2006a) Coupling energy metabolism with a mechanism to support brain-derived neurotrophic factor-mediated synaptic plasticity. Neurosci. 139, 1221–1234.

Vaynman, S. and Gomez-Pinilla, F. (2006b) Revenge of the "sit": how lifestyle impacts neuronal and cognitive health through molecular systems that interface energy metabolism with neuronal plasticity. J. Neurosci. Res. 84, 699–715.

Vaynman, S.S., Ying, Z., Yin, D. and Gomez-Pinilla, F. (2006c) Exercise differentially regulates synaptic proteins associated to the function of BDNF. Brain Res. 1070, 124–30.

Vaynman, S., Ying, Z. and Gomez-Pinilla, F. (2004) Hippocampal BDNF mediates the efficacy of exercise on synaptic plasticity and cognition. Eur. J. Neurosci. 20, 2580–2590.

Vaynman, S., Ying, Z. and Gomez-Pinilla, F. (2003) Interplay between BDNF and signal trans-duction modulators in the regulation of the effects of exercise on synaptic-plasticity. Neurosci. 122, 647–57.

Vicario-Abejon, C., Collin, C., McKay, R.D.G. and Segal, M. (1998) Neurotrophins induce forma-tion of functional excitatory and inhibitory synapses between cultrual hippocampalm neurons, J. Neurosci. 18, 7256–7271.

Vidal-Puig, A., Solanes, G., Grujic, D., Flier, J.S. and Lowell, B.B. (1997) UCP3: an uncoupling protein homologue expressed preferentially and abundantly in skeletal muscle and brown adi-pose tissue. Biochem. Biophys. Res. Commun. 235, 79–82.

Waitzkin, L. (1966) A survey for unknown diabetics in a mental hospital. II. Men from age fifty. Diabetes 15, 164–172.

Weaver, I.C., Cervoni, N., Champagne, F.A., D'Alessio, A.C., Sharma, S., Seckl, J.R., Dymov, S., Szyf, M. and Meaney, M.J. (2004) Epigenetic programming by maternal behavior. Nat. Neu-rosci. 7, 847–854.

Weaver, I.C., Grant, R.J. and Meaney, M.J. (2002) Maternal behavior regulates long-term hippocampal expression of BAX and apoptosis in the offspring. J. Neurochem. 82, 998–1002.

Wendorf, M. and Goldfine, I.D. (1991) Archaeology of NIDDM. Excavation of the "thrifty" genotype. Diabetes 40, 161–165.

Winokur, A., Maislin, G., Phillips, J.L. and Amsterdam, J.D., (1988) Insulin resistance after oral glucose tolerance testing in patients with major depression. Am. J. Psychiatry 145, 325–330.

Worm, N. (2002) Syndrome X oder ein Mammut auf den Teller. Mit Steinzeit-Diät det Wohl stands Falle. Systemed-Verlag, Lünen.

Wren, A.M., Seal, L.J., Cohen, M.A., Brynes, A.E., Frost, G.S., Murphy, K.G., Dhillo, W.S., Ghatei, M.A. and Bloom, S.R. (2001) Ghrelin enhances appetite and increases food intake in humans. J. Clin. Endocrinol. Metab. 86, 5992.

Wu, A., Ying, Z. and Gomez-Pinilla, F. (2004) The interplay between oxidative stress and brain-derived neurotrophic factor modulates the outcome of a saturated fat diet on synaptic plasticity and cognition. Eur. J. Neurosci. 19, 1699–1707.

Xu, B., Goulding, E.H., Zang, K., Cepoi, D., Cone, R.D., Jones, K.R., Tecott, L.H., Reichardt, L.F. (2003) Brain-derived neurotrophic factor regulates energy balance downstream of melanocortin-4 receptor. Nat. Neurosci. 6, 736–742.

Yacoubian, T.A. and Lo, D.C. (2000) Truncated and full-length TrkB receptors regulate distinct modes of dendritic growth. Nat. Neurosci. 3, 342–349.

Yeo, G.S., Hung, C.-C.C., Rochford, J., Keogh, J., Gray, J., Sivaramakrishnan, S., O'Rahilly, S. and Farooqi, I.S. (2004) A de novo mutation affecting human TrkB associated with severe obesity and developmental delay. Nat. Neurosci. 7, 1187–1189.

Yin, J.C., DelVecchio, M., Zhou, H. and Tully, T. (1995) CREB as a memory modulator: induced expression of a dCREB2 activator isoform enhances long-term memory in Drosophila. Cell 81, 107–115.

Yin, J.C., Wallach, J.S., DelVecchio, M., Wilder, E.L., Zhou, H., Quinn, W.G. and Tully, T. (1994) Induction of a dominant negative CREB transgene specifically blocks long-term memory in Drosophila. Cell 79, 49–58.

Yin, J.C. and Tully, T. (1996) CREB and the formation of long-term memory. Curr. Opin. Neurobiol. 6, 264–268.

Ying, S., Futter, M., Rosenblum, K., Webber, M.J., Hunt, S.P., Bliss, T.V. and Bramham, C.R. (2002) Brain-derived neurotrophic factor induces long-term potentiation in intact adult hippocampus: requirement for ERK activation coupled to CREB and upregulation of Arc synthesis. J. Neurosci. 22, 1532–1540.

Zafra, F., Hengerer, B., Leibrock, J., Thoenen, H. and Lindholm, D. (1990) Activity dependent regulation of BDNF and NGF mRNAs in the rat hippocampus is mediated by non-NMDA glutamate receptors. EMBO J. 9, 3545–3550.

Zeineh, M.M., Engel, S.A., Bookheimer, S.Y. (2000) Application of cortical unfolding techniques to functional MRI of the human hippocampal region. NeuroImage. 11, 668–683.

Zhao, W.Q., Chen, H., Quon, M.J., Alkon, D.L. (2004) Insulin and the insulin receptor in experimental models of learning and memory. Eur. J. Pharmacol. 490, 71–81.

Zhou, Z., Hong, E.J., Cohen, S., Zhao, W.N., Ho, H.Y., Schmidt, L., Chen, W.G., Lin, Y., Savner, E., Griffith, E.C., Hu, L., Steen, J.A., Weitz, C.J. and Greenberg, M.E. (2006) Brain-specific phosphorylation of MeCP2 regulates activity-dependent BDNF transcription, dendritic growth, and spine maturation. Neuron 52, 255–269.

Chapter 18
CREB Responsive Transcription and Memory Formation

Thomas C. Tubon, Jr. and Jerry C.P. Yin

Abstract Inhibitors of protein synthesis block memory formation when they are acutely delivered around the time of behavioral training. This requirement for *de novo* synthesis of proteins includes a prerequisite for gene transcription, since inhibitors of RNA polymerase II display similar effects. These observations, together with the strong biochemical and genetic evidence in *Aplysia* and *Drosophila*, led to experiments testing the importance of the cAMP-Response Element (CRE) and its Binding protein (CREB) in long-term memory and synaptic plasticity. In this chapter, we will review the molecular biology of *CREB* genes, before summarizing the work that demonstrates CREB is an important factor in memory formation. We will then address the more complex issue of why this requirement remains controversial. Drawing from emerging work in *Drosophila*, we will discuss the complexity of *CREB* gene expression and how revealing the molecular mechanisms that underlie CREB activity may provide insights and resolutions to the earlier experimental discrepancies.

18.1 Molecular Organization of CREB

The requirement for cAMP-responsive transcription evolved very early in cells, since there is evidence of this pathway in yeast and *Dictyostelium* (Hintermann and Parish 1979; Peters, Cammans, Smit, Spek, van Lookeren Campagne and Schaap, 1991; Shemarova 2005). The importance of this molecular pathway is made apparent by the evolutionary conservation across the animal kingdom of one of the key regulating molecules in this signaling cascade, - CREB. The earliest known homolog of what is now recognized as the prototypical CREB gene was identified in hydra (Chera, Kaloulis, and Galliot 2007). In invertebrates, a single PKA-dependent CREB gene appears to be the rule, and it has been well studied in *Aplysia*, *Drosophila*, *Apis mellifera*, and *Lymnaea* (Yin and Tully 1996; Bartsch, Casadio, Karl, Serodio, and Kandel 1998; Eisenhardt, Friedrich, Stollhoff, Muller, Kress, and Menzel 2003; Ribeiro, Serfozo, Papp, Kemenes, O'Shea, Yin, Benjamin, and Kemenes 2003; Sadamoto, Sato, Kobayashi, Murakami, Aonuma, Ando,

S. M. Dudek (ed.), *Transcriptional Regulation by Neuronal Activity.*
© Springer 2008

Fujito, Hamano, Awaji, Lukowiak, Urano, and Ito 2004). In vertebrates, the single CREB precursor gene appears to have duplicated and diverged during evolution, producing a multi-gene family. In rodents and humans, there are three members in the CREB gene family (CREB, CREM and ATF-1) (Meyer and Habener 1993). All CREB genes have the general structure that is summarized in Figure 1A. The highly conserved basic region-leucine zipper (bZIP) domain is located near the carboxy-terminal end of the protein, and confers dimerization and DNA binding properties on the protein. Dimerization of CREB molecules can occur within and between CREB family members to form both homodimers and heterodimers via the bZIP domain. In the middle of the protein, there is a region of high conservation (the P-box or KID domain) that contains numerous amino acid residues that are target substrates for various kinases such as protein kinase A (PKA), protein kinase C (PKC), casein kinase II (CKII), Calcium-calmodulin-dependent protein kinase II (CamKII), ataxia-telangiectasia-mutated (ATM) protein kinase, and glycogen synthase kinase 3 (GSK3) (Johannessen and Moens 2007; Shanware, Trinh, Williams, and Tibbetts 2007). The most widely studied phosphorylation site on CREB resides within this region at amino acid residue Serine 133 (S133). Although many target phosphorylation sites within the P-box have been identified, S133 is unique among these sites because its phosphorylation is necessary for interaction with known histone acetyltransferases, CREB Binding Protein (CBP) or its relative p300 (Chrivia, Kwok, Lamb, Hagiwara, Montminy, and Goodman 1993; Kwok, Lundblad, Chrivia, Richards, Bachinger, Brennan, Roberts, Green, and Goodman 1994). The addition of a phosphoryl group to S133 and binding to CBP/P300 is thought to induce a structural change in CREB that promotes gene transactivation through subsequent recruitment of RNA polymerase II. The functional roles for

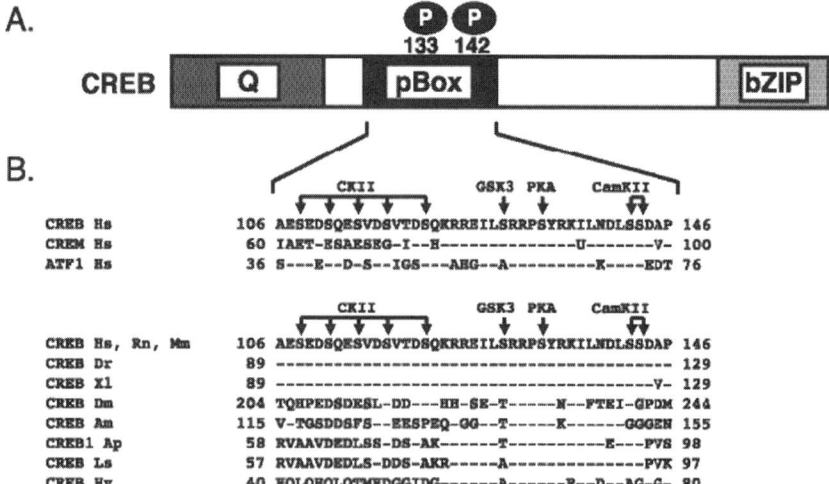

Fig. 1 Structural Organization of CREB: (a) General Structural Features of CREB. (b) Sequence comparison of the P-box region. Hs, *Homo sapiens*; Rn, *Rattus norvegicus*; Mm *Mus musculus*; Dr, *Danio rerio*; Xl *Xenopus laevis*; Dm, *Drosophila melanogaster*; Am; *Apis mellifera*; Ap, *Aplysia californica*, Ls *Lymnaea stagnalis*; Hv *Hydra vulgaris*.

two additional phosphorylation sites located within the CREB P-box have also been described at Serine 142 (S142) and Serine 143 (S143) (Mayr and Montminy 2001; Gau, Lemberger, von Gall, Kretz, Le Minh, Gass, Schmid, Schibler, Korf, and Schutz 2002; Kornhauser, Cowan, Shaywitz, Dolmetsch, Griffith, Hu, Haddad, Xia, and Greenberg 2002). Interestingly, phosphorylation of Ser142 by CaMKII inhibits the formation of CREB-CBP complex, thereby preventing the activation of CREB-CBP dependent downstream target genes. Transgenic mice harboring a mutation in the S142 site (S142A) display a down-regulation in the expression of circadian clock genes, *c-Fos* and *mPer1* (Gau *et al.*, 2002). The concept that CREB phosphorylation at serine residues at positions 133, 142, and 143 may lead to differences in gene expression programs is further supported by the observation that the triply phosphorylated protein is required in neurons for effective depolarization-induced gene transcription (Kornhauser *et al.*, 2002). These studies illustrate how distinct combinations of posttranslational modifications within the P-box of CREB can direct the expression of specific programs of gene expression.

A number of experimental observations have been described where CREB functionally interacts with other transcriptional co-activators, independently of S133 phosphorylation (Fimia, De Cesare, and Sassone-Corsi 1999; Conkright, Canettieri, Screaton, Guzman, Miraglia, Hogenesch, and Montminy 2003). These interactions are likely not based on physical contact between the different co-activators and the CREB P-box region, but rather through the carboxy-terminal bZIP domain. Historically, however, the presence of the S133 site has been one of the defining characteristics of CREB family genes and proteins. In the single *Drosophila CREB* gene (*dCREB2 or dCREB-A*), the bZIP domain is highly conserved between flies and humans (*Homo sapiens* CREB amino acids 265-324; *Drosophila melanogaster* dCREB2 amino acids 268-327 (81% identity 52/64), while the P-box region is less conserved (34% identity 14/41). However, it is clear from a comparison of the primary sequences for CREB from multiple species that what remains conserved are the numerous phosphorylation sites that act as specific substrates for PKA, PKC, CKII, CamKII, and GSK3. A comparison of these sites is shown in Figure 1B, revealing the evolutionary conservation of these substrate target sites for phosphorylation among CREB homologs.

In addition to multiple phosphorylation events, there have been several reports of other modifications occurring on CREB, including O-GlcNAc-glycosylation, SUMOylation, and redox-based modifications (Goren, Tavor, Goldblum, and Honigman 2001; Comerford, Leonard, Karhausen, Carey, Colgan, and Taylor 2003; Khidekel and Hsieh-Wilson 2004). We have evidence for each of these occurring on various dCREB2 protein isoforms, but the functional consequences of these protein modifications are still unclear.

One of the unforeseen, and neglected effects of heavy post-translational modifications is an effect on the sensitivity and affinity of CREB-antibody reagents in protein detection. Clearly, we observe that dCREB2 protein isoforms are modified extensively, with a subset of post-translational modifications that are known to differentially affect the affinity of almost all of our 14 CREB-specific antibodies. These antibodies are generally classifiable into two groups, and members of the first recognize certain modified forms, but not others, and vice versa. Thus, based on immuno-affinity Western Blot detection, we have been able to identify distinct CREB antibody pools that detect the presence of several CREB isoforms that vary in both post-translational modifications and primary amino acid composition. Although this analysis of antibody reagents has unveiled valuable molecular tools for investigating dCREB2 proteins, the use of these reagents dramatically increases the dimensionality involved in interpreting results from both western and immnocytochemical experiments. To date, similar analysis of antibody reagents for the detection of mammalian CREB isoforms and post-translational modifications has not been performed.

All *CREB* genes, including *dCREB2*, also undergo extensive alternative splicing (Ruppert, Cole, Boshart, Schmid, and Schutz 1992; Yin, Wallach, Wilder, Klingensmith, Dang, Perrimon, Zhou, Tully, and Quinn 1995b; Girardet, Walker, and Habener 1996; Walker, Girardet, and Habener 1996; Sanborn, Millan, Meistrich, and Moore 1997; Yang, Lanier, and Kraig 1997; Daniel and Habener 1998; Pietruck, Xie, Sharma, Meuser, and Palmer 1999a; Pietruck, Xie, Sharma, Meuser, and Pierce Palmer 1999b; Huang, Zhang, Lu, Yin, Xu, Wang, Zhou, and Sha 2004). Figure 2 shows the known transcripts generated from the *dCREB2* gene locus. Across all species, the different splice products can be catalogued into four patterns. The first type of alternative splicing results in the inclusion or exclusion of coding regions. These regions can vary in size from dCREB2's 4 amino acid exon 4, to CREM's exon Ia, which spans 400 nucleotides and encodes a fragment of CREM's DNA-binding domain. There are also alternative splices into the middle of the gene that introduce frame shift mutations resulting in premature termination of translation. This type of alternative splicing results in a carboxy-terminal truncation of the full-length protein. In some cases, it may

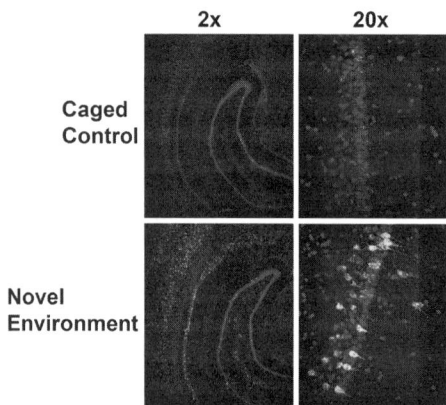

Fig. 2 Dramatic induction of Arc mRNA expression by a brief exposure of a rat to a novel environment. The upper panel shows both DAPI and *Arc* RNA staining, detected using FISH, of a region of ventral hippocampus from a caged control rat taken with both a 2x and 20x objective. The lower panel shows both DAPI and *Arc* RNA staining from a rat that was exposed for 5 min to a novel environment, and then returned to its home cage for 25 min. For each the upper and lower panels, the 2x field encompasses a large portion of the hippocampus and areas of the neocortex, and the 20x field shows the CA1 region of hippocampus. Please note the dramatic increase of *Arc* RNA staining in the "novel environment" section as compared to the "caged control" in all fields of the hippocampus and in the neocortex

also result in the re-initiation of translation from a downstream ATG codon. For *dCREB2*, the –r and –s splice forms likely represent examples of this, although their amino-terminal regions were not characterized. The third type of splicing occurs on the mammalian *CREM* gene, where an entire alternative bZIP region is substituted in at the C-terminus of the protein. Not all of these types of splicing have been documented in neurons, but they are very abundant in certain tissues such as testis. In testis, the regulation of splicing is also complex, with both temporal and signal-dependent influences (Waeber, Meyer, LeSieur, Hermann, Gerard, and Habener 1991). The last type of splicing occurs in the non-coding parts of the mRNA (Sanborn *et al.,* 1997). This has not been extensively characterized, and its functional consequences remain poorly understood.

Although the transcriptional regulation of *CREB* genes has not been studied in detail, there are adequate reports of regulation of transcription, and examples of alternative promoters, to warrant closer scrutiny. There is evidence of transcriptional regulation of one of the predominant promoters on the mammalian *CREB* gene (Meyer, Waeber, Lin, Beckmann, and Habener 1993), and there is clear indication of cAMP-responsive up regulation of the *ICER* promoter on mammalian *CREM* (Lalli, Lee, Lamas, Tamai, Zazopoulos, Nantel, Penna, Foulkes, and Sassone-Corsi 1996). At face value, these reports suggest that changes in cAMP levels can both increase and decrease cAMP-responsive transcription, probably being gated by the absolute levels of change, its duration, the other signaling pathways that are co-activated, and the cell type. In addition to these effects of cAMP, the first knock out of the α and Δ isoforms on the mouse *CREB* gene resulted in up regulation of the cryptic β-specific promoter located in intron 3 (Blendy, Kaestner, Schmidt, Gass, and Schutz 1996). In *Drosophila*, the underlying mechanisms that regulate transcription of the *dCREB2* gene have not been characterized.

18.2 CREB and Memory Formation – Causal Intervention

The first hint that CREB-responsive transcription might be important for long-lasting plasticity involved experiments in *Aplysia* long-term facilitation (LTF), a cellular model for memory formation (Dash, Hochner, and Kandel 1990). Double-stranded CRE-element oligonucleotides were injected into ganglia, resulting in a blockade of long-, but not short-lasting facilitation. Presumably, the oligonucliotide-containing CRE sites titrated out endogenous *Aplysia* CRE-binding protein(s), preventing activation and transcription of necessary endogenous downstream target genes. Injection of control, scrambled oligonucleotides did not affect LTF, thus demonstrating the molecular specificity of the disruption. These results suggested, but did not directly demonstrate, that CREB family proteins were the most likely targets of the CRE-oligonucleotide titration.

The direct demonstration of CREB's involvement in memory formation required molecular cloning and characterization of the *Drosophila dCREB2* gene, which encodes a "blocker" (naturally occurring dominant negative) and "activator" protein isoforms (Yin, Wallach, Del Vecchio, Wilder, Zhou, Quinn, and

Tully 1994; Yin, Del Vecchio, Zhou, and Tully 1995a; Yin *et al.,* 1995b). In transgenic flies, the expression of these dCREB2 isoforms were placed under the control of the heat-shock promoter, thus allowing very rapid and tight control over protein induction, while maintaining the transgenes in an "off" configuration prior to behavioral training. When flies containing the "blocker" transgene were shifted from normal temperature to 37°C, the transgene produced enough inhibiting protein to specifically disrupt long-term memory (LTM) formation of an olfactory avoidance behavioral paradigm. Transgene induction did not affect learning, or the formation of another form of consolidated memory (ARM), demonstrating the behavioral specificity of the intervention. Remarkably, induction of the dCREB2 activator transgene accelerated the formation of LTM, producing asymptotic levels of LTM after only a single training trial. (This experimental result will be discussed in greater detail in a later section of this chapter). Together, these loss- and gain-of-function experiments argued persuasively for the participation of dCREB2-responsive transcription in the formation of LTM.

In *Aplysia*, the identification and characterization of the *CREB1* gene revealed that it encodes both CRE-responsive activators and inhibitors (Bartsch *et al.,* 1998), while the CREB2 gene codes for an inhibitor (Bartsch, Ghirardi, Skehel, Karl, Herder, Chen, Bailey, and Kandel 1995). When antibodies raised against CREB2 were injected into the nuclei of co-cultured sensory-motor neurons, they enhanced the formation of long- but not short-term facilitation. In these studies, the "relief of inhibition" observed in *Aplysia* complemented earlier observations in the *Drosophila* "gain-of-function" experiments, providing compelling support for the role of CREB in LTM. Subsequently, it was shown that injection of activated CREB1a protein into the nuclei would also accelerate LTF (Bartsch *et al.,* 1998). Taken together, the *Drosophila* and *Aplysia* experiments provided compelling evidence that CREB activity was needed for both memory formation and long-term facilitation. These experiments also demonstrated that enhancing CREB's activity was sufficient to accelerate these processes. Recently, correlational data supporting the role of CREB in LTM has been reported in the *Lymnaea* model system (Azami, Wagatsuma, Sadamoto, Hatakeyama, Usami, Fujie, Koyanagi, Azumi, Fujito, Lukowiak, and Ito 2006).

18.3 CREB and Memory Formation in Mammals

Contemporaneous with the invertebrate work, an insertion mutation was made in the 5' coding region of the mouse *CREB* gene, thus truncating the full-length α and Δ protein isoforms (Blendy *et al.,* 1996). These proteins are well-characterized activators of CRE-mediated transcription. However, removal of the α and Δ isoforms led to the upregulation of the β isoform, normally a minor, barely detectable species. Initial experiments using this mutant mouse produced data in complete agreement with the story emerging from invertebrate studies. The CREB αΔ mouse was compromised for long-, but not short-term, memory of cued and contextual fear conditioning, and displayed behavioral deficits in the Morris Water Maze (MWM)

(Bourtchuladze, Frenguelli, Blendy, Cioffi, Schutz, and Silva 1994; Morris, Garrud, Rawlins, and O'keefe 1982). Although there was up regulation of the β isoform, the interpretation was that there was a net decrease in activator potential when CREB αΔ isoforms were removed. Interestingly, these behavioral deficits could be overcome if the inter-trial training interval was altered (Kogan, Frankland, Blendy, Coblentz, Marowitz, Schutz, and Silva 1997).

18.4 CREB and Memory Formation – Correlation

The serine 133-phosphoacceptor site is the target of a myriad of "upstream" signal transduction systems, including the important PKA, CaMK II/IV, MEK and PKC pathways (Johannessen and Moens 2007). It has been well established that phosphorylation of this residue is a necessary step in the activation of transcription, at least in most situations, since phosphorylation is needed for productive interaction with CREB Binding Protein (CBP), a co-activator with histone acetylase activity. The development of a phospho-specific antibody allowed correlations to be made between S133 phosphorylation and memory formation (Ginty, Kornhauser, Thompson, Bading, Mayo, Takahashi, and Greenberg 1993). This activity-dependent event has been correlated with many different types of neuronal activity, including memory formation. Impey and colleagues extended this type of correlational analysis through the use of transgenic CRE reporter mice (Impey, Smith, Obrietan, Donahue, Wade, and Storm 1998). They first used hippocampal slice physiology to validate their mice, showing that the reporter, and phosphorylation of CREB at serine 133, were responsive to stimuli that produced transcription-dependent Long Term Potentiation (LTP); the best cellular model for synaptic activity associated with memory formation. They then utilized this tool to identify memory-dependent cells in the hippocampus when animals were trained with hippocampal-dependent behavioral tasks. Activation of the reporter occurred in parallel with S133 phosphorylation. Together with the CREB αΔ and invertebrate behavioral experiments, these correlations formed a coherent picture of CREB's role during memory formation and long-lasting plasticity.

18.5 Conflicting Data

In spite of the wealth of causal and correlational observations, doubts have surfaced about the importance of CREB-dependent events in long-term plasticity and memory formation. Transgenic mice harboring targeted deletions in two CREB isoforms (α and Δ; *CREB αΔ* mice) were generated and used to demonstrate the role of CREB in memory formation (Bourtchuladze *et al.,* 1994). In these seminal studies using the *CREB αΔ* mice, several behavioral paradigms were employed to determine the effect of altered CREB levels on hippocampal- and amygdalar-dependent memory formation. To minimize possible confounding effects from the genetic background, mice carrying the *CREB* $\alpha\Delta$ mutation were backcrossed into the C57BL/6 (B6) and 129/SvJ (129) backgrounds. They were then crossed together, producing transgenic animals that were homozygous for the CREB $\alpha\Delta$ deletion, but containing a random hybrid genetic background from the B6 and 129 mouse strains (Bourtchuladze *et al.,* 1994). In contextual fear conditioning (an amygdalar- and hippocampal-dependent task), these mice displayed normal memory 30 minutes post-training, but were impaired at 60 minutes and 24 hours after training. Similarly, in cued fear conditioning, an amygdalar-dependent task, these animals displayed normal memory at 1 hour post training, but had deficits in memory at the 2 and 24-hour time points. These studies argued that removal of the $\alpha\Delta$ isoforms produce deficits in long-term memory in fear conditioning paradigms. Subsequent studies demonstrated impairment in hippocampal-dependent memory of the MWM, as well as deficits in social recognition, and social transmission of food preferences (Kogan, Frankland, and Silva 2000).

In a similar set of experiments, Gass and colleagues generated *CREB* $\alpha\Delta$ deletion mice in a hybrid F1 breeding scheme involving C57/Bl6 and FVB/N backgrounds (Gass, Wolfer, Balschun, Rudolph, Frey, Lipp, and Schutz 1998). These F1 *CREB* $\alpha\Delta$ mice were homozygous for the *CREB* $\alpha\Delta$ deletions, but maintained hybrid vigor because of their mixed genetic background. Unexpectedly, these *CREB* $\alpha\Delta$ mice displayed no obvious impairment in contextual and cued fear conditioning. Spatial memory was assayed using the MWM where two different training regimens were employed. The mutant mice did not show deficits in either protocol, one of which was originally used (and produced behavioral deficits) (Kogan *et al.,* 1997), and one that is a second standard protocol used in the Lipp laboratory (Wolfer, Muller, Stagliar, and Lipp 1997). In addition, the F1 *CREB* $\alpha\Delta$ hybrid mice did not show a deficit in the social transmission of food preference, another behavioral paradigm that previously showed CREB-dependence. In contrast, mice carrying one *CREB* $\alpha\Delta$ allele and one *CREB* null allele (in which all of the CREB isoforms were disrupted, *CREBcomp*) displayed a marked deficit in memory in both the MWM and in fear conditioning.

To complicate matters further, Graves and co-workers analyzed CREB $\alpha\Delta$ mutants in an F1 B6/129 hybrid background (Graves, Dalvi, Lucki, Blendy, and Abel 2002). They demonstrated that the *CREB* $\alpha\Delta$ B6/129 F1 hybrids have deficits in short- and long-term memory of contextual and cued fear conditioning. In contrast, based on performance in the hidden platform MWM, these animals displayed normal hippocampal function.

Clearly, these studies involving the *CREB* $\alpha\Delta$ mutant mice highlight the contribution and impact that genetic backgrounds can make to the experimental outcome in behavioral paradigms. These observations are further complicated by the possibility that attenuated *CREB* gene expression is subject to compensatory upregulation of other CREB isoforms or family members. Most notably, the *CREBcomp* mutant which contains a *CREB* $\alpha\Delta$ allele and a *CREBnull* allele show broad defects in learning and memory tasks, perhaps from attenuated expression of compensating CREB isoforms (Gass *et al.,* 1998).

In an effort to clarify the issues surrounding the *CREB* $\alpha\Delta$ mutant mice, the Schutz lab generated transgenic mice to test the effects of totally removing CREB (Balschun, Wolfer, Gass, Mantamadiotis, Welzl, Schutz, Frey, and Lipp 2003). LoxP sites were inserted flanking exon 10 of the *CREB* locus, and these mice were crossed with mice that contained the αCaM kinase promoter driving expression of the CRE recombinase. This promoter expresses exclusively in forebrain regions beginning about 2 weeks after birth, allowing for spatially-restricted changes in CREB expression during later post-natal development. The resulting "floxed" mice express CREB without its C-terminal bZIP (dimerization and DNA binding) domains. Presumably, CREB-responsive transcription is more severely affected in these mice, and will be referred to as "nulls". The mice were in a C57/Bl6 background, and they were crossed to $\alpha\Delta$ mice in the FVB/N background, producing heterozygous mice with one *CREB* $\alpha\Delta$ chromosome and one *CREB* null chromosome (but in a hybrid genetic background). The progeny mice showed modest behavioral deficits in the MWM and fear conditioning, but were normal in the social transmission of food assay. However, detailed analysis of the performance in the MWM revealed that the behavioral deficit was due to a lack of flexibility in MWM strategy, rather than impairment in memory formation. In searching for the platform, these mutant mice used a thigmotactic strategy, preferring to spend most of their time hugging the walls of the MWM. Furthermore, there were no deficits in L-LTP or LTD in hippocampal slices from the mutant mice. In fact, the only clear-cut behavioral deficit in these heterozygous mice was an effect on conditioned taste aversion. In summary, *CREB* knock out mice have produced conflicting results in different labs, and the replacement of one $\alpha\Delta$ chromosome with a "floxed" chromosome produced mice with very little behavioral and physiological impairment.

Where does all of this conflicting data leave us? Across all invertebrates, the experimental data seems consistent and clear, yet in mice the evidence appears to be awash in paradoxical observations. Although the genetic background is indeed a critical factor in the CREB $\alpha\Delta$ mice studies, we believe that the biggest single factor that determines the clarity of the data is whether the experimental intervention on *CREB* gene expression is acute or chronic. All of the invertebrate approaches utilize acute interventions, whereas all described experiments to date involving transgenic knock-outs are chronic, and, arguably, lead to the apparent disparity in behavioral observations as described above. Chronic interventions are obviously more susceptible to compensatory changes, some of which may be attributable to differences in genetic background.

18.6 Acute versus Chronic

Why is the difference between acute and chronic disruption important? One of the major problems with mouse knock out studies in general, and the CREB $\alpha\Delta$ and null mutations in particular, is that the chronic nature of the genetic manipulation means that the targeted genes and their proteins are missing prior to behavioral training, especially during development. In highly dynamic, interactive gene networks, the demand for survival and viability during development, including early postnatal periods, selects for compensating mechanisms that can bypass normally required gene products that are missing or mutant. In the original *CREB* $\alpha\Delta$ hypomorphic mice, there is molecular understanding of some of the compensation that occurs in response to removing two isoforms. The *CREB* α△ mutants retain an estimated 10–20% of residual CREB activity, which may partly explain their resilience to deficits in behavioral paradigms (Blendy *et al.,* 1996; Maldonado, Blendy, Tzavara, Gass, Roques, Hanoune, and Schutz 1996). Despite the targeted disruption of *CREB* α△, these mice display no obvious defects in growth or development. When the insertion was made in an early coding exon, usage of an internal, downstream in-frame ATG codon was increased, producing more than normal amounts of the CREB β isoform. This protein has all of the same properties of the CREB α△ isoforms, including CRE binding and cAMP-responsive transcriptional activation, except that it is a slightly weaker activator. This intragenic compensation suggests that the mammalian *CREB* gene locus somehow can autoregulate itself, and that removing the normally predominant CREB α△ isoforms results in up regulation of the β isoform (Blendy *et al.,* 1996). In addition, there is evidence of intergenic compensation, with an accompanying increase in CREM protein isoforms (Hummler, Cole, Blendy, Ganss, Aguzzi, Schmid, Beermann, and Schutz 1994). CREB, CREM and ATF-1 are members of a superfamily of transcriptional activators that share many, if not most, of the same molecular properties. The detectable increases in both CREM activator and blocker isoforms, coupled together with the increase in the CREB β activator isoform suggests that the compensating mechanisms are not straightforward. Simplistically, one might have expected just an increase in activator isoforms, since the CREB α△ isoforms function as transcriptional activators.

The second generation null mutation mice are more compromised, since homozygous mice die from various systemic problems, including ones that have postnatal onset (e.g. respiratory problems, neuronal development and degeneration, dwarfism, reproductive issues, immune system problems) (Mantamadiotis, Lemberger, Bleckmann, Kern, Kretz, Martin Villalba, Tronche, Kellendonk, Gau, Kapfhammer, Otto, Schmid, and Schutz 2002). These mice die prematurely, indicating that *CREB* is an essential gene required for normal development and full adult viability. It should not be surprising, then, if these mice have compensatory mechanisms to allow somewhat normal development and early adult function. Although detailed molecular evidence for compensation has not been described, we will return to this issue at the end of this Chapter.

18.7 Mammalian Acute Intervention Experiments

If, as we hypothesize, differences between acute and chronic intervention are responsible for some of the experimental discrepancies, then one prediction is that mammalian experiments done with acute intervention should show a requirement for CREB during memory formation. Three different types of acute intervention have been used to disrupt or augment CREB activity, and each of these has shown clear effects on memory formation. The first approach is to inject oligonucleotides directly into particular brain tissue. This approach has the advantage that there is both temporal and anatomical specificity in the delivery of the oligonucleotides. Since the oligonucleotides are injected shortly before behavioral training, there is very little time for molecular compensation to occur. Guzowski and colleagues injected antisense, or scrambled control, oligonucleotides into the dorsal hippocampus prior to behavioral training and produced specific effects on long-, but not short-, term memories (Guzowski and McGaugh 1997). These effects were shown to specifically affect memory formation, and not hippocampal function in general. Antisense, but not scrambled, oligonucleotides injected 6 or 20 hours prior to training impaired 48-hour memory, but not acquisition in the MWM. However, if these were injected 6 days prior to training, there was no effect, demonstrating that the oligonucleotides are not generally cytotoxic. In addition, rats with impaired MWM performance could re-learn the task, suggesting that their hippocampi remain functional. Although the detailed mechanism(s) of disruption is(are) not known, it is presumed that some antisense-based mechanism inactivated *CREB* mRNA and its role in gene expression. Lamprecht and coworkers demonstrated similar behavioral effects on long-, but not short-term amygdalar involvement in conditioned taste aversion, and extended their analysis to show biochemical effects of the antisense, but not control, oligonucleotides on CREB protein levels (Lamprecht, Hazvi, and Dudai 1997). Most recently, Zhang showed that antisense oligonucleotides can decrease total and phosphorylated CREB levels in the olfactory bulb during aversive learning (Zhang, Okutani, Inoue, and Kaba 2003).

The second general approach has been to use viral vectors to deliver inhibiting or activating forms of CREB into neurons. This strategy also has a high degree of anatomical and temporal control over the intervention, minimizing the chances for compensation. Josselyn and colleagues used Herpes Simplex Virus to package and deliver CREB into the amygdala of rats, and showed a specific enhancement of fear potentiated startle (Josselyn, Shi, Carlezon, Neve, Nestler, and Davis 2001). A massed training paradigm, which normally does not produce long-lasting memory, could produce robust memory if the animals were augmented with virally-delivered CREB. This effect was dependent upon CREB activity, explicitly paired training, and showed both anatomical and time-course requirements consistent with specific effects on long-lasting, but not short-term, memory. Moreover, Mouravlev and colleagues have used somatic gene transfer techniques to elevate CREB expression in a subset of hippocampal neurons, thereby preventing the aging-related decrease in long-term memory observed in control animals (Mouravlev, Dunning, Young, and During 2006). Viral delivery methods have also been used to disrupt memory formation of recognition memory and social transmission of food preference (Brightwell, Smith, Countryman, Neve, and Colombo 2005; Warburton, Glover, Massey, Wan, Johnson, Bienemann, Deuschle, Kew, Aggleton, Bashir, Uney, and Brown 2005). Significantly, these studies have used different viral-based systems to deliver the activating or inhibiting forms of CREB, thus arguing against viral-specific effects.

Finally, cutting-edge genetic methodologies have been used to acutely intervene with CREB function. Two different inducible systems, doxycycline and tamoxifen, have been adopted to disrupt CREB and test for effects on memory formation (Kida, Josselyn, de Ortiz, Kogan, Chevere, Masushige, and Silva 2002; Bozon, Kelly, Josselyn, Silva, Davis, and Laroche 2003; Josselyn, Kida, and Silva 2004). Although the kinetics of induction for the doxycycline system is slow, behavioral effects can be measured in the striatum. The tamoxifen-inducible CREB-estrogen receptor fusion protein is activated within hours, and shows effects in cued and contextual fear conditioning paradigms, recognition memory, and conditioned taste aversion.

In summary, all acute interventions in the rodent systems show effects of CREB on memory formation. These interventions can result in enhancing or disrupting effects, seem to be independent of the type of behavior, and are achievable using very different experimental methodologies. Collectively, all of the behavioral data on mammalian CREB forces us to focus on the issue of compensation, which can occur during chronic interventions, but is less of a factor with acute strategies.

18.8 Compensation

What are the molecular bases for differences between acute and chronic intervention? Can these mechanisms shed light upon why there are inter-laboratory behavioral differences with the *CREB* $\alpha\Delta$ mice? How do we account for the lack of behavioral effects when the second-generation *CREB* null mice are analyzed? The molecular analysis of the *CREB* $\alpha\Delta$ mice already tells us that both intragenic as well as intergenic molecular compensatory mechanisms occur upon the removal of these two protein isoforms. In the context of the genomic *CREB* locus, the increase in *CREB* β mRNA results from the usage of a cryptic promoter located in the intron prior to exon 4. The CREB β protein isoform is generated by translational initiation from an ATG codon that is normally located prior to the intron-exon 4 junction, within the "intron". These events emphasize how strong pressure resulting from the removal of the *CREB* α and Δ isoforms produces dramatic molecular compensatory mechanisms. The molecular mechanisms that upregulate CREM protein isoforms are not characterized, nor is it clear what other compensatory changes occur in upstream or downstream signaling pathways. What is clear, though, is that changes to the equilibrium of the CREB/CREM/ATF-1 family produce dramatic compensatory events. Why do acute interventions yield more consistent behavioral results? Molecular analysis of the *Drosophila* gene might shed some light on this question.

When transgenic flies containing the inducible activator protein are expressed, the induced protein increases but reaches a plateau level within 2 hours. Once this plateau is achieved, the protein expression level remains constant. Most importantly, direct *in vivo* measurements of transcription activity show that the kinetics of increased CRE-responsive transcription is very rapid (within 2 hours) and short-lived (less than 2 hours) in its duration (T. Tubon, unpublished data). These experiments demonstrate that the induced activator protein is only transiently activated, even though its levels remain elevated for many hours. Conversely, when the transgenic dCREB2-b blocker isoform is induced, time-course Western analysis shows rapid (within 2 hours) up regulation of activator isoforms. Together, these experiments suggest very strongly that the *dCREB2* gene is autoregulatory, and that this self-correcting system is very rapid. These results are logically identical to what was described in the *CREB* $\alpha\Delta$ mutant mouse, where CREB β and various CREM protein isoforms were upregulated. However, they are sobering because of their rapid kinetics of compensation, demonstrating that even acute disruptions of the CREB equilibrium will produce compensatory changes if the delay between intervention and analysis extends beyond the narrow temporal window of efficacy. Most acute behavioral interventions may be more consistent in their consequences because behavioral training usually occurs within hours of the intervention, not allowing a new, compensated "equilibrium" to be stably established. Knock out mice clearly have ample time to establish a new "equilibrium" both intra- and inter-genically prior to behavioral interrogation.

The second point comes from the analysis of the *CREB* $\alpha\Delta$ mice, where compensation can be widespread, involving at least the *CREB* and *CREM* genes. How the *CREB/CREM/ATF-1* genes "talk" to each other, and measure the net activator to blocker ratio, is unknown. However, it is likely that basic cellular signaling pathways, such as cAMP and calcium are involved, since *CREB* and *CREM* transcription, alternative splicing and protein activation all respond to these signals. It seems most likely, then, that inter-laboratory differences in behavioral results are at least partially due to a combination of factors including variation in behavioral protocols, housing, and genetic background differences, each of which is capable of affecting signaling pathways, their corresponding genetic interactions, and the behavioral outcomes. Crabbe and coworkers demonstrate in a very well controlled study, that even when apparatus, test protocols, and many environmental variables are controlled, there are inter-laboratory behavioral differences using the same strains of mice (Crabbe, Wahlsten, and Dudek 1999). Although the magnitude of the differences varied, they were seen with both inbred strains and one null mutant mouse when subjected to a battery of 6 different behavioral tests. Thus, it is possible that a whole set of "environmental" and genetic background factors can affect the "set-point" of cAMP and calcium-responsive signaling.

But what about the second generation null mice? Why would a more severe mutation produce negligible effects? Once again, work on the *Drosophila* gene may provide an some resolution. The *dCREB2* gene has only one characterized genomic null mutation known as S162 (Eberl, Perkins, Engelstein, Hilliker, and Perrimon 1992). This X-linked mutation produces lethality in males and homozygous females. The S162 mutation contains a single base pair change just upstream of the basic region and leucine zipper. This substitution results in a stop codon, and the termination of translation. At a low frequency ($\sim$0.5%), mutant "escaper" male flies can be recovered. These flies still contain the S162 mutation, but have somehow bypassed the removal of most of the dCREB2 protein isoforms. Molecular characterization of protein products in the escaper flies shows that there are indeed aberrant dCREB2 proteins produced (Fig. 3A) (Belvin, Zhou, and Yin 1999). One product likely results from the C-terminal truncation predicted to result from premature termination of translation. This product is detectable in both mutant escaper (male) flies, and in heterozygous females. The other protein is smaller than the predicted truncation product, and is only detectable in homozygous escaper flies. The molecular characteristics of this protein are currently under investigation.

But do these products have functional consequences? One way to test the relative "toxicity" of S162 is to compare its effects versus that of a deletion that removes many genes on both sides of the *dCREB2* locus. When the appropriate genetic crosses are carried out, the *S162* mutation produces fewer progeny than

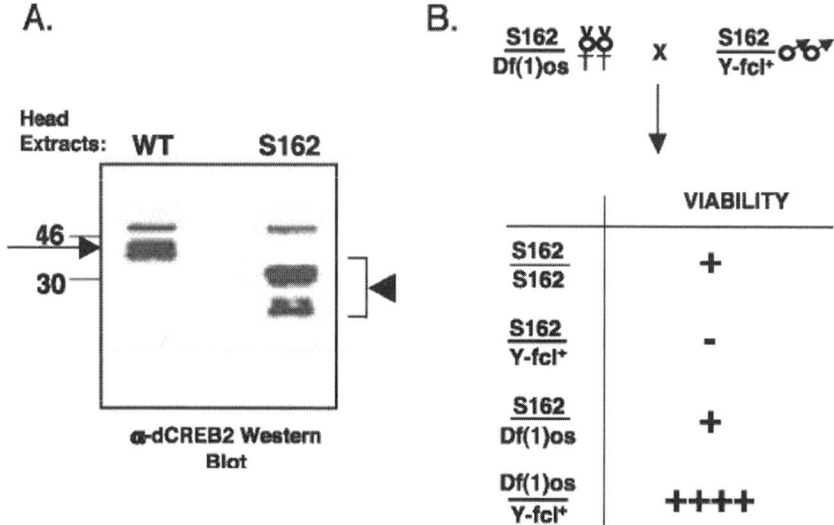

Fig. 3 dCREB2 S162 dominant-negative. (a) Western blot analysis of extracts from wild-type males and *S162* escaper males. dCREB2 bands are visualized at approximately 38–40kD, as indicated by the arrow. The S162 mutant forms are indicated by the arrowhead. (b) Crossing *S162/Df(1)os* virgin females to *S162/Y-fcl+* males produces no viable *S162/Y-fcl+* male progeny, and primarily *Df(1)os/Y-fcl+* males. *S162*, premature stop codon mutant in the *dCREB2* gene; *Df(1)os*, deficiency covering the genomic *CREB* locus ; *Y-fcl+*, Y-chromosome translocation and duplication of the region covering the *dCREB2* locus.

a genomic deficiency that removes many genes, including *dCREB2* (E. Drier, personal communication) (Fig. 3B). This result is strong genetic evidence that the S162 mutation actually produces harmful (dominant negative) products, not just a lack of full-length products. Currently, we do not know which (or both) of the two types of products detectable on western blots have dominant negative properties. The stop codon in S162 is located just N-terminal to the basic region and leucine zipper. Thus, the truncated protein is most likely missing these domains, but would contain N-terminal amino acid residues, including the moderately conserved P-box or KID domain. This is the part of the protein that contains a large number of conserved phosphorylation sites. Perhaps a truncated protein with this region is capable of titrating important kinases and their signaling partners. Alternatively, activity-dependent changes due to the aberrant dCREB2 protein product may be the culprit, such as modifications in DNA-binding or interactions with other transcriptional co-factors such as CBP/p300.

Our present understanding of the *S162* mutation suggests that the second-generation null mice may have similar issues. The null mutation results from floxed removal of the bZIP domain, the exact piece of the dCREB2 protein that is removed in the S162 mutant background. This should produce the same truncated "aberrant" proteins that are seen in the fly S162 mutant, which would probably have "dominant negative" properties. Thus the null mice are probably missing normal CREB isoforms, as well as expressing aberrant, dominant negative proteins. Therefore, the starting and ending points for compensation will differ from those of the *CREB* $\alpha\Delta$ mice. It is likely that the actual mechanisms of compensation in the null mice will be considerably more complex, owing to the involvement of more effector molecules and biologically-dependent pathways. In lieu of such complexity, one fruitful avenue might be to explore alternative splicing mechanisms on the *CREB* gene and in the null mice, with special attention towards an analysis of CREB isoforms that may be expressed at markedly lower levels in wild-type animals.

18.9 The Drosophila dCREB2 Story

Given this complexity in the mammalian systems, what clues about CREB's involvement in memory formation might result from studies in evolutionarily "simpler" systems? The first lesson is that the single genes in the invertebrate systems are very complex. In both *Aplysia* and *Drosophila*, there are multiple RNA transcripts and protein isoforms with "unusual" (relative to the simplified mammalian cartoon of CREB) subcellular localizations. Secondly, most, if not all, of the complex post-translational modifications have been conserved. Thirdly, the work in honeybees begins to hint at the evolutionary forces that must have accompanied speciation (Eisenhardt *et al.*, 2003). The honeybee gene produces many alternatively spliced isoforms, most of which remain functionally uncharacterized. It seems likely that the honeybee represents one of the last species prior to gene duplication and concomitant expansion in the number of genes in the *CREB* gene family. Finally, as suggested initially (Yin *et al.*, 1995a), it seems most likely that the ancestral founding gene in this family is actually *CREM*. This hypothesis is based on the general organization of the genes, the patterns of alternative splicing that produce activator species, and the complex subcellular localization of different protein isoforms (Fenaroli, Vujanac, De Cesare, and Zimarino 2004), all of which show more in common between *Drosophila dCREB2* and *CREM* than with *CREB*. From this perspective, if *CREB* is indeed the later evolved gene, it may have evolved with some special "niches" to fill, such as accelerating long-lasting plasticity and memory formation in general, both of which arguably occur faster in mammals than flies.

In *Drosophila*, even with the (ultimate) acute intervention tool (the rapidly inducible heat-shock promoter), recently reported differences have muddied the previous behavioral data (Perazzona, Isabel, Preat, and Davis 2004). The original *Drosophila*-based papers showing dCREB2-mediated enhancement of memory formation utilized transgenic flies (C28 and C30) where an activator isoform was placed under control of the heat-shock promoter (Yin *et al.*, 1995a). Recently, these results were challenged by Perazzona and colleagues on a number of bases: a) the C28 transgene contains a sequence rearrangement early in its coding region, leading to a frame shift and subsequent stop codon at amino acid residue 78, b) the failure to detect an inducible protein and c) inability of the authors to repeat behavioral enhancement of *dCREB2-a* using these flies. In 1996, Marianne Bienz first noticed that the C28 gene, when programmed into reticulocyte lysates, did not produce a protein of the predicted size of 40kD. This led promptly to DNA sequencing of the original *dCREB2* C28 transgene, and subsequent identification of the sequence rearrangement. However, a number of experimental observations that followed provided compelling evidence that the protein produced from C28 was most likely a truncated, biologically active dCREB2 isoform. First, in reticulocyte lysate, using the original *CREB* C28 transcript as an RNA template, a shorter protein was produced that disappeared when an in-frame, downstream ATG (methionine codon) to TAA (stop codon) mutation was introduced in the transcript. Second, we observed an increase in CRE-DNA binding activity when C28 was induced, in addition to a distinct shift in the mobility of CRE-DNA bound complexes following C28 induction (T. Tubon, H. Zhou, unpublished results). Third, all of the parallel experiments that were published, demonstrating that C28 made a PKA-responsive transcriptional activator when transfected into F9 cells, were done with the same rearranged cDNA (Yin *et al.*, 1995b). Contrary to earlier published results, Perazzona and colleagues were unable to show an enhancement in memory using C28 in the Pavlovian odor-avoidance learning and memory behavioral paradigm. Interestingly, in these studies, induced expression of two independent UAS-dCREB2-a transgenic lines corrected for the original mutation were profoundly deleterious, leading to 99% lethality at day 7 following induction, suggesting a critical requirement for tightly regulated expression of CREB during development.

The report by Perazzona and colleagues has prompted a more detailed study and re-evaluation of the role of dCREB2 in memory formation. In our ongoing studies, we have returned to this issue with increasingly more sensitive reagents, generating new insights into the role of dCREB in memory and a deep appreciation of the intricacies underlying dCREB2 gene expression and activity. Our present course of study has lead to strong evidence supporting the original claims of behavioral enhancement. Detailed molecular analysis reveals that translation initiation from an internal, downstream ATG occurs in the C28 transgene, and that the protein produced functions as originally reported (i.e. a PKA- and CRE-responsive transcriptional activator). Precedence for this type of translational regulation has been observed for mammalian *CREB*, in the expression of the CREB β isoform (Blendy *et al.*, 1996), as well as in the alternative ATG selection reported in the *CREM* gene (Gellersen, Kempf, Sandhowe, Weinbauer, and Behr 2002). Consistent with the mammalian CREB homolog, the dCREB2 protein is indeed heavily modified, thus partially accounting for the previous difficulty in its detection. The development of new antibodies to dCREB2 and the recent finding that these antibody reagents differentially recognize dCREB2 isoforms that harbor various posttranslational modifications has provided invaluable tools to detect and validate the dCREB2 activator. With previous antibodies, it was not possible to detect the activator protein initiated from the downstream ATG codon. Over the years, the frequency of behavioral enhancement of LTM with a p-value of less than 0.05 has been observed in about one-third of the experiments. We do not fully understand the experimental parameters that contribute to inconsistent enhancement, but place more valuation on enhancement when it occurs, than on the times when it does not.

Most importantly, this information has prompted us to make a second generation of transgenic *dCREB2* activator flies that induce the shorter protein. These flies show a narrow temporal window following heat-shock protein induction during which transcriptional activation occurs. The narrow temporal window of transcriptional activity likely accounts for a restricted time frame during which behavioral training must occur for enhancement of memory formation. Experiments to prove this point are underway. Finally, this shorter protein has been identified as an endogenous wild-type dCREB2 transcriptional activator in *Drosophila*, whose expression is dependent on various *dCREB2* mRNA molecules that initiate translation from an internal start site. Thus, the original reports on dCREB2's involvement in memory formation seem correct, but also illustrate the complexity of this gene family, even when there is only a single member in the fly.

From this discussion, it should be clear that there is no single experimental system that has all of the advantages that are necessary for a complete understanding of this gene family and its role in memory formation. However, an integration of behavioral, cellular and molecular information across the various species helps in clarifying some of the experimental issues. Continued molecular analysis will undoubtedly provide more information that further clarifies and extends our understanding of this complex gene family's role in this very complex process.

18.10 Looking to the Future

When the CREB behavioral papers first were published, there was optimism that the identification of this nuclear convergence point would allow the assignment of different patterns of stimulation to defined molecular events, particularly on CREB. For example, since repetitive spaced, but not massed, training produces *CREB*-dependent, long-term memory, it was felt that a molecular dissection of the CREB nuclear endpoint would be possible, and this would facilitate working "backwards" towards cellular signaling pathways. Similarly, it seemed that the patterns of stimulation that produce long-term depression, long-term potentiation or late-LTP should differentially affect CREB's modification status. Although this view was overly optimistic, it now seems that closure on all of dCREB2's post-translational modifications is near. These modifications now need to be placed into their cellular and subcellular contexts, and this will then complete the description of the "code" of changes that are possible. Finally, connecting the particular set of changes with their behavioral and physiological stimulations will then be possible.

References

Azami, S., Wagatsuma, A., Sadamoto, H., Hatakeyama, D., Usami, T., Fujie, M., Koyanagi, R., Azumi, K., Fujito, Y., Lukowiak, K. and Ito, E. (2006) Altered gene activity correlated with long-term memory formation of conditioned taste aversion in Lymnaea. J. Neurosci. Res. 84, 1610–1620.
Balschun, D., Wolfer, D.P., Gass, P., Mantamadiotis, T., Welzl, H., Schutz, G., Frey, J.U. and Lipp, H.P. (2003) Does cAMP response element-binding protein have a pivotal role in hippocampal synaptic plasticity and hippocampus-dependent memory? J. Neurosci. 23, 6304–6314.

Bartsch, D., Casadio, A., Karl, K.A., Serodio, P., and Kandel, E.R. (1998) CREB1 encodes a nuclear activator, a repressor, and a cytoplasmic modulator that form a regulatory unit critical for long-term facilitation. Cell 95, 211–223.

Bartsch, D., Ghirardi, M., Skehel, P.A., Karl, K.A., Herder, S.P., Chen, M., Bailey, C.H. and Kandel, E.R. (1995) Aplysia CREB2 represses long-term facilitation: relief of repression converts transient facilitation into long-term functional and structural change. Cell 83, 979–992.

Belvin, M.P., Zhou, H., Yin, J.C. (1999) The Drosophila dCREB2 gene affects the circadian clock. Neuron 22, 777–787.

Blendy, J.A., Kaestner, K.H., Schmid, W., Gass, P. and Schutz, G. (1996) Targeting of the CREB gene leads to up-regulation of a novel CREB mRNA isoform. EMBO J. 15, 1098–1106.

Bourtchuladze, R., Frenguelli, B., Blendy, J., Cioffi, D., Schutz, G. and Silva, A.J. (1994) Deficient long-term memory in mice with a targeted mutation of the cAMP-responsive element-binding protein. Cell 79, 59–68.

Bozon, B., Kelly, A., Josselyn, S.A., Silva, A.J., Davis, S. and Laroche, S. (2003) MAPK, CREB and zif268 are all required for the consolidation of recognition memory. Philos. Trans. R. Soc. Lond. B. Biol. Sci. 358, 805–814.

Brightwell, J.J., Smith, C.A., Countryman, R.A., Neve, R.L. and Colombo, P.J. (2005) Hippocampal overexpression of mutant CREB blocks long-term, but not short-term memory for a socially transmitted food preference. Learn. Mem. 12, 12–17.

Chera, S., Kaloulis, K. and Galliot, B. (2007) The cAMP response element binding protein (CREB) as an integrative HUB selector in metazoans: Clues from the hydra model system. Biosys tems 87, 191–203.

Chrivia, J.C., Kwok, R.P., Lamb, N., Hagiwara, M., Montminy, M.R. and Goodman, R.H. (1993) Phosphorylated CREB binds specifically to the nuclear protein CBP. Nature 365, 855–859.

Comerford, K.M., Leonard, M.O., Karhausen, J., Carey, R., Colgan, S.P. and Taylor, C.T. (2003) Small ubiquitin-related modifier-1 modification mediates resolution of CREB-dependent responses to hypoxia. Proc. Natl. Acad. Sci. USA 100, 986–991.

Conkright, M.D., Canettieri, G., Screaton, R., Guzman, E., Miraglia, L., Hogenesch, J.B. and Montminy, M. (2003) TORCs: transducers of regulated CREB activity. Mol. Cell. 12, 413–423.

Crabbe, J.C., Wahlsten, D. and Dudek, B.C. (1999) Genetics of mouse behavior: interactions with laboratory environment. Science 284, 1670–1672.

Daniel, P.B. and Habener, J.F. (1998) Cyclical alternative exon splicing of transcription factor cyclic adenosine monophosphate response element-binding protein (CREB) messenger ribonucleic acid during rat spermatogenesis. Endocrinology 139, 3721–3729.

Dash, P.K., Hochner, B. and Kandel, E.R. (1990) Injection of the cAMP-responsive element into the nucleus of Aplysia sensory neurons blocks long-term facilitation. Nature 345, 718–721

Eberl, T., Perkins, L.A., Engelstein, M., Hilliker, A.J., Perrimon, N. (1992) Genetic and developmental analysis of polytene section 17 of the X chromosome of Drosophila melanogaster. Genetics 130, 569–583.

Eisenhardt, D., Friedrich, A., Stollhoff, N., Muller, U., Kress, H. and Menzel, R. (2003) The AmCREB gene is an ortholog of the mammalian CREB/CREM family of transcription factors and encodes several splice variants in the honeybee brain. Insect. Mol. Biol. 12, 373–382.

Fenaroli, A., Vujanac, M., De Cesare, D. and Zimarino, V. (2004) A small-scale survey identifies selective and quantitative nucleo-cytoplasmic shuttling of a subset of CREM transcription factors. Exp. Cell Res. 299, 209–226.

Fimia, G.M., De Cesare, D., Sassone-Corsi, P. (1999) CBP-independent activation of CREM and CREB by the LIM-only protein ACT. Nature 398, 165–169.

Gass, P., Wolfer, D.P., Balschun, D., Rudolph, D., Frey, U., Lipp, H.P. and Schutz, G. (1998) Deficits in memory tasks of mice with CREB mutations depend on gene dosage. Learn. Mem. 5, 274–288.

Gau, D., Lemberger, T., von Gall, C., Kretz, O., Le Minh, N., Gass, P., Schmid, W., Schibler, U., Korf, H.W. and Schutz, G. (2002) Phosphorylation of CREB Ser142 regulates light-induced phase shifts of the circadian clock. Neuron 34, 245–253.

Gellersen, B., Kempf, R., Sandhowe, R., Weinbauer, G.F. and Behr, R. (2002) Novel leader exons of the cyclic adenosine 3', 5'-monophosphate response element modulator (CREM) gene, transcribed from the promoters P3 and P4, are highly testis-specific in primates. Mol. Hum. Reprod. 8, 965–976.

Ginty, D.D., Kornhauser, J.M., Thompson, M.A., Bading, H., Mayo, K.E., Takahashi, J.S. and Greenberg, M.E. (1993) Regulation of CREB phosphorylation in the superchiasmatic nucleus by light and a circadian clock. Science 260, 238–241.

Girardet, C., Walker, W.H. and Habener, J.F. (1996) An alternatively spliced polycistronic mRNA encoding cyclic adenosine 3',5'-monophosphate (cAMP)-responsive transcription factor CREB (cAMP response element-binding protein) in human testis extinguishes expression of an internally translated inhibitor CREB isoform. Mol. Endocrinol. 10, 879–891.

Goren, I., Tavor, E., Goldblum, A. and Honigman, A. (2001) Two Cysteine residues in the DNA-binding domain of CREB control binding to CRE and CREB-mediated gene expression. J. Mol. Biol. 313, 695–709.

Graves, L., Dalvi, A., Lucki, I., Blendy, J.A. and Abel, T. (2002) Behavioral analysis of CREB alphadelta mutation on a B6/129 F1 hybrid background. Hippocampus 12, 18–26.

Guzowski, J.F. and McGaugh, J.L. (1997) Antisense oligodeoxynucleotide-mediated disruption of hippocampal cAMP response element binding protein levels impairs consolidation of memory for water maze training. Proc. Natl. Acad. Sci. USA 94, 2693–2698.

Hintermann, R. and Parish, R.W. (1979) The intracellular location of adenylyl cyclase in the cellular slime molds Dictyostelium discoideum and Polysphondylium pallidum. Exp. Cell Res. 123, 429–434.

Huang, X., Zhang, J., Lu, L., Yin, L., Xu, M., Wang, Y., Zhou, Z. and Sha, J. (2004) Cloning and expression of a novel CREB mRNA splice variant in human testis. Reproduction 128, 775–782.

Hummler, E., Cole, T.J., Blendy, J.A., Ganss, R., Aguzzi, A., Schmid, W., Beermann, F. and Schutz, G. (1994) Targeted mutation of the CREB gene: compensation within the CREB/ATF family of transcription factors. Proc. Natl. Acad. Sci. USA 91, 5647–5651.

Impey, S., Smith, D.M., Obrietan, K., Donahue, R., Wade, C. and Storm, D.R. (1998) Stimulation of cAMP response element (CRE)-mediated transcription during contextual learning. Nat. Neurosci. 1, 595–601.

Johannessen, M. and Moens, U. (2007) Multisite phosphorylation of the cAMP response element-binding protein (CREB) by a diversity of protein kinases. Front. Biosci. 12, 1814–1832.

Josselyn, S.A., Kida, S. and Silva, A.J. (2004) Inducible repression of CREB function disrupts amygdala-dependent memory. Neurobiol. Learn. Mem. 82, 159–163.

Josselyn, S.A., Shi, C., Carlezon, W.A., Jr., Neve, R.L., Nestler, E.J. and Davis, M. (2001) Long-term memory is facilitated by cAMP response element-binding protein overexpression in the amygdala. J. Neurosci. 21, 2404–2412.

Khidekel, N. and Hsieh-Wilson, L.C. (2004) A 'molecular switchboard'–covalent modifications to proteins and their impact on transcription. Org. Biomol. Chem. 2, 1–7.

Kida, S., Josselyn, S.A., de Ortiz, S.P., Kogan, J.H., Chevere, I., Masushige, S. and Silva, A.J. (2002) CREB required for the stability of new and reactivated fear memories. Nat. Neurosci. 5, 348–355.

Kogan, J.H., Frankland, P.W. and Silva, A.J. (2000) Long-term memory underlying hippocampus-dependent social recognition in mice. Hippocampus 10, 47–56.

Kogan, J.H., Frankland, P.W., Blendy, J.A., Coblentz, J., Marowitz, Z., Schutz, G. and Silva, A.J. (1997) Spaced training induces normal long-term memory in CREB mutant mice. Curr. Biol. 7, 1–11.

Kornhauser, J.M., Cowan, C.W., Shaywitz, A.J., Dolmetsch, R.E., Griffith, E.C., Hu, L.S., Haddad, C., Xia, Z. and Greenberg, M.E. (2002) CREB transcriptional activity in neurons is regulated by multiple, calcium-specific phosphorylation events. Neuron 34, 221–233.

Kwok, R.P., Lundblad, J.R., Chrivia, J.C., Richards, J.P., Bachinger, H.P., Brennan, R.G., Roberts, S.G., Green, M.R. and Goodman, R.H. (1994) Nuclear Protein CBP is a coactivator for the transcription factor CREB. Nature 370, 223–226.

Lalli, E., Lee, J.S., Lamas, M., Tamai, K., Zazopoulos, E., Nantel, F., Penna, L., Foulkes, N.S. and Sassone-Corsi, P. (1996) The nuclear response to cAMP: role of transcription factor CREM. Philos. Trans. R. Soc. Lond. B. Biol. Sci. 351, 201–209.

Lamprecht, R., Hazvi, S. and Dudai, Y. (1997) cAMP response element-binding protein in the amygdala is required for long- but not short-term conditioned taste aversion memory. J. Neurosci. 17, 8443–8450.

Maldonado, R., Blendy, J.A., Tzavara, E., Gass, P., Roques, B.P., Hanoune, J. and Schutz, G. (1996) Reduction of morphine abstinence in mice with a mutation in the gene encoding CREB. Science 273, 657–659.

Mantamadiotis, T., Lemberger, T., Bleckmann, S.C., Kern, H., Kretz, O., Martin Villalba, A., Tronche, F., Kellendonk, C., Gau, D., Kapfhammer, J., Otto, C., Schmidt, W. and Schutz, G. (2002) Disruption of CREB function in brain leads to neurodegeneration Nat. Genet. 31, 47–54.

Mayr, B. and Montminy, M. (2001) Transcriptional regulation by the phosphorylation-dependent factor CREB. Nat. Rev. Mol. Cell. Biol 2, 599–609.

Meyer, T.E. and Habener, J.F. (1993) Cyclic adenosine 3',5'-monophosphate response element binding protein (CREB) and related transcription-activating deoxyribonucleic acid-binding proteins. Endocr. Rev. 14, 269–290.

Meyer, T.E., Waeber, G., Lin, J., Beckmann, W. and Habener, J.F. (1993) The promoter of the gene encoding 3',5'-cyclic adenosine monophosphate (cAMP) responsive element binding protein contains cAMP response elements: evidence for positive autoregulation of gene transcription. Endocrinology 132, 770–780

Morris, R.G., Garrud, P., Rawlins, J.N. and O'Keefe, J. (1982) Place navigation impaired in rats with hippocampal lesions. Nature 297, 681–683.

Mouravlev, A., Dunning, J., Young, D. and During, M.J. (2006) Somatic gene transfer of cAMP response element binding protein attenuates memory impairment in aging rats Proc. Natl. Acad. Sci. 103, 4705–4710.

Perazzona, B., Isabel, G., Preat, T. and Davis, R.L. (2004) The role of cAMP response element-binding protein in Drosophila long-term memory. J. Neurosci. 24, 8823–8828.

Peters, D.J., Cammans, M., Smit, S., Spek, W., van Lookeren Campagne, M.M. and Schaap, P. (1991) Control of cAMP-induced gene expression by divergent signal transduction pathways. Dev. Genet. 12, 25–34.

Pietruck, C., Xie, G.X., Sharma, M., Meuser, T. and Palmer, P.P. (1999a) Alternative exon splicing of cyclic AMP response element-binding protein in peripheral sensory and sympathetic ganglia of the rat. Life Sci. 65, 2205–2213.

Pietruck, C., Xie, G.X., Sharma, M., Meuser, T. and Pierce Palmer, P. (1999b) Multiple splice patterns of cyclic AMP response element-binding protein mRNA in the central nervous system of the rat. Brain Res. Mol. Brain Res. 69, 286–289.

Ribeiro, M.J., Serfozo, Z., Papp, A., Kemenes, I., O'Shea, M., Yin, J.C., Benjamin, P.R. and Kemenes, G. (2003) Cyclic AMP response element-binding (CREB)-like proteins in a molluscan brain: cellular localization and learning-induced phosphorylation. Eur. J. Neurosci. 18, 1223–1234.

Ruppert, S., Cole, T.J., Boshart, M., Schmid, E. and Schutz, G. (1992) Multiple mRNA isoforms of the transcription activator protein CREB: generation by alternative splicing and specific expression in primary spermatocytes. EMBO J. 11, 1503–1512.

Sadamoto, H., Sato, H., Kobayashi, S., Murakami, J., Aonuma, H., Ando, H., Fujito, Y., Hamano, K., Awaji, M., Lukowiak, K., Urano, A. and Ito, E. (2004) CREB in the pond snail Lymnaea stagnalis: cloning, gene expression, and function in identifiable neurons of the central nervous system. J. Neurobiol. 58, 455–466.

Sanborn, B.M., Millan, J.L., Meistrich, M.L. and Moore, L.C. (1997) Alternative splicing of CREB and CREM mRNAs in an immortalized germ cell line. J. Androl. 18, 62–70.

Shanware, N.P., Trinh, A.T., Williams, L.M. and Tibbetts, R.S. (2007) Coregulated ATM and casein kinase sites modulate CREB-coactivator interactions in response to DNA damage. J. Biol. Chem. 282, 6283–6291.

Shemarova, I.V. (2005) [cAMP-PKA signal pathway in the lower eukaryotes]. Tsitologiia 47, 296–310.

Waeber, G., Meyer, T.E., LeSieur, M., Hermann, H.L., Gerard, N. and Habener, J.F. (1991) Developmental stage-specific expression of cyclic adenosine 3',5'-monophophate response element-binding protein CREB during spermatogenesis involves alternative exon splicing. Mol. Endocrinol. 5, 1418–1430.

Walker, W.H., Girardet, C. and Habener, J.F. (1996) Alternative exon splicing controls a translational switch from activator to repressor isoforms of transcription factor CREB during spermatogenesis. J. Biol. Chem. 271, 20145–21050.

Warburton, E.C., Glover, C.P., Massey, P.V., Wan, H., Johnson, B., Bienemann, A., Deuschle, U., Kew, J.N., Aggleton, J.P., Bashir, Z.I., Uney, J. and Brown, M.W. (2005) cAMP responsive element-binding protein phosphorylation is necessary for perihinal long-term potentiation and recognition memory. J. Neurosci. 25, 6296–6303.

Wolfer, D.P., Muller, U., Stagliar, M. and Lipp, H.P. (1997) Assessing the effects of the 129/Sv genetic background on swimming navigation learning in transgenic mutants: a study using mice with a modified beta-amyloid precursor protein gene. Brain Res. 771, 1–13.

Yang, L., Lanier, E.R. and Kraig, E. (1997) Identification of a novel, spliced variant of CREB that is preferentially expressed in the thymus. J. Immunol. 158, 2522–2525.

Yin, J.C. and Tully, T. (1996) CREB and the formation of long-term memory. Curr Opin Neurobiol 6, 264–268.

Yin, J.C., Del Vecchio, M., Zhou, H. and Tully, T. (1995a) CREB as a memory modulator: induced expression of a dCREB2 activator isoform enhances long-term memory in Drosophila. Cell 81, 107–115.

Yin, J.C., Wallach, J.S., Wilder, E.L., Klingensmith, J., Dang, D., Perrimon, N., Zhou, H., Tully, T., Quinn, W.G. (1995b) A Drosophila CREB/CREM homolog encodes multiple isoforms, including a cyclic AMP-dependent protein kinase-responsive transcriptional activator and antagonist. Mol. Cell. Biol 15, 5123–5130.

Yin, J.C., Wallach, J.S., Del Vecchio, M., Wilder, E.L., Zhou, H., Quinn, W.G. and Tully, T. (1994) Induction of a dominant negative CREB transgene specifically blocks long-term memory in Drosophila. Cell 79, 49–58.

Zhang, J.J., Okutani, F., Inoue, S. and Kaba, H. (2003) Activation of the cyclic AMP response element-binding protein signaling pathway in the olfactory bulb is required for the acquisition of olfactory aversive learning in young rats. Neuroscience 117, 707–713.

Chapter 19
Dynamic Transcription of the Immediate-Early Gene *Arc* in Hippocampal Neuronal Networks: Insights into the Molecular and Cellular Bases of Memory Formation

John F. Guzowski, Ting Nie, and Teiko Miyashita

Abstract The activity-regulated cytoskeletal-associated protein (*Arc*) is an immediate-early gene (IEG) that is dynamically regulated by neuronal activity. IEGs encode a diverse range of proteins including regulatory transcription factors, structural and signal transduction proteins, growth factors, proteases, and enzymes [reviewed in (Lanahan and Worley, 1998)]. Moreover, several IEGs have been shown to be required for long-lasting synaptic plasticity and memory consolidation processes [reviewed in (Guzowski, 2002)]. Of the IEGs investigated in learning and memory, *Arc*, also referred to as *Arg3.1* (activity-regulated gene 3.1), has been of particular interest because of its tight experience-dependent regulation in behaviorally defined neural networks, its mRNA transport to and expression in activated synapses, its capacity for modification of synaptic function, and its critical role in memory consolidation. This chapter provides an overview of the research on *Arc*'s properties, putative functions, and regulation at cellular and network levels.

19.1 The Cloning and Early Characterization of *Arc*

Arc was cloned and characterized in 1995 by two groups of investigators identifying genes activated by neural activity. The gene was called the activity-regulated cytoskeletal associated protein by Worley and colleagues [*Arc*; (Lyford et al., 1995)] and activity-regulated gene 3.1 by Kuhl and colleagues [*Arg3.1*; (Link et al., 1995)]. (Please note that from this point on, we will simply use the term *Arc* to refer to this gene.) The investigations were prompted by evidence that long-term synaptic plasticity believed to underlie memory consolidation requires protein synthesis and *de novo* transcription (Goelet et al., 1986; Lanahan and Worley, 1998). Because of their rapid activity-regulated expression, IEGs were logical candidates to encode proteins involved in synaptic modification. Using differential cloning techniques following maximal electroconvulsive seizure (MECS), an *Arc/Arg3.1* cDNA of approximately 3,000 bp was identified by both the Worley and Kuhl groups (Link et al., 1995; Lyford et al., 1995). The *Arc* primary transcript contains two small introns (in the 3' UTR) and there is only one known splice variant. The open reading frame is contained in a single exon (exon 1) and encodes a 396 amino acid protein that migrates at approximately 55 kDa on SDS-PAGE (Lyford et al., 1995).

Arc is expressed at detectable levels in rats starting from about postnatal day 12, reaches maximal levels at postnatal day 21, and remains at steady levels for the rest of life (Lyford et al., 1995). In terms of amino acid sequence homology, *Arc* is not strongly related to any other protein and only bears some homology with the actin-binding protein α-spectrin, a cytoskeletal component. Northern blot analysis of *Arc* mRNA expression revealed a low basal level and a dramatically elevated level in the hippocampus and neocortex following MECS (Link et al., 1995; Lyford et al., 1995). This increase was blocked by systemic injection of NMDA antagonist MK-801, demonstrating that *Arc* transcription is regulated by excitatory synaptic activity via NMDA receptor activation. Moreover, unilateral intraocular injection of tetrodotoxin showed that the basal levels of *Arc* seen in the visual cortex were not constitutive gene expression, but rather highly dependent on neuronal activity / input (Lyford et al., 1995).

19.2 Dendritic Targeting of *Arc* mRNA and Protein

Arc has attracted considerable attention in memory research due to the finding that its mRNA translocates to dendrites and active synapses (Steward and Schuman, 2001), the site of plasticity. This feature of *Arc* is unique among all known IEGs expressed in neurons (Wallace et al., 1998). *In situ* hybridization of *Arc* mRNA in MECS and LTP brains revealed localization of signal in both the cell body layers and dendritic regions (e.g., the molecular layer of the dentate gyrus) (Wallace et al., 1998; Steward and Worley, 2002). This pattern of staining supports the assertion that *Arc* mRNA is transported from soma to dendrites. Studies by Steward and Worley further demonstrated that *Arc* mRNA can be targeted specifically to active synapses, in an NMDA receptor dependent manner (Steward et al., 1998; Steward and Worley, 2001). Transport of *Arc* mRNA to active synapses was observed through high-frequency stimulation in specific entorhinal cortex (EC) projections that terminate at specific layers of the molecular layer of the dentate gyrus. Depending on the projections stimulated, *Arc* mRNA localized specifically at the activated region of the molecular layer of the dentate gyrus. MECS produced even distribution of *Arc* throughout the molecular layer within 1 hr of stimulation. A Delivery of a specific perforant path stimulation triggered a prominent labeling of *Arc* in the active layer and depleted the mRNA from nonactivated regions. When an NMDA receptor antagonist was applied to the hippocampus through a micropipette, *Arc* induction as well as its specific targeting was completely blocked in the area of infusion.

The time course of *Arc* mRNA distribution following MECS is demonstrated in Fig. 1. Of particular interest is the observation that *Arc* RNA signal is exclusively in the cell body layer at 15 min, and distributed widely through the molecular layer and dendritic regions of CA1 within another 15 min (at the 30 min time point). This rapid translocation from cell body to distal dendritic regions is consistent with recent reports of dynamic trafficking of *Arc* RNA granules in live cell imaging studies (Dynes and Steward, 2007).

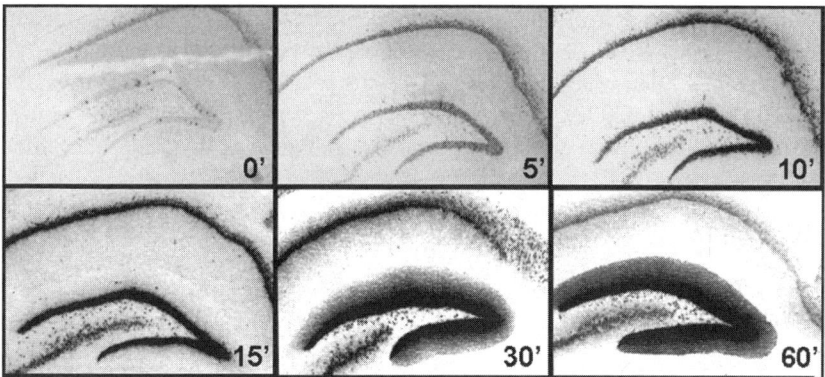

Fig. 1 Time course of *Arc* mRNA expression following maximal electroconvulsive seizure (MECS) in rat hippocampus. *Arc* RNA staining is indicated by dark color. Please note that in the caged control (0') there are low levels of *Arc* RNA staining and single *Arc*+ cells can be identified in the dentate gyrus. Also note how *Arc* RNA staining is restricted to the cell layers of the dentate gyrus and hippocampus proper (CA1 and CA3) at 10' and 15' after MECS. Within just an additional 15' (30' section) *Arc* mRNA is then found in dendritic regions of the dentate gyrus (e.g., molecular layer) and pyramidal cell layers of hippocampus. *Arc* mRNA was detected with a digoxigenin labeled riboprobe and a chromagenic substrate

In addition to this rapid translocation of *Arc* mRNA, earlier studies had shown that nascent *Arc* primary transcript could be detected at the genomic alleles within 2 min following an electrical or behavioral stimulus (Guzowski et al., 1999). Moreover, the presence of perinuclear / cytoplasmic *Arc* mRNA could be detected within 10-15 min. This rapid processing of *Arc RNA*, like that of other IEGs (such as c-fos and zif268), is due to the fact that the mRNA is derived from a short primary transcript ($\sim$3.4 kb) containing only two short introns. Together, these features allow rapid (within $\sim$30 min), and very precise targeting of new *Arc* mRNA to active synapses following a plasticity inducing stimulus or learning event in an animal.

Arc protein, like *Arc* mRNA, is also found in neuronal dendrites. (It should also be noted that *Arc* protein is also located in the soma and, occasionally, in the nuclei of neurons.) The use of post-embedding immunogold electron microscopy revealed that Arc protein is specifically expressed in excitatory dendritic spines of the activated layer of the dentate gyrus following LTP-producing stimulation of the perforant path (Moga et al., 2004), extending the earlier light microscopic observations of Steward and colleagues. Further support for a specific targeting of *Arc* protein to synapses was the finding that *Arc* was a component of the NMDA receptor multiprotein complex (Husi et al., 2000), although the functional significance of this interaction is not known at present. The presence of *Arc* mRNA in dendrites, which are known to contain ribosomes and other components of the translational machinery (Steward and Levy, 1982), indicates that *Arc* protein might be translated locally. Consistent with this idea, *Arc* targeting studies demonstrated that *Arc* immunostaining was localized in the same specific layer where *Arc* mRNA was present (Steward et al., 1998; Steward and Worley, 2001). Further support for the dendritic translation of *Arc* mRNA comes from the finding that *Arc* translation could be induced by addition of brain derived neurotrophic factor (BDNF) in isolated synaptoneurosome preparations and that the effect was blocked by the NMDA receptor antagonist MK-801 (Yin et al., 2002). Thus, *Arc* can achieve synapse specificity by virtue of localization of its mRNA to dendrites and the presence of dendritic translational machinery.

19.3 A Critical Role for *Arc* in Neural Plasticity and Memory Consolidation

Because of its regulation by neural activity *in vivo*, its regulation by LTP inducing stimuli, and its dendritic localization, *Arc* was an excellent candidate for a protein to be involved with synaptic plasticity and long-term memory formation. The first study to address the role of *Arc* in synaptic plasticity used intrahippocampal infusions of antisense oligodeoxynucleotides (ODNs) to inhibit local *Arc* protein expression reversibly, and examine the effect of this inhibition on the induction / maintenance of long-term potentiation (LTP) and learning and memory consolidation in the Morris water maze task (Guzowski et al., 2000). In the first experiment, *Arc* antisense ODNs were delivered to one hippocampus and scrambled ODNs to the other 1.5 h before induction of LTP in the perforant path of rats chronically implanted with bilateral infusion cannulae (which also functioned as the recording electrode). Initially, the magnitude of early LTP did not differ between *Arc* ODN-treated and scrambled ODN-treated groups. However, long-lasting LTP in the *Arc* antisense ODN-treated hemispheres decayed more rapidly than scrambled ODN hemispheres, with a significant difference detected as early as 4 h after induction. In biochemical control studies, *Arc* antisense ODNs reduced hippocampal expression of Arc protein by 60% relative to control scrambled ODN-treated hippocampi. These studies showed that blocking *Arc* protein expression did not affect the induction of LTP, but impaired the maintenance of long-lasting LTP in the rat hippocampus.

Using the same approach, the effects of inhibiting hippocampal Arc protein expression on learning and memory were assessed in spatial reference learning in the water maze (Guzowski et al., 2000). *Arc* antisense or scrambled control ODNs were infused bilaterally into the hippocampus 3 hr before training sessions. The infusions of antisense or scrambled ODNs did not influence acquisition of this task, whereas the *Arc* antisense ODNs produced a significant impairment in spatial memory on a retention probe test, given 2 days later, as compared to the controls. In a separate experiment, *Arc* antisense or scrambled control ODNs were infused into the hippocampus immediately after two training sessions of the spatial water maze task, with the two session separated by just 2 min. In addition, another group was given post-training infusion of the *Arc* antisense ODNs 8 hours after training. In this study, the animals given the scrambled ODN immediately after training and those given the *Arc* antisense ODNs 8 hours after training performed indistinguishably in the 48 hour retention probe test, showing spatial bias for the training location.

By contrast, the rats given the *Arc* antisense ODN immediately after training were strongly impaired relative to the other two groups. As will be seen below, the expression of *Arc* RNA and protein are tightly coupled to recent experience, and it was predicted that the antisense infusion, delayed by 8 hours, would not have an effect on 48 hour test performance. These results show that *Arc* expression in the hippocampus, immediately after training, is critical for consolidation of long-term memory for spatial learning.

Considerable experimental evidence indicates that the basolateral complex of the amygdala (BLA) modulates memory storage by faciliting neuroplastic change in other brain regions, including the hippocampus (Guzowski and McGaugh, 1997; McGaugh, 2000). It has been shown, for example, that BLA neuronal activity can modulate hippocampal LTP (Ikegaya et al., 1994, 1995). Given that *Arc* plays a critical role in the maintenance of hippocampal LTP, McIntyre et al. tested the hypothesis that BLA neuronal activity influenced learning induced *Arc* gene expression in the hippocampus (McIntyre et al., 2005). Infusions of the β-adrenoreceptor agonist, clenbuterol, into the BLA immediately after training on an inhibitory avoidance task enhanced memory tested 48 h later. The same dose of clenbuterol significantly increased Arc protein levels in the dorsal hippocampus 45 min after training. Additionally, posttraining intra-BLA infusions of a memory-impairing dose of lidocaine (a local anesthetic that blocks neuronal action potentials) significantly reduced *Arc* protein levels in the dorsal hippocampus at this same time point. Interestingly, increases or decreases in Arc protein levels, associated with intra-BLA clenbuterol or lidocaine, were not accompanied by similar alterations in the levels of *Arc* mRNA 30 min after training. These data are consistent with the notion that amygdala modulation of plasticity related *Arc* protein expression in efferent brain regions occurs at a posttranscriptional level. McIntyre et al. then used *Arc* antisense ODNs to directly block training induced *Arc* protein expression in the hippocampus. This manipulation significantly impaired 48 h retention performance, providing support for the hypothesis that the BLA influences long-term memory, at least in part, by modulating plasticity in the hippocampus.

Recently, Kuhl and colleagues described the effect of germline knockout of the *Arc* gene in mice (Plath et al., 2006). These mice are completely deficient of the *Arc* open reading frame, but are viable and generated in the expected Mendelian ratios. Brains appeared histologically normal, and quantitative analysis of dendritic branching and spine density of CA1 neurons showed no differences between the *Arc* knockouts and normal littermates. Additionally, the *Arc* knockout animals showed normal baseline neurotransmission, AMPA and NMDA synaptic conductances, and paired-pulse facilitation. Thus, developmental loss of *Arc* did not lead to gross physiological abnormality or alteration of neuronal development / function. In contrast, *Arc* knockout mice were profoundly deficient in long-term plasticity. In both *in vivo* perforant path / dentate gryus LTP and *in vitro* Schaffer collateral / CA1 LTP, *Arc* knockout mice (or hippocampal slices derived from them) exhibited an interesting phenotype. In both instances, *Arc* knockout preparations showed an enhanced early- and impaired late- LTP. In both forms of LTP, responses of the *Arc* knockouts returned to baseline levels within 60 to 90 min, coinciding with the time course of *Arc* protein expression. Furthermore, *Arc* knockouts showed impairments both in the initial induction of Schaffer collateral / CA1 long-term depression (LTD) and its maintenance, with knockouts returning to baseline levels within 40 to 60 min. Thus, *Arc* knockout mice cannot stably modify synaptic weights (up or down) in hippocampal networks.

Consistent with the strong deficits in long-term synaptic plasticity, *Arc* knockout mice have impaired long-term memory in several behavioral tasks, including the spatial version of the Morris water maze task, cued and contextual fear conditioning, conditioned task aversion, and object recognition (Plath et al., 2006). In the tasks where measures of short-term memory and acquisition were possible, the *Arc* knockout mice were not different from littermate controls. Thus, as concluded from the earlier *Arc* antisense studies (Guzowski et al., 2000), *Arc* does not play a role in learning or short-term memory, but is essential for the consolidation of long-term memories.

19.4 Reported Cellular Functions of *Arc*

Given the critical role for *Arc* in the maintenance of long-term plasticity and memory consolidation, there has been much interest in determining the cellular function, or functions, of *Arc* in neurons. In studies of transfected primary neurons, Fujimoto and colleagues (Fujimoto et al., 2004) suggested a role for *Arc* in destabilizing the cytoskeleton through interactions with microtubule associate protein 2 (MAP2). Such interactions could be important for dendritic remodeling associated with long-term alterations of synaptic efficacy. While intriguing, these results are very preliminary and need much further testing before this mechanism can be supported. In another study, it was suggested that *Arc* interacts with calcium-calmodulin kinase II (CaMKII), and that this interaction (defined by *in vitro* studies) promotes neurite outgrowth in cultured neuroblastoma cells (Donai et al., 2003). In a recent study, *Arc* protein was shown to co-localize (using immunofluorescence) with GluR4 receptors and co-precipitate with actin and GluR4 in turtle brain (Mokin et al., 2006).

While the above studies are all consistent with *Arc* interacting with a number of different dendritic proteins (in NMDA receptor multiprotein complexes, with MAP2, with CaMKII, and with actin and GluR4), they are largely correlative or were in heterologous cell systems. Moreover, these studies lacked a strong mechanistic functional dissection. Recently, Worley and colleagues identified a novel cellular pathway function for *Arc*. Initially, yeast two-hybrid screens were used to identify potential protein interaction partners of *Arc*. Two partners were identified that implicate *Arc* in the neural endocytotic pathway—dynamin 2 and endophilin 3. Through a series of well designed cell biological studies, Chowdury et al. provided strong evidence that through its interactions with dynamin 2 and endophilin 3, *Arc* regulates the endocytosis of AMPA receptors (Chowdhury et al., 2006). Shepherd et al. extended these studies to show that *Arc* plays a critical role in homeostatic synaptic scaling of AMPA receptors in cultured hippocampal and cortical neurons (Shepherd et al., 2006). In normal neuronal cultures blocking neural activity with tetrodotoxin decreases *Arc* expression and leads to upregulation of surface expression of AMPA receptors. By contrast, increasing neural activity in normal cultures with picrotoxin, a GABA antagonist, increases *Arc* expression and decreases surface expression of AMPA receptors. Neuronal cultures from *Arc* knockout mice lose the ability to modulate surface AMPA receptor expression in both tetrodotoxin and picrotoxin containing media. Although these data are consistent with the idea that *Arc* controls surface AMPA receptor expression, they do not prove it. To more directly test this hypothesis, Chowdury et al. showed that transfection of wild type *Arc* into *Arc* knockout neurons rescues the synaptic scaling phenotype, and that *Arc* protein mutants that are deleted in the region necessary for interaction with endophilin 3 do not (Chowdhury et al., 2006). The hypothesis that *Arc* regulates surface AMPA receptor equilibrium by controlling endocytosis was also supported in a study using hippocampal organotypic slice cultures (Rial Verde et al., 2006).

So what can we make of these several studies on *Arc* function? At present, we suggest that it is too early to tell. As detailed in a review covering these three studies implicating *Arc* in regulating AMPA receptor endocytosis and trafficking (Chowdhury et al., 2006; Rial Verde et al., 2006; Shepherd et al., 2006), many questions remain to be resolved in reconciling whole animal electrophysiological, plasticity, and memory studies with the cell biological data (Tzingounis and Nicoll, 2006). Although the cell biological data implicating *Arc* in regulating AMPA receptor endocytosis is compelling, it is hard to reconcile with the LTP phenotype of the knockout mice and *Arc* antisense ODN injected rats, for example. In this instance, it is not clear how endocytosis of AMPA receptors by an *Arc*-dynamin-endophilin complex would be involved in stabilizing a recent increase in synaptic efficacy (LTP), which is believed to be, at least in part, due to the insertion of AMPA receptors (Malinow and Malenka, 2002). From the perspective of memory consolidation and storage, it is attractive to think that *Arc* could act as a critical molecule for maintaining homeostatic synaptic scaling in hippocampal and neocortical networks. By virtue of its tight coupling to neural activity and its ability to provide synapse specific targeting, *Arc* would be well suited to serve such a function.

19.5 Regulation of *Arc* Gene Expression: Behavioral and Cellular Studies.

The notion that experience-dependent *Arc* gene expression is critical for the consolidation of newly learned experiences is supported by many studies showing that *Arc* RNA and protein expression are dramatically increased by behavioral experience [e.g., (Guzowski et al., 2001a; Dickey et al., 2004; Burke et al., 2005; Chawla et al., 2005; Ramirez-Amaya et al., 2005; Zhang et al., 2005; Zou and Buck, 2006; Han et al., 2007)]. To date, there are at least 40 publications showing upregulation of *Arc* in response to learning, memory retrieval, or stimulus presentation in rodents. Given the diversity of behavioral tasks and brain regions examined, it is not practical to include a complete review of this literature in this chapter. Suffice it to say that *Arc* gene expression is dramatically regulated by behavioral experience in many regions of the brain including (but not limited to) the neocortex, the hippocampus, the amygdala, the olfactory bulb, and the striatum. An example of this dramatic and rapid induction of *Arc* transcription in response to exposure to a novel environment is shown in Fig. 2.

The issue of what cell types express *Arc* following behavioral induction was addressed using double label fluorescence *in situ* hybridization and immunohisto-chemistry. In an extensive characterization of *Arc* expression following exposure to a novel environmental context, it was demonstrated that *Arc* was expressed exclusively in α calcium-calmodulin kinase II (αCaMKII) expressing neurons in the hippocampus, neocortex, and dorsal striatum (Vazdarjanova et al., 2006). In each of these brain regions, *Arc* protein was not detected in GFAP+ astrocytes, further supporting the conclusion that in intact, healthy brain, *Arc* is expressed exclusively in neurons. In the hippocampus and neocortex, *Arc* was not detected in GAD65/67+ cells of rats exposed to the novel context, but only in αCaMKII+ cells. By contrast, *Arc* was detected in a small population of GAD65/67+ cells following MECS, indicating the capacity for *Arc* to be induced in some interneurons, although under non-physiological conditions. Together, these results show that in normal brain *Arc* is expressed in principal glutamatergic neurons of the hippocampus and neocortex, consistent with a role for *Arc* in the modification of dendritic spine function. In the dorsal striatum a slightly different picture emerged—*Arc* was induced in GAD65/67+ cells. These cells, however, are also αCaMKII+, as the principal neurons of the striatum are GABAergic. Thus, *Arc* is expressed in αCaMKII+

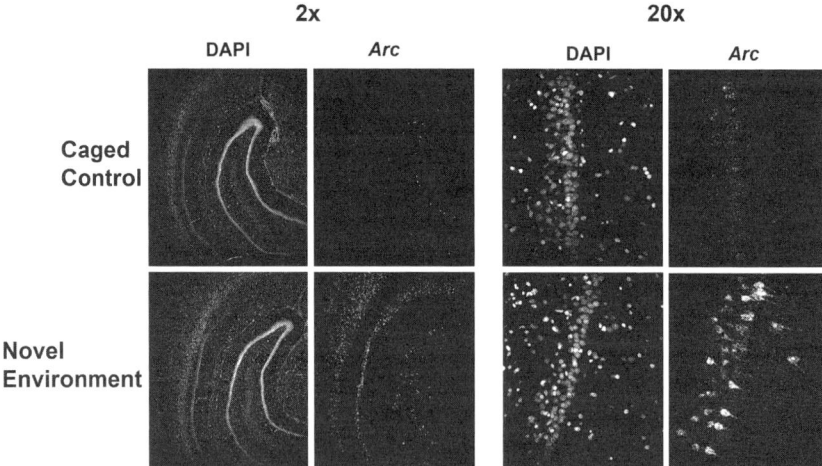

Fig. 2 Dramatic induction of *Arc* mRNA expression by a brief exposure of a rat to a novel environment. The upper panel shows both DAPI and *Arc* RNA staining, detected using FISH, of a region of ventral hippocampus from a caged control rat taken with both a 2x and 20x objective. The lower panel shows both DAPI and *Arc* RNA staining from a rat that was exposed for 5 min to a novel environment, and then returned to its home cage for 25 min. For each the upper and lower panels, the 2x field encompasses a large portion of the hippocampus and areas of the neocortex, and the 20x field shows the CA1 region of hippocampus. Please note the dramatic increase of *Arc* RNA staining in the "novel environment" section as compared to the "caged control" in all fields of the hippocampus and in the neocortex. (*See* Color Plate 13).

principal neurons in the forebrain, independent of whether they are GABAergic or glutamatergic. The tight association of *Arc* expression with αCaMKII+ supports the hypothesis that Arc and αCaMKII act as "plasticity partners" to promote functional and/or structural synaptic modifications that accompany learning.

To date, the exact signal transduction pathway(s) and transcription factors controlling experience-dependent *Arc* transcription is not known. "Prototypical" IEG promoters, such as those for c-*fos* and *zif268*, are known to contain both CRE and SRE consensus sequences within the proximal portion of the promoter, upstream of the TATA box (Herdegen and Leah, 1998). Both CREs and SREs bind transcription factors responsive to changes in calcium and cAMP, and are thought to be critical promoter elements for IEG promoters. In the c-*fos* promoter, both the calcium/cAMP response element and the SRE are critical for the induction of a reporter gene in transgenic mice following kainic acid induced seizures (Robertson et al., 1995), suggesting a synergistic requirement for both transcription factors binding sites. Interestingly, *Arc* does not contain a consensus CRE within its proximal promoter (up to base -1737 relative to the transcript start site), but does contain consensus SREs. Recently it was shown that mice containing a germline knockout of SRF (which binds the SRE) cannot upregulate *Arc* or other IEGs (including c-fos, zif268, FosB, or egr2) in response to electroconvulsive shock (ECS) in the dentate gyrus (Ramanan et al., 2005). Interestingly, the SRF knockouts could upregulate expression of another IEG, *Homer 1a*, which is co-expressed with *Arc* by behavioral activation (Vazdarjanova et al., 2002), in response to ECS (Ramanan et al., 2005). This data provide positive evidence that SRF mediates activity regulated transcription of *Arc* and other IEGs.

The foregoing results, and the lack of a consensus CRE within the proximal promoter of *Arc*, do not, however, rule out a role for CREB in the regulation of *Arc*. In the studies by Waltereit and colleagues, treatment of primary neurons with forskolin (a direct activator of adenyl cyclase) activated transcription of the endogenous *Arc* gene, but in transfection assays forskolin did not activate the reporter gene driven by the 1737 bases upstream of the *Arc* start site (Waltereit et al., 2001). These data are consistent with the idea that functional CRE sites influencing cAMP mediated *Arc* transcription lie outside of the proximal promoter region. This finding is supported further in a genome wide chromatin immunoprecipitation screen of genes regulated by the phosphorylation of CREB in PC12 cells, following treatment with forskolin (Impey et al., 2004). In this study, *Arc*, which is strongly upregulated by forskolin in PC12 cells, was identified as a member of the "CREB regulon". Stated simply, CREs in the vicinity of the *Arc* start site were occupied with phospho-CREB following forskolin treatment, and correlated with the induction of *Arc* mRNA expression. Like many of the genes regulated by forskolin, the CRE sites identified with *Arc* were in non-canonical positions. Thus, it remains possible that *Arc* might be regulated, like at least some other IEGs, by CREB-CRE interactions.

Lastly, existing evidence supports a role for PKA and MAPK/ERK in *Arc* transcriptional regulation. The increase of *Arc* RNA expression in cultured hippocampal neurons in response to forskolin was blocked by inhibitors of both PKA and MAPK/ERK (Waltereit et al., 2001). Additionally, studies by Bramham and colleagues also point to a role for MAPK/ERK in the regulation of *Arc* expression

(Messaoudi et al., 2002; Ying et al., 2002). In these studies, infusion of BDNF into the dentate gryus of anesthetized rats leads to the gradual development of a macro-molecular synthesis-dependent LTP (Messaoudi et al., 2002). This BDNF-induced LTP is correlated with the increased phosphorylation of MAPK/ERK and CREB, and the upregulation of *Arc*, but not another IEG, zif268. Co-infusion of BDNF plus an U0126, an inhibitor of MEK (the kinase immediately upstream of ERK), prevented the induction of LTP, phoshphorylation of ERK and CREB, and induc-tion of *Arc* mRNA expression (Ying et al., 2002). The findings of a requirement for MAPK/ERK in the activation of *Arc* transcription are also consistent with the idea that the SREs of the *Arc* promoter may be critical for its regulation—past studies have shown that Elk-1, part of the ternary complex acting at SREs, is phosphorylated by MAPK/ERK (Xia et al., 1996).

19.6 Studying IEG Transcription in Single Neurons using RNA Fluorescence *in situ* Hybridization

While developing a sensitive fluorescence *in situ* hybridization (FISH) method to study co-regulation of different IEG mRNAs in neurons following behavioral acti-vation, my colleagues and I made a serendipitous discovery. The principle finding was that both primary (unprocessed) RNA transcript in the nucleus (at the genomic alleles) and mature mRNA in the cytoplasm of specific IEGs could be detected with the use of sensitive FISH and confocal microscopy. Time course analyses using a temporally precise strong electrical stimulus, MECS, showed that *Arc* RNA transcript was detected within 1-2 minutes at intense intranuclear foci (INF) and that these INF disappeared within ~15 minutes (Guzowski et al., 1999). Subsequently, from ~15-45 min after the stimulus, a prominent labeling was observed in cyto-plasmic and dendritic regions. After that, *Arc* mRNA could be observed principally in dendritic regions. Other well characterized IEGs, such as zif268 and c-fos, show similar dynamics of transcription and mRNA processing, although these transcripts remain within the cytoplasm and do not translocate to dendrites. Several control experiments (including RNase pretreatment of tissue sections, use of sense ribo-probes, and combined use of intron and exon probes) have supported the conclusion that these INF represent the sites of transcription at the genomic alleles for specific IEGs.

This time course of appearance / disappearance of *Arc* signal described above is similar for neurons in many regions of the brain, including the pyramidal cell layers of the hippocampus, the neocortex, amygdala, and striatum. One region that displays different kinetics is the granule cell layer of the dentate gyrus. In these neurons, the appearance of *Arc* transcription foci is as rapid, but transcription continues to persist for greater than 4 and less than 8 hours after MECS. Thus, in granule cells *Arc* mRNA levels reach much higher levels as compared to CA1 and CA3 pyramidal neurons, due to this prolonged period of active transcription.

The kinetics of *Arc* mRNA appearance / disappearance were then examined using natural stimuli—a brief (5 min) exposure to a novel environmental context. In initial

(Guzowski et al., 1999) and then follow-up studies (Vazdarjanova et al., 2002), a maximum proportion of *Arc* transcriptionally active cells was detected in rats sacrificed immediately after the 5 min exploration session. Within 16 min after removal from the environment, the proportion of *Arc* INF-positive cells was back to baseline levels. Thus, a brief "burst" of *Arc* transcription accompanies learning about a novel environment. *Arc* mRNA was then detected in the cytoplasm and dendrites ~20–60 min after removal from environment (Guzowski et al., 1999; Vazdarjanova et al., 2002).

The above basic findings into the transcriptional regulation of *Arc* (and other IEGs) in intact, behaving animals are significant for two reasons. First, these are simply observations of what occurs in the brain—not the result of an *in vitro* system or the response to an experimental stimulus. As such, these data necessarily constrain the list of putative mechanisms controlling IEG transcription in hippocampal neurons *in vivo*. In the next section entitled "A model of the neural and molecular events leading to the activation of *Arc* transcription in hippocampal neurons" we will consider this topic in detail. Second, by defining the temporal dynamics of IEG RNA expression, starting from the appearance of nuclear transcription foci and ending as a cytoplasmic/dendritic signal, it is possible to infer the activity history of individual neurons in a brain tissue section for different time intervals before the animal's death using FISH and confocal microscopy. A practical application of this approach to imaging activity in neural networks engaged during distinct behaviors is described in the following paragraph.

The ability to define the time of activation of single neurons based on the subcellular distribution of *Arc* RNA suggested that it might be possible to expose rats to two different behavioral experiences, separated by a rest interval, and visualize the population of cells activated during each experience. We termed this new imaging approach "cellular compartment analysis of temporal activity by fluorescence *in situ* hybridization", or "catFISH". In concept, catFISH provided an alternative means to address whether activation of IEG transcription was specifically related to information processing or merely associated with increased behavioral arousal, motor activity, stress, etc, typically associated with different learning paradigms (Guzowski et al., 2001b). With this approach, neural activity "maps" for two discrete experiences by a single animal can be visualized and compared. The within-subject comparisons provided by catFISH contrast strongly with "standard approach" for IEG imaging, in which changes in the proportion of activated neurons are compared across different groups of animals (between-subjects). With the between-subjects design of standard IEG experiments, a brain area can be inferred to be specifically associated with a behavioral task only if a different percentage of cells are active as compared to the "learning control" group. With catFISH, even when two experiences activate the same number of neurons in a given brain region, it is possible to determine the degree to which the two experiences activated the same population or two distinct neuronal populations. Such information about the ensemble representation for two distinct behaviors greatly expands the use of IEGs for mapping the degree to which the active cell populations for the two experiences were the same or different.

catFISH was first validated as an imaging technique by examining neuronal activation patterns in the CA1 region of the hippocampus as rats were exposed sequentially to either the same environment twice or to two distinct environments (Guzowski et al., 1999). This study, and several subsequent studies, demonstrated that the activation of *Arc* transcription in hippocampal pyramidal neurons is information content specific (Guzowski et al., 1999; Guzowski et al., 2004; Guzowski et al., 2005; Guzowski, 2006; Guzowski et al., 2006). catFISH cell imaging has also been used to examine the dynamics of information processing with the hippocampus, by comparing the ensemble activity of CA3 and CA1 neurons following defined perturbations of environmental context (Guzowski et al., 2004; Vazdarjanova and Guzowski, 2004). These past studies of *Arc* activation in CA1 and CA3 are consistent, both qualitatively and quantitatively, with single unit recording studies of complex spiking activity of hippocampal neurons. Our current working model, described in more detail in the following section, is that the expression of a hippocampal firing field, characterized by a burst of action potentials, is a sufficient signal to activate transcription of *Arc* and probably other similarly sensitive IEGs such as zif268 and Homer 1a (Guzowski et al., 1999; Vazdarjanova et al., 2002; Guzowski et al., 2005; Guzowski, 2006; Guzowski et al., 2006). In addition to these studies in hippocampus, *Arc* catFISH has been used to investigate neuronal ensemble responses in parietal (Burke et al., 2005) and olfactory (Zou and Buck, 2006) cortices. Moreover, Petrovich and colleagues used a variant of *Arc* catFISH [*Arc*/Homer 1a catFISH; (Vazdarjanova et al., 2004; Guzowski et al., 2005)] and tract tracing methods to study the influence of amygdalar and prefrontal cortical inputs to the lateral hypothalamus in driving conditioned potentiation of feeding (Petrovich et al., 2005).

19.7 A Model of the Neural and Molecular Events Leading to the Activation of *Arc* Transcription in Hippocampal Neurons

The finding that the transcription of the IEG *Arc* (and others such as *Homer 1a*) is tightly linked to the neural activity associated with the expression of a "place field" in hippocampal neurons has several important implications for understanding hippocampal function in memory and provides a point of convergence for other lines of research in this area. First, studies by Kentros and colleagues have shown that the stabilization of place field "maps" of CA1 neurons requires NMDA receptor function (Kentros et al., 1998) and protein synthesis (Agnihotri et al., 2004). In these two studies, the NMDA antagonist CPP or the protein synthesis inhibitor anisomycin did not affect the formation of place field maps or their stability at short intervals (i.e., 1 hr) when rats where exposed to a novel environment, but prevented the long-term stabilization of these maps. This NMDA receptor- and protein synthesis-dependent stabilization was evidenced when the rats were re-introduced to the environment 24 hours later and a new map was activated. In vehicle treated controls, hippocampal maps for the same environment were stabile at this 24 hour interval. Interestingly, neither of these drugs influenced the stability of map activation in familiar environments. Thus, stabilization of a place cell network representation of a new experience requires NMDA receptor activation and protein synthesis. These findings are consistent with the notion that expression of IEG proteins such as *Arc* and *H1a*, which require NMDA receptor activation for transcriptional activation (Lyford et al., 1995; Brakeman et al., 1997) and an intact translation apparatus for functional protein expression, is critical for the stabilization of place field maps. This hypothesis is supported by the finding that inhibition of *Arc* protein expression can block the maintenance phase of long-term potentiation (without affecting its induction) and impair long-term memory in a spatial task (without affecting learning or short-term memory) (Guzowski et al., 2000).

Recent experiments have helped further define the requirements for *Arc* activation by physiological experience. Fletcher et al. showed that experience-dependent upregulation of *Arc* mRNA in hippocampus was completely blocked by fornix lesions, which also impaired spatial learning and memory (Fletcher et al., 2006). By contrast, selective depletion of cholinergic inputs to the hippocampus did not affect learning-dependent *Arc* mRNA induction and spared spatial learning and memory (Fletcher et al., 2007). In both manipulations, "place cell" firing was intact in hippocampal neurons, but fornix lesions eliminated hippocampal theta rhythm and the selective cholinergic depletion only reduced the magnitude of, and did not eliminate, hippocampal theta (Lee et al., 1994). From these two studies it was concluded that hippocampal theta constitutes a critical co-factor for hippocampal plasticity and experience-dependent *Arc* induction (Fletcher et al., 2007). Additionally, the contribution of NMDA receptors and voltage-dependent calcium channels to the *in vivo* activation of *Arc* transcription was examined (Guzowski et al., 2001c). In these studies, systemic injection of drugs that block NMDA receptors (CPP; 10 mg/kg, i.p.) or voltage dependent calcium channels (verapamil; 10 mg/kg, i.p.) were given to rats before exposure to a novel environment. The effect of these drugs on the appearance of *Arc* transcription foci in CA1 neurons was determined and compared to vehicle treated rats. These studies showed that CPP completely blocked experience-dependent *Arc* transcription and verapamil had no effect. Importantly, this dose of CPP was used in single unit recording studies of CA1 neurons previously and shown to not affect basic *in vivo* firing parameters of CA1 neurons, but did block the stabilization of place field maps (Kentros et al., 1998) and experience-dependent place field expansion (Ekstrom et al., 2001).

In considering the molecular mechanisms governing *Arc* transcription *in vivo* in hippocampal neuronal networks, the rapidity of *Arc* transcriptional activation [nascent transcript detected within 2 min; (Guzowski et al., 1999)] rules against a relatively slow synapse to nucleus mechanism (Adams and Dudek, 2005). Moreover, the transient nature of *Arc* transcription [cessation of new transcription within 16 min; (Vazdarjanova et al., 2002)] rules against a requirement for alterations in chromatin structure or histone modifications, which are typically seen occurring not before 1 hour after behavioral training (Levenson et al., 2004; Chwang et al., 2006). This rapidity of transcriptional activation / shut-off suggests that the *Arc* promoter is "primed" with necessary trans-acting factors to receive a rapid calcium mediated signal following synaptic activation or neuronal firing. In terms of mechanism, two distinct possibilities can be envisioned. First, action potentials could serve as the critical event, leading to the opening of somatic voltage-dependent calcium channels and rapid invasion of calcium to the nucleus (Adams and Dudek, 2005). Second, calcium entry through synaptic NMDA receptors could activate inositol 1,4,5 triphosphate receptors (InsP3-R), leading to a regenerative "calcium wave" within the endoplasmic reticulum to the nuclear membrane. Both mechanisms would be extremely rapid, and lead to increased nuclear calcium. From there, free nuclear calcium could bind nuclear localized calmodulin, and then activate calcium-calmodulin kinase IV (CaMKIV) to phosphorylate transcription factors such as CREB and SRF to activate transcription of *Arc* and other IEGs (Kang et al., 2001; Limback-Stokin et al., 2004; Ramanan et al., 2005). The data showing an absolute requirement for NMDA receptor function for *in vivo* activation of *Arc* transcription in hippocampal neurons (Lyford et al., 1995; Guzowski et al., 2001c) is most readily consistent with the "calcium wave" hypothesis of Berridge (Berridge, 1998). Interestingly, in our studies using systemic injection of the NMDA antagonist CPP (Guzowski et al., 2001c), *Arc* transcription was not affected in overlying neocortex, despite the pronounced inhibition in hippocampus. These findings raise the possibility that in neocortical neurons *Arc* transcriptional regulation could be driven by the action potential mediated mechanism proposed by Adams and Dudek (Adams and Dudek, 2005). As there is considerable evidence supporting each model, it would not be too surprising that different neuronal populations, and potentially the same population under different neural states, could use distinct mechanisms for mediating calcium regulated neuronal gene transcription.

19.8 Conclusions

In the 12 years since its discovery, *Arc* has proven an important gene for both understanding the molecular and cellular bases of brain plasticity underlying such functions as memory, and has also been important for monitoring neuronal processing dynamics within and across different brain systems. As illustrated by the recent experiments suggesting a role for *Arc* in AMPA receptor trafficking, further molecular studies of *Arc* will help define the very nature of the plastic changes critical for memory consolidation and perhaps memory storage, and may help lead us into considering new directions on our thinking about plasticity and memory. Moreover, the precise dynamics of *Arc* transcription in brain will continue to provide a powerful tool for dissecting the signal transduction pathways critically regulating this process and for monitoring brain systems interactions during different cognitive processes. In sum, *Arc* will continue to serve as a Rosetta Stone to aid in deciphering the intricacies of the brain, at synaptic, cellular, and systems levels.

Acknowledgments Portions of the research described in this chapter were funded by the National Institutes of Health Grant MH060123 to JFG.

References

Adams, J.P. and Dudek, S.M. (2005) Late-phase long-term potentiation: getting to the nucleus. Nat. Rev. Neurosci. 6, 737–743.

Agnihotri, N.T., Hawkins, R.D., Kandel, E.R. and Kentros, C. (2004) The long-term stability of new hippocampal place fields requires new protein synthesis. Proc. Natl. Acad. Sci. USA 101, 3656–3661.

Berridge, M.J. (1998) Neuronal calcium signaling. Neuron 21, 13–26.

Brakeman, P.R., Lanahan, A.A., O'Brien, R., Roche, K., Barnes, C.A., Huganir, R.L. and Worley, P.F. (1997) Homer: A protein that selectively binds metabotropic glutamate receptors. Nature 386, 284–288.

Burke, S.N., Chawla, M.K., Penner, M.R., Crowell, B.E., Worley, P.F., Barnes, C.A. and McNaughton, B.L. (2005) Differential encoding of behavior and spatial context in deep and superficial layers of the neocortex. Neuron 45, 667–674.

Chawla, M.K., Guzowski, J.F., Ramirez-Amaya, V., Lipa, P., Hoffman, K.L., Marriott, L.K., Worley, P.F., McNaughton, B.L. and Barnes, C.A. (2005) Sparse, environmentally selective expression of Arc RNA in the upper blade of the rodent fascia dentata by brief spatial experience. Hippocampus 15, 579–586.

Chowdhury, S., Shepherd, J.D., Okuno, H., Lyford, G., Petralia, R.S., Plath, N., Kuhl, D., Huganir, R.L. and Worley, P.F. (2006) Arc/Arg3.1 interacts with the endocytic machinery to regulate AMPA receptor trafficking. Neuron 52, 445–459.

Chwang, W.B., O'Riordan, K.J., Levenson, J.M. and Sweatt, J.D. (2006) ERK/MAPK regulates hippocampal histone phosphorylation following contextual fear conditioning. Learn. Mem. 13, 322–328.

Dickey, C.A., Gordon, M.N., Mason, J.E., Wilson, N.J., Diamond, D.M., Guzowski, J.F. and Morgan, D. (2004) Amyloid suppresses induction of genes critical for memory consolidation in APP + PS1 transgenic mice. J. Neurochem. 88, 434–442.

Donai, H., Sugiura, H., Ara, D., Yoshimura, Y., Yamagata, K. and Yamauchi, T. (2003) Interaction of Arc with CaM kinase II and stimulation of neurite extension by Arc in neuroblastoma cells expressing CaM kinase II. Neurosci. Res. 47, 399–408.

Dynes, J.L. and Steward, O. (2007) Dynamics of bidirectional transport of Arc mRNA in neuronal dendrites. J. Comp. Neurol. 500, 433–447.

Ekstrom, A.D., Meltzer, J., McNaughton, B.L. and Barnes, C.A. (2001) NMDA receptor antagonism blocks experience-dependent expansion of hippocampal "place fields". Neuron 31, 631–638.

Fletcher, B.R., Calhoun, M.E., Rapp, P.R. and Shapiro, M.L. (2006) Fornix lesions decouple the induction of hippocampal arc transcription from behavior but not plasticity. J. Neurosci. 26, 1507–1515.

Fletcher, B.R., Baxter, M.G., Guzowski, J.F., Shapiro, M.L. and Rapp, P.R. (2007) Selective cholinergic depletion of the hippocampus spares both behaviorally induced Arc transcription and spatial learning and memory. Hippocampus 17, 227–234.

Fujimoto, T., Tanaka, H., Kumamaru, E., Okamura, K. and Miki, N. (2004) Arc interacts with microtubules/microtubule-associated protein 2 and attenuates microtubule-associated protein 2 immunoreactivity in the dendrites. J. Neurosci. Res. 76, 51–63.

Goelet, P., Castellucci, V.F., Schacher, S. and Kandel, E.R. (1986) The long and the short of long-term memory–a molecular framework. Nature 322, 419–422.

Guzowski, J.F. (2002) Insights into immediate-early gene function in hippocampal memory consolidation using antisense oligonucleotide and fluorescent imaging approaches. Hippocampus 12, 86–104.

Guzowski, J.F. (2006) Immediate early genes and the mapping of environmental representations in hippocampal neural networks. In: R. Pinaud, L.A. Tremere, (Eds.), Immediate early genes in sensory processing, cognitive performance, and neurological disorders. Springer, New York, pp. 159–176.

Guzowski, J.F. and McGaugh, J.L. (1997) Interaction of neuromodulatory systems regulating memory storage. In: M. Decker and J.D. Brioni, (Eds.), *Alzheimer's Disease: Molecular Aspects and Pharmacological Treatments*. Wiley-Liss. pp 37–61.

Guzowski, J.F., Knierim, J.J. and Moser, E.I. (2004) Ensemble Dynamics of Hippocampal Regions CA3 and CA1. Neuron 44, 581–584.

Guzowski, J.F., McNaughton, B.L., Barnes, C.A. and Worley, P.F. (1999) Environment-specific expression of the immediate-early gene *Arc* in hippocampal neuronal ensembles. Nat. Neurosci. 2, 1120–1124.

Guzowski, J.F., Setlow, B., Wagner, E.K. and McGaugh, J.L. (2001a) Experience-dependent gene expression in the rat hippocampus after spatial learning: A comparison of the immediate-early genes *Arc*, *c-fos*, and *zif268*. J. Neurosci. 21, 5089–5098.

Guzowski, J.F., McNaughton, B.L., Barnes, C.A. and Worley, P.F. (2001b) Imaging neural activity with temporal and cellular resolution using FISH. Curr. Opin. Neurobiol. 11, 579–584.

Guzowski, J.F., Houston, F.P., Worley, P.F. and Barnes, C.A. (2001c) Experience-dependent Arc expression in hippocampal neurons: Role of NMDA receptors and voltage-dependent calcium channels. In: Society for Neuroscience Annual Meeting. San Diego, CA.

Guzowski, J.F., Timlin, J.A., Roysam, B., McNaughton, B.L., Worley, P.F. and Barnes, C.A. (2005) Mapping behaviorally relevant neural circuits with immediate-early gene expression. Curr. Opin. Neurobiol. 15, 599–606.

Guzowski, J.F., Lyford, G.L., Stevenson, G.D., Houston, F.P., McGaugh, J.L., Worley, P.F. and Barnes, C.A. (2000) Inhibition of activity-dependent arc protein expression in the rat hippocampus impairs the maintenance of long-term potentiation and the consolidation of long-term memory. J. Neurosci. 20, 3993–4001.

Guzowski, J.F., Miyashita, T., Chawla, M.K., Sanderson, J., Maes, L.I., Houston, F.P., Lipa, P., McNaughton, B.L., Worley, P.F. and Barnes, C.A. (2006) Recent behavioral history modifies coupling between cell activity and Arc gene transcription in hippocampal CA1 neurons. Proc. Natl. Acad. Sci. USA 103, 1077–1082.

Han, J.H., Kushner, S.A., Yiu, A.P., Cole, C.J., Matynia, A., Brown, R.A., Neve, R.L., Guzowski, J.F., Silva, A.J. and Josselyn, S.A. (2007) Neuronal competition and selection during memory formation. Science 316, 457–460.

Herdegen, T. and Leah, J.D. (1998) Inducible and constitutive transcription factors in the mammalian nervous system: control of gene expression by Jun, Fos and Krox, and CREB/ATF proteins. Brain Res. Brain Res. Rev. 28, 370–490.

Husi, H., Ward, M.A., Choudhary, J.S., Blackstock, W.P. and Grant, S.G. (2000) Proteomic analysis of NMDA receptor-adhesion protein signaling complexes. Nat. Neurosci. 3, 661–669.

Ikegaya, Y., Saito, H. and Abe, K. (1994) Attenuated hippocampal long-term potentiation in basolateral amygdala-lesioned rats. Brain Res. 656, 157–164.

Ikegaya, Y., Saito, H. and Abe, K. (1995) Requirement of basolateral amygdala neuron activity for the induction of long-term potentiation in the dentate gyrus in vivo. Brain Res. 671, 351–354.

Impey, S., McCorkle, S.R., Cha-Molstad, H., Dwyer, J.M., Yochum, G.S., Boss, J.M., McWeeney, S., Dunn, J.J., Mandel, G. and Goodman, R.H. (2004) Defining the CREB regulon: a genome-wide analysis of transcription factor regulatory regions. Cell 119, 1041–1054.

Kang, H., Sun, L.D., Atkins, C.M., Soderling, T.R., Wilson, M.A. and Tonegawa, S. (2001) An important role of neural activity-dependent CaMKIV signaling in the consolidation of long-term memory. Cell 106, 771–783.

Kentros, C., Hargreaves, E., Hawkins, R.D., Kandel, E.R., Shapiro, M. and Muller, R.V. (1998) Abolition of long-term stability of new hippocampal place cell maps by NMDA receptor blockade. Science 280, 2121–2126.

Lanahan, A. and Worley, P. (1998) Immediate-early genes and synaptic function. Neurobiol. Learn. Mem. 70, 37–43.

Lee, M.G., Chrobak, J.J., Sik, A., Wiley, R.G. and Buzsaki, G. (1994) Hippocampal theta activity following selective lesion of the septal cholinergic system. Neuroscience 62, 1033–1047.

Levenson, J.M., O'Riordan, K.J., Brown, K.D., Trinh, M.A., Molfese, D.L. and Sweatt, J.D. (2004) Regulation of histone acetylation during memory formation in the hippocampus. J. Biol. Chem. 279, 40545–40559.

Limback-Stokin, K., Korzus, E., Nagaoka-Yasuda, R. and Mayford, M. (2004) Nuclear calcium/calmodulin regulates memory consolidation. J. Neurosci. 24, 10858–10867.

Link, W., Konietsko, U., Kauselmann, G., Krug, M., Schwanke, B., Frey, U., Kuhl, D. (1995) Somatodendritic expression of an immediate-early gene is regulated by synaptic activity. Proc. Natl. Acad. Sci. USA 92, 5734–5738.

Lyford, G.L., Yamagata, K., Kaufmann, W.E., Barnes, C.A., Sanders, L.K., Copeland, N.G., Gilbert, D.J., Jenkins, N.A., Lanahan, A.A. and Worley, P.F. (1995) Arc, a growth factor and activity-regulated gene, encodes a novel cytoskeleton-associated protein that is enriched in neuronal dendrites. Neuron 14, 433–445.

Malinow, R. and Malenka, R.C. (2002) AMPA receptor trafficking and synaptic plasticity. Annu. Rev. Neurosci. 25, 103–126.

McGaugh, J.L. (2000) Memory–A Century of Consolidation. Science 287, 248–251.

McIntyre, C.K., Miyashita, T., Setlow, B., Marjon, K.D., Steward, O., Guzowski, J.F. and McGaugh, J.L. (2005) Memory-influencing intra-basolateral amygdala drug infusions modulate expression of Arc protein in the hippocampus. Proc. Natl. Acad. Sci. USA 102, 10718–10723.

Messaoudi, E., Ying, S.W., Kanhema, T., Croll, S.D. and Bramham, C.R. (2002) Brain-derived neurotrophic factor triggers transcription-dependent, late phase long-term potentiation in vivo. J. Neurosci. 22, 7453–7461.

Moga, D.E., Calhoun, M.E., Chowdhury, A., Worley, P., Morrison, J.H. and Shapiro, M.L. (2004) Activity-regulated cytoskeletal-associated protein is localized to recently activated excitatory synapses. Neuroscience 125, 7–11.

Mokin, M., Lindahl, J.S. and Keifer, J. (2006) Immediate-early gene-encoded protein Arc is associated with synaptic delivery of GluR4-containing AMPA receptors during in vitro classical conditioning. J. Neurophysiol. 95, 215–224.

Petrovich, G.D., Holland, P.C. and Gallagher, M. (2005) Amygdalar and prefrontal pathways to the lateral hypothalamus are activated by a learned cue that stimulates eating. J. Neurosci. 25, 8295–8302.

Plath, N., Ohana, O., Dammermann, B., Errington, M.L., Schmitz, D., Gross, C., Mao, X., Engelsberg, A., Mahlke, C., Welzl, H., Kobalz, U., Stawrakakis, A., Fernandez, E., Waltereit, R., Bick-Sander, A., Therstappen, E., Cooke, S.F., Blanquet, V., Wurst, W., Salmen, B., Bosl, M.R., Lipp, H.P., Grant, S.G., Bliss, T.V., Wolfer, D.P. and Kuhl, D. (2006) Arc/Arg3.1 is essential for the consolidation of synaptic plasticity and memories. Neuron 52, 437–444.

Ramanan, N., Shen, Y., Sarsfield, S., Lemberger, T., Schutz, G., Linden, D.J. and Ginty, D.D. (2005) SRF mediates activity-induced gene expression and synaptic plasticity but not neuronal viability. Nat. Neurosci. 8, 759–767.

Ramirez Amaya, V., Vazdarjanova, A., Mikhael, D., Rosi, S., Worley, P.F. and Barnes, C.A. (2005) Spatial exploration-induced Arc mRNA and protein expression: evidence for selective, network-specific reactivation. J. Neurosci. 25, 1761–1768.

Rial Verde, E.M., Lee-Osbourne, J., Worley, P.F., Malinow, R. and Cline, H.T. (2006) Increased expression of the immediate-early gene arc/arg3.1 reduces AMPA receptor-mediated synaptic transmission. Neuron 52, 461–474.

Robertson, L.M., Kerppola, T.K., Vendrell, M., Luk, D., Smeyne, R.J., Bocchiaro, C., Morgan, J.I. and Curran, T. (1995) Regulation of c-fos expression in transgenic mice requires multiple interdependent transcription control elements. Neuron 14, 241–252.

Shepherd, J.D., Rumbaugh, G., Wu, J., Chowdhury, S., Plath, N., Kuhl, D., Huganir, R.L. and Worley, P.F. (2006) Arc/Arg3.1 mediates homeostatic synaptic scaling of AMPA receptors. Neuron 52, 475–484.

Steward, O. and Levy, W.B. (1982) Preferential localization of polyribosomes under the base of dendritic spines in granule cells of the dentate gyrus. J. Neurosci. 2, 284–291.

Steward, O. and Schuman, E.M. (2001) Protein synthesis at synaptic sites on dendrites. Annu. Rev. Neurosci. 24, 299–325.

Steward, O. and Worley, P.F. (2001) Selective targeting of newly synthesized Arc mRNA to active synapses requires NMDA receptor activation. Neuron 30, 227–240.

Steward, O. and Worley, P. (2002) Local synthesis of proteins at synaptic sites on dendrites: role in synaptic plasticity and memory consolidation? Neurobiol. Learn. Mem. 78, 508–527.

Steward, O., Wallace, C.S., Lyford, G.L. and Worley, P.F. (1998) Synaptic activation causes the mRNA for the IEG Arc to localize selectively near activated postsynaptic sites on dendrites. Neuron 21, 741–751.

Tzingounis, A.V. and Nicoll, R.A. (2006) Arc/Arg3.1: linking gene expression to synaptic plasticity and memory. Neuron 52, 403–407.

Vazdarjanova, A. and Guzowski, J.F. (2004) Differences in hippocampal neuronal population responses to modifications of an environmental context: evidence for distinct, yet complementary, functions of CA3 and CA1 ensembles. J. Neurosci. 24, 6489–6496.

Vazdarjanova, A., McNaughton, B.L., Barnes, C.A., Worley, P.F. and Guzowski, J.F. (2002) Experience-dependent coincident expression of the effector immediate-early genes *Arc* and *Homer 1a* in hippocampal and neocortical neuronal networks. J. Neurosci. 22, 10067–10071.

Vazdarjanova, A., Ramirez-Amaya, V., Insel, N., Worley, P.F., Guzowski, J.F. and Barnes, C.A. (2004) Behavior induces expression of the plasticity-related immediate-early gene Arc in excitatory and inhibitory CaMKII - positive neurons. In: Society for Neuroscience 34th Annual Meeting. San Diego, CA.

Vazdarjanova, A., Ramirez-Amaya, V., Insel, N., Plummer, T.K., Rosi, S., Chowdhury, S., Mikhael, D., Worley, P.F., Guzowski, J.F. and Barnes, C.A. (2006) Spatial exploration induces ARC, a plasticity-related immediate-early gene, only in calcium/calmodulin-dependent protein kinase II-positive principal excitatory and inhibitory neurons of the rat forebrain. J. Comp. Neurol. 498, 317–329.

Wallace, C.S., Lyford, G.L., Worley, P.F. and Steward, O. (1998) Differential intracellular sorting of immediate early gene mRNAs depends on signals in the mRNA sequence. J. Neurosci. 18, 26–35.

Waltereit, R., Dammermann, B., Wulff, P., Scafidi, J., Staubli, U., Kauselmann, G., Bundman, M. and Kuhl, D. (2001) Arg3.1/Arc mRNA induction by Ca2+ and cAMP requires protein kinase A and mitogen-activated protein kinase/extracellular regulated kinase activation. J. Neurosci. 21, 5484–5493.

Xia, Z., Dudek, H., Miranti, C.K. and Greenberg, M.E. (1996) Calcium influx via the NMDA receptor induces immediate early gene transcription by a MAP kinase/ERK-dependent mechanism. J. Neurosci. 16, 5425–5436.

Yin, Y., Edelman, G.M. and Vanderklish, P.W. (2002) The brain-derived neurotrophic factor enhances synthesis of Arc in synaptoneurosomes. Proc. Natl. Acad. Sci. USA 99, 2368–2373.

Ying, S.W., Futter, M., Rosenblum, K., Webber, M.J., Hunt, S.P., Bliss, T.V. and Bramham, C.R. (2002) Brain-derived neurotrophic factor induces long-term potentiation in intact adult hippocampus: requirement for ERK activation coupled to CREB and upregulation of Arc synthesis. J. Neurosci. 22, 1532–1540.

Zhang, W.P., Guzowski, J.F. and Thomas, S.A. (2005) Mapping neuronal activation and the influence of adrenergic signaling during contextual memory retrieval. Learn. Mem. 12, 239–247.

Zou, Z. and Buck, L.B. (2006) Combinatorial effects of odorant mixes in olfactory cortex. Science 311, 1477–1481.

Subject Index

Printed in the United States of America.